# 30th International Symposium on Space Terahertz Technology (ISSTT 2019)

AA005002

Gothenburg, Sweden
15-17 April 2019

ISBN: 978-1-7138-0322-5

**Printed from e-media with permission by:**

Curran Associates, Inc.
57 Morehouse Lane
Red Hook, NY 12571

**Some format issues inherent in the e-media version may also appear in this print version.**



International Symposium on Space Terahertz Technology
c/o NRAO Headquarters
520 Edgemont Road
Charlottesville, VA 22903-2475
USA

Phone: (434) 296-0254
Fax: (434) 296-0278

mbishop@nrao.edu

**Additional copies of this publication are available from:**

Curran Associates, Inc.
57 Morehouse Lane
Red Hook, NY 12571 USA
Phone: 845-758-0400
Fax: 845-758-2633
Email: curran@proceedings.com
Web: www.proceedings.com

# ISSTT 2019

## 30th International Symposium on Space Terahertz Technology

# Proceedings Book

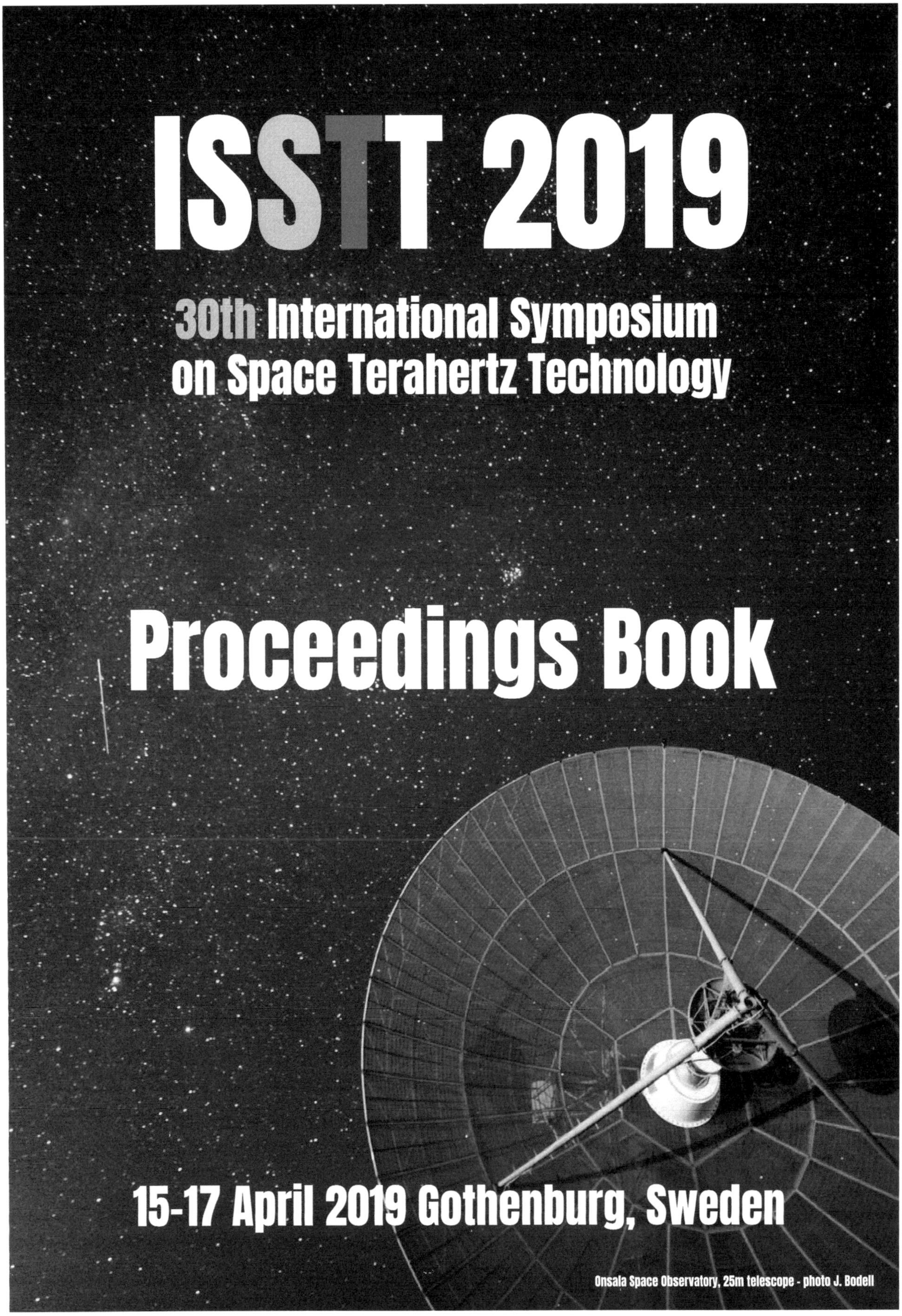

## 15-17 April 2019 Gothenburg, Sweden

Onsala Space Observatory, 25m telescope - photo J. Bodell

# Contents

# Sponsors

We greatly appreciate the support from our industrial sponsors:

## Strategic Sponsor:

Virginia Diodes, Inc.
www.vadiods.com

## Contributing Sponsors:

Keysight Technologies
www.keysight.com

Rohde & Schwarz
www.rohde-schwarz.com

Wasa Millimeter Wave AB
www.wmmw.se

# Committees

The **Local Organizing Committee (LOC)** was responsible for the planning and organizing activities. LOC also decided on accepting of the 6 late contributions in the Poster Session. The LOC consists of the following persons affilated with the Chalmers University of Technology, Gothenburg, Sweden:

Belitsky Victor (Chair)
Desmaris Vincent
Pavolotsky Alexey
Sjögren Paulina
Stake Jan

The LOC can be reached at isstt2019info@gmail.com

The **Scientific Advisory Committee (SAC)** members were responsible for reviewing the abstracts and making recommendations on acceptance and format of presentation for each abstract. This Committee will also decide on the place and dates for the next ISSTT2020, proceedings publication policy and the Best PhD Student Contribution Award committee.

The SAC members are:

| | |
|---|---|
| Andrey Baryshev | University of Groningen, the Netherlands |
| Victor Belitsky | Chalmers University of Technology, Sweden |
| Brian Ellison | Rutherford Appleton Laboratory, UK |
| Jian-Rong Gao | SRON, the Netherlands |
| Gregory Goltsman | MSPU, Russia |
| Christopher Groppi | Arizona State University, USA |
| Jeffrey Hessler | Virginia Diodes Inc., USA |
| Heinz-Wilhelm Hübers | DLR, Germany |
| Boris Karasik | Jet Propulsion Laboratory, USA |
| Valery Koshelets | IRE RAS, Russia |
| Alain Maestrini | Paris Observatory, France |
| Hiroshi Matsuo | NAOJ, Japan |
| Imran Mehdi | Jet Propulsion Laboratory, USA |
| Patricio Mena | University of Chile, Chile |
| Netty Honingh | University of Cologne, Germany |
| Christophe Risacher | IRAM, France |
| Sheng-Cai Shi | Purple Mountain Observatory, China |
| Jan Stake | Chalmers University of Technology, Sweden |
| Edaward Tong | Smithsonian Astrophysics Observatory, USA |
| Yoshinory Uzawa | NAOJ, Japan |
| Ghassan Yassin | University of Oxford, UK |

# 2019 30th Insternational Symposium on Space Terahertz Technology (ISSTT 2019)

April 15 – 17, 2019, Gothenburg, Sweden

## Technical Program[1]

[1]In the Technical Program, it is only the submitting author names along with the titles of the contributions, which are listed. The full lists of authors could be found in the corresponding paper.

# Monday, April 15, 2019

## 08:45 – 09:00 Welcome

## 09:00 – 10:20 Session I. Instruments, Devices and Technologies for Small Satellites

***Chair: Vincent Desmaris***

09:00 – 09:20 Goutam Chattopadhyay - Planetary/Cometary Submillimeter-Wave Instruments on Ultra-Small Platforms. Page 19

09:20 – 09:40 Maria Alonso del Pino - Fly's Eye Lens Phased Array for Submillimeter-Wave Space Instruments. Page 20

09:40 – 10:00 Jonathan Hoh - Development of an Integrated Dual-Band Schottky Receiver in the Terahertz Regime for Use in Cubesat Systems. Page 21

10:00 – 10:20 Christine P. Chen - Design and Fabrication of Silicon Stacked Architecture for 2.06 THz Receiver Front End. Page 22

## 10:50 - 11:30 Invited talk I

**Donal Murtagh**, *Chalmers University of Technology* - Mm and sub-mm spectroscopy in atmospheric science.

## 11:30 - 12:30 Session II. Schottky Receivers and Technologies

***Chair: Jan Stake***

11:30 – 11:50 Diego Moro-Melgar - Reliability and Reproducibility of Discrete Schottky Diodes-Based Sources up to 370 GHz. Page 26

11:50 – 12:10 Jeanne Treuttel - Development of Room-Temperature Schottky Diode Technology for applications in the Tera-Hertz ranges. Page 27

12:10 – 12:30 Karl Jacob - Radiometric Performance of the 530 to 625 GHz Receiver Unit of the Submillimetre Wave Instrument on JUICEs. Page 28

## 13:50 - 14:30 Invited talk II

**Karl-Friedrich Schuster**, *Institut de Radioastronomie Millimétrique* - General Development Strategies for Millimeter-wave Astronomy and historic and current approaches at IRAM.

## 14:30 - 15:30 Session III. SIS Receivers and Mixers

***Chair: Christopher Groppi***

14:30 – 14:50 Raymond Blundell - A 1.3 mm Superconductor Insulator Superconductor Mixer Receiver with 40 GHz Wide Instantaneous Bandwidth. Page 33

14:50 – 15:10 Takafumi Kojima - Performance of a 275-500 GHz SIS mixer with 3-22 GHz IF. Page 34

15:10 – 15:30 Wenlei Shan - Experimental Study of a Monolithic Planar-integrated Dual Polarization Balanced SIS Mixer. Page 35

## 16:00 - 17:00 Session IV. THz sources

***Chair: Imran Mehdi***

16:00 – 16:20 Bertrand Thomas - Digitally tunable 150 GHz Local Oscillator chian for the Submillimeter Wave Instrument onboard the ESA JUICE mission. Page 37

16:20 – 16:40 Jose V. Siles - High-power broad-band room-temperature 2.46-2.70 THz LO sources to enable high-spectral resolution mapping of HD and [NII]. Page 38

16:40 – 17:00 Nickolay Kinev - Superconducting flux-flow oscillator as the terahertz external local oscillator for heterodyne receiving. Page 39

## 17:00 - 19:30 Poster Session

### MM and Microwave Passive and Active Components

P1-1 Cristian López - Design and implementation of a broadband and compact 90-degree waveguide twist with simplified layout. Page 42

P1-2 Daniel Montofre - Study and Development of two Low-Cost and Easy-Construction Horn Antennas for Astronomy Applications. Page 44

P1-3 Jie Hu - Design of a Silicon-based 160 – 320GHz tanh-profile wide-band Corrugated Horn. Page 46

P1-4 Cristian Lopez - Broadband Waveguide-to-Substrate Transition Using a Unilateral Etched Finline Structure. Page 47

P1-5 Hawal Rashid - Compact Wideband Passive and Active Component Chips for Radio Astronomy Instrumentation. Page 50

P1-6 Isaac Lopez-Fernandez - Compact Cryogenic Wide-Band Balanced Amplifiers with Superconducting 90° Hybrids for the IF of Submillimeter-Wave SIS Mixer. Page 57

P1-7 Patricio Mena - Modelling dielectric losses in microstrip traveling-wave kinetic-inductance parametric amplifiers. Page 63

P1-8 Vincent Desmaris - Characterization of GaN-based Low Noise Amplifiers at Cryogenic Temperatures. Page 67

**THz Optics and Devices**



**THz Sources**

**Systems**



**Antennas and Telescopes**



# Tuesday, April 16, 2019

## 08:45 - 10:05 Session V. SIS Devices and Receivers

***Chair: Christophe Risacher***

## 13:50 - 14:30 Invited talk III

**Leonardo Testi**, *European Organisation for Astronomical Research in the Southern Hemisphere* - The ALMA 2030 Development Roadmap: science goals and instrument development vision.

## 11:30 - 12:30 Session VI. Future Missions and Projects - I
***Chair: Patricio Mena***

11:15 – 11:35 Paul Goldsmith - A Space Mission to Probe the Trail of Water. Page 172

11:35 – 11:55 Christopher Groppi - First Generation Heterodyne Instrumentation Concepts for the Atacama Large Aperture Submm/mm Telescope. Page 173

11:55 – 12:15 Andrei Smirnov - Millimetron Space Observatory: progress in the development of payload module. Page 180

## 13:30 - 14:10 Invited talk IV

**Paola Caselli**, *Max-Planck-Institute for Extraterrestrial Physics* - Astrochemistry at the dawn of star and planet formation.

## 14:10 - 15:10 Session VII. THz Optics and Antennas
***Chair: Hiroshi Matsuo***

14:10 – 14:30 Richard Hills - Wide-Field Designs for Off-Axis Telescopes: Application to the Optics of CCAT-prime. Page 183

14:30 – 14:50 Andrey Baryshev - In Flight Measurements System of Millimetron Primary Mirror Surface. Page 184

14:50 – 15:10 Jose Silva - Far-field beam pattern technique for high pointing accuracy characterization of GUSTO HEB mixer arrays. Page 185

## 15:40 - 17:20 Session VIII. HEBs and KIDs
***Chair: Gregory Goltsman***

15:40 – 16:00 Yuan Ren - Mid-infrared heterodyne receiver based on a superconducting hot electron bolometer and a quantum cascade laser. Page 187

16:00 – 16:20 Akira Kawakami - 2 THz Hot Electron Bolometer Mixer using a Magnetic Thin Film. Page 188

16:20 – 16:40 Changyun Yoo - Demonstration of a TACIT Heterodyne Detector at 2.5 THz. Page 191

16:40 – 17:00 Tess Skyrme - Understanding dissipative behaviour in superconducting microresonators over a wide range of readout power. Page 192

17:00 – 17:20 Eduard Driessen - Increased multiplexing of kinetic-inductance detector arrays by post- characterization adaptation of the individual detectors. Page 193

### 17:40 - 18:20 Session IX. Future Missions and Projects - II
***Chair: Valery Koshelets***

17:40 – 18:00 Hiroshi Matsuo - Prospects of High Angular Resolution Terahertz Astronomy from Antarctica. Page 195

18:00 – 18:20 Viacheslav Vdovin - New stage of the Suffa Submm Observatory in Uzbekistan Project. Page 196

## Wednesday, April 17, 2019

### 08:45 - 10:05 Session X. Future Missions and Projects - III
***Chair: Edward Tong***

08:45 – 09:05 Jose V. Siles - COMETS – Comets Observation & Mapping Enhanced THz Spectrometer at 210-580 GHz: Objectives and Development Status. Page 203

09:05 – 09:25 Martina Wiedner - The Origins Space Telescope and the Heterodyne Receiver HERO. Page 204

09:25 – 09:45 Christopher Groppi - The Terahertz Intensity Mapper (TIM): a Next-Generation Experiment for Galaxy Evolution Studies. Page 208

09:45 – 10:05 Satoshi Ochiai - Study for proposal of SMILES-2 to JAXA M-class mission. Page 216

### 10:35 - 11:15 Invited talk V

**Susanne Aalto**, *Chalmers University of Technology* - Molecules as probes of galaxy evolution - exploring the hidden growth of galaxies.

### 11:15 - 12:35 Session XI. QCL THz Sources
***Chair: Heinz-Wilhelm Hübers***

11:15 – 11:35 Marc Mertens - A Double-Metal QCL with Backshort Tuner. Page 221

11:35 – 11:55 Martin Wienold - Frequency tuning of terahertz quantum-cascade lasers by optical excitation. Page 222

11:55 – 12:15 Till Hagelschuer - A compact 4.7-THz source based on a high-power quantum-cascade laser with a back-facet mirror. Page 223

12:15 – 12:35 Yuner Gan - 81-beam supra-THz local oscillator by a phase grating and a quantum cascade. Page 224

## 13:45 - 15:25 Session XII. Radars, Systems, Backend
*Chair: Sheng-Cai Shi*

13:45 – 14:05 Bernd Klein - Digital high-resolution wide-band Fast Fourier Transform Spectrometer. Page 226

14:05 – 14:25 Ken Cooper - Validation Measurements of Humidity Profiling in Rain Using a 170 GHz Differential Absorption Radar. Page 227

14:25 – 14:45 Theodore Reck - Cold-Source Noise Temperature Measurements with a Vector Network Analyzer Frequency Extender at WR-6.5. Page 228

14:45 – 15:05 Gabriel Santamaria Botello - On the Comparison Between Low Noise Amplifiers and Photonic Upconverters for Millimeter and Terahertz Radiometry. Page 229

15:05 – 15:25 David Monasterio - A broadband down-conversion module for the extended W-Band. Page 233

## 16:05 - 17:25 Session IV. New Devices and Technologies
*Chair: Jian-Rong Gao*

16:05 – 16:25 Sergey Cherednichenko - Quantum transport at Dirac point enables graphene for terahertz heterodyne astronomy. Page 236

16:25 – 16:45 Hajime Ezawa - Design and Evaluation of SIS Photon Detectors at Terahertz Frequencies. Page 237

16:45 – 17:05 Wen Zhang - Near infrared photon detectors using titanium-based superconducting transition-edge sensors. Page 238

17:05 – 17:25 Andrey Pankratov - On-chip refrigerator integrated into a photon-noise-limited detector for high-performance Cosmology missions. Page 239

# Proceeding Contributions

# Monday, April 15, 2019

# Session I. Instruments, Devices and Technologies for Small Satellites

# Planetary/Cometary Submillimeter-Wave Instruments on Ultra-Small Platforms

G. Chattopadhyay[1], M. Alonso delPino[1], C. Jung-Kubiak[1], T. Reck[2], J. Siles[1], C. Lee[1], and A. Tang[1]

CubeSats are shoe box size satellites with low-mass, low available power, and until recently had limited communication and instrumentation capability. The size of the CubeSats are referred in the unit of 'U', where one 'U' is a cube with 10 cm x 10 cm x 10 cm dimensions.

Until recently, the development of CubeSats and its related instrumentation was primarily confined to undergraduate university research. However, in recent years, the national space agencies have been actively looking into CubeSats and SmallSats as useable platforms to supplement main missions as well as using them for standalone scientific missions. In some cases, they are being used to provide a communication relay platform for the missions where direct to earth (DTE) links are not always feasible. A case in point is the Mars Cube One (MarCO) [1] where two CubeSats were used to provide communications with the main space craft and ground control during high risk maneuvers and critical events during the recent Insight landing.

In general, a small 6U will have approximately 2U available (the rest of the space is used for solar cells for power generation, star tracker for guidance, attitude control, and other electronics) for low-power, low-mass, yet highly capable scientific payload. These CubeSat based flights will not only enable advancing proof of concept instruments to higher technology readiness level (TRL) by flying them in relevant environment, but also will allow to have multiple targeted flights with scientific data returns. Due to budget constraints, flying large missions are becoming more and more challenging. CubeSat and SmallSat platforms will provide the avenue to more scientific missions as they are much cheaper than large missions.

Developing scientific payloads CubeSat platform poses a host of challenges. First, the instrument needs to be highly compact due to the lack of available space. Second, it has to be ultra-low power due to the severe restrictions on DC power availability. And finally, one has to be innovative in the design of the antennas as traditional high gain reflector antennas (for scientific payload as well as for data communication) are not practical. Design and development of aperture deployable antennas and other innovative structures are gaining a lot of attention in this regard.



[1] Jet Propulsion Laboratory (JPL), California Institute of Technology, Pasadena, CA 91109 USA.

[2] Was at JPL, now at Virginia Diodes Inc., Charlottesville, VA 22902, USA.

We are currently developing fully functional submillimeter-wave standalone scientific instruments as technology demonstrations on 6U CubeSats. Specifically, we are developing a low-mass and low-power 500-600 GHz high-resolution spectrometer instrument on a 6U CubeSat platform capable of remotely measuring water isotopes on comets. The instrument's 18 cm diameter aperture consists of a novel low-profile leaky-wave lens based antenna with waveguide feed integrated on the CubeSat. A microelectromechanical system (MEMS) based calibration switch [2] is integrated with the receiver front-end along with low-power CMOS based backend spectrometer circuits. This allows the overall instrument mass and power to be in the range ideal for implementation on ultra-small platforms such as CubeSats. Fig.1 shows the schematic of the concept.

In this presentation, we will provide the design and implementation of the 500-600 GHz spectrometer instrument with details of innovative packaging solutions, antenna technology, and low-power backend solutions suitable for CubeSat and SmallSat platforms.

The research described herein was carried out at the Jet Propulsion Laboratory, California Institute of Technology, Pasadena, California, USA, under contract with National Aeronautics and Space Administration.

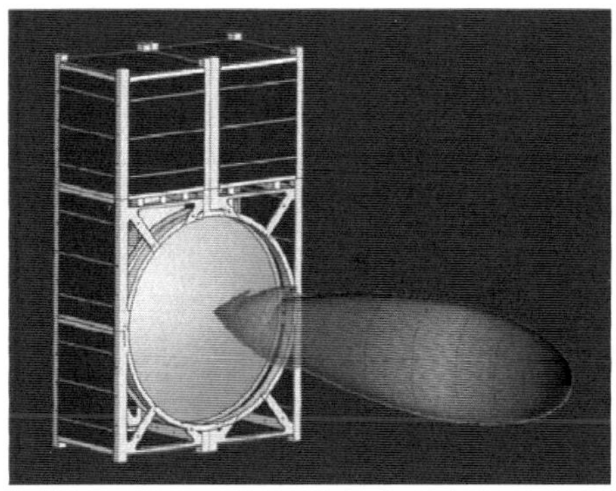

Fig. 1: Conceptual schematic of a submillimeter spectrometer instrument on a CubeSat Platform.

## References

[1] S. Asmar et al., "Mars Cube One (MarCO) Shifting the Paradigm in Relay Deep Space Operations," in *the Proceedings of the 14th International Conference on Space Operations (SpaceOps 2016)*, Daejeon, Korea, May 2016.

[2] T. Reck et al., "A 700-GHz MEMS Waveguide Switch," *IEEE Trans. THz Sci. Tech.*, vol. 6, no. 4, pp. 641-643, July 2016.

NOTES:

# Fly's Eye Lens Phased Array for Submillimeter-Wave Space Instruments

M. Alonso-delPino[1], S.Bosma[2], C. Jung-Kubiak[1], G. Chattopadhyay[1] and N. Llombart[2].

Next planetary missions to Mars or Venus require the development of submillimeter-wave heterodyne instruments to enable limb-sounding measurements that characterize the temperature and composition of gases of their atmosphere. At these frequencies, the main approach to perform such a large field of view is by using bulky mechanical scanners on the quasi-optical system. But the amount of volume, mass and power required for this mechanical scanning is impractical for most of these planetary missions.

In this contribution, we present a sparse phased array based on a fly's eye lens antenna that performs wide angle beam scanning using a piezo-electric motor. An array of silicon lenses is linearly translated from its central position of the antenna creating the steering of the main beam. The required displacement of the lens with this approach is significantly reduced compared to free standing lenses; for subwavelength applications, the maximum translation can be in the order of a few millimeters, which can still be covered by a piezo electric motor. This architecture combines a low mechanical complexity and a greatly reduced number of active elements compared to a fully sampled array. The grating lobes on the sparse array are attenuated by using a very directive element pattern (a lens antenna of a low f-number), an approach similar to limited-scan arrays. However, on the contrary of limited-scan arrays, this architecture can reach large steering angles thanks to the steering of the element pattern and the array factor.

In this work, we have studied the scanning properties of the leaky wave feed and its integration with a piezo-electric motor at 550 GHz in [1]. Furthermore, we have improved the leaky wave feed by removing the iris and adding a dielectric layer in-between the lens and the air cavity to improve the aperture efficiency and the bandwidth [2] compared to [1]. Simulated results of the array show aperture efficiencies greater than 80% for a bandwidth of 20%. Scanning angles of +/-20 degrees can be achieved with a gain loss lower than 1.5dB.

We are currently developing an array prototype of 36.7dB gain and a scanning range of +/-20 degrees at 550GHz, shown in Figure 1. The fly's eye lens array is fabricated using laser micromachining and is coated with Parylene to minimize the internal reflections of the silicon lenses. The rest of silicon wafers that define the leaky-wave feed and support the lens are processed using the DRIE process developed in [3]. The silicon wafers are sustained on a metal block fixture that integrates the piezo-electric actuator and can integrate the receiver front-end. The piezo-electric actuator is a commercial miniature translation stage based on piezoelectric inertia that is able to achieve a travel range of 12 mm with a 1 nm sensor resolution at 10 mm/s. The details on the design, fabrication and integration of the proposed fly's eye lens antenna array will be presented in the conference.

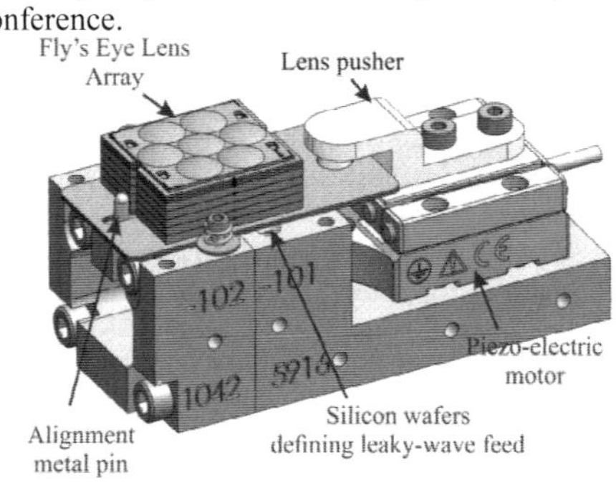


Fig. 1. Sketch of the fly's eye lens phased array prototype at 550 GHz with integrated piezo electric motor. The aperture diameter of each element is around 5.13mm.



A part of the work was carried out at the Jet Propulsion Laboratory (JPL), California Institute of Technology, under a contract with the National Aeronautics and Space Administration. A part of this work is supported by the European Union through the ERC Starting Grant LAA-THz-CC (639749).

[1] Jet Propulsion Laboratory, California Institute of Technology, Pasadena, CA. 91109 USA

[2] Terahertz Sensing group, Technical University of Delft, The Netherlands.

#### References

[1] M. Alonso-delPino, C. Jung-Kubiak, T. Reck, N. Llombart and G. Chattopadhyay, "Beam Scanning of Silicon Lens Antennas Using Integrated Piezomotors at Submillimeter Wavelengths," in *IEEE Transactions on Terahertz Science and Technology*, vol. 9, no. 1, pp. 47-54, Jan. 2019.

[2] S. Bosma, M. Alonso-delPino, D. Blanco, N. Llombart, "Fly's Eye Lens Phased Array for Submillimeter Wavelengths " submitted to 11th *Eur. Conf. Antennas Propag.*, Apr. 2019.

[3] C. Jung-Kubiak et al., "A Multistep DRIE Process for Complex Terahertz Waveguide Components," in *IEEE Transactions on Terahertz Science and Technology*, vol. 6, no. 5, pp. 690–695, 2016.

NOTES:

# Development of an Integrated Dual-Band Schottky Receiver in the Terahertz Regime for Use in Cubesat Systems

Jonathan R. Hoh[1], Jose V. Siles[2], and Christopher E. Groppi[1]

Molecular clouds are the coldest form of matter found in the universe and are the heart of the star formation process. Reaching temperatures as low as a few Kelvin, these vast regions are often called "star nurseries" due to their unique environment which allows for the collapse of mass into nascent stars. Molecular clouds are typically comprised of little more than a collection of molecular hydrogen ($H_2$) and trace elements such as silicon and carbon found in intermittent dust grains. Due to the extremely low temperatures in these regions, observing molecular clouds is a difficult task that relies on the radiation produced by low-energy rotational transitions.

Water vapor ($H_2O$) is a key tracer of star formation and is of particular interest in learning about the origins of the Earth and our own solar system. Since the ground-state rotational transition of ($H_2O$) at 556 GHz (0.5 mm) only requires 27 Kelvin of thermal energy between its upper and lower states, it provides an excellent probe of the excitation conditions of the cold gas surrounding birthing stars. This in combination with significant self-absorption due to the high optical depth of the transition provides a complex ($H_2O$) profile which unveils information about the molecular cloud such as density, temperature, velocity structure, and even geometry.

While ($H_2O$) is an extremely important spectral line for demystifying molecular clouds, detecting and mapping these signatures is a difficult task. Most notably, the 556 GHz line is completely saturated when looking out from Earth due to the water vapor in our atmosphere. Astronomers have attempted to make observations of this spectral line from balloons launched high above the ground, but even the small amount of water in the atmosphere at this altitude makes observations difficult. The only way to truly avoid atmospheric occlusion in this band is to observe on a satellite in space, which was the primary driver of missions such as SWAS, Odin, and Herschel. However, full-sized satellite missions such as these are prohibitively expensive and are usually are not focused specifically on the 556 GHz line. In order to push further into ($H_2O$) observations of molecular clouds, technology must be developed to allow for specialized cubesat or smallsat missions which will make observations in this spectral band economically feasible without trading away spatial or spectral resolution.

Here we present the completed design for a low-mass, low-power, highly integrated Schottky diode based coherent receiver system suitable for deployment on cubesats or other smallsat platforms. The current state of coherent Schottky diode receivers are still far too large and consume too much power to be considered for use on these smaller platforms. Through use of a novel packaging system, we have condensed JPL designs for both a modular 520-600 GHz receiver and a 1040-1200 GHz receiver into a single integrated receiver and mixer block. This combined block is pumped by a single local oscillator (LO) and has been shown to have significantly smaller volume and power consumption than the current state-of-the-art while maintaining the noise temperature of its larger counterparts.

Furthermore, we will discuss and present the designs of a Gaussian-optic thermal break which will separate the Schottky diodes from the heat dissipation caused by the LO chain feeding the mixers. We will then present some results of tests and measurements done to the manufactured integrated receiver. Finally, we will conclude with a discussion of future plans for the system and its promising potential for use in cubesat interferometry systems.


## References

*Please use this section to provide references (up to 3 references)*

[1] Ashby, M. L., Bergin, E. A., Plume, R., Carpenter, J. M., Neufeld, D. A., Chin, Melnick, G. J. (2000). An Analysis of Water Line Profiles in Star Formation Regions Observed by the [ITAL] Submillimeter Wave Astronomy Satellite. The Astrophysical Journal, 539(2)

[2] Bally J. (1986). Interstellar Molecular Clouds. Science, 232(4747), 185-193

[3] Snell, R. L., Howe, J. E., Ashby, M. L., Bergin, E. A., Chin, G., Erikson, N. R., Melnick, G. J. (2000). Water Abundance in Molecular Cloud Cores. The Astrophysical Journal, 539(2)



[1] School of Earth and Space Exploration, Arizona State University, 781 S Terrace Rd, Tempe, AZ, USA.

[2] Jet Propulsion Laboratory, California Institute of Technology, 4800 Oak Grove Drive, Pasadena, CA, USA.


NOTES:

# Design and Fabrication of Silicon Stacked Architecture for 2.06 THz Receiver Front End

Christine P. Chen, Darren Hayton, Robert Lin, Cecile Jung-Kubiak, Jose Siles, Joseph Lee, Maria Alonso, Alex Peralta, and Imran Mehdi

***Abstract*— By designing a Si three-dimensional stack, a reproducible process for iterative-use and accurate assembly at 2.06 THz is created. Low noise temperature is measured at 2.06 THz with a metal block at room temperature. The Si block system created via multi-step DRIE process is described. The mask layout and design considerations are discussed. Design for fine assembly, as well as process specifications of side wall smoothness and accurate etch depths of the μm-scale waveguides are essential to optimal sub-harmonic mixing performance. Variability tolerances are described, with future explorations covered.**

***Index Terms*—Aerospace components, Silicon, Space technology, Submillimeter wave propagation, System integration, Wafer scale integration**

## I. Introduction

NASA'S earth science directorate includes as a focus understanding the atmospheric dynamics in Earth's upper atmosphere [1]. Schottky diode mixers operating at several THz have historically been a stable method of passive sensing in space, with prior mission deployments [2]. The capability to leverage silicon (Si) fabrication [3] and integrate the local oscillator (LO) and intermediate frequency (IF) signal onto a compact Si micro-machined package has the potential to introduce new features and design paradigms, including multi-functional arrays for compact systems.

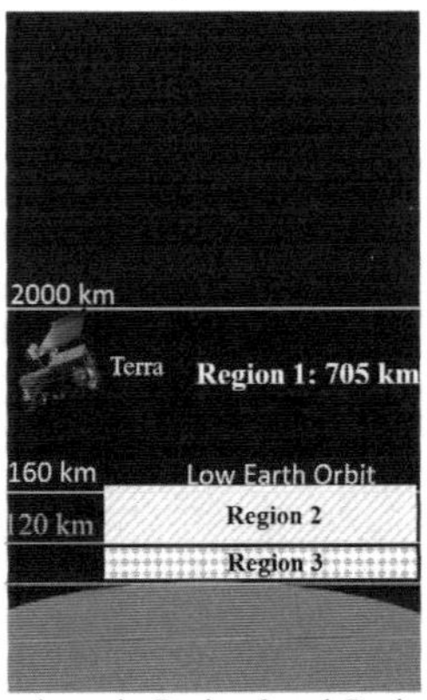


Fig. 1. Target science in Region 2 and Region 3, while prior missions have occurred in Region 1, like Terra. $O_2$ will be measured in Region 3, which is <100 km above Earth. Meanwhile, dynamics including those in Region 2 will be measured via Doppler shift and OI emission.

A heterodyne receiver front end is being developed to perform measurements at 2.06 THz, where a neutral oxygen [OI] line exists. By measuring this atmospheric feature, thermos-spheric models can be created for understanding space weather and its impact on earth climate. Recently, laboratory measurements have been performed with traditional metal machined blocks to validate design. The system currently consists of a LO chain with a frequency synthesizer, followed by a gallium nitride (GaN) power amplifier and a solid-state-based tripler chain up to 1.03 THz. At 63.9 K ambient temperature, the mixer was measured to have a double sideband (DSB) noise temperature of 8186 K at 2.0034 THz.

This paper introduces a Si stacked architecture designed to replace the metal blocks, specifically the interface housing the low-parasitic 2.06 THz Schottky diode mixer. Si micro-machining allows for desirable features such as accuracy and reproducibility. While the Si fabrication is utilized for the last stage of this LO chain, the scheme is such that it could be extendable to combine with lower frequency mixers due to the modularity of each of these parts. The Si block houses the waveguide structure, IF and bias boards, as well as the interconnection to the electrical output feeds. Considerations for this hybrid Si and metal interface include having a reusable interface assembly between the metal and Si block and having an interchangeable Si block for different interface design. Compared to Computer Numerical Controller (CNC) milling, lithographical features can result in better precision. Furthermore, by having the vertical waveguide coupling, the design envisions a compact coupling in of the RF signal. With high-resistivity Si, it is estimated that losses are comparable with metal-machining.

## II. Background

Silicon micromachining is used to create the fine waveguide features in the three-dimensional stack. The fabrication process consists of two main steps. In order to etch straight side walls, reducing ohmic losses, multiple $SiO_2$ masks are used. This type of process allows for thicker structures and multiple-levels for different feature heights [4]. The recipe used in this particular paper was first developed and described in [3] at Jet Propulsion Laboratory and adapted to be used for this work. The main steps

C. P. Chen was supported by the NPP Fellowship, administered by USRA.

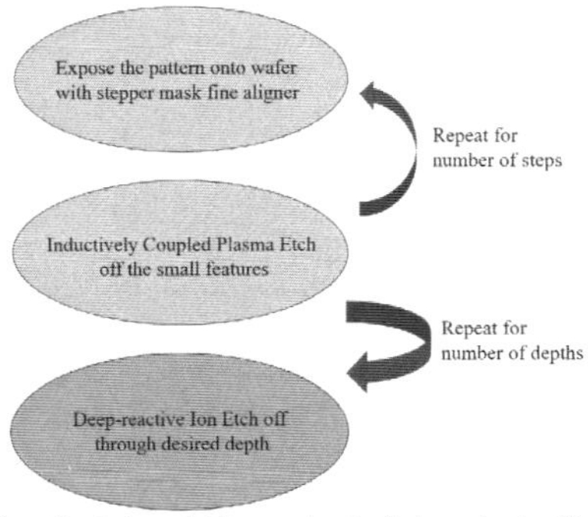


Fig. 2. In order to have high-aspect ratio feature size in silicon, deep-reactive ion etching is used. This consists of a three-step cycle of etching with $SF_6$ and passivation with $C_4F_8$.

are illustrated in Fig. 2, where a three-step process for deep-reactive ion etching is performed. Negative photoresist is lithographically patterned on the 4-inch Si wafer. With the pre-determined selectivity of the process, inductively coupled plasma (ICP) etch is initially performed on the $SiO_2$. This step allows for finer precision over final target thicknesses. Then, deep-reactive ion etching (DRIE) is done, creating the desired depths of the Si waveguide. Each etch creates the vertical waveguide features. Waveguide depths are measured using a profilometer, in order to ensure accuracy of process, which will be discussed further in Sec. IV.

## III. Methodology and Measurement Results

The Si stack is designed to be embedded into a LO chain

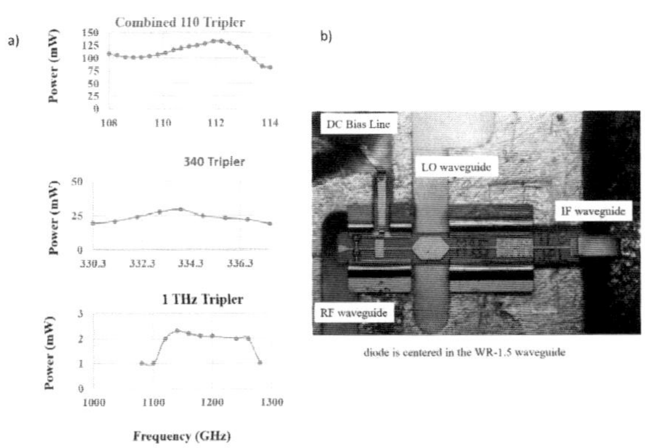


Fig. 3. a) The diagrams on the left are frequency sweeps at each stage of the multiplier chain. b) The diagram on the right is the top down view of device layout.

depicted in Fig. 4. The GaAs Schottky diode mixer is a technology optimized for lower parasitics and high-frequency mixing [5]. The chain is composed of a sequence of Schottky-diode based multiplier blocks, all the way up to 1 THz. Each component of this chain is characterized, as shown in Fig. 3a. The LO chain, with the experimental setup shown in Fig. 4, was measured to capture the power across frequency at each step of the multiplier chain. The key desired metric for this

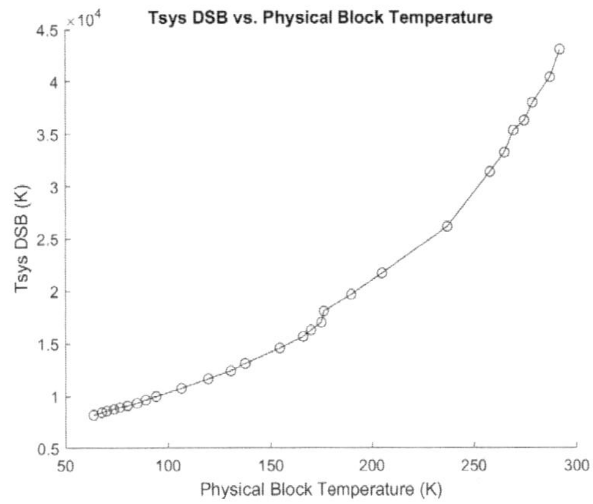


Fig. 5. Double side band (DSB) noise temperature measurement of the Schottky diode mixing in a metal-machined waveguide block structure at 2.003 THz.

room temperature LO source is generating enough power, relative to existing sources, to enable the 2.06 THz subharmonic mixing, which has been developed in prior works [6-8]. Alignment tolerances become tighter as frequency is scaled. In the 2.06 THz subharmonic mixer, the interface between the 1 and 2 THz blocks are crucial. Fine assembly will dictate the performance of the mixer.

The noise temperature measurement with the metal block is shown in Fig. 5, a world-record value of 8186 K measured at 63 K. With such a development of the chain, the Si micro-machined part will be introduced. This will minimize the cost and time to delivery. It will also miniaturize the assembly, integral for flight purposes. In the next section, Si micromachining is examined for the 2.06 THz Si architecture.

## IV. Silicon Integration Results and Discussion

As frequency scales, the size of the waveguide and device anode miniaturize, increasing complexity for assembly. The considerations become much tighter for 2.06 THz mixing.

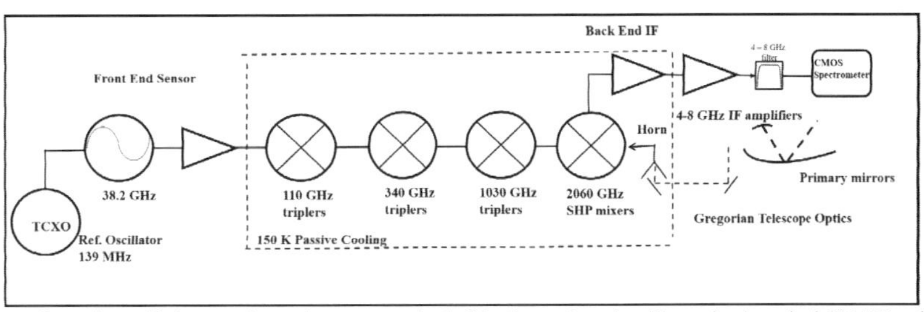


Fig. 4. Flight instrument configuration, with the experimental setup comprised of the frontend portion, illustrating how the 2.003 THz mixer is being measured

Diode alignment at the center of the waveguide is essential. Performance can deviate from simulated behavior when device is not coupling the quasi-TEM mode signal as expected in the center of the waveguide [9]. With a 40 μm pitch, a misalignment of 20 μm is already 13% of the wavelength.

Mechanically, the contact pads as seen in Fig. 3b are effectively sandwiched between the top and bottom blocks. Where these regions are relatively located will influence the overall soundness of assembly. Beyond assembly expertise, design considerations for optimized integration are integral for alignment of such tight tolerance. This makes integration design all the more essential to an optimized process and optimal performance.

The Si blocks were designed based on prior experience with high-frequency alignment. A key challenge for this type of integrated architecture is to perform reliable testing given sub-millimeter constraints. With the features being etched onto two pieces of 350-μm Si wafer, the Si block is designed to be

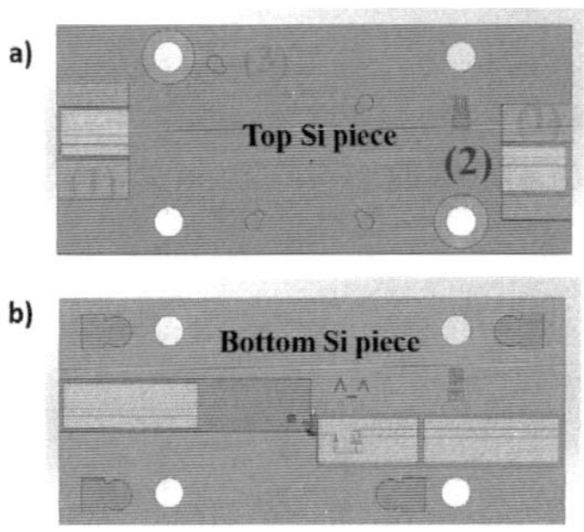


Fig. 6. The (a) top and (b) bottom Si pieces, with several places denoted. (1) quartz-based IF boards, (2) indirect monitor of coupling through relative power shifts, and (3) interfaces to block via pins. During assembly, the two pieces are stacked on top of each other.

assembled with fine tolerance by incorporating both prior metal block assembly strategies and a novel optical alignment scheme. Using a previous mechanism of fine alignment by placing Si pins within etched cavities between each of the interfaces [3], the chain can be assembled sequentially. Crucial also to optimal performance is fabrication accuracy. By measuring the waveguide features across wafer, accuracy and uniformity are verified during the fabrication process. The vertical waveguide, which bridges the two wafers, must have fine fabrication alignment in order to have any mismatch <1 μm order.

From Fig. 6, the top and bottom of the Si stack is shown. These are aligned via the alignment marks indicated. Prior work in this area of alignment have yielded H-plane signal losses at 500-650 GHz to 0.08 dB/mm at 600 GHz, which overlaps with the idealized waveguide loss and is only 0.02 dB/mm greater than a metal machined waveguide [10]. In order to attain such high precision, on the fabrication side, losses due to sidewall roughness are minimized, and accuracy as compared to vertical waveguide lengths simulated are ensured, both possible per the steps described in Sec. II.

Considerations for methods of extracting the IF line out are explored, including with smaller, integrated I/O. More compact integration of IF line is possible at the expense of a slightly higher power loss over SMA output, with room for improvement.

Once these pieces of the Si stack are aligned, the signal is designed to be coupled out via a Si microlens antenna [11]. Novel alignment protocols allow for alignment monitoring prior to and during measurement.

## V. Conclusion

Future THz instruments may be strengthened in the arena of multi-pixel arrays and integrated subsystems through exploring and refining new fabrication technologies [12]. The scheme presented in this paper will help to enable these larger-scale array architectures with reliability at a much smaller stack area of around 400 mm$^2$ area than previously demonstrated. Furthermore, the next steps in realizing a 2.06 THz mixer for THz limb sounding in the lower earth atmosphere are developed.

## Acknowledgment

The authors are at the NASA/Jet Propulsion Laboratory, California Institute of Technology, Pasadena, CA, 91109, with C. P. Chen under the NPP program, which is administered by the Universities Space Research Association under contract with NASA.

## References

[1] "Atmospheric Composition" https://science.nasa.gov/earth-science/programs/research-analysis/atmospheric-composition, 2018.
[2] I. Mehdi, "THz Instruments for Space Exploration," *Proceedings of 2017 Asia Pacific Microwave Conference*, 2017.
[3] C. Jung-Kubiak, *et al.* "A Multistep DRIE Process for Complex Terahertz Waveguide Components," *IEEE Transactions on Terahertz Science and Technology*, vol. 6, no. 5, Sept., 2016.
[4] Y. Mita, *et al.* "Embedded-Mask-Methods for mm-scale multi-layer vertical/slanted Si structures," *Proceedings IEEE Thirteenth Annual International Conference on Micro Electro Mechanical Systems*, 2000.
[5] I. Mehdi, *et al.* "THz Diode Technology: Status, Prospects, and Applications," Proceedings of the IEEE, 2017.
[6] J. Treuttel, *et al.* "A 2 THz Schottky solid-state heterodyne receiver for atmospheric studies," SPIE, 2016.
[7] P. Siegel, "Terahertz Technology," IEEE Transactions on Terahertz Science and Technology, Vol. 50, No. 3, 2002.
[8] P. Siegel, "2.5-THz GaAs monolithic membrane-diode mixer," IEEE Transactions on Terahertz Science and Technology, Vol. 47, Is. 5, 1999.
[9] A. Maestrani, "Multi-Anode Frequency Triplers at Sub-Millimeter Wavelengths," trs.jpl.nasa.gov, 2005.
[10] T. J. Reck, *et al.* "Measurement of Silicon Micromachined Waveguide Components at 500-750 GHz," IEEE Transactions on Terahertz Science and Technology, Vol. 4, No. 1, 2014.
[11] M. Alonso, "Development of Silicon Micromachined Microlens Antennas at 1.9 THz," IEEE Transactions on Terahertz Science and Technology, Vol. 7, No. 2, 2017.
[12] I. Mehdi, "THz Instruments for Space Exploration," APMC, 2017.

# Monday, April 15, 2019

# Session II. Schottky Receivers and Technologies

# Reliability and Reproducibility of Discrete Schottky Diodes-Based Sources up to 370 GHz

*D. Moro-Melgar, O. Cojocari, I. Oprea, M. Hoefle & M. Rickes.

*ACST GmbH, Hanau, 63457, Germany*
*Contact: diego.moro-melgar@acst.de

***Abstract***— We present a set of two different THz sources covering the 275-360 GHz frequency range average output power of 14-15 dBm. This power is obtained without any power combining technique allowing a significant reduction in the complexity of the source and the number of components. The presented sources are based on ACST high power frequency doublers covering the 135-185 GHz and 270-370 GHz frequency ranges based on novel Schottky-varactor discrete diode structures and assembly process. These varactor diodes integrate a CVD-diamond substrate to increase the thermal dissipation capabilities and provide higher power handling capabilities to the doubler chip. The first source presented in this work cover the frequency band proposed for the submillimeter wave instrument (SWI) in the ESA's mission JUICE-SWI, and it demonstrates high reproducibility and reliability. The D and G-band doublers can safely handle up to 600 mW input power to provide 160-180 mW output power. However, the D-band doubler (135-160 GHz) doubler has demonstrated 250 mW output power using 1 W input power without failure. The Y-band doublers can safely handle up to 140 mW input power to provide 30-40 mW output power. An overview of the experimental results compared to simulated predictions are discussed here.

## 1. Results

Nowadays the market volumes for submm-wave sources and receivers are very limited. Moreover, specification requirements may vary from one particular application to another. Under these conditions, the use of discrete diode structures for frequency multipliers and mixers is of significant advantages in comparison to monolithically-integrated MMIC diode circuits, concerning flexibility and price. The experimental results of two different manufactured THz sources are plotted in Fig. 1. The first source operates in the 275-315 GHz frequency range, and it consists of an active multiplication chain (AMC) to reach E-band (60-90 GHz) and two multiplication stages at 135-160 GHz and 270-320 GHz. The second source is a scaled version of the first one to operate in the 315-360 GHz frequency range. The second source consists of an AMC to reach E-band and two multiplication stages at 155-185 GHz and 310-370 GHz. Commercial 71-76 GHz and 81-86 GHz high power amplifiers are used in source one and two, respectively, to build up the setups. The maximum power achieved by these sources is between 14-16 dBm with >15 % bandwidth. These sources are defined using discrete Schottky diodes without any power combined technique in the multiplication chain. These discrete diodes-based sources provide state-of-the-art performance previously defined only by MMIC sources [1].

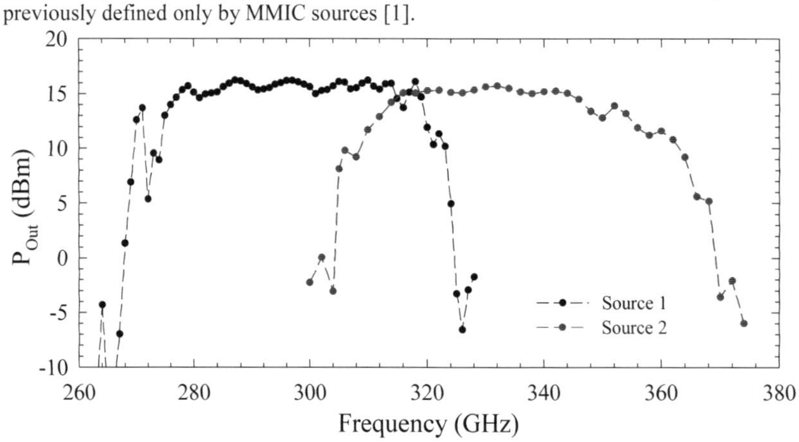


Fig. 1. Output power provided by two different THz sources developed at ACST.

[1] D. Moro-Melgar, O. Cojocari, and I. Oprea. "High Power High-Efficiency 270-320 GHz Source Based on Discrete Schottky Diodes." In *2018 15th European Radar Conference (EuRAD)*, pp. 337-340. IEEE, 2018.

# Development of Room-Temperature Schottky Diode Technology for applications in the Tera-Hertz range

J. Treuttel[1], L. Gatilova[1], J. Valentin[1], A. Maestrini[1], Y. Jin[2], A. Cavanna[2], C. Goldstein[3], P. DeMaagt[4]

Since 2006 the LERMA-Observatoire de Paris in close collaboration with C2N has made a great progress in the development of the French technology of THz electronic components based on Schottky diodes. By bringing together the unique knowledge and skills of both laboratories, we have developed submillimeter devices at 300GHz, 600GHz and 1.2 THz, with state-of-the-art performances. These devices are selected today for the Submillimetre Wave Instrument (SWI) of the JUICE planetary probe, ESA's first class L mission. The progress made over the last years and our future work on the device miniaturization and increasing working frequency will be discussed in this presentation.

The design of MMIC Schottky based circuits are made by LERMA-Observatoire de Paris. The growth of the especially dedicated MBE epitaxial GaAs structures and the manufacturing of the Schottky-based circuits are performed in the clean room of C2N by LERMA and C2N staff members. We have developed the approach using direct E-beam writing for all fabrication steps, allowing a great flexibility of design, excellent anode definition and perfect alignment between the different lithography steps. An important effort has been made during the last two years in order to improve the fabrication process. A dedicated metal evaporator has been acquired in order to improve the ohmic and Schottky contact. The result of this several-months study was the production of diodes with excellent characteristics and very good homogeneity across the wafer.

In 2015, LERMA demonstrated state-of-the art results on its first MMIC subharmonic mixer at 600GHz. The record low average noise temperature of 750K has been measured at 150K in 520-620 GHz range (with the minimum value of 550K at 560 GHz) [1]. The mixer was used in our laboratory for molecular spectroscopy measurements. It was obtained that our compact 600 GHz room-temperature receiver provides the same order of sensitivity and frequency accuracy as InSb HEB spectrometer [2].

In 2016-2017 we realized our first 1200 GHz receiver for SWI. We obtained a receiver noise temperature two times better than specified for the mission.

Current work focuses on the development of a bias-able 2THz sub-harmonic mixer based on a preliminary design made for ESA R&D study [4].

This work was supported by CNES, ESA and the LABEX ESEP.

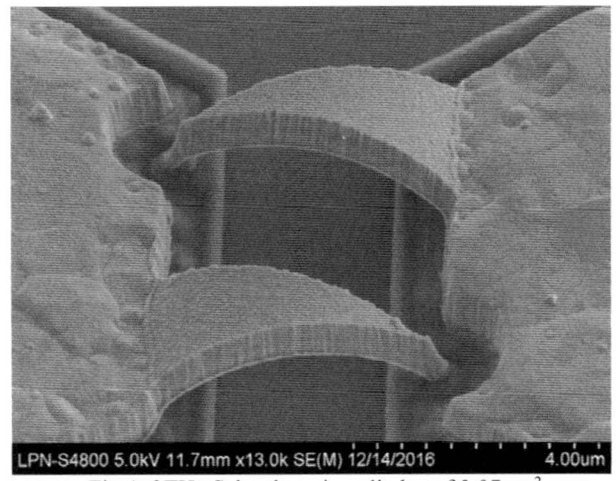


Fig.1. 2THz Schottky mixer diodes of 0.07μm$^2$

### REFERENCES

[1] J. Treuttel, L. Gatilova, A. Maestrini et al., "A 520–620-GHz Schottky receiver front-end for planetary science and remote sensing with 1070 K–1500 K DSB noise temperature at room temperature," *IEEE Trans. Terahertz Sci. Technol.*, vol. 6, no. 1, pp. 148-155, Jan. 2016.

[2] A. Pienkina, "Theory of traveling-wave optical laser Heterodyne Receiver for Interstellar Molecular Laboratory Spectroscopy," *PhD Thesis,* University of Lille 1, march 2018.

[3] A. Maestrini, L. Gatilova, J. Treuttel ,Y. Jin, A. Cavanna, D. Moro Melgar, T. Vacelet, A. Féret, S. Caroopen, G. Gay, F. Dauplay, J-M. Krieg, P. De Maagt, C. Goldstein, "1080-1280GHz Schottky Receiver for JUICE-SWI with 1600-2600K DSB Receiver Noise Temperature" *29th IEEE International Symposium on Space THz Technology (ISSTT2018)*, Pasadena, CA, USA, March 26-28, 2018

[4] Funding Program ESA-ITT1/ 8271 / NL /MH

[1] LERMA, Observatoire de Paris, PSL Research University, CNRS, UMR 8112, Sorbonne. Universités UPMC. Paris 06, F-75014 Paris, France

[2] Center for Nanoscience and Nanotechnology (C2N), 10 Boulevard Thomas Gobert, 91120 Palaiseau, France

[3] Centre National d'Etudes Spatiales (CNES), 18 avenue Edouard Belin, F-31401 Toulouse cedex 9, France

[4] European Space Agency (ESA-ESTEC), Keplerlaan 1, PO Box 299, NL-2200 AG Noordwijk, The Netherlands.

NOTES:

# Radiometric Performance of the 530 to 625 GHz Receiver Unit of the Submillimetre Wave Instrument on JUICE

Karl Jacob[1,*], M. Kotiranta[1], P. Sobis[2], A. Emrich[2], V. Drakinskiy[3], J. Stake[3], A. Maestrini[4], J. Treuttel[4], B. Thomas[5], M. Philipp[5], P. Hartogh[6], and A. Murk[1]

[1]*Institute of Applied Physics, University of Bern, CH-3012, Bern, Switzerland*

[2]*OMNISYS Instruments AB, SE-421 32, Västra Frölunda, Sweden*

[3]*Chalmers University of Technology, 41296, Göteborg, Sweden*

[4]*LERMA, Observatoire de Paris, 75014, Paris, France*

[5]*Radiometer Physics GmbH, Meckenheim, Germany*

[6]*Max Planck Institute for Solar System Research, D-37077, Göttingen, Germany*

*Contact: karl.jacob@iap.unibe.ch.

***Abstract*— The upcoming Submillimeter Wave Instrument on the JUICE spacecraft is a passive radiometer/spectrometer instrument with two heterodyne receivers which are independently tunable in the frequency bands 530 to 625 GHz and 1080 to 1275 GHz. It will study Jupiter's atmosphere as well as the atmospheres and surface properties of the Galilean moons. This work presents the results of first radiometric tests with a prototype of the 600 GHz receiver. In this context, the baseline ripples caused by the internal calibration target have been characterized using two conical prototypes with a linear and an exponential absorber coating profile. A significant reduction of the baseline ripple amplitude has been measured with the target having the exponential cone profile. The spectroscopic baseline has been characterized for various frequency steps when applying frequency switching as an alternative calibration mode. At some operating frequencies a very flat switching baseline has been measured for frequency throws up to 90 MHz, while at other frequencies significant spectral distortions are measured even with a step size of 22.5 MHz. The first radiometric tests of the sideband gain ratio with a passive Fourier Transform Spectroscopy method demonstrate the general applicability in the 530 to 625 GHz band.**

## I. INTRODUCTION

The JUpiter ICy moons Explorer (JUICE) is a L-class mission of the European Space Agency (ESA) to investigate Jupiter and the Galilean satellites Ganymede, Callisto and Europa. The launch of the spacecraft is scheduled for 2022 and the scientific payload consists of 10 instruments. One of these instrument is the Submillimetre Wave Instrument (SWI). SWI is a passive heterodyne radiometer/spectrometer instrument that is going to observe the atmospheres and surface properties of the Jovian objects with two orthogonally polarized receivers which can be independently tuned in the frequency ranges from 530 to 625 GHz and from 1080 to 1275 GHz [1]. The two double sideband (DSB) receivers are based on sub-harmonic mixers which will be passively cooled to a temperature of about 140K to improve the sensitivity of the instrument. The back-end of SWI includes two Chirp Transform Spectrometers (CTS), as well as two Auto Correlation Spectrometers (ACS) and two continuum channels. SWI will be calibrated with views to cold space and using an internal conical blackbody calibration target (CHL) which acts as the hot reference [2]. A planar flip mirror can be activated to allow the 600GHz and the 1200GHz receivers to view the CHL. This internal calibration target should exhibit a homogeneous temperature distribution, as well as an emissivity close to unity. The coherent backscattering $S_{11}$ of the internal calibration target leads to standing waves between the calibration target and the receivers, that cause periodic baseline ripples in the calibrated spectra. This limits the absolute accuracy of the calibration and complicates the evaluation of the spectra. Since the flip mirror mechanism has a relatively slow switching time and a limited lifetime, alternative calibration modes need to be considered to overcome gain drifts of the receivers. Hence, it is also planned to apply frequency switching as an alternative calibration mode that requires a less frequent calibration with the CHL. A flat spectroscopic baseline is necessary over the entire intermediate frequency (IF) band and large frequency throws are preferable for the data evaluation. Additionally, an accurate analysis of the observed spectral lines requires a precise knowledge of the sideband gain ratio of the double sideband mixers. Therefore, determining the receiver characteristics before the launch is crucial. This paper first presents the design of the 530 to 625 GHz receiver unit of SWI and then reports the results of the first radiometric measurements obtained with a receiver prototype.

## II. RECEIVER UNIT DESIGN

A view from the bottom into the receiver unit box together with the optical beam path is illustrated in Figure 1. More detailed information about the optical design and the components can be found in [3]. The 600 GHz receiver chain consists of an E-band

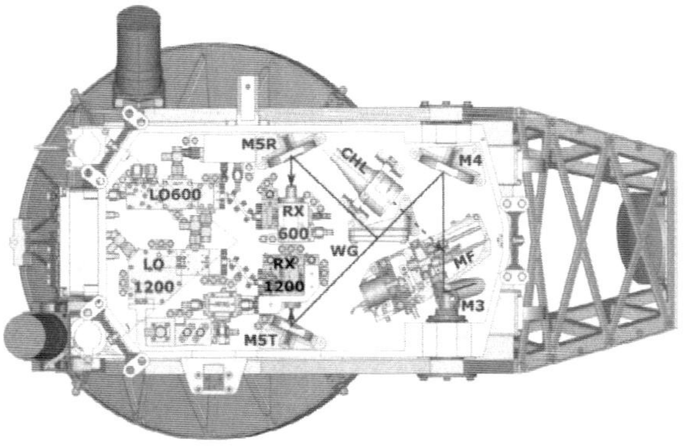


*Figure 1. Bottom view of the receiver unit box with optical beam path.*

tripler and an E-band power amplifier (MPA) from Radiometer Physics GmbH (RPG) in Meckenheim, Germany. The elements are followed by a chain of two varactor diode multipliers, a 140 GHz doubler from RPG and a 280 GHz doubler from LERMA in Paris, France. The local oscillator (LO) delivers a maximum power between 4 and 10 mW at ambient temperature operation across the entire SWI frequency range from 265 to 312.5 GHz. The sub-harmonic DSB mixer has been developed by Omnisys Instruments AB in Västra Frölunda, Sweden. The mixer block consists of a broadband GaAs Schottky membrane diode and a commercial low noise amplifier (LNA) that has been designed by the company Low Noise Factory, which are both optimized for cold temperature operation. The mixer is based on Terahertz Monolithically Integrated circuits (TMIC) and cryogenic InP High Electron Mobility Transistor (HEMT) LNA MMIC from Chalmers University of Technology. Both chips are integrated into a single block including the bias connections for the LNA in order to reduce the size and mass. The mixer block with the LNA will be passively cooled to a temperature of about 150 K in order to improve the signal-to-noise ratio by a thermal strap that is connected to an external cold space radiator. In addition, the last doubler is cooled in order to improve the efficiency and extend its lifetime. Since the other components (LO600 in Fig. 1) of the receiver unit will be operated at the temperature of the spacecraft of about 220 K, the cooled components (RX600 in Fig. 1) are insulated by an additional titanium waveguide. More information about the receiver unit design can be found in [4].

## III. Radiometric Characterization

### *A. Standing Waves to Calibration Target*

The baseline ripples, which are caused by a small but non-zero reflectivity of the CHL, have been determined in combination with a prototype of the 600 GHz receiver unit integrated into a breadboard model of the receiver unit optics and a prototype of the CHL. Two prototypes of the CHL with different geometries have been designed and manufactured, one with a commonly used linear and one with an exponential profile of the absorber coating to reduce the reflectivity of the CHL as described in [5]. The IF signal has been measured with the prototype of the SWI

*Figure 2. Baseline ripple determined with the linear CHL prototype.*

ACS, resolving a bandwidth of 4.4 GHz with 256 channels. The measurement setup is calibrated with a Y-factor measurement using a liquid nitrogen and an ambient temperature reference. The differences of two calibrated spectra, one of them measured with an axial movement of the target of 1/4λ, is calculated to make sure that only the standing waves caused by the test object are observed. At a fixed central frequency and distance to the calibration target, the double sideband detection of the standing wave leads to destructive and constructive superposition in the lower (LSB) and the upper sideband (USB) in the IF spectrum. Because the absolute position of the target is not clearly defined and the location at which the reflection at the target occurs is difficult to identify, measurements at only two positions are not sufficient to determine the amplitude in the worst case, where the resulting baseline ripples in the two sidebands superimpose constructively. Therefore, the resulting spectra are measured at two additional target positions, which are shifted by 1/8 λ to the first set of measurements. The amplitude in the worst case is calculated with the root sum square (RSS) of both amplitudes. Figure 2 depicts the baseline ripples determined with the linear CHL prototype at the central frequency of 590 GHz. Additional measurements with a commercially available TK-Ram absorber from Thomas Keating Ltd., UK, and a state-of-the-art pyramidal calibration target have been performed for a comparison. The pyramidal target is the on-board calibration target of the ice cloud imager (ICI) instrument as part of the second generation of the meteorological operational satellites (MetOP-SG) [6]. All results are summarized in Table 1. The amplitudes determined with both CHL prototypes and the ICI target are significantly smaller in comparison to the TK-Ram absorber. The worst case amplitude of the linear prototype is a factor of four higher compared to the worst case amplitude of the exponential cone.

*Table 1. Extracted standing wave amplitudes of the measured targets.*

| Target | **ΔTB [K] (0/8 – 2/8)λ** | **ΔTB [K] (1/8 – 3/8)λ** | ΔTB[K] RSS |
|---|---|---|---|
| SWI CHL lin. | **0.78** | **0.11** | 0.79 |
| SWI CHL exp. | **0.10** | **0.16** | 0.19 |
| MetOp-SG ICI | **0.43** | **0.15** | 0.46 |
| TK-RAM | **9.1** | **8.7** | 12.6 |

### *B. Frequency Switching Baseline*

The frequency switching method will to be implemented as one of the SWI calibration modes and, therefore, the quality of the difference spectra has been investigated with the SWI 600 GHz receiver unit prototype and the prototype of the SWI ACS. The

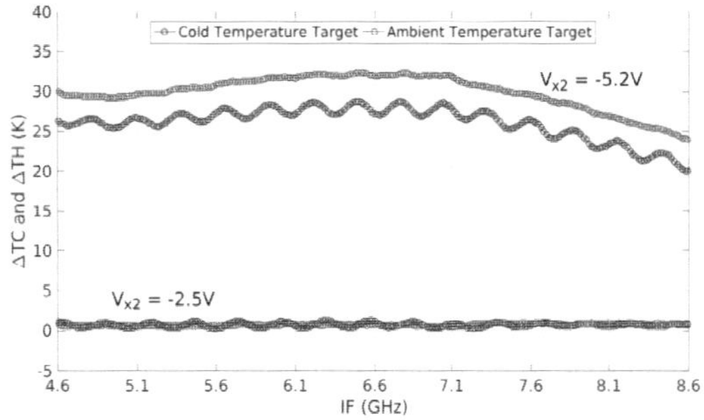


*Figure 3. Switching baseline at 590 GHz when looking onto a liquid nitrogen and a room temperature target for two bias voltages of the last LO stage with the same LO power and for a stepsize of 22.5 MHz.*

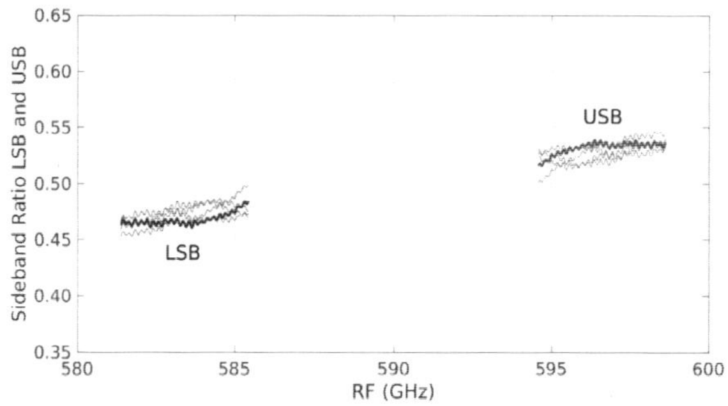


*Figure 4. Fitted sideband ratio at 590 GHz with different path length. of the MPI. The results obtained with the distance showing the highest condition number of the coefficient matrix is shown with the thick line.*

minimum frequency step of the SWI synthesizer is 1.875 MHz and with the harmonic number of 24, this results in the total step of the subharmonic LO of 22.5 MHz. The measurements were carried out by applying a similar frequency step and multiples of the step size using a laboratory synthesizer. The beam of the receiver was periodically directed onto a liquid nitrogen and a conical ambient temperature target for one second and the LO frequency was switched at each position. As an example Figure 3 shows the switching baseline at a LO frequency of 295 GHz for different bias settings of the last LO stage with the same LO output power and for the smallest stepsize of 22.5 MHz. In both cases the receiver has been optimized for the minimum noise temperature by changing the bias voltage setting of the 280GHz doubler according to a pre-determined look-up table. At a LO frequency of 295 GHz the optimum is reached at two different bias voltages while maintaining the bias voltage setting of the remaining LO components. The bias voltages -2.5 V and -5.2 V result in a similar LO output power of about 1.8 mW. While an almost flat switching baseline is observed at a voltage of -2.5 V, the switching baseline at a voltage -5.2 V shows significant spectral baseline distortions. In the second case a more complex calibration scheme would be needed in which the bias settings are also changed at each LO frequency step. The periodic ripple on the cold target are an artifact due to a standing wave from reflections at the liquid nitrogen surface. This result shows that the voltage settings cannot be selected solely on the basis of an optimal noise temperature if frequency switching will be used.

### C. Sideband Gain Ratio Measurements

The frequency dependent sideband gain ratio of the 600 GHz receiver prototype has been measured with a scanning Martin Puplett Interferometer (MPI). The used polarizing dual-beam interferometer consists of three polarizing wire grids, a fixed rooftop reflector, and a second rooftop reflector mounted on a translation stage. Behind the MPI a flip mirror is used to switch between an ambient temperature and a liquid nitrogen target. From the resulting oscillations of the Y-factor with varying path length difference in the two interferometer arms, the frequency response can be determined with a Fourier transformation. In contrast to similar test setups using a single power detector with relative wide IF bandwidth, the measurement setup uses the prototype of the SWI ACS. This allows a more detailed Fourier transform of the interferogram, that results in a much higher frequency resolution and reveals more details on the variability of the sideband response [7]. In a first approximation of the sideband ratio between the USB and LSB the resulting beating pattern can be fitted with two cosine functions of a periodicity corresponding to the LO and IF frequencies. The amplitude and phase of the two cosines is derived using a linear least squares fitting algorithm. The ratio of these amplitudes correspond to the sideband ratio, whereas the fitted phase is only needed to compensate the uncertainty of the zero path length position. The detailed description of the used setup, which has been placed in front of the SWI receiver unit optics, and the data analysis can be found in [7]. Figure 4 gives an example of the fitted sideband response at a central frequency of 590 GHz at room temperature operation. The minimum path length used in the fit were fixed to a value where the interference pattern of the lowest IF shows at least one zero-crossing of the envelope of the beating pattern. As a quality criterion the condition number of the coefficient matrix was determined, which indicates the highest value with 12 mm path length. The result of the corresponding fit is plotted with the thick line. The fast periodic ripple on the traces are an artifact due to standing waves in the measurement setup. The main reasons for these standing waves are the reflections at the liquid nitrogen surface of the cold reference load, as well as the alignment errors and other imperfections of the MPI elements.

## IV. Conclusions

Radiometric measurements of the baseline ripples caused by the CHL demonstrate a reduction of the standing wave amplitude with the exponential prototype. On the basis of these test results the geometry with an exponential cone profile has been selected as the internal calibration target for SWI. Frequency switching as alternative calibration mode, which requires a less frequent gain calibration with the CHL, was investigated, showing very promising results with an almost flat switching baseline at the smallest possible stepsize of the SWI synthesizer and even at larger frequency steps. However, these tests have demonstrated that suitable bias voltages of the LO chain components need to be selected to avoid large baseline distortions. The presented FTS test method allowed to determine the frequency dependent sideband gain ratio, enabling the possibility of a pre-launch calibration of the sideband response over the full IF bandwidth and a wide range of LO settings. The verification of the FTS technique with gas cell measurements is planned in the future and a more compact, vacuum-compatible setup for cryogenic sideband gain ratio tests at 600 and 1200 GHz is in construction.

## ACKNOWLEDGMENTS

The work at the University of Bern has been funded through the Swiss National Science Foundation under Grant No. 200020-165744 and the ESA PRODEX program.

## REFERENCES

[1] P. Hartogh et al., "The Submillimetre Wave Instrument on JUICE," in Proc. EPSC 2013, London, United Kingdom.

[2] K. Jacob et al., "Design of the calibration target for SWI on JUICE," in Proc. 2016 41st International Conference on Infrared, Millimeter, and Terahertz Waves, 2016, Copenhagen, Denmark.

[3] M. Kotiranta et al., "Optical design and analysis of the Submillimeter-Wave Instrument on JUICE," IEEE Transactions on Terahertz Science and Technology, vol. 8, no. 6, pp. 588-595, 2018.

[4] K. Jacob et al., "Characterization of the 530 GHz to 625 GHz SWI receiver unit for the Jupiter mission JUICE," in Proc. 36th ESA Antenna Workshop 2015, Nordwijk, Netherlands.

[5] K. Jacob, A. Schröder, and A. Murk, "Design, manufacturing, and characterization of conical blackbody targets with optimized profile," IEEE Transactions on Terahertz Science and Technology, vol. 8, no. 1, pp. 76-84, 2018.

[6] A. Schröder et al., "Electromagnetic design of calibration targets for MetOp-SG microwave instruments," IEEE Transactions on Terahertz Science and Technology, vol. 7, no. 6, pp. 677-685, 2017.

[7] M. Renker et al., "Sideband characterization and atmospheric observations with various 340 GHz heterodyne receivers," Review of scientific instruments, vol. 85, no. 8, pp. 084706-1-10, 2014.

# Monday, April 15, 2019

# Session III. SIS Receivers and Mixers

# A 1.3 mm Superconductor Insulator Superconductor Mixer Receiver with 40 GHz Wide Instantaneous Bandwidth

Raymond Blundell[1], Robert Kimberk[1], Edward Tong[1] Paul Grimes[1], Nathan Hagos[1], Lingzhen Zeng[1]

The majority of SIS mixers at millimeter and submillimeter wavelengths have been designed for radioastronomy applications where low-noise and high sensitivity are of paramount importance. These include double sideband, single sideband, and 2 SB mixer receivers. Historically, the development of low noise IF amplifiers has enabled the instantaneous frequency coverage of these receivers to be extended from about 500 MHz or so available bandwidth using L Band amplifiers, centered typically at 1.5 GHz, to 8 GHz and beyond. For example, many of the receivers incorporated into the Atacama Large Millimeter Array provide IF output from 4 – 12 GHz, which is similar to that of receivers operating at the IRAM 30 m radio telescope and those at the SMA, which now provide IF output across the 4 – 16 GHz frequency range. However, in all of these recent designs, IF output below 4 GHz is not generally processed.

In this paper, we present an SIS mixer receiver with an IF bandwidth of 20 GHz, which also makes use of the frequency range below 4 GHz with low noise. To achieve this, we use an IF diplexer with two outputs, the first is used to couple IF from ~ 100 MHz to 4 GHz to a commercially available low-noise SiGe amplifier, and the second couples IF above 4 GHz to a separate commercially available low noise amplifier via a wideband circulator produced in-house[1]. In this way, the instantaneous frequency response of the receiver extends over an almost continuous 40 GHz band, with only a small gap, about 100 MHz wide, centered at the local oscillator frequency.

Referring to figure 1, the receiver noise performance at low IF is similar to that at intermediate IF. However, at high IF, the receiver noise increases as a result of increased IF noise contribution, which is partly due to non-optimal mixer tuning and partly due to higher IF noise. In addition, a standing wave with period of about 1 GHz occurs at high IF. A more optimized mixer tuning coupled with improved circulator performance is expected to reduce this.

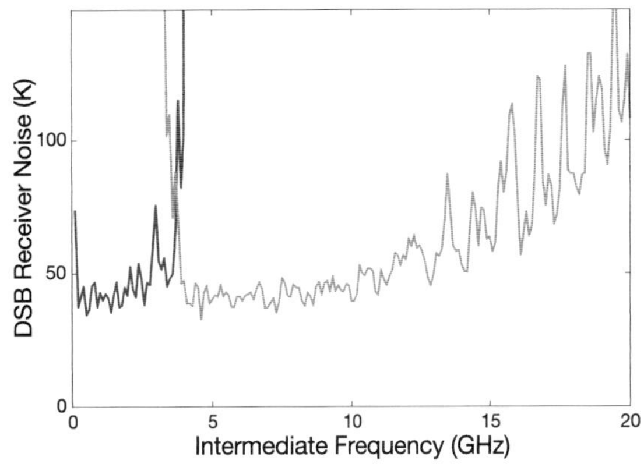


Fig. 1. Double Side Band receiver noise as a function of IF measured at an LO frequency of 225.5 GHz: the low frequency IF chain shows good receiver performance to ~ 4 GHz, whereas the high frequency IF extends operation to ~ 20 GHz.

This is the first demonstration of a truly wide-band SIS mixer, which when coupled to an ortho-mode transducer could be used to perform wide band dual polarization observations. Alternatively, incorporating separate local oscillators (40 GHz apart) to feed the mixers, one could observe an 80 GHz instantaneous bandwidth, to facilitate spectral line surveys or to enable accurate determination of the redshift of high-z targets without the need for multiple frequency tunings.

## References

[1] L. Zeng, C. E. Tong, R. Blundell, P. K. Grimes and S. N. Paine, "A Low-Loss Edge-Mode Isolator with Improved Bandwidth for Cryogenic Operation," *IEEE Trans. Microwave Theory and Techniques* vol. 66, no. 5, pp. 2154-2160, May 2018. 10.1109/TMTT.2018.2799574.

[1] Smithsonian Astrophysical Observatory,
60 Garden Street, Cambridge, MA02138, USA.

NOTES:

# Performance of a 275-500 GHz SIS mixer with 3-22 GHz IF

Takafumi Kojima[1], Matthias Kroug, Kazunori Uemizu, Yasutaka Niizeki, Akihira Miyachi, Keiko Kaneko, Yoshinori Uzawa,

The increase of the instantaneous bandwidth of low-noise heterodyne receivers is a key aspect for creating new prospects in radio astronomy at millimeter and submillimeter wavelengths. For example, instruments with wider intermediate frequency (IF) bandwidth would offer better sensitivity and multiline spectral observations without changing the local oscillator (LO) frequency.

Wideband technology with respect to radio frequency (RF) is also beneficial because the possibility to cover wide RF ranges with the same receiver offers new science cases, mostly related to accurate observations of multiple spectral lines with the same receiver calibrations. In addition, the wideband RF technology allows us to reduce the total number of receivers to cover a particular RF range, and thus, to simplify maintenance and operation of telescopes.

So far, we have independently studied and developed wideband RF and IF technologies. The double sideband (DSB) receiver implementing wideband RF components showed about 2 to 3 times the quantum noise over the RF 275-500 GHz with the IF band of 4-8 GHz [1]; The SIS mixer-preamplifier module based on high current density SIS junctions demonstrates low-noise and flat gain over the 3–18-GHz IF range at local oscillator frequencies of 400-480 GHz [2].

Our aim in this study is to develop a wideband RF and IF receiver technology with the same SIS mixer. We have designed and fabricated a wideband RF and IF SIS mixer combining two technologies. As shown in Fig. 1, the preliminary result showed DSB receiver noise temperature below 3 times the quantum noise for the 3-22 GHz IF over the entire LO frequencies. Moreover, the noise temperature averaged over 3-22 GHz was almost comparable with the one for 4-8 GHz. In the symposium, we will present the current status and latest result of the wideband RF and IF SIS mixer.

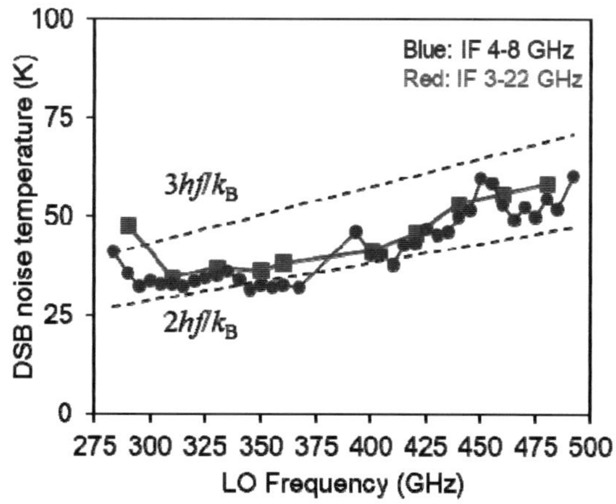


Fig. 1. Measured DSB noise temperatures averaged over 4-8 GHz and 3-22 GHz IF. While an isolator between the SIS mixer and cryogenic amplifier (CLNA) was used in the 4-8 GHz IF, the SIS mixer and CLNA was directly connected without using the isolator in the 3-22 GHz IF.


REFERENCES

[1] T. Kojima, M. Kroug, A. Gonzalez, K. Uemizu, K. Kaneko, A. Miyachi, Y. Kozuki, and S. Asayama, "275–500-GHz Wideband Waveguide SIS Mixers," IEEE Transactions on Terahertz Science and Technology, vol. 8, no. 6, pp. 638-646, Nov. 2018.

[2] T. Kojima, M. Kroug, K. Uemizu, Y. Niizeki, H. Takahashi and Y. Uzawa, "Performance and Characterization of a Wide IF SIS-Mixer-Preamplifier Module Employing High-J c SIS Junctions," IEEE Transactions on Terahertz Science and Technology, vol. 7, no. 6, pp. 694-703, Nov. 2017


[1] National Astronomical Observatory of Japan, Osawa 2-21-1, Mitaka, Tokyo Japan.

NOTES:

# Experimental Study of a Monolithic Planar-integrated Dual Polarization Balanced SIS Mixer

Wenlei Shan[1], Shohei Ezaki[1], Akihira Miyachi[1], Takafumi Kojima[1], and Yoshinori Uzawa[1]

Coherent radio astronomical observation delivers precise chemical and kinematic information of celestial objects, which incoherent observation does not convey. However, the complexity of coherent receiver frontends imposes a limit to the number of pixels arrayed in the focal plane of a radio telescope and results in a narrow field of view, which is usually much less than the available field of view constrained by the radio telescope optics.

We have been developing compact focal plane heterodyne detector arrays with SIS mixers for wide field-of-view astronomical observation. Then central idea to achieve improved compactness is the integration of multiple pixels in a monolithic silicon chip. We have constructed the conceptual technical solution to implement this idea in our previous works and carried out proof-of-concept study to proof the feasibility with a prototype receiver [1-3]. Although important experimental evidences were obtained, which strongly support the feasibility of the concept, a complete experimental investigation had not been available.

The prototype single-pixel SIS mixer chip has most of the key features of the concept. The mixer is designed to operate at ALMA Band 4 frequencies. The mixer chips are fabricated from silicon-on-insulator (SOI) wafers, which are locally thinned to 6 μm thick membranes on which probes lie to couple the LO and the signal from the waveguides (see the insets in Fig. 1). An orthogonally placed polarization-sensitive probe is adopted for signal coupling with polarization separation. For each polarization a balanced mixing configuration is adopted, which is a representative of various complex circuitries that can be potentially integrated.

In this presentation we will report a complete experimental study of the prototype receiver in the full frequency band. According to the measurement results, the cross-polarization is lower than -20 dB for any frequency in the band. The major reason for the cross talk between the two polarizations will be discussed. The noise rejection ratio is better than 15 dB as shown in Fig. 1. This relatively high noise rejection ratio is believed to be an advantage of good balance of the circuit achieved by photolithography technique. With the balanced mixer we measured the LO noise, which is difficult to be measured with other means. The receiver noise is found to be lower than 40 K in the frequency band and does not noticeably depend on the frequency.

[1] National Astronomical Observatory of Japan, Tokyo, 181-8588, Japan.

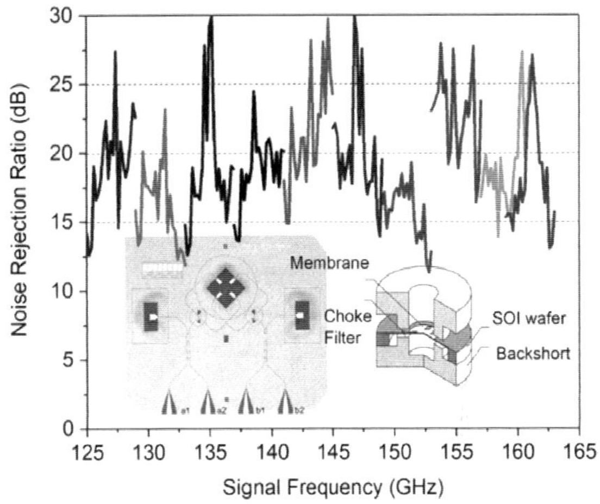


Fig. 1. The noise rejection ratio of the balanced SIS mixer as a function of the signal frequency. The insets show the image of the front side of the mixer chip and a schematic drawing of the membrane waveguide probe as well as the mounting method.

## References

[1] Wenlei Shan, Shohei Ezaki, Shinichiro Asayama, Takashi Noguchi, and Satoru Iguchi, "Planar-integration of Arrayed SIS Receiver Frontends," *ISSTT 2018*, Pasadena, CA, US, Mar. 2018.

[2] W. Shan, S. Ezaki, J. Liu, S. Asayama, T. Noguchi, S. Iguchi, "Planar superconductor-insulator-superconductor mixer array receivers for wide field of view astronomical observation," *Proc. SPIE 10708*, Millimeter, Submillimeter, and Far-Infrared Detectors and Instrumentation for Astronomy IX, 1070814 , July 2018.

[3] Wenlei Shan, Shohei Ezaki, Jie Liu, Shinichiro Asayama, and Takashi Noguchi, "A New Concept for Quasi-Planar Integration of Superconductor-Insulator-Superconductor Array Receiver Front Ends," *IEEE Trans. on Terahertz Sci. and Tech.*, vol. 8, no. 4, 472-474, 2018.

NOTES:

# Monday, April 15, 2019

# Session IV. THz sources

# Digitally tunable 150 GHz Local Oscillator chain for the Submillimeter Wave Instrument onboard the ESA JUICE mission

B. Thomas[1], M. Philipp[1], K. Lichius[1], M. Trasatti[1], A. Maestrini[2], E.P. Miettinen[3] and P. Hartogh[3]

The Submillimeter Wave Instrument (SWI) is one of the 10 scientific instruments selected as payload onboard the JUpiter ICy moons Explorer (JUICE) satellite [1], the first large-class mission in ESA's Cosmic Vision 2015-2025 programme. SWI will perform atmospheric remote sensing of Jupiter's middle atmosphere and the exosphere of the Galilean moons [2].

SWI features two heterodyne channels centered around 600 GHz and 1200 GHz frequencies, relying on passively cooled Schottky diodes receivers [3]. Part of the Local Oscillator operates nevertheless at close to ambient temperature.

RPG is developing the 150 GHz Local Oscillator chains for both 600 GHz and 1200 GHz channels. It uses a combination of discrete Schottky Varactor diodes (from Teratech Ltd.) and amplifier MMICs (from ADI/Hittite). Although the LO chain for the 600 GHz channel is relatively straightforward with a cascaded E-band tripler, E-band Medium Power Amplifier and 150 GHz doubler chain, the 1200 GHz LO chain relies on power-combining techniques already described in [4]. A view of this power-combined LO chain is shown in Fig.1.

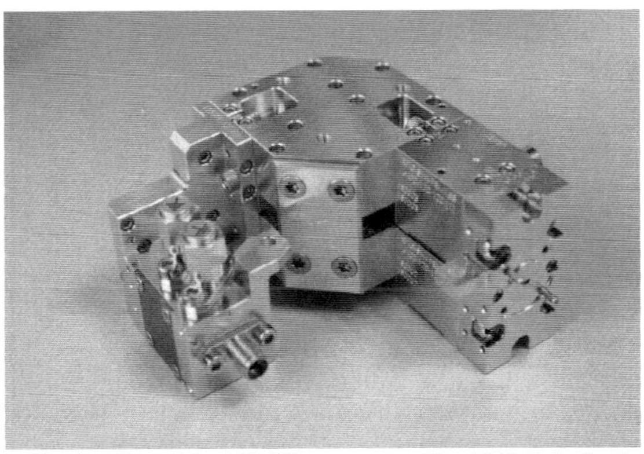

Fig. 1. View of the 150 GHz power-combined LO chain for the 1200 GHz EM channel. Outer dimensions are 70x74x38.5 mm$^3$

This chain operates between 136 GHz and 158 GHz and includes internal voltage regulation and digital control of the output power. The main challenge in developing the EM LO chain was to integrate a DC circuit controllable electronically with 12 bits of TTL lines in order to tune the amount of LO power from +13dBm to +19dBm. Each amplifier and doubler module includes an independent fixed DC bias voltage, and 3 bits of control. The tripler operates in self-biasing mode.

The components and LO chains have been designed for broad bandwidth, wide power tuning capability, low DC power consumption and maximum integration of the control electronics, featuring radiation hard parts and additional spot shielding to increase resilience against the harsh Jupiter radiation environment.

Extensive test results of the EM LO chains for both 600 and 1200 GHz channels will be presented, as well as the FM development status and qualification approach.

### References

[1] ESA JUICE web portal. Website: http://sci.esa.int/juice/

[2] P. Hartogh el al, "The Submillimeter Wave Instrument on JUICE", proceedings of the 29th Internal Symposium on Space Terahertz Technology – ISSTT, Pasadena, CA, USA, March 26-28, 2018.

[3] A. Maestrini el al, "The 1200GHz receiver front-end of the Submillimeter Wave Instrument of ESA Jupiter Icy moons Explorer", proceedings of the 8th ESA Workshop on millimeter-wave technology and applications – ESA-ESTEC, Noordwijk, The Netherlands, December 10-12, 2018.

[4] B. Thomas et al, "Compact power-combined LO sources for SWI onboard JUICE", proceedings of the 36th ESA Antenna Workshop on Antennas and RF Systems for Space Science, ESTEC, Noordwijk, the Netherlands, Oct. 6-9, 2015.

[1] Radiometer Physics GmbH, 53340, Meckenheim, Germany
[2] Observatoire de Paris, LERMA, 75014, Paris, France.
[3] Max Planck Institute for Solar System Research, 37077, Göttingen, Germany.

NOTES:

# High-power broad-band room-temperature 2.46-2.70 THz LO sources to enable high-spectral resolution mapping of HD and [NII]

Jose V. Siles[1], Jonathan H. Kawamura[1], Robert H. Lin[1], Choonsup Lee[1], Imran Mehdi[1], Paul F. Goldsmith[1] and Jorge Pineda[1].

Understanding the structure and dynamics of ionized gas in star forming regions is critical for exploring the effect of radiative feedback of massive stars into the interstellar medium. Stellar feedback is a key driver of galaxy evolution, and the ionized gas is an important tracer of these processes. A particularly powerful probe of the ionized gas is the far-infrared fine-structure lines of ionized nitrogen. These lines are not hindered by dust extinction and can readily propagate through the plane of the Galaxy. The intensity ratio of the [NII] 122 μm (2.46 THz) and 205 μm (1.46 THz) lines is a key diagnostic of the electron density of the gas, and therefore can be used in the position-position-velocity space to determine the 3D distribution of ionized gas densities in star forming regions. This information is crucial for understanding the process of star formation and the effect that stellar feedback has in the regulation of star formation in galaxies. However, high-spectral resolution observations of the [NII] 122 μm have not been yet performed due to the lack of powerful local oscillator sources and the difficulty of performing observations at SOFIA flight altitudes. Future balloon-borne and space-born platforms (e.g. OST) will require high-power LO sources able to enable array receivers in the 2.5 THZ range. These LO sources need to be able to operate at room-temperature and be efficient in order to minimize the dc power consumption. Broadband operation between 2.46 THZ and 2.7 THz is required to cover the important HD 1-0 line. HD is considered the best tracer of the gas mass in protoplanetary disks, which is currently not well-constrained.

Current state-of-the-art room-temperature 2.5-2.7 THz sources (JPL) produce only 3-14 μW in the 2.5 THz frequency range [1], with powers levels under 3 μW at the [NII] and HD frequencies. The 2.7 THz receiver of GREAT on board SOFIA features a LO source with output power levels <3.5 μW in this frequency range [2] (Virginia Diodes), barely enough to pump a single Hot-Electron-Bolometer (HEB) based receiver unless the source is cryogenically cooled. None of these sources provide enough LO power to enable observation of the [NII] 122 μm line, and they are far from being able to allow array receivers in this frequency range.

In order to overcome this issue, we planned to update one of the x3x3x3 2.7 THz Schottky diode based frequency multiplied LO chains reported in [1] with (i) new high-performance 300 GHz and 900 GHz tripler designs based on the successful JPL on-chip power-combined topology used in the THz sources presented in [3], and (ii) a novel biasable 2.4-2.7 THz tripler design to replace the former unbiased 2.5 THz tripler design used in [1]. The goal is to provide output power levels in excess of 30 μW to enable high-spectral resolution mapping of HD and [NII], similar to the power levels of the recently demonstrated JPL 16-pixel 1.9 THz source and the 4-pixel 1.9 THz source successfully flown on-board the Stratospheric Terahertz Observatory for [CII] mapping. Preliminary results with only the new 900 GHz tripler stage added are shown in Fig. 1. A factor of four improvement in power (up to 20 μW) is already achieved together with a considerable increase in bandwidth. The injected power at W-band has not been increased so the improvement is completely due to the higher performance of the new multiplier designs. In the conference, we will present these new designs and further tests showing the additional improvement once the rest of stages are updated with these new designs.

A new compact architecture design, significantly smaller than the one in [1] will also be presented together with a preliminary design of a 16-pixel LO source at 2,7 THz.

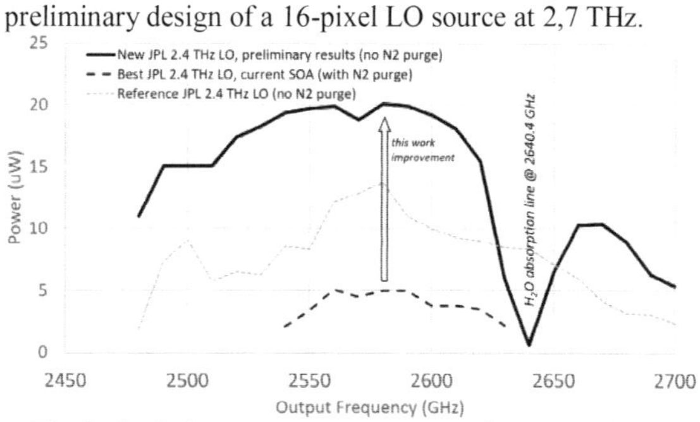


Fig. 1. Preliminary room-temperature performance of the new JPL 246-2.7 THz LO source (solid line) consisting of the reference JPL 2.4 THz LO source (dashed line) updated with on-chip power-combined frequency triplers. Current state-of-the-art results from [2], under N2 purge to remove the impact of the water absorption line, are also plotted (dotted line).

[1] Jet Propulsion Laboratory, California Institute of Technology, Pasadena, CA 91109.

*The research was carried out at the Jet Propulsion Laboratory, California Institute of Technology, under a contract with the National Aeronautics and Space Administration.* 

### References

[1] C. A. Duran, et al., "4GREAT: A Multiband Extension of GREAT from 490 GHz to 2.7 THz," *28th International Symposium on Space THz Technology*, March 2017.

[2] A. Maestrini, et al., "Design and characterization of a room temperature all-solid-state electronic source tunable from 2.48 to 2.75 THz", *IEEE Trans. on THz Science and Technology*, Vol. 2, No. 2, pp. 177-185, March 2012.

[3] J. V. Siles, et al., "A New Generation of Room Temperature Frequency Multiplied Sources With up to 10× Higher Output Power in the 160 GHz – 1.6 THz Range," *IEEE Trans. on THz Science and Technology*, Vol. 8, No. 6, pp. 596-604, November 2018.

NOTES:

# Superconducting Flux-Flow Oscillator as a Terahertz External Local Oscillator for Heterodyne Receiving

Nickolay V. Kinev, Kirill I. Rudakov, Lyudmila V. Filippenko, Mikhail Yu. Fominskiy, Andrey M. Baryshev, and Valery P. Koshelets

***Abstract*—We have developed, fabricated and tested a terahertz (THz) source based on the long Josephson junction acting as the flux-flow oscillator (FFO). The FFO was integrated to the lens antenna on a single chip providing the THz emission to open space. We used the double slot type of antenna coupled to the semielliptical silicon lens to form a narrow output beam. The impedance and the emission properties of the antenna were numerically simulated. The experimental samples were fabricated with the FFO based on Nb/AlN/NbN superconductor–insulator–superconductor (SIS) trilayers with a current density of 11 kA/cm$^2$ and a gap voltage of about 3.55 mV. The output radiation was studied by a THz spectrometer based on the SIS receiver with high spectral resolution, the signal has been observed in the range of 440-700 GHz with the signal-to-noise ratio up to 55 dB. A feasibility of the phase locking by using the harmonic mixer has been discussed.**

A flux-flow oscillator (FFO) based on the unidirectional flow of magnetic vortexes (fluxes) in a long Josephson junction is a promising solution of local source for heterodyne receiving in terahertz (THz) region [1],[2]. Its frequency tuning range up to 100% of central frequency is unique among other types of THz sources. The FFO output power of about 0.1-1 μW gained in the peak with a linewidth of ~40 kHz by using the phase-locking loop (PLL) is sufficient for heterodyne requirements. Up to now, the FFO has been implemented as the local oscillator utilizing the on-chip integration with the receiving SIS-mixer at the on-board superconducting receiver for the TELIS balloon instrument [3],[4].

In this study, we have coupled the FFO output radiation to a transmitting lens antenna and developed a compact THz oscillator emitting to open space. The idea of the oscillator is presented in Fig. 1. The cryogenic module with the oscillator can be installed on the same cold plate as a detector and operate as a heterodyne source. Recently we tested this concept using also a double slot antenna but with another type of the excitation and another topology of the antenna [5],[6].

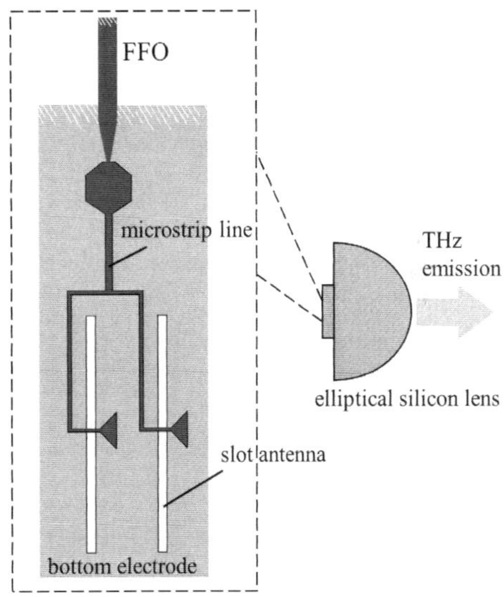


Fig. 1. On the left – layout of the planar structure with the FFO coupled by a microstrip line to the slot antenna; on the right – the scheme of the chip containing a microcircuit placed at the focus of the silicon lens. The scheme on the right is not to scale.

The numerical simulations were performed using a specialized microwave 3D-modeling software. The impedance of the antenna and the power emitted to open space are presented in Fig. 2. According to simulations, the antenna operating range is 350-550 GHz defined by a power level of emission to open space higher than 0.5 of the total output FFO power. After this, the batch of the samples containing the FFO with dimensions 16×400 μm$^2$ based on Nb/AlN/NbN trilayer and the slot antenna based on Nb film with a thickness of ~200 nm was fabricated. The current density of SIS trilayer on the batch is about 11 kA/cm$^2$, which corresponds to parameter

The numerical simulations, fabrication, and experimental study of the samples were supported by the Russian Science Foundation (Project No. 17-79-20343). The technological maintenance for fabrication of the tunnel structures was carried out at the Kotel'nikov IREE RAS within the framework of the state task.

N. V. Kinev, L. V. Filippenko, M. Yu. Fominskiy, V. P. Koshelets are with the Kotel'nikov Institute of Radio Engineering and Electronics RAS, Moscow 125009, Russia. (e-mail: nickolay@hitech.cplire.ru).

K. I. Rudakov is with the Kotel'nikov Institute of Radio Engineering and Electronics RAS, Moscow 125009, Russia, and also with the Kapteyn Astronomical Institute, University of Groningen, 9712 CP Groningen, The Netherlands.

A. M. Baryshev is with the Kapteyn Astronomical Institute, University of Groningen, 9712 CP Groningen, The Netherlands.

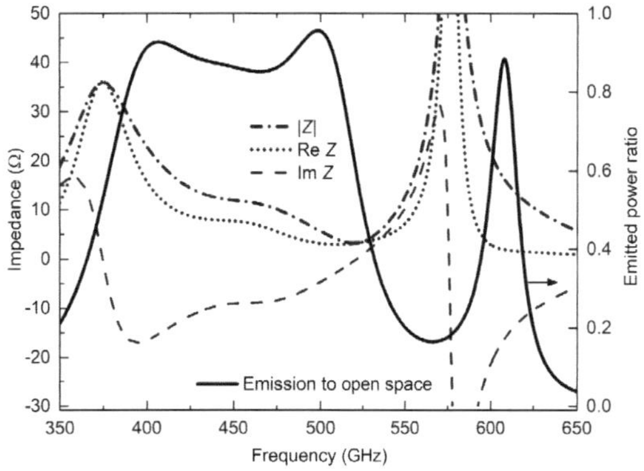


Fig. 2. Input impedance Z of the antenna at the point of connection to the impedance transformer (the real part – dot line, the imaginary part – dashed line, and the absolute value – dash-dot line), and the emitted power (solid line) normalized to the total output FFO power.

$R_n \times A \sim 20\ \Omega\ \mu m^2$, and the quality factor $R_j/R_n$ is about 30, where $R_n$ is the normal-state resistance, $R_j$ is the sub-gap resistance and $A$ is the area of the junction.

The microchip with "FFO & antenna" integrated circuit was mounted on the back flat surface of a semielliptical silicon lens, so that the center of the antenna was located in the far focus of the lens. For proper FFO operation, the chip was also located inside the magnetic shields. The module was installed in the liquid helium cryostat opposite an output THz window with IR filters. The measurements of the emission to open space were carried out using a THz spectrometer based on the superconducting integrated receiver (SIR) [2],[3] utilizing another FFO as a local oscillator and located in a separate cryostat opposite the "emitting" cryostat. The experimental setup is simpler than discussed in [6] (see Fig. 4 in [6]) due to the harmonic mixer is not used in the present paper.

The emission of the FFO in the range from 440 GHz up to 700 GHz, available for the SIR operation, was studied by the SIR at the intermediate frequency (IF) range of 4 - 8 GHz. The results for recorded spectra at some selected frequencies are presented in Fig. 3. The signal-to-noise ratio for the recorded

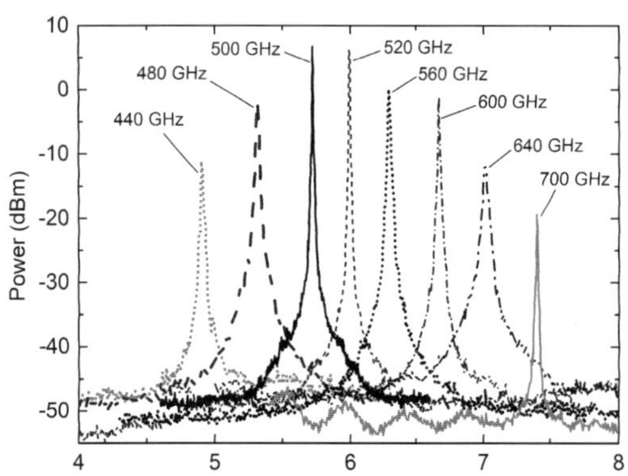


Fig. 3. Emission spectra of the FFO recorded using the SIR by a spectrum analyzer with the resolution bandwidth of 1 MHz. The frequency of emission is pointed for each curve.

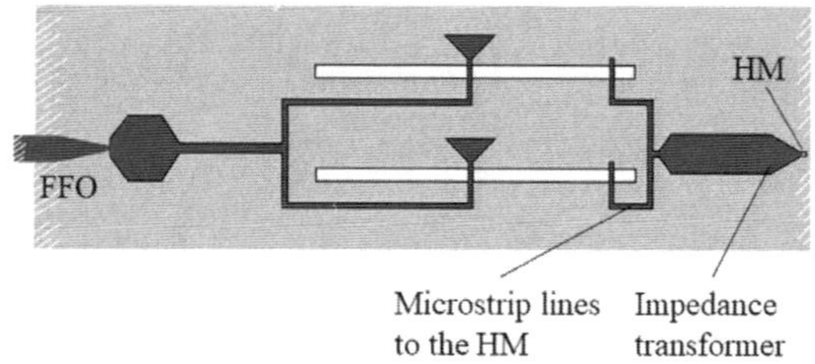


Fig. 4. The idea of embedding the HM into the FFO-based integrated structure shown in Fig. 1 not changing the geometry of the slot antenna.

spectral lines is up to 55 dB (see solid and short-dash curves in Fig. 3 for 500 GHz and 520 GHz, respectively), the spectral linewidth is estimated to be 2 - 15 MHz. We had no ability to get a higher accuracy for the shape and the linewidth definition since the oscillator has not been locked. Note that the actual upper border of the studied operating range is at least 150 GHz higher than the range defined by 0.5 level in simulations, and the signal-to-noise ratio at 700 GHz is as high as 25 dB. The frequencies below 440 GHz are out of the receiver operating range.

For practical applications as a heterodyne, the FFO should be phase locked. For this, the harmonic mixer (HM) based on the SIS junction and the PLL are commonly used [1],[2],[3],[6]. We are planning to embed the HM into integrated "FFO & antenna" structure using two symmetric microstrip lines that connect to a single line with an impedance transformer (see Fig. 4); at the output edge of the transformer the HM will be located. This idea is to be elaborated and tested.

## References


[1] V. P. Koshelets and S. V. Shitov, "Integrated superconducting receivers," *Supercond. Sci. Technol.*, vol. 13, no. 5, pp. R53–R69, 2000, DOI: 10.1088/0953-2048/13/5/201.

[2] V. P. Koshelets, P. N. Dmitriev, M. I. Faley, L. V. Filippenko, K. V. Kalashnikov, N. V. Kinev, O. S. Kiselev, A. A. Artanov, K. I. Rudakov, A. de Lange, G. de Lange, V. L. Vaks, M. Y. Li, and H. B. Wang, "Superconducting Integrated Terahertz Spectrometers," *IEEE Trans. Terahertz Sci. Technol.*, vol. 5, no. 4, pp.687–694, Jun. 2015, DOI: 10.1109/TTHZ.2015.2443500.

[3] G. Lange, D. Boersma, J. Dercksen, P. Dmitriev, A. B. Ermakov, L. V. Filippenko, H. Golstein, R. Hoogeveen, L. de Jong, A. V. Khudchenko, N. V. Kinev, O. S. Kiselev, B. Kuik, A. de Lange, J. Rantwijk, A. S. Sobolev, M. Y. Torgashin, E. Vries, P. A. Yagoubov, and V. P. Koshelets, "Development and characterization of the superconducting integrated receiver channel of the TELIS atmospheric sounder," *Supercond. Sci. Technol.*, vol. 23, no. 4, pp. 045016-1–045016-8, Mar. 2010, DOI: 10.1088/0953-2048/23/4/045016.

[4] O. Kiselev, M. Birk, A. Ermakov, L. Filippenko, H. Golstein, R. Hoogeveen, N. Kinev, B Kuik, A. Lange, G. Lange, P. Yagoubov, and V. Koshelets, "Balloon-Borne Superconducting Integrated Receiver for Atmospheric Research," *IEEE Trans. Appl. Supercond.*, vol. 21, no. 3, pp. 612–615, Jun. 2011, DOI: 10.1109/TASC.2010.2091712.

[5] N.V. Kinev, K.I. Rudakov, L.V. Filippenko, A.M. Baryshev, and V.P. Koshelets. "Wideband Josephson THz flux-flow oscillator integrated with the slot lens antenna and the harmonic mixer," *EPJ Web Conf.*, vol. 195, pp. 02003-1–02003-2, Nov. 2018, DOI: 10.1051/epjconf/201819502003.

[6] N.V.Kinev, K.I. Rudakov, L.V. Filippenko, A.M. Baryshev, and V.P. Koshelets, "Flux-flow Josephson oscillator as the broadband tunable terahertz source to open space," *J. Appl. Phys.*, vol. 125, no. 15, pp. 151603-1–151603-7, Mar. 2019, DOI: 10.1063/1.5070143.

# Monday, April 15, 2019

## Poster Session

# Design and Implementation of a Broadband and Compact 90-degree Waveguide Twist with Simplified Layout

C. Lopez[1], V. Desmaris[1], D. Meledin[1], A. Pavolotsky[1] and V. Belitsky[1]

***Abstract*— We report on a novel wideband single step 90-degree twist with a tolerant geometry, hence allowing for a simplified manufacturing process achieved through direct milling. Experimental verification shows a return loss of less than 20 dB over the 140-220 GHz band, which constitutes a 44% fractional bandwidth. Furthermore, the insertion loss is comparable with a continuous twist for the same band. The performance, compactness and fabrication tolerance insensitivity make it very suitable for use in various waveguide systems from cm to mm wavelength range.**

***Index Terms*—waveguides, waveguide twist, simplified manufacturing process, single step twist.**

## I. Introduction

WAVEGUIDE twists are an essential interconnection part in many millimetre and sub-millimetre systems, especially for modern polarization sensitive THz receivers [1]. Among the different types, the 90-degree twist is the most common one. As a consequence, it has been largely studied and many implementations can be found in the literature [2-7]. Nevertheless, the main solutions are the continuous rotation and single step twist. The first approach is frequently adopted in a commercially available twists. Although the smooth rotation guarantees low insertion loss and minimizes the reflection over larger bandwidths, it requires length of several wavelengths and implies complicated and time-consuming fabrication processes. As compactness is often a key aspect for instrumentation receiver systems, most of the research in this field has focused on a single step twist. There have been numerous studies to investigate step twist with cross sections based on corner cut waveguides [2]. More complicated geometries with multiple sharp corners or ridge waveguides [3,4,5] have also been proposed. These geometries usually introduce additional cuts that maximize the bandwidth of the twist. Nonetheless, the performance of such twists is frequently rather sensitive to their geometrical dimensions, hence requiring tight fabrication tolerances. Therefore, simple fabrication techniques such as milling often are not applicable. For instance, in [6], the design proposed in [7] is implemented using micromachining techniques.

[1] Group for Advanced Receiver Development, Chalmers University of Technology, SE412 96, Gothenburg, Sweden.

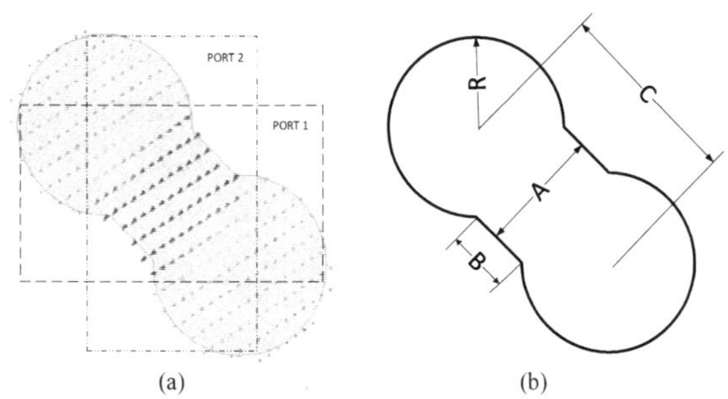


Fig. 1. WR-5 Twist. (a) Cross Section view of twist and waveguide ports. Electric field illustration of the twist dominant mode. (b) Design of WR-5 waveguide Twist: R= 420μm, A= 560 μm, B= 294 μm and C= 920 μm. The thickness is 500 um.

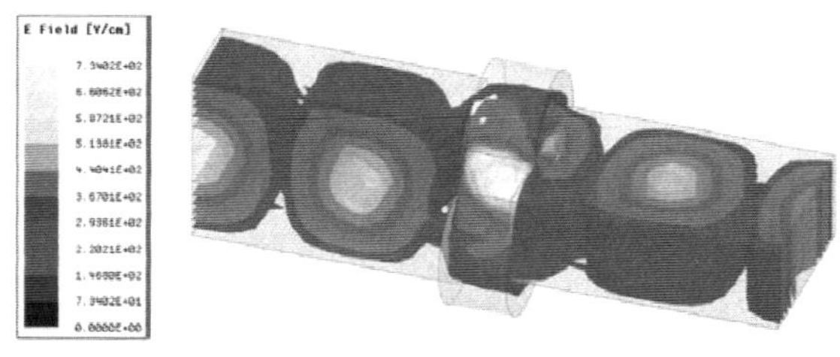


Fig. 2. Visualization of 90 ° EM field rotation. The TE10 mode is transformed into TE01 through the dominant mode of the twist.

In this paper, we present a novel single step 90-degree twist with more tolerant geometry, hence allowing for a simplified manufacturing process.

## II. Design and Fabrication

The proposed waveguide twist is depicted in Fig. 1. It comprises two circular waveguides with radius R interconnected by a rectangular waveguide defined by the parameters A and B. The distance between the centres of the circular waveguides is defined by C. The thickness is approximately a quarter of the guided wavelength inside the structure at the centre frequency, i.e. 500 μm for our designed Twist.

The twist transforms the $TE_{10}$ mode into $TE_{01}$ through the dominant mode of the structure. This mode allows the rotation of the polarization from 0 to 45 degrees inside the twist, and

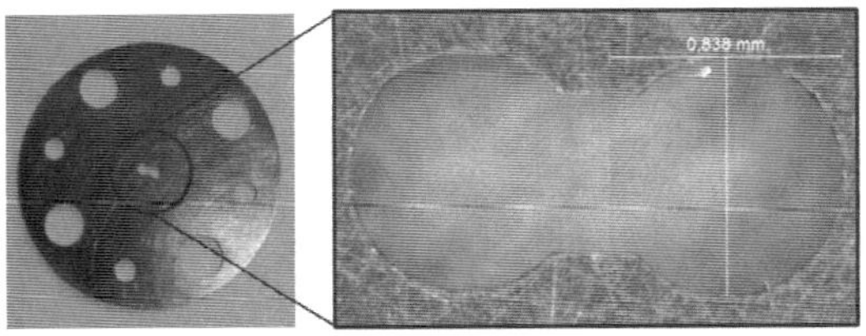

Fig. 3. Photograph of fabricated twist and detail of structure.

finally to 90 degrees as it is shown in Fig. 2. The shape of the twist allows easy fabrication using milling or combination of drilling and milling.

The waveguide twist has been optimized using the full-wave 3D simulator Ansys HFSS aiming frequency range 140-220 GHz. Tolerance analysis has shown that a maximum deviation of 10μm in each parameter could be allowed for the twist operating in this frequencies.

The test structure was fabricated of tellurium copper through direct milling and illustrated in Fig. 3.

## III. MEASURED PERFORMANCE

The fabricated twist was characterized using VNA frequency extension modules (VDI Inc. extension modules and a Keysight PNA-X). A SOLT calibration was applied.

The simulated and measured results are compared in Fig.4. The return loss is below -20dB over the whole band, which implies a 44% fractional bandwidth. It can be seen that return loss measurements show good agreement with simulation.

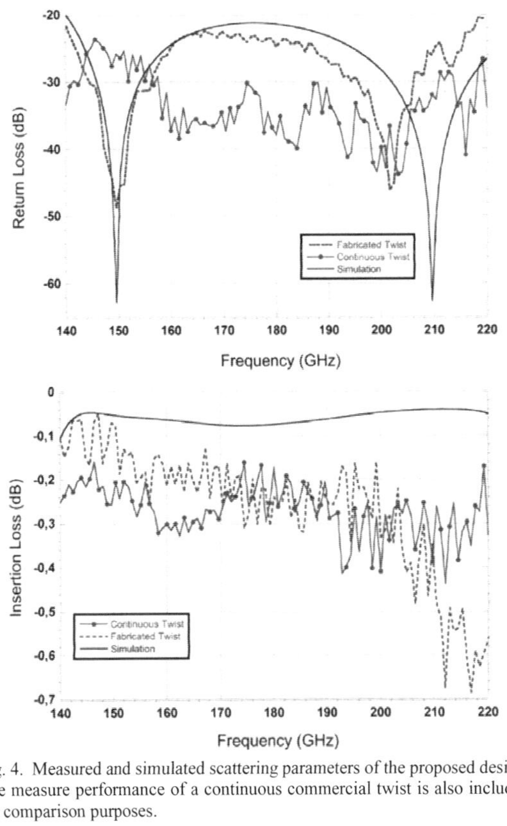


Fig. 4. Measured and simulated scattering parameters of the proposed design. The measure performance of a continuous commercial twist is also included for comparison purposes.

However, the predicted resonance at 210 GHz seems to be shifted down by 8 GHz due to fabrication inaccuracy. Regarding the insertion loss, it is less than 0.4dB between 140 GHz and 200 GHz. However, it rapidly degrades for frequencies above 200 GHz. It is clear from the graph that the fabricated twist presents additional losses that were not predicted by the simulation. We proposed that this phenomenon can be due to fabrication tolerances. Misalignment of the twist with respect to the input and output waveguides could excite non-dominant modes inside the twist cavity. Therefore, energy transfer between modes is promoted and the overall insertion loss is increased. Simulations initially confirm this hypothesis. Nevertheless, further investigation is required.

For the sake of comparison, a commercial 4mm long continuous twist fabricated through electroforming was measured using the same setup. It is easy to see that the overall insertion loss of the continuous twist is in the same range as our design for almost the entire band.

## IV. CONCLUSION

A novel 90° single step twist with a simple and compact geometry has been presented. It has been shown that the design is well-suited for fabrication through standard milling techniques. Experimental results have shown a 44% fractional bandwidth with return loss better than -20dB. Although the measured transmission loss is similar to a continuous twist, it is higher than expected. A possible explanation could be that the excitation of higher order modes caused additional loss. However, further research is needed in order to fully describe the complete set of modes inside the twist and its relation with structure alignment.

## ACKNOWLEDGMENTS

The research leading to these results has received funding from the European Commission Seventh Framework Programme (FP/2007-2013) under grant agreement No 283393 (RadioNet3).

## REFERENCES

[1] Belitsky et al, "SEPIA – a new single pixel receiver at the APEX telescope", Astronomy and Astrophysics, vol. 612, A23, 2018.

[2] Wu, J. M., Zhou, D. F., Lei, X., Gao, J., & Hu, H. M. , "Design of a compact broadband 90 waveguide twist based on double-corner-cut square slots.", International Journal of Antennas and Propagation, 2017.

[3] Asao, H., Hiraiwa, G., & Katayama, A, "A compact 90-degree twist using novel ridged waveguide for integrated waveguide subsystems.", 2006 European Microwave Conference, pp.1185-1188, IEEE, September 2006.

[4] Sun, M., Xu, Y., Peng, T., Shang, Y., & Zhen, Q.,"Design of a compact and broadband 90-degree waveguide twist.",*2018 IEEE International Vacuum Electronics Conference (IVEC)* , pp.203-204, IEEE, April 2018.

[5] Cano, Juan Luis; Mediavilla Sanchez, Angel, "Arbitrary-angle single-step waveguide twist for quasi-octave bandwidth performance", Progress In Electromagnetics Research, vol 162, pp. 73-79, 2018.

[6] Chen, L., Arsenovic, A., Stanec, J. R., Reck, T. J., Lichtenberger, A. W., Weikle, R. M., & Barker, N. S. (2011). "A Micromachined Terahertz Waveguide 90 Twist." *IEEE Microwave and Wireless Components Letters*, vol *21*(5), pp. 234-236,2011

[7] Litvinov, V. R., Rud, L. A., & Sverdlenko, E. A. "Compact 90° twists into rectangular waveguides", *Radioelectronics and Communications Systems*, vol. *53*, pp. 162-166, 2010.

# High-Performance Smooth-Walled Antennas for THz Frequency Range: Design and Evaluation

D. Montofré, R. Molina, A. Khudchenko, R. Hesper, A. Baryshev, N. Reyes and F.P. Mena

***Abstract*—Traditionally, corrugated conical horn antennas have been the main choice for use in astronomical receivers in the range of millimeter and submillimeter waves. They present low cross-polar level and high coupling efficiency into the fundamental Gaussian mode. However, this type of antenna is difficult to manufacture, inevitably increasing its price and extending the production process. In this work, we present two kinds of feed horn antennas, aimed for use in the ALMA Band-6 frequency range (211-275 GHz), which can be fabricated in a much simpler way with conventional machining tools. Specifically, we present the design and performance comparison of smooth-walled spline-profile horns in two geometries, diagonal and conical. Optimization of the designs has been made by means of an algorithm that allowed us to obtain models whose electrical and mechanical characteristics make them competitive when compared with corrugated horns. Our simulations have shown a good cross-polar performance with levels below −20dB and gaussicity above 96%. These properties make this type of horns an option at the time of choosing a feed system for cutting-edge astronomical applications**

***Index Terms*—spline-profile, diagonal, conical horn, millimeter/submillimeter wave.**

## I. Introduction

CORRUGATED conical horn antennas have been, traditionally, the main choice when it comes to developing instruments for astronomy applications in the range of millimeter and submillimeter waves (including part of the THz frequency spectrum) [1]. They have been selected due to their excellent characteristics. In particular, their coupling to the fundamental Gaussian mode (in short *Gaussicity*) is around 98% and cross-polar level lower than -30 dB [2]. Nonetheless, this sort of antennas is difficult to manufacture, inevitably extending the production process and, therefore, increasing their price. There are other options to corrugated horns, e.g., conical and pyramidal horns. However, they lack symmetry in the radiation pattern and, additionally, pyramidal horns present astigmatism [3].

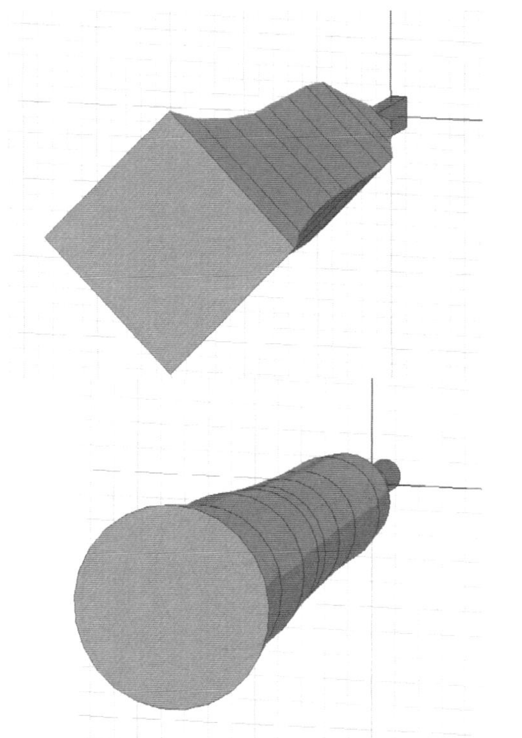

Fig. 1. The two horns under study. (Top) Diagonal-spline horn connected to a rectangular waveguide. (Bottom) Conical horn connected to a circular waveguide. Both designs were optimized in order to reach the goals described in table 1.

In this work, we present the comparison of two types of smooth-walled spline-profile horn antennas, diagonal and circular (figure 1). This work has been focused on the frequency range of ALMA Band 6 (211-275 GHz), currently one of the most interesting observation bands for astronomers, with the image of HL-Tau [4] and the first image ever taken of a Black Hole

D. Montofré is with the Department of Electrical Engineering, Universidad de Chile, Santiago, Chile and with the Kapteyn Institute of Astronomy, University of Groningen, The Netherlands. (e-mail: d.a.montofre@rug.nl).

R. Molina is with the Millimeter Wave Laboratory and the Department of Astronomy, Universidad de Chile, Santiago, Chile (e-mail: rocio.molina@raig.uchile.cl).

A.~Khudchenko, R. Hesper, A.M.~Baryshev, are with the Kapteyn Astronomical Institute, University of Groningen, Groningen 9747 AD, The Netherlands (e-mail: A.Khudchenko@sron.nl; R.Hesper@astro.rug.nl; A.M.Baryshev@astro.rug.nl)

N. Reyes is with the Department of Electrical Engineering, Universidad de Chile, Santiago, Chile and with the Max-Planck Institute for Radio Astronomy, Bonn, Germany

F.P. Mena is with the Millimeter Wave Laboratory and the Department of Electrical Engineering, Universidad de Chile, Santiago, Chile. (e-mail: fpmena@uchile.cl)

Event Horizon [5] as examples of its major achievements. Simulations and measurements have shown cross-polar levels lower than -20 dB, sidelobes below -25 dB in the E and H cardinal planes, and a Gaussicity around 96%, in agreement with previous reports [6].

## II. Horn Designs

The starting point to design both horn antennas is the profile of the walls, which defines the features of the radiation pattern. We have followed a similar methodology as described in [7]. In order to simplify the machining even further, we have used straight lines to connect nodes instead of a cubic spline curve Once the initial wall shape is defined, the next step is creating the volume. On the one hand, the diagonal-spline horn is created by rotating the 1-D wall profile in 90°, then the both profiles are linked using a straight line. The result is profiles connected by segmented flat rectangular surfaces. Then, the final volume is obtained just mirroring the resulting surface, by means of selecting the proper symmetry axis. On the other hand, the conical-spline is created by applying a 360° rotational sweep to the initial 1-D wall profile. Both horn geometries are shown in figure 1. Diagonal-spline and conical-spline horn designs are fed using a rectangular (0.5-mm×1.0-mm) and a circular (diameter = 1.30-mm) waveguide, respectively. This election was made to facilitate the subsequent machining process.

## III. Performance Evaluation of Designs

Figure 2 shows the simulated radiation pattern of the diagonal and conical spline horns at their three key analysis frequencies. Both horns show a very good symmetry between E, D and H planes for levels above -20 dB. This means that both horns will produce a beam with good circular shape or, equivalently, they present low beam ellipticity. The cross-polar levels are lower than -20 dB for both designs over the entire bandwidth. However, the conical-spline design shows a better performance at all frequencies, reaching levels close to -30 dB. This better cross-polar performance correlates with lower levels of sidelobes in D plane. For the diagonal-spline geometry, sidelobe level is below -40 dB in the E and H-plane. D plane shows higher level of sidelobes with values above -30 dB for 211 and 243 GHz, and a level slightly below -20 dB for 275 GHz. For the conical-spline geometry, on the other hand, sidelobe level is around 30 dB in the E, D and H plane for 211 and 243 GHz. A difference is observed at 275 GHz, where sidelobe level is slightly below -20 dB in the H-plane. This difference in sidelobes level at different cardinal planes results in the diagonal-spline design having a more symmetrical beam shape, but at expenses of having higher levels of cross-polar component.

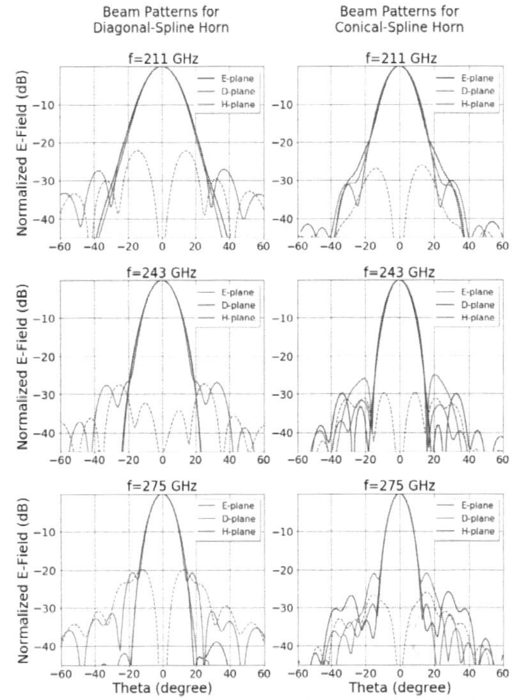


Fig. 2. Comparison of radiation pattern for diagonal-spline (left column) and conical-spline horns for their key analysis frequencies. Co-polar curves are shown in solid line, and cross-polar curves for D plane are shown in dashed line. Diagonal-spline horn it has lower level of sidelobes in E and H planes and a wider beam width. On the other hand, conical-spline horn exhibits lower sidelobes in D-plane, which correlates with lower levels of cross-polar for all the frequencies.

## IV. Conclusion

We have designed and analyzed two smooth-walled spline-profile horns. The simulated radiation patterns have shown a very good performance with a Gaussicity over 96%, cross-polar lower than -20 dB over the entire frequency range. We deem that the simulated performance for both antennas meets the requirements to be used in astronomical applications.

## References

[1] J.F. Johansson, "A Comparison of Some Feed Types", Multi-feed System for Radio Telescopes, ASP Conference Series, vol. 75, 1995

[2] A. Ludwig, "The definition of cross polarization", IEEE Transactions on Antennas and Propagation, vol. 21, no. 1, pp. 116-119, 1973

[3] E Muelhdorf, "The Phase Center of Horn Antennas", IEEE Trans-actions on Antennas and Propagation, vol. ap-18, no. 6, 1970.

[4] ALMA Partnership, "The 2014 ALMA Long Baseline Campaign: First Results from High Angular Resolution Observations toward the HL Tau Region", The Astrophysics Journal Letters, vol. 8, no. 1,2015

[5] Event Horizon Telescope Collaboration, "First M87 Event Horizon Telescope Results. VI. The Shadow and Mass of the Central Black Hole", The Astrophysical Journal Letters, vol. 875, no. 1, 2019.

[6] H.J. Gibson et.al., "A Novel Spline-Profile Diagonal Horn Suitable for Integration into THz Split-Block Components", IEEE Transactions on Terahertz Science and Technology, vol. 7, no. 6, 2017

[7] C. Granet et.al., "A Smooth-Walled Spline-Profile Horn as an Alternative to the Corrugated Horn for Wide Band Millimeter-Wave Applications", IEEE Transactions on Antennas and Propagation, ,vol. 52, no. 3, 2004

# Design of a Silicon-based 160~320GHz tanh-profile wideband Corrugated Horn

Jie Hu[1, 2], Zheng Lou[2], Hao-tian Zhu[3], Wei-tao Lv[2], Dong Liu[2] and Sheng-Cai Shi[2]

Abstract: This paper presents the design and simulation a silicon-based 160~320GHz wideband corrugated horn. The horn is formed by stacking 30 gold-coated silicon platelets [1]. The corrugation of the horn is formed by photography and deep reactive ion etching (DRIE). The tanh profile is chosen to make the horn compact as compact as possible [2]. The corrugated horn is simulated by a home-made mode matching script. The simulated beam patterns across the frequency band are shown in Fig. 1. It shows great symmetry and low sidelobe and cross-polarization level, which are below -35dB and -20dB respectively. The S11 is also below -20dB across the frequency band. The effect of the rectangular to circular waveguide transformation has also been taken into account.

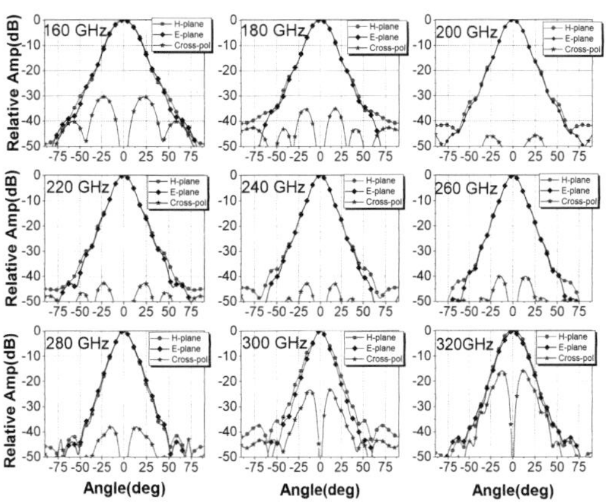


Fig. 1. Simulated far-field pattern of the corrugated horn. The black curves are E-plane, the red curves are H-plane and the blue curves are cross-polarization in D-plane.

REFERENCES

[1] J. Britton, K. W. Yoon, J. A. Beall, D. Becker, H. M. Cho, G. C. Hilton, M. D. Niemack, and K. D. Irwin, "Progress toward corrugated feed horn arrays in silicon," *AIP Conference Proceedings,* vol. 1185, no. 1, pp. 375-378, 2009.

[2] J. E. McKay, D. A. Robertson, P. A. S. Cruickshank, R. I. Hunter, D. R. Bolton, R. J. Wylde, and G. M. Smith, "Compact wideband corrugated feedhorns with ultra-low sidelobes for very high performance antennas and quasi-optical systems," *IEEE Trans. Antennas Propag.,* vol. 61, no. 4, pp. 1714-1721, 2013.

[1] APC, University Paris Diderot, Paris, 75013, France.
[2] Purple Mountain Observatory, 210032, Nanjing, China. .
[3] LERMA, Observatorie de Paris, Paris, 75014, France.

NOTES:

# Broadband Waveguide-to-Substrate Transition Using a Unilateral Etched Finline Structure

C. Lopez[1], V. Desmaris[1], D. Meledin[1], A. Pavolotsky[1] and V. Belitsky[1]

*Abstract*—**We present a novel broadband waveguide-to-substrate transition that aims to be used for broadband mixer design. The transition consists of a unilateral finline structure with etched substrate between the fins. This particular feature reduces the overall insertion loss and facilitates matching with the waveguide. The transition is designed of a thin silicon substrate covered by a superconducting niobium thin layer. An auxiliary gold layer situated on top of the Nb-layer provides grounding for the fins and facilitates a simple mounting process in the split-block waveguide mount. In order to compare simulations with measurements, a back-to-back arrangement was designed and simulated using HFSS in the 211-373 GHz frequency band. The back-to-back simulation results show an insertion loss of less than 0.3 dB in the whole band. Furthermore, a fractional bandwidth of 55% with a return loss better than 15 dB is achieved.**

*Index Terms*—**Unilateral Finline, Broadband Waveguide to Substrate Transition.**

## I. Introduction

SINCE THz active components are produced using thin film technology, the transition from a waveguide to substrate plays an essential role in the performance of any THz system. Both a good impedance match and ease of fabrication are both fundamental features required in the design of such transitions. Mounting accuracy is also of high importance.

For decades, one of the most popular structures in the THz receivers field has been waveguide E-probes. Although this solution achieves large fractional bandwidths [1,2], its performance is rather sensitive to its position inside the waveguide. As alternative solution, unilateral finline structures [3] are more tolerant in terms of accuracy of their positioning in the waveguide, yet, this approach results in problems related to impedance matching between the waveguide and the substrate, which affects its performance over large operational bandwidths. Other approaches that make use of transmission lines different from microstrip and slotline have been proposed over the years, such as the presented in [4]. Nevertheless, none of these structures is able to address simultaneously all the outlined requirements.

We report on a novel broadband waveguide-to-substrate transition, which employs a unilateral finline structure for prospective use in a broadband mixer design.

[1] Group for Advanced Receiver Development, Chalmers University of Technology, SE412 96, Gothenburg, Sweden.

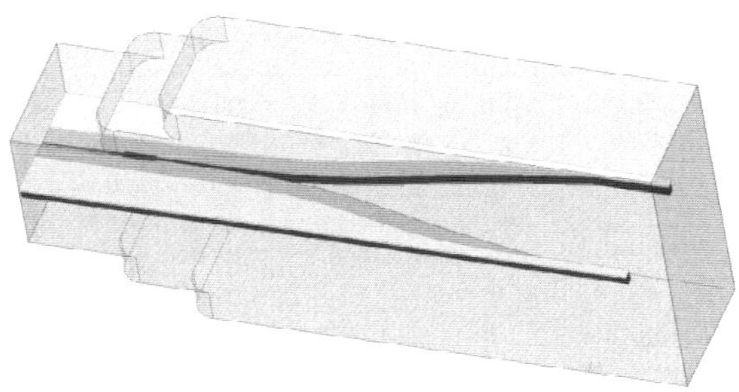

Fig. 1. CAD model of the proposed waveguide-to-substrate transition. The whole structure is centered in the split-plane of a rectangular waveguide (split-block configuration). The layer legend: gold (yellow), niobium (orange) and silicon (black).

## II. Design and Simulation

The matching between the high impedance of the full-height waveguide and a slotline is accomplished in two stages. The first stage employs a unilateral etched finline structure, while a 2-section slotline Chebyshev transformer is implemented to finally reach the desired slotline impedance of 60 Ohm. A CAD model of the proposed structure is shown in Fig. 1.

Although several different methods of unilateral finline design have been reported [5,6], none have been formulated for etched finlines. Therefore, we suggest an alternative approach to define the finline profile. Using Ansys HFSS, the impedance and guided wavelength for finline sections of different widths was investigated and mapped. This data was employed to create a 4-step Chebyshev transformer. Finally, a spline curve was drawn through the centre of each step. The same procedure was applied for the slotline Chebyshev transformer.

The transition is designed on a 30 μm silicon substrate covered by a layer of 400 nm thick superconducting niobium. The fins gradually shrink into a slotline defined by the niobium layer. The substrate between the fins has been removed in order to enable a smoother impedance transition between the waveguide and the substrate and avoid dielectric material loss in the weveguide. A 5 μm thick gold layer deposited over the niobium film extends 200 μm beyond the substrate in each direction and serves as beam-leads. This way, upon closing the waveguide, the gold layer is clamped between the two waveguide halves, providing grounding for the fins and facilitating the mounting process. Additionally, the waveguide width is gradually reduced in 2 steps to obtain subcritical dimensions. These steps are introduced late in the transition, i.e. when the field has already concentrated inside the fins. Because of this, the waveguide shrinking have almost no influence over the impedance of the transition but provide good isolation against unwanted waveguide modes.

In order to compare simulations with measurements, a back-to-back arrangement was design and simulated using HFSS in the 211-373 GHz frequency band. The simulation results are depicted in Fig. 2. An insertion loss less than 0.3 dB and a return loss better than 15 dB is achieved over the whole band. This result implies a fractional bandwidth of 55%. It is important to note that with the intention of facilitating mounting and improving the mechanical strength of the assembly, silicon tips have been introduced in the back to back transition as it is depicted in Fig. 3. These tips extend from the main substrate and fit inside cavities located in the lower end of the split block.

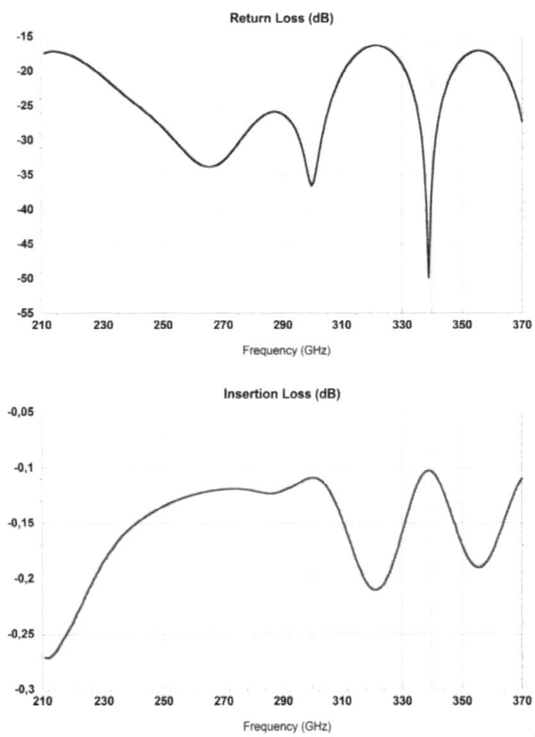


Fig. 2. Simulated scattering parameters for the back to back arrangement. The return loss is below 15 dB and the insertion loss is better that 0.3dB in the whole band.

## III. FABRICATION

The fabrication of the back-to-back transition was made in-house. The devices were processed using SOI wafer with a 30 um thick device layer. First, an Nb thin film with a thickness of 400 nm was deposited by DC magnetron sputtering. The first lithography step served to pattern the finline structure and the central slotline transformer. A thin 100 nm sputtered aluminum film was employed as hard mask to protect the underling Nb layer from the subsequent dry etch process. In a second lithography step a resist patch was created to preserve the silicon layer that outlines the slotline. Next, 5 μm of the unprotected silicon were anisotropically etched with help of Bosch process. This guarantees an accurate definition of the etched finline structure in the first microns, which are the most critical for the device performance. The 5 μm gold beamleads were created by photolithography and electroplating process. A Ti/Au bilayer was used for the electroplating seed.

For backside processing, the chip was mounted upside down on a transparent 4-inch sapphire wafer using a release layer and adhesive layer. The next step consists on etching the thick silicon handle layer using the buried $SiO_2$ layer as etch stop. The $SiO_2$ was later striped away in order to allow the final lithography step on top of the remaining 30 μm thick silicon layer. Backside lithography over a thick photo resist provided an etch mask for the subsequent anisotropic etching that completed the device definition. Finally, the adhesive layer and the release layer were removed in solvent, freeing the samples. SEM pictures of the devices are shown in figure 3.

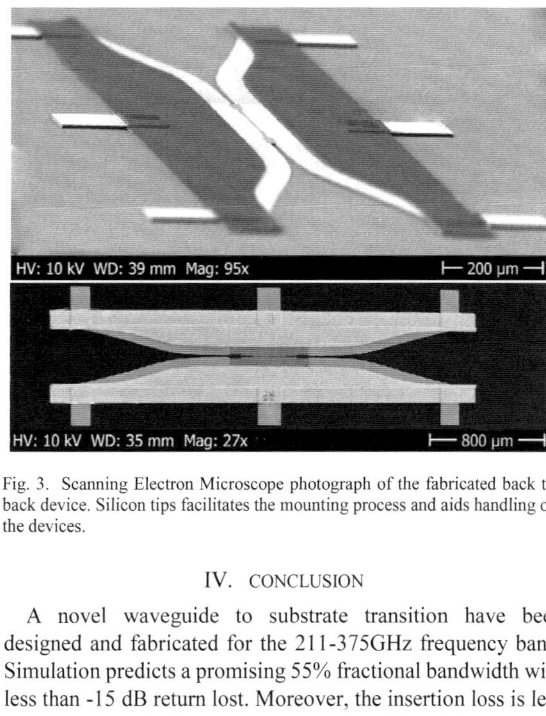


Fig. 3. Scanning Electron Microscope photograph of the fabricated back to back device. Silicon tips facilitates the mounting process and aids handling of the devices.

## IV. CONCLUSION

A novel waveguide to substrate transition have been designed and fabricated for the 211-375GHz frequency band. Simulation predicts a promising 55% fractional bandwidth with less than -15 dB return lost. Moreover, the insertion loss is less than 0.3 dB. A back to back transition have been fabricated and experimental verification is in progress.

## ACKNOWLEDGEMENTS

The research leading to these results has received funding from the European Commission Seventh Framework Programme (FP/2007-2013) under grant agreement No 283393 (RadioNet3).

## REFERENCES


[1] Risacher, C., Vassilev, V., Pavolotsky, A., & Belitsky, V. "Waveguide-to-microstrip transition with integrated bias-T.", *IEEE Microwave and Wireless Components Letters*, vol *13*, pp. 262-264, 2003.

[2] J. W. Kooi, G. Chattopadhyay, S. Withington, F. Rice, J. Zmuidzinas, C. Walker, & G. Yassin, "A full-height waveguide to thin-film microstrip transition with exceptional RF bandwidth and coupling efficiency," International Journal of Infrared and Millimeter Waves, pp.261-284, 2003.

[3] Tan, Boon Kok, and Ghassan Yassin, "Unilateral finline transition at THz frequencies," Twentieth International Symposium on Space Terahertz Technology", 2009.

[4] Leong, Kevin MKH, et al. "A 340–380 GHz integrated CB-CPW-to-waveguide transition for sub millimeter-wave MMIC packaging.", *IEEE Microwave and Wireless Components Letters* 19.6, pp 413-415, 2009.

[5] North Chris, Ghassan Yassin, and Paul Grimes. "Rigorous analysis and design of finline tapers for high performance millimetre and submillimetre detectors.",Seventeenth International Symposium on Space Terahertz Technology, 2006.

[6] Schieblich, Christian., J. K. Piotrowski, and J. H. Hinken. "Synthesis of optimum finline tapers using dispersion formulas for arbitrary slot widths and locations.", IEEE transactions on microwave theory and techniques 32.12, pp.1638-1645, 1984

# Compact Wideband Passive and Active Component Chips for Radio Astronomy Instrumentation

H. Rashid[1], D. Meledin[1,2], V. Desmaris[1,2], E. Sundin[2], and V. Belitsky[1 2]

*Abstract*— In the 2SB receivers, the rejection ratio is determined among other factors by the performance of the quadrature hybrids. We have developed and fine-tuned components that intended to be used in the IF chain and meet the requirements for new generation of compact wideband 2SB receivers. We present here the design and characterization of three multi-section compact wideband 3 dB quadrature couplers (coupled line coupler – Lange coupler–coupled line coupler). Specifically, the miniaturized 3-section hybrid chip made using thin-film technology utilizes gold plated transmission lines and air bridges to connect the fingers of the Lange coupler (middle section The hybrids were designed to have the amplitude and phase imbalance better than 0.6 dB and ±3° respectively over a 3.5-12.5 GHz and 4-16 GHz frequency bands. Experimental verification of the assembly at 293 K and 4 K shows very good agreement between the measurements and simulations. Additionally, we demonstrate the suitability of the miniature hybrid chips, by implementing them in a balanced amplifier designed using a modular approach, which demonstrated promising results.

## I. Introduction

The demand for wideband millimeter-submillimeter wave multi-pixel receivers is continuously increasing. So far, only DSB multi-pixel heterodyne systems have been demonstrated [1]. Implementing of wideband sideband separating mixers into multi-pixel design is hindered however, by the fact that the number and the size of passive and active components (LNA) increases, which leads to unpractical receiver pixel footprint. A typical sideband separating receiver consists of among others, RF and IF quadrature hybrids, bias-Ts, isolators and low noise amplifiers, see Fig. 1. One of the key figures of merit in sideband separating receivers is the image rejection ratio, which is mainly dependent on the overall amplitude and phase impedance of the receiver chain [2]. This sets tough requirements on both the RF and IF quadrature hybrids.

In order to shrink the wideband receiver pixel footprint, compact and wideband passive and active device are required. One way to reduce the size is to use substrates with high dielectric constant and to integrate the bias-T with the wideband IF hybrid [2]. Furthermore, by using a balanced LNA topology [3] (employing the wideband IF hybrids) would yield a both compact and wideband LNA with sufficient return loss, which practically removes the need for having an isolator in the receiver chain [2]. Hence, the importance of wideband quadrature hybrid with excellent amplitude and phase imbalance performance is vital [4-5]. This way, only two devices, an integrated 2SB mixer including bias-T followed by balanced IF LNA are required instead of four, which saves space considerably and would improve the performance. Furthermore, it is essential for integration and fabrication purpose that the circuits are planar.

To meet these goals, both compact wideband IF (one octave) and RF circuits has been developed and demonstrated [4-6]. Some of these wideband circuits have successfully been integrated into ALMA Band 5 receivers [6] or in the SEPIA receiver [7], which demonstrated image rejection ratios typically better than 15 dB, which is a significant improvement [6]. However, to meet the future demands of wideband radio astronomy receivers (IF bandwidth of two octaves or more), a new generation of circuit with bandwidth and enhanced performance have been designed.

In the next section, two compact wideband hybrids are presented, one operating at 3.5-12.5 GHz, and one at 4-16 GHz. Measurements of the manufactured circuits show exceptional performance at both room and cryogenic temperatures.

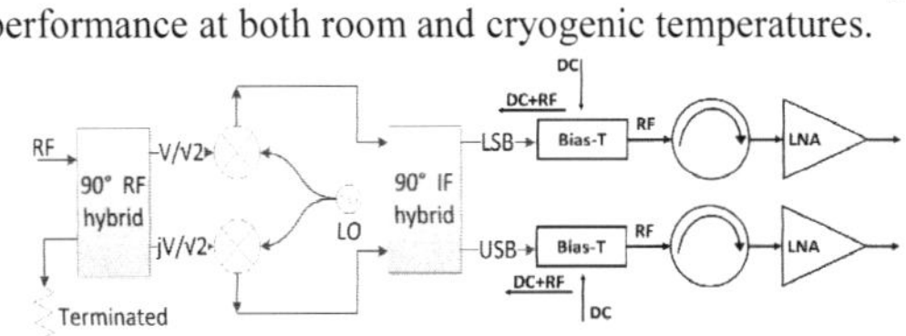


Fig. 1. Schematic of a sideband separating receiver

## II. High Performance Wideband Hybrids

### A. Hybrid Topology

The requirement on the amplitude and phase imbalance was discussed in [5], where the maximum allowed imbalance was 0.6 dB and ±3° respectively over one octave fractional bandwidth. However, future receivers require at least the same imbalance performance over two octaves or more. Hybrids employing single-section topologies cannot provide the required amplitude imbalance for bandwidths of greater than one octave at best [8, 9]. Therefore, multi-section layout must be used to meet the requirements upcoming

[1] Milliwave, Gothenburg, Sweden.

[2] Group for Advanced Receiver Development, Chalmers University of Technology, SE412 96, Gothenburg, Sweden.

wideband/multipixel receivers. A multi-section layout of three sections have been used here, since three sections is enough to meet the requirements, and the same time keep the circuit compact.

*B. Substrate selection*

The possibility to use a substrate with reasonably high dielectric permittivity allows reaching compact design with minimum insertion loss; the latter depending on dielectric loss and conducting loss in the transmission lines. The miniature hybrid chip allows it to be integrated into virtually any sideband separating mixer, which is especially advantageous for multi-pixel receivers or low noise balanced amplifier layouts. Moreover, due to the multi-section design, it would be advantageous to use a substrate with high dielectric constant in order to minimize the overall chip dimensions. Therefore, Alumina substrate was selected as it exhibits excellent mechanical properties, low loss tangent, good thermal conductivity, and a dielectric constant of 9.6 at the design frequency.

*C. Three section hybrids*

The proposed design employs a symmetrical three-section coupler. The even mode characteristic impedance for each stage of the multi-section coupler can be found in tables for a given equi-ripple across the band [10]. Two hybrids with different bandwidths were designed. The design procedure in both cases are the same. First step is to calculate the coupling coefficients of each section of the hybrid. The center section has always the highest coupling coefficient, while the adjacent sections are rather weakly coupled. Therefore, the first and the third sections are realized as a coupled line coupler structure whereas the middle section coupler employs a Lange coupler. Through these choices of couplers, the final structure becomes planar and the hybrid dimensions are completely determined by photolithography process. The miniaturized 3-section hybrid chip are made using thin-film technology, which utilizes gold plated transmission lines and air bridges to connect the fingers of the Lange coupler (middle section). A schematic of the three-section hybrid is illustrated in Fig. 2.

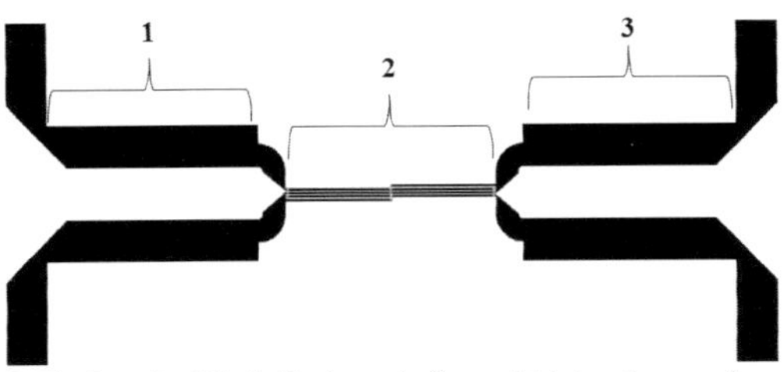


Fig. 2: The 3-section IF hybrid schematic, first and third sections are the coupled line couplers whereas the second section is the Lange coupler.

The initial dimensions for the multi-section coupler were calculated with Keysight ADS [10] LineCalc, using the coupling coefficients as described above. The complete hybrid structure was then optimized with Keysight Momentum [11]. The total length and width of the hybrid are 18.3 mm and 5 mm respectively. Fig.3 and Fig.4 shows the simulated performance of the 3.5-12.5 GHz and 4-16 GHz hybrid with excellent amplitude imbalance performance. Moreover, by employing superconducting Nb lines as described in [12] the performance can be further improved by removing unwanted capacitive coupling if the Lange coupler.

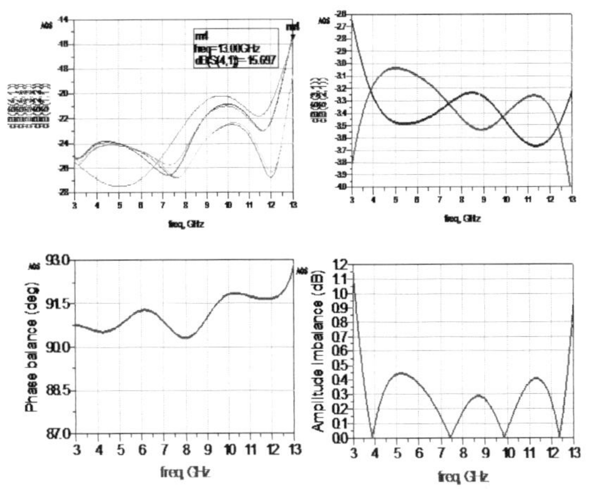


Fig. 3: Simulated performance of the 3.5-12.5 GHz hybrid. Top left: Reflection and isolation plot (dB). Top right: Though and coupled plot (dB). Bottom left: Phase balance (deg). Bottom right: Amplitude imbalance (dB).

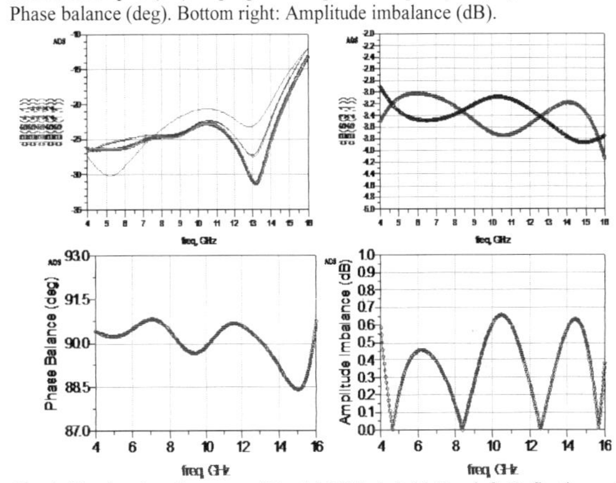


Fig. 4: Simulated performance of the 4-16 GHz hybrid. Top left: Reflection and isolation plot (dB). Top right: Though and coupled plot (dB). Bottom left: Phase balance (deg). Bottom right: Amplitude imbalance (dB).

## III. Hybrid Performance

*A. Characterization setup*

The fabricated chips were mounted in a housing featuring 4 K-type connectors, which was then installed in a cryostat for cooling down the samples down to 4K, using a close cycle machine. 4-port S-parameters measurements were carried out at room temperature (293 K) and at cryogenic temperatures (~4 K) using a Rohde&Schwarz ZVA40 vector network analyser (VNA). Owing to the lack of calibration standards at cryogenic temperatures, the S-parameters measurements at 4K were conducted using a special calibration procedure for cryogenic temperature, as described in details in [6]. The measured performance of the hybrids is shown in Fig. 5-9.

*B. 3.5-12.5 GHz hybrid measured at 293 K*

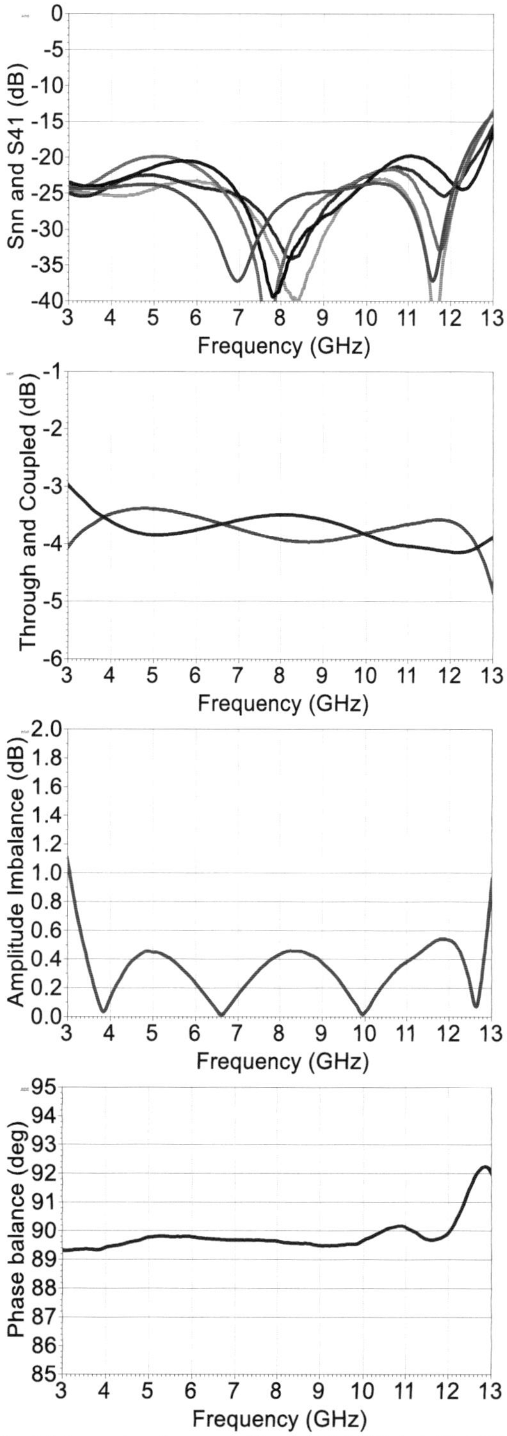


Fig. 5: Measured performance of the 3.5-12.5 GHz hybrid at 293 K. Figures from top to bottom represent: Reflection and isolation plot (dB); Though and coupled plot (dB); Amplitude imbalance (dB); Phase balance (deg.)

*C. 3.5-12.5 GHz hybrid measured at 4 K*

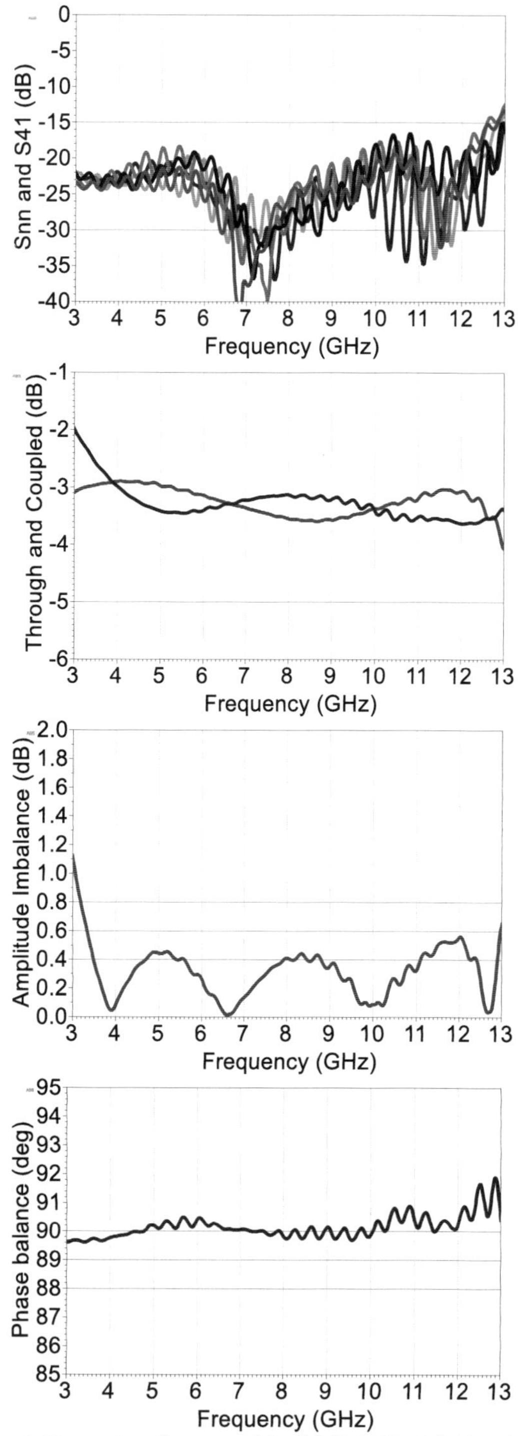


Fig. 6: Measured performance of the 3.5-12.5 GHz hybrid at 4 K. Figures from top to bottom represent: Reflection and isolation plot (dB); Though and coupled plot (dB); Amplitude imbalance (dB); Phase balance (deg.).

*D. 4-16 GHz hybrid Measured at 293 K*

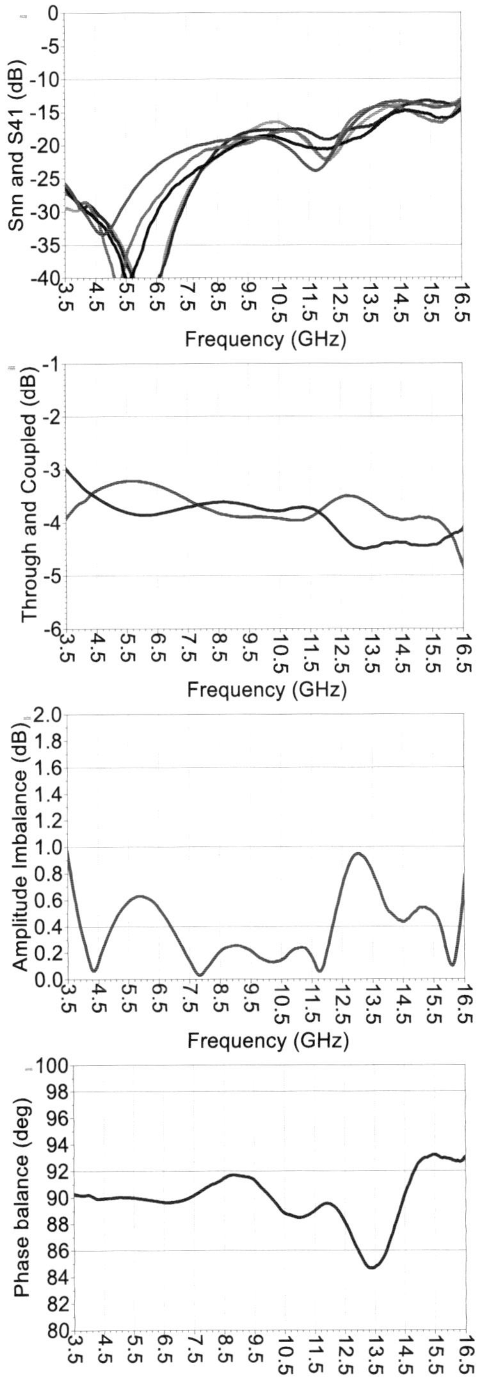


Fig. 7: Measured performance of the 4-16 GHz hybrid at 293 K. Figures from top to bottom represent: Reflection and isolation plot (dB); Though and coupled plot (dB); Amplitude imbalance (dB); Phase balance (deg.)

*E. 4-16 GHz hybrid Measured at 293 K*

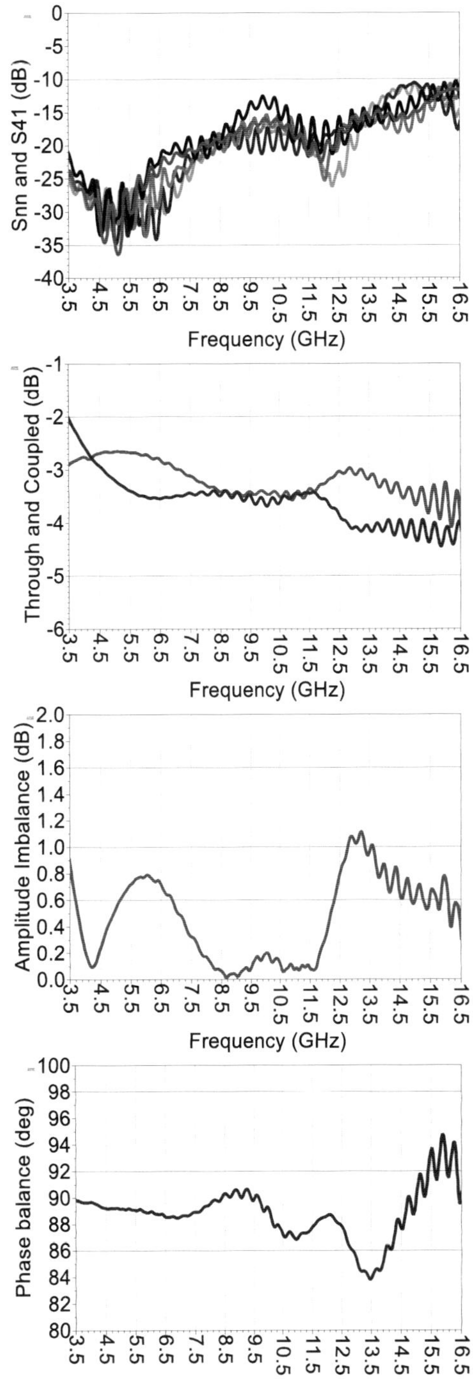


Fig. 8: Measured performance of the 4-16 GHz hybrid at 4 K. Figures from top to bottom represent: Reflection and isolation plot (dB); Though and coupled plot (dB); Amplitude imbalance (dB); Phase balance (deg.)

*F. 4-12 GHz superconducting hybrid measured at 4K*

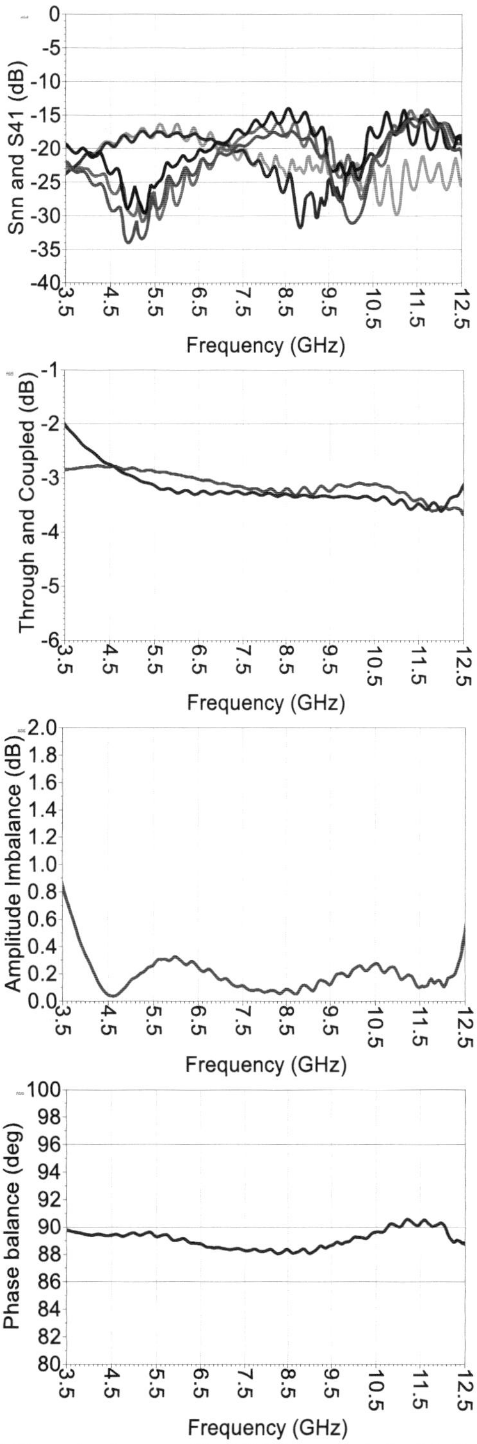


Fig. 9: Measured performance of the 4-12 GHz superconducting hybrid at 4 K. Figures from top to bottom represent: Reflection and isolation plot (dB); Though and coupled plot (dB); Amplitude imbalance (dB); Phase balance (deg.).

*G. The hybrids' performance*

Comparing Fig. 5 to 6, it can be seen that the amplitude and phase imbalance 3.5-12.5 GHz hybrid are not affected by the change in temperature, even though the insertion loss is clearly improved when the devices are operated at 4K, most likely due to the increase of conductivity of the metallic microstrip lines when cooling down the device.

Despite of a slight over-coupling, we observe a similar relative independence of the amplitude and phase imbalance as well as an improvement of the insertion loss for the 4-16 GHz hybrids, as show on as shown on Fig. 7 and Fig. 8. However for this particular hybrid, we observe a discrepancy in return loss between the measurement results Fig. 7 and the simulation in Fig. 3 is probably due to due mounting or/and fabrication

The 4-12 GHz hybrid made of superconducting transmission lines shows measured 4-12 GHz hybrid with almost ideal performance and is very suitable for integration into 2SB THz receivers based on superconducting mixers.

## IV. Balanced Amplifier Characterization

*A. Practical implementation*

An advantageous use of the miniaturize chips would be their implementation in balanced amplifiers, eventually used in the IF amplification chain of cryogenic receivers. Therefore, we constructed a modular balanced amplifier using 2 single end 4-16 GHz LNF-LNC4_16B amplifiers from Low Noise Factory [13] and two wideband miniature 3dB 90 degrees hybrids: one 4-12 GHz hybrid at the input and one 4-16 Ghz hybrid at the output (Fig. 10). Obviously, the modular approach is not optimum, since it introduced many interfaces and parasitics originating from the connection between the different elements of the amplifier. Yet it is a good test bed for evaluating the hybrids' performance.

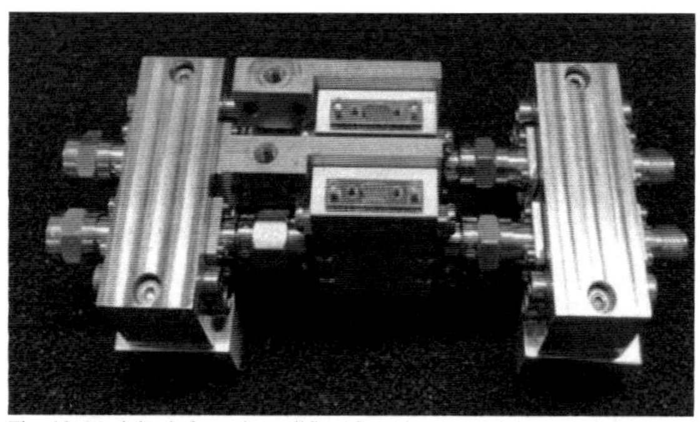

Fig. 10. Modular balanced amplifiers featuring two miniature hybrids and two single-end amplifiers.

*B. Characterization setup*

The noise performance of the LNAs were measured in a cryogenic system comprising a cryostat equipped with a 2-stage close cycle-machine, which allows cooling down to 4 K (Fig. 10). The noise temperature measurements were performed by the standard Y-factor technique and using the cold attenuator method similarly to [3], with an Agilent N4000A ENR noise diode and an Agilent MXA N9020A spectrum analyzer with noise figure measurement option. The input and output reflection factors were measured after removal of the cold attenuator using a Rohde&Schwarz ZVA40, employing a specific calibration procedure described in [6] and biasing the individual amplifiers under the exact same conditions as during the noise measurements.

### C. *Characterization results*

The noise temperature and gain of the balanced amplifier over the 4-12 GHz frequency band are presented in Fig. 11, whereas the input and output reflection factors are plotted in Fig 12.

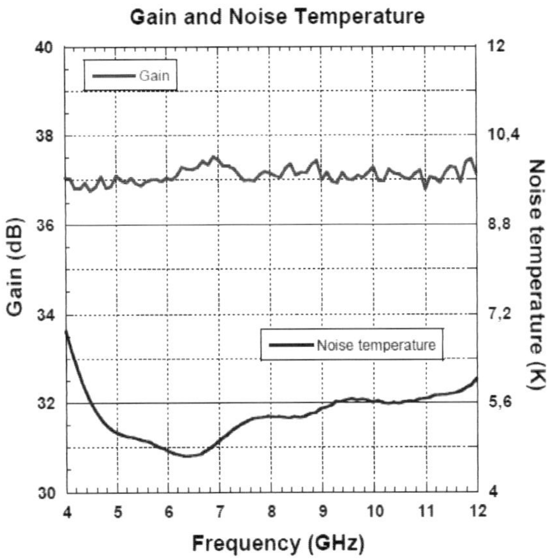


Fig. 11. Gain and Noise temperature of the Modular balanced amplifier.

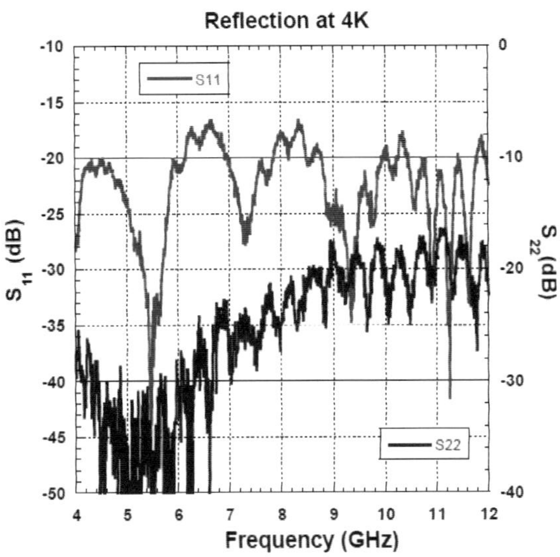


Fig. 12. Gain and Noise temperature of the Modular balanced amplifier.

The results show an improvement of about 5-7 dB in terms of return loss of the balanced amplifier as compared to the single-end amplifiers, yet without noticeable deterioration of the gain performance, which is consistent with our expectations. However, the noise temperature was raised by about 1K on average over the 4-12 GHz frequency range. This is likely to be ascribed to the inevitable parasitic effects introduced by the modular design with its different internal and external (SMA) interconnections and the fact that the hybrids used in this experiment are made of Gold transmission lines, which is not completely lossless even at cryogenic temperatures.

## V. Conclusions

In this work, we report on the design and characterization of a 3.5-12.5 GHz and 4-16 GHz quadrature 3 dB directional couplers. The compact size of the hybrid chip allows it to be integrated into virtually any sideband separating (2SB) mixer operating at room temperature or cryogenic temperatures and is furthermore especially advantageous for multi-pixel 2SB receivers or balanced amplifier layouts.

The hybrid was fabricated using in-house thin-film technology on Alumina substrate and uses air bridges to inter-connect the fingers of the Lange coupler. The four-port S-parameters measurement showed very good agreement with the simulation results.

Furthermore, the good performance measured on a modular balanced amplifier constructed using the miniaturized hybrid chips let us to believe that with further integration of the chips in the amplifier designs even more broadband and compact IF LNAs could be realised for future use in radioastronomy receivers.

## Acknowledgment

The authors would like to acknowledge Sven-Erik Ferm for his help with the fabrication of the mechanical fixtures to for the measurement tests.

The research leading to these results has received funding from the European Commission Seventh Framework Programme (FP/2007-2013) under grant agreement No 283393 (RadioNet3).

## References

[1] C. E. Groppi and J. H. Kawamura, "Coherent Detector Arrays for Terahertz Astrophysics Applications," *Terahertz Science and Technology, IEEE Transactions on,* vol. 1, pp. 85-96, 2011.

[2] I. Malo-Gomez, J. D. Gallego-Puyol, C. Diez-Gonzalez, I. Lopez-Fernandez, and C. Briso-Rodriguez, "Cryogenic Hybrid Coupler for Ultra-Low-Noise Radio Astronomy Balanced Amplifiers," *Microwave Theory and Techniques, IEEE Transactions on,* vol. 57, pp. 3239-3245, 2009

[3] E. Sundin, A. Ermakov, V. Desmaris, H. Rashid, D. Meledin, M. Bylund and V. Belitsky, "Cryogenic IF Balanced LNAs Based on Superconducting Hybrids for Wideband 2SB THz Receivers", *Proceedings of the ISSTT2017*, Cologne, Germany, 2017.

[4] H. Rashid, D. Meledin, V. Desmaris, and V. Belitsky, "Novel Waveguide 3 dB Hybrid With Improved Amplitude Imbalance," *Microwave and Wireless Components Letters, IEEE,* vol. 24, pp. 212-214, 2014.

[5] H. Rashid, D. Meledin, V. Desmaris, A. Pavolotsky, and V. Belitsky, "Superconducting 4-8 GHz Hybrid Assembly for 2SB

Cryogenic THz Receivers," *Terahertz Science and Technology, IEEE Transactions on,* vol. 4, pp. 193-200, 2014.

[6] V. Belitsky *et al.*, "ALMA Band 5 receiver cartridge: Design, performance, and commissioning", *Astronomy and Astrophysics*, vol. 611, A68, 2018.

[7] V. Belitsky *et al.*, "SEPIA – a new single pixel receiver at the APEX telescope", *Astronomy and Astrophysics*, vol. 612, A23, 2018.

[8] B. Billade, H. Rashid, V. Desmaris, and V. Belitsky, "Superconducting 4 - 8 GHz IF Hybrid for Low Noise mm-Wave Sideband Separation SIS Receiver," *Microwave and Wireless Components Letters, IEEE,* vol. 22, pp. 589-591, 2012.

[9] H. Rashid, P. Y. Aghdam, D. Meledin, V. Desmaris, and V. Belitsky, "Wideband Planar Hybrid With Ultralow Amplitude Imbalance," *IEEE Microwave and Wireless Components Letters,* vol. 27, pp. 230-232, 2017.

[10] E. G. Cristal and L. Young, "Theory and Tables of Optimum Symmetrical TEM-Mode Coupled-Transmission-Line Directional Couplers," *Microwave Theory and Techniques, IEEE Transactions on,* vol. 13, pp. 544-558, 1965.

[11] ADS. *www.keysight.com.*

[12] V. Belitsky *et al.*, "Superconducting microstrip line model studies at millimetre and sub-millimetre waves". *International Journal of Infrared and Millimeter Waves*. 27. Pp. 809-834, 2006

[13] https://www.lownoisefactory.com/products/cryogenic/4-16-ghz/

.

# Compact Cryogenic Wide-Band Balanced Amplifiers with Superconducting 90° Hybrids for the IF of Submillimeter-Wave SIS Mixers

R. I. Amils, I. López-Fernández, J. D. Gallego, *Member, IEEE*, V. Desmaris, V. Belitsky, D. Meledin, E. Sundin, C. Diez, I. Malo

***Abstract*—The pressing demand for mm-wave cryogenic radio astronomy receivers with increasing instantaneous bandwidth has spurred interest in more complex intermediate frequency amplifier configurations, like the balanced amplifier, as the traditional options have increasing difficulties to comply either with the noise or the input matching specifications. This solution is typically penalized by the slight increment in noise produced by the quadrature hybrid losses. We propose a balanced amplifier using a novel 3 dB quadrature hybrid coupler design with superconducting lines. The prototype unit built for the 4-12 GHz band integrates in the same module the hybrid coupler chips, the low noise amplifiers and the bias circuitry. The mechanical design allows for an independent testing of the individual amplifiers. The average noise temperature is 4.4 K, only 0.3 K more than the average of its amplifiers. The input reflection improves more than 10 dB. This compact balanced amplifier is also compared with a non-integrated version, showing an improvement in noise and reflection. It has been tested in a complete mm-wave receiver, with advantage over other IF schemes.**

***Index Terms*—Cryogenic balanced amplifier, submillimeter wave receivers, superconducting hybrid coupler, low noise, radio astronomy.**

## I. Introduction

THE new generations of THz receivers for radio astronomy demand increasingly wider instantaneous bandwidths [1], [2]. The requirements for the Intermediate Frequency (IF) cryogenic amplifiers are extremely difficult to meet, as larger fractional bandwidths force a trade-off between noise temperature and input reflection matching. The mismatch between the mixer, usually a superconducting-insulating-superconducting (SIS) junction and the low noise amplifier (LNA) poses many practical problems, like high ripples in the band or mixer instability. Two approaches to address this issue have been attempted in the past: (1) the use of input cryogenic isolators and (2) the implementation a balanced amplifier configuration. Both transfer the input reflection demands from the LNA to another additional component.

The first solution was employed, for example, in bands 5 and 9 of the Atacama Large Millimeter Array (ALMA) [3], which represent a standard for mm-wave receiver architectures. Its main disadvantages are, on the one hand, the insertion losses of the isolator, which could degrade the noise temperature by about 3 K for the 4-12 GHz ALMA band (at 15 K ambient temperature), and on the other hand, the limited fractional bandwidth of ferrite isolators. The performance of the available isolators in the ALMA band degrades at the edges, and it is very difficult to procure (or design) isolators for larger fractional bands.

The balanced solution is implemented in some radio astronomical receivers [4]. It is more cumbersome and expensive, as it requires two quadrature 3 dB cryogenic hybrid couplers and two LNAs. Although it suffers from the same loss problem, with a good hybrid design the results are better in noise (a penalty of 1.5 K in the aforementioned band) and reflection than with the input isolator. Moreover, the potential bandwidth is greater, making this solution expandable to future wider band receivers.

In this work, we propose using superconducting hybrid couplers [5] in the balanced amplifier to minimize the insertion losses and thus, the noise penalty of this configuration. Furthermore, we accomplish the integration of the pair of hybrids and amplifiers in the same block (what will be called henceforth a "compact balanced amplifier") in order to overcome the size drawback and to improve its matching and noise performance by eliminating the connectors between the balanced amplifier components.

The classical ALMA 4-12 GHz band, where adequate LNA and hybrid designs were already available, was selected for this implementation. Even though cryogenic amplifiers with acceptable input reflection levels have been recently developed for this band [6], the balanced configuration is superior in input matching, noise band ripple (due to the uncorrelation between input and output noise waves, see [4]), redundancy against failures in the field and even linear dynamic range. These advantages are increasingly determining

The work done at Yebes Observatory was supported in part by the European Union's Horizon 2020 research and innovation programme under grant agreement 73562 (RadioNet).

R. I. Amils, I. López-Fernández, J. D. Gallego and C. Diez are with Observatorio de Yebes, Centro de Desarrollos Tecnológicos del Instituto Geográfico Nacional (IGN), 19141 Yebes, Spain (e-mail i.lopez@gmail.com).

V. Desmaris, V. Belitsky, D. Meledin and E. Sundin are is with the Group for Advanced Receiver Development (GARD) of Chalmers University of Technology, SE-412 96 Gothenburg, Sweden (e-mail: vincent.desmaris@chalmers.se).

as the fractional bandwidth is further expanded.

The characteristics of the LNAs and the quadrature hybrids used are analyzed in sections II and III. The balanced amplifiers are presented in section IV. A first non-compact (modular) approach was assembled and tested, both in the labs and in a complete receiver with promising results. Finally, a prototype of a compact balanced amplifier was demonstrated. The measurement system and the results are analyzed in sections V and VI.

## II. Low Noise Amplifiers

The low noise amplifiers used for this work are based on the well stablished design used in the 4-12 GHz IF of ALMA bands 5 and 9. A description can be found in [3]. It is implemented in microstrip hybrid technology, with single chip transistors for each of the three stages. For this application, InP HEMT devices were placed in all stages to optimize noise and reduce power consumption. In particular, the first stage input network was tuned to fit a 150 × 0.1 µm gate Diramics[1] transistor with state-of-the-art noise performance, developed in collaboration with ETH[2] (see Fig. 1). The original ALMA design was conceived to be used in combination with a cryogenic input isolator, hence no effort was put on the optimization of the input return losses, which makes it an excellent candidate to test the benefits a balanced configuration.

The LNAs used in the balanced amplifier yield a noise temperature around 4 K (measured at 5 K ambient temperature). Their performances are given in section V.

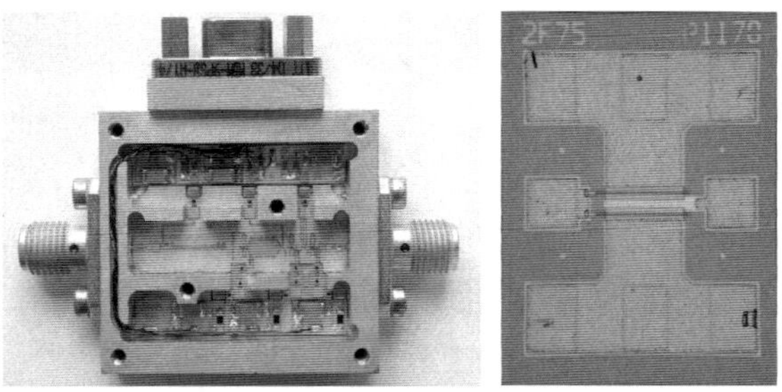

Fig. 1. On the left, ALMA LNA on which this development is based. On the right, Diramics InP transistor with two 75×0.1 µm gate fingers used in the first stage to obtain the best noise.

## III. Quadrature Hybrid Coupler

The superconducting 4-12 GHz compact hybrids are fabricated on alumina and employ the three-sections planar design suggested in [5]. The central section with a Lange coupler is placed between two sections with coupled line couplers. The Lange coupler provides the highest coupling coefficient while the other two sections are loosely coupled.

The whole design features superconducting transmission lines to minimize the insertion loss, have better control of the transmission line geometry and ease the manufacturing of the devices. The hybrids are produced using thin-film microfabrication technology that combines photolithography and dry etching processes for the formation of superconducting Nb transmission lines, and galvanically plated air bridges to connect the coupler fingers in the middle section. The overall size of the miniature hybrid chip is as small as 27x9 mm and is currently limited only by the pitch required between the through and coupled ports, in order to conveniently interface with the existing LNA structures.

The performance of the hybrids has been verified on a single witness hybrid sample that was selected from the same wafer as one of the three units used for this work. The S-parameters of the witness hybrid have been measured at 4K, using a 4-port VNA connected to a cryostat, similarly to [7]. The results show excellent performance of the hybrids in terms of insertion loss, return loss, as well as amplitude and phase imbalance (see Fig. 3).

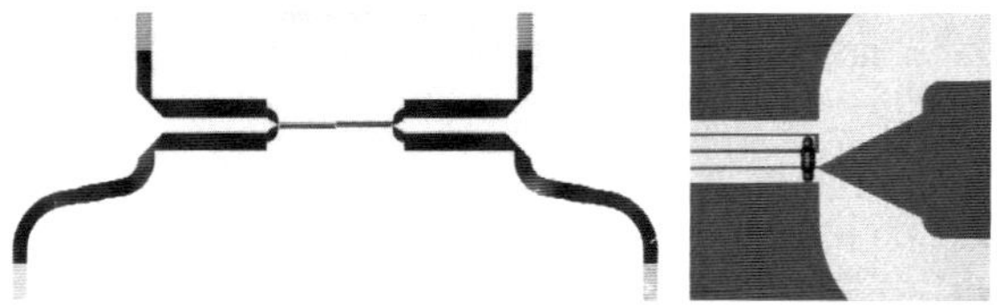

Fig. 2. On the left, photograph of the hybrid chip, 27×9 mm. On the right (in inverted colors), detail of the Lange coupler section and air bridge.

## IV. Balanced Amplifier Design and Fabrication

### *A. Modular balanced LNA*

Two identical LNAs as described in section II were connected to two quadrature hybrid modules to test the performance of a balanced configuration (see Fig. 4) and compare it with the compact version. The connectorized hybrid modules are made of CuTe and include a prior version of the chip described in section III, with slightly worse input reflection [7].

Similar modular balanced amplifiers had been tested in the past at Yebes Observatory [2], but not with superconducting lines. The validity of this approach was demonstrated before proceeding with the fabrication of the compact balanced version by including this amplifier on a complete 300 GHz receiver, where an advantage in noise terms over the pre-existing isolator solution was measured [7].

### *B. Compact balanced LNA*

The integration of the balanced amplifier into a single module was attempted as a proof of concept to validate its potential advantages. In this first attempt no effort was put on the size reduction of the module (one of the benefits of the integration). As shown in Fig. 5, the amplifier comprises two specular versions of the individual LNA, sharing only the drain bias circuit cavity. No changes in the microstrip matching circuits were made. Both DC connectors were

[1] Diramics AG, Hinterbergstrasse 28, 6312 Steinhausen, Switzerland (http://diramics.com/).

[2] Millimeter-Wave Electronics Laboratory, Eidgenössische Technische Hochschule (ETH), Zürich

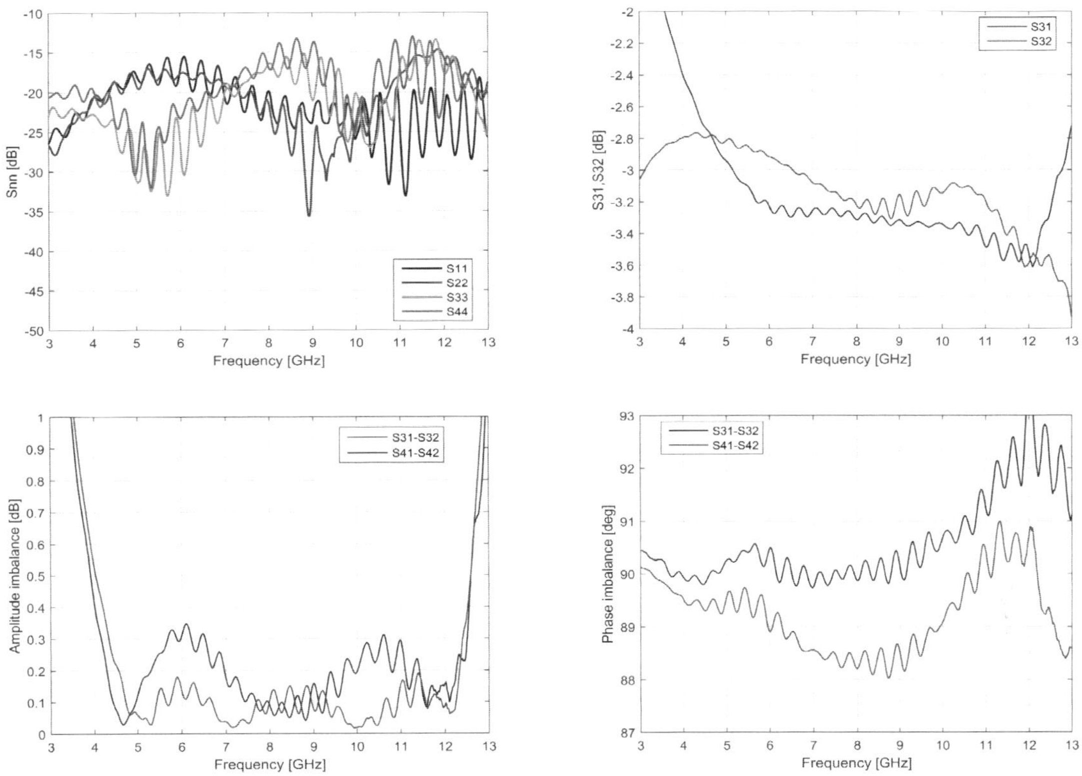


Fig. 3. Performance of the superconducting hybrid. From left to right, top to bottom, return losses, insertion losses, amplitude imbalance and phase imbalance.

preserved to simplify the design.

The interface of the microstrip line with the hybrid chip consists of a coaxial hermetic seal (glass bead) with a stress relief contact on both ends to avoid cracking of the soldering due to differential thermal contraction between pin and line.

One key advantage of this design is the possibility of performing measurements of each amplifier independently. This can be achieved before attaching the hybrid blocks to the dual amplifier chassis, by simply placing 2.9 mm coaxial connectors on the outer end of the glass beads pins. These measurements (1) allow a direct comparison of the single ended vs the balanced configurations and (2) facilitate the tuning of the individual amplifiers.

All mechanical parts (amplifier chassis, hybrid blocks and lid) were machined in CuTe at Chalmers GARD workshop and gold plated at Yebes Observatory. The overall dimensions without connectors are 53×52.4×12.5 mm.

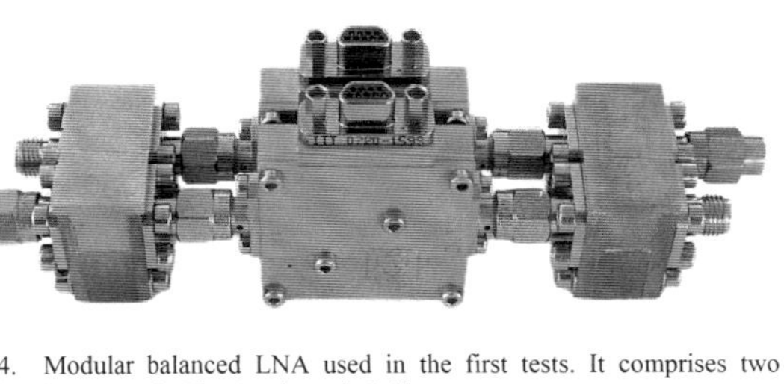

Fig. 4. Modular balanced LNA used in the first tests. It comprises two individual LNAs as in Fig. 1 and two hybrids.

## V. Results

### *A. Measuring system*

Cryogenic noise measurements were performed at Yebes Observatory. Two different setups were used to characterize the amplifiers.

The validation and tuning measurements of the LNAs included in the compact balance amplifier were done in a 15 K Dewar using the cold attenuator method. A HP N8975A noise figure meter was used in combination with an Agilent N4002A noise source, and an in-house-built 15 dB attenuator specially designed in a quartz substrate [8] for broadband cryogenic operation and accurate temperature readings. The main reason for using this setup was the advantage it gives for tuning an amplifier in terms of speed (due to the fast-switching noise diode) and the possibility of measuring cryogenic S parameters simultaneously.

The balanced amplifiers with superconducting hybrids ($T_c$ around 9 K) were measured in a 5 K Dewar using the controlled temperature load method. In this case, the noise source is broadband very low reflection coaxial termination connected directly to the LNA input inside the cryostat. It is based on a GaAs chip featuring a 50 Ω load and an integrated

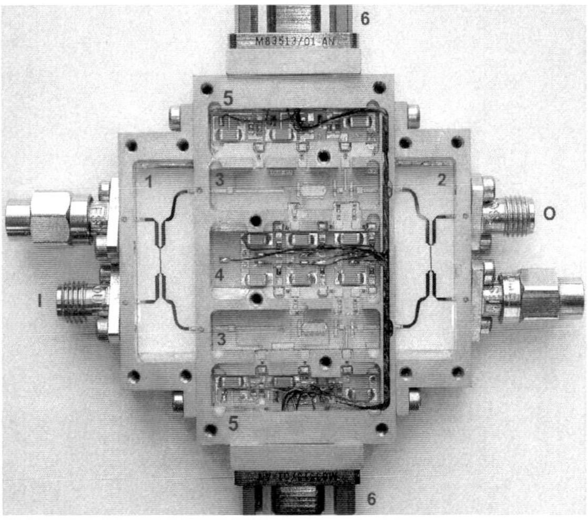


Fig. 5. Compact balanced LNA, sized 53×52.4×12.5 mm without connectors. Input (I) and output (O) are 2.9 mm coaxial connectors. 50 Ω loads are connected to the other hybrid ports. Note the input (1) and output (2) hybrid chips assembled into detachable modules, the specular RF cavities (3) as in the original LNAs of Fig. 1, the common gate bias cavity (4), the drain bias cavities (5) and DC connectors (6) of each amplifier. Hybrid blocks can be substituted by 2.9 mm connectors enabling direct measurements of the individual LNAs.

heating resistor and temperature sensor, fabricated by IAF[3] [9]. Its temperature can be controlled in closed loop using the heater and the sensor, and the temperature of the cold plate is controlled as well by an independent loop (allowing measurements at different temperatures). A photograph of the setup inside the Dewar is shown in Fig. 6.

The final measurements of the individual LNAs were also taken in this system at 5 K to provide a homogeneous comparison with the balanced LNA. Data at 15 K were also obtained, showing very good agreement with the measurements taken in the 15 K Dewar with the other method.

Cryogenic S parameter data was measured at 5 K in the same Dewar using a Keysight PNA-X N5247A and Keysight N4694A electronic calibration kit used in place of the DUT at ambient temperature inside the Dewar. Stainless steel access lines are supposed invariant with temperature. The loss change with temperature of the flexible cable connecting the DUT with the output line is subtracted from the S11 and S21 and its residual phase variations upon cooling are eliminated by time domain gating.

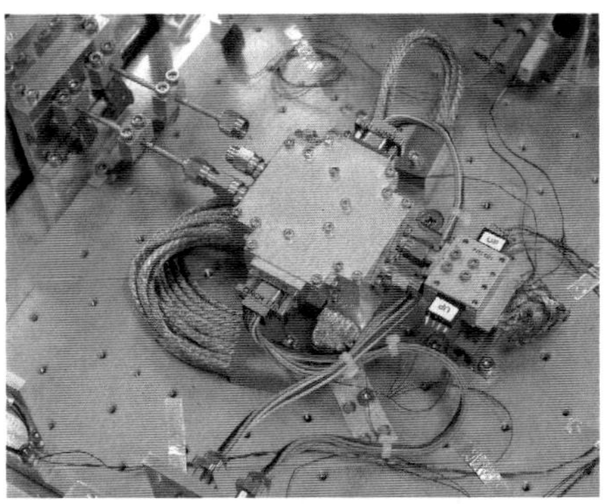

Fig. 6. Noise measurement setup in the 5 K Dewar. A cryogenic heatable load [9] is connected to the input of the compact balanced amplifier.

[3] Fraunhofer Institute for Applied Solid State Physics, Tullastrasse 72, 79108 Freiburg, Germany (https://www.iaf.fraunhofer.de/)

*B. Experimental results*

Comparisons between a modular balanced amplifier and an isolator in this band had been presented previously [4], [7], showing an advantage in noise in favor of the balanced option. In this section we will focus only on the measurements of the balanced amplifiers.

Figure 7 shows the noise and gain curves of the compact balanced amplifier, its individual LNAs (measured before the assembly of the hybrid blocks) and the modular amplifier. The balanced amplifier noise is almost bounded between the noise of its amplifiers. It is remarkable that the average noise in the band is 4.4 K, only 0.3 K more than both LNAs averaged, which demonstrates the importance of the low losses of the superconducting hybrid lines and the close integration of hybrid and LNA. The noise degradation in the compact balanced amplifier noise with respect to a single-ended is

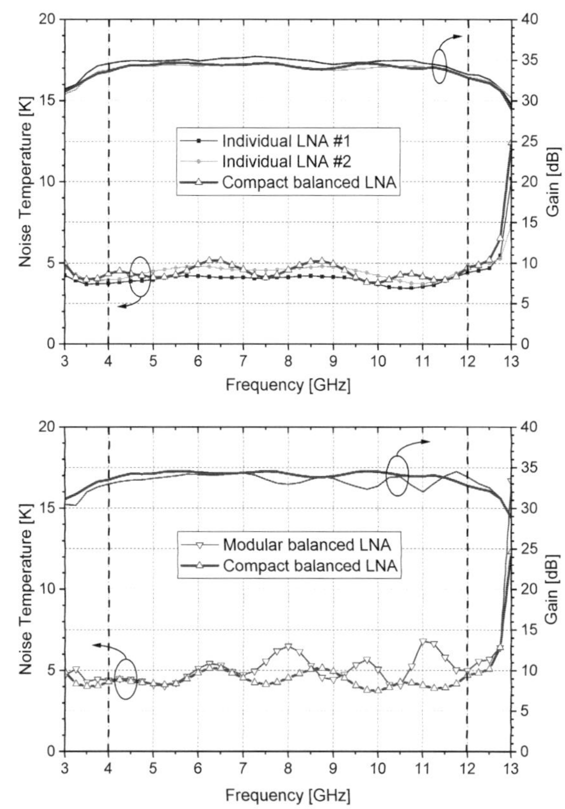


Fig. 7. Noise and gain curves of the compact balanced LNA compared to its individual LNAs (up) and the modular balanced LNA (down), measured at 5 K.

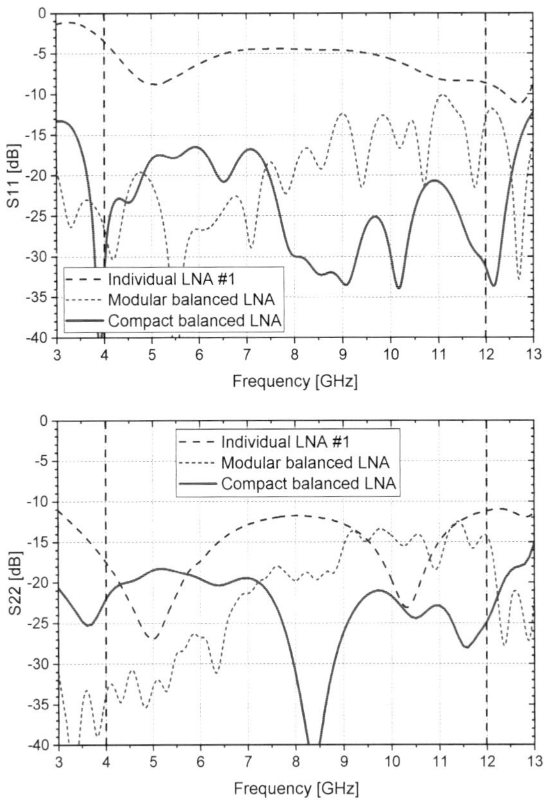


Fig. 8. Input (up) and output (down) reflection of the compact balanced LNA (solid red) compared to one individual LNA (dashed blue) and the modular balanced LNA (short dashed green), measured at 5 K

higher probably due to the losses and mismatches in the connectors, and to the slightly more reflective hybrids used. This is even clearer in the noise (and gain) ripple, produced by stationary noise waves in the balanced amplifier. While the compact balanced amplifier presents less than 1 K peak to peak, the modular balanced ripple is 2.7 K. Note that the amplifier is being measured in 50 Ω setup in which the single-ended amplifiers noise is quite flat. In a real non-matched environment, the amplifiers will be exposed to the mixer reactive load and the system noise and gain ripple (due in this case to the stationary waves between the mixer and the LNA) will be much higher with the non-balanced amplifier.

The input and output return losses of the three configurations are shown in Figure 8. As expected, the balanced configuration improves the single amplifier reflection, especially at the input which is much more difficult to match for a wide band low noise amplifier. The variation is more than 10 dB between the single-ended and the compact balanced versions. The improvement of the compact balanced amplifier with respect to the modular version is more evident in the higher end of the band.

The results collected in this section were obtained at 5 K ambient temperature and with the same bias for the amplifiers. They are summarized in Table I. The bias used optimizes the performance (noise, gain flatness and output reflection) of a single-ended amplifier, while keeping a discrete power dissipation (7.5 mW). It was replicated in the balanced amplifiers to ensure that the results were perfectly comparable. The power dissipation of a balanced amplifier is then roughly twice that of one individual LNA, and this is not a minor issue for certain applications with very restricted cooling power. However, a significant reduction of this power consumption is possible in a balanced amplifier by retuning the bias of the last amplifying stages: the output return losses are not affected as they are dominated by the hybrid reflection. Preliminary tests showed that a 25% reduction is easily attainable (11.5 mW overall) with only a minor impact in the gain flatness (less than 1 dB increase).

TABLE I
PERFORMANCE OF THE DIFFERENT LNA CONFIGURATIONS TESTED

| @ 5 K, 4-12 GHz | **Single ended** [a] | **Modular balanced** [b] | **Compact balanced** |
|---|---|---|---|
| Average Noise (pp) (K) | 4.1 (1.0) | 5.1 (2.7) | 4.4 (1.4) |
| Average Gain (pp) (K) | 34.4 (2.0) | 33.5 (2.5) | 34.1 (1.7) |
| IRL max. (dB) | -3.5 | -10 | -16.4 |
| ORL max. (dB) | -11.2 | -12.3 | -18.3 |
| Power dissipation [c] (mw) | 7.5 | 15 | 15 |

[a] Values presented are the average of both LNAs used in the compact balanced.
[b] Individual amplifiers are not the same as in the compact balanced, but of identical design and transistor batches. Hybrids differ slightly in its design.
[c] Balanced amplifier measured with the same bias as single amplifiers. Further power dissipation reduction is possible.

At the time of this writing, the compact balanced LNA has already been measured in a 300 GHz receiver at GARD with excellent results which will be made available soon.

## VI. CONCLUSION

A cryogenic compact balanced low noise amplifier has been designed, fabricated and critically compared with its single-ended amplifiers and with a similar non-integrated balanced LNA. This unit benefits from excellent 3 dB quadrature hybrid coupler and LNAs designs, as well as Diramics InP 0.1×150 μm transistors. The superconducting lines negligible losses and the tight integration are responsible for the very low noise increment of just 0.3 K (measured at 5 K) with respect to the individual LNAs and for the reduced noise ripple of less than 1.5 K pp in the 4-12 GHz band. As expected from a balanced configuration, the input return loss is drastically improved and lies well below 15 dB. Such performance makes this type of amplifiers excellent candidates for the IF of wide band mm-wave receivers and demonstrates its potential for wider bands, for which other alternatives are not possible (like cryogenic isolators) or degrade significantly the performance (like single-ended amplifiers).

Additional improvements of the compact balanced amplifier design are under study, such as the simplification of the bias circuitry, the integration of a mixer bias T, a further reduction in size or the extension of the bandwidth to higher frequencies.

## References


[1] J. Carpenter, D. Iono, L. Testi, N. Whyborn, A. Wootten, N. Evans, "The ALMA Development Road Map," ALMA Observatory, Jun. 18, 2018. [Online]. Available: https://www.almaobservatory.org/wp-content/uploads/2018/07/20180712-alma-development-roadmap.pdf

[2] P. Grimes, R. Blundell, S. Paine, C.-Y. E. Tong, L. Zeng, "Next generation receivers for the submillimeter array," *Proc. SPIE 9914, Millimeter, Submillimeter, and Far-Infrared Detectors and Instrumentation for Astronomy VIII*, 991424, Jul. 19, 2016.

[3] I. López-Fernández, J. D. Gallego Puyol, C. D. González and A. B. Cancio, "Development of Cryogenic IF Low-Noise 4-12 GHz Amplifiers for ALMA Radio Astronomy Receivers," *2006 IEEE MTT-S Int. Microwave Symp. Dig.*, San Francisco, CA, 2006, pp. 1907-1910.

[4] I. Malo-Gómez, J. D. Gallego-Puyol, C. Diez-Gonzalez, I. López-Fernández and C. Briso-Rodriguez, "Cryogenic Hybrid Coupler for Ultra-Low-Noise Radio Astronomy Balanced Amplifiers," *IEEE Trans. Microwave Theory Tech.*, vol. 57, no. 12, pp. 3239-3245, Dec. 2009.

[5] H. Rashid, D. Meledin, V. Desmaris, A. Pavolotsky, V. Belitsky, "Superconducting 4–8 GHz Hybrid Assembly for 2SB Cryogenic THz Receivers", *IEEE Transactions on Terahertz science and Technology*, 4 (2) pp. 193-200. 2014.

[6] J. Schleeh et al., "Cryogenic LNAs for SKA band 2 to 5," *2017 IEEE MTT-S International Microwave Symposium (IMS)*, Honololu, HI, 2017, pp. 164-167.

[7] D. Meledin, V. Desmaris, E. Sundin, H. Rashid, A. Pavolotsky, J. D. Gallego, I. López-Fernández, C. Diez, V. Belitsky, "A 1 mm SIS Receiver Utilizing Different IF Configurations," presented at the *30th International Symposium on Space Terahertz Technolgy*, Gothemburg, Sweden, Apr. 15-17, 2019.

[8] J. L. Cano, N. Wadefalk and J. D. Gallego-Puyol, "Ultra-Wideband Chip Attenuator for Precise Noise Measurements at Cryogenic Temperatures," *IEEE Trans. Microwave Theory Tech.*, vol. 58, no. 9, pp. 2504-2510, Sept. 2010.

[9] D. Bruch, R. I. Amils, J. D. Gallego, M. Seelmann-Eggebert, B. Aja, F. Schafer, C. Diez, A. Leuther, M. Schlechtweg, O. Ambacher, I. Kallfass, "A Noise Source Module for In-Situ Noise Figure Measurements From DC to 50 GHz at Cryogenic Temperatures," *IEEE Microwave and Wireless Components Letters*, vol. 22, no. 12, pp. 657-659, Dec. 2012.

[10] V. Belitsky, *et al.*, "Facility Heterodyne Receiver for the Atacama Pathfinder Experiment Telescope", *IRMMW-THz 2007* Cardiff, UK, 2007.

# Modelling dielectric losses in microstrip traveling-wave kinetic-inductance parametric amplifiers

D. Valenzuela, F.P. Mena, and J. Baselmans.

***Abstract*— The development of travelling-wave kinetic inductance parametric amplifiers has introduced a promising new technology in radio astronomy [1]. In fact, it has the potential to reach near quantum-limited noise on a wider bandwidth, larger dynamic range and higher operation frequency when compared with current technologies. While the first implementation of a TKIPA used a coplanar waveguide for the transmission line [1], in this work, we present simulation studies using a microstrip line. One of the main motivations is to be able to obtain a 50-Ω line suitable for readily connection with other components. Therefore, reducing the reflections produced by an impedance mismatch between the ports. In this work we present the design of two different 50-Ω microstrip engineered transmission lines for parametric amplification using the four- and three-wave mixing effects. Additionally, we have included losses in the model. We have simulated their dispersion relation, and the parametric gain with pump frequencies at 7 and 14 GHz for the four- and three-wave mixing effects, respectively. We demonstrate that the three-wave mixing process requires less pumping power for obtaining the same gain with the added advantage of not having a stop band within the operation range. Furthermore, our simulations demonstrate that three-wave mixing is less affected by losses on the transmission line.**

***Index Terms*— Filters, microwave, parametric, superconductors.**

## I. Introduction

HIGH electron mobility transistors have been especially useful in astronomy. Amplifiers based on it are reliable, stable, and can work in an octave or higher bandwidth without increasing the noise significantly. However, their added noise reach between 10-20 times the quantum limit, depending on the operation frequency. Recently, a new kind of amplifiers, namely the traveling-wave kinetic inductance parametric amplifiers (TKIPA), have been the focus of attention due to its promising properties. The TKIPA makes use of the non-linear kinetic inductance of a superconductor, which is constructed as transmission line, to produce parametric amplification. In this process, amplification occurs when a weak signal, $f_s$, is fed into the line in the presence of a strong pump signal, $f_p$. Since the kinetic inductance depends quadratically with current, interaction of four photons in the form of four wave mixing (FWM) is expected. Conservation of energy requires that an additional idle signal, $f_i$, is created according to

$$f_s + f_i = 2f_p.$$

Unwanted harmonics of the pump and target signal are suppressed by constructing the transmission line as a periodic-structure filter, which results in a concentration of power in the desire bandwidth. While the first implementation of a TKIPA has used a coplanar waveguide for the transmission line [1], in

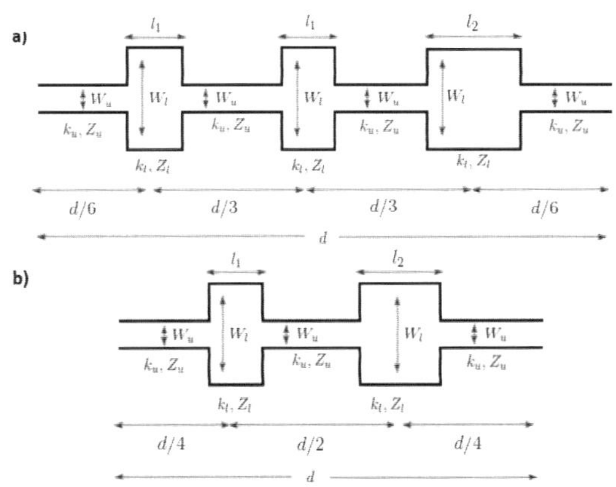


Fig. 1. Unit cells of the periodic filter used to implement the parametric amplifier. The geometrical parameters $d$, $W_u$, $W_l$, $l_1$, and $l_2$ were optimized as to select the desired operation frequency, and location and strength of the stop bands. The widths of the lines, $W_u$ and $W_l$, define their propagation constants, $k_u$ and $k_l$, and characteristic impedances, $Z_u$ and $Z_l$. a) Cell supporting FWM. b) Cell supporting TWM.

This work was supported by Conicyt through grants Fondecyt 1180700 and Basal AFB-170002. We also thank AWR for granting the use of its licenses and the EE Dep. of U. Chile for its travelling grant *Programa de Ayuda de Viajes*.

D. Valenzuela is with the Electrical Engineering Department, Universidad de Chile, Santiago, Chile.

F. P. Mena is with Electrical Engineering Department and with the Millimeter Wave Lab, Universidad de Chile, Santiago, Chile (e-mail: fpmena@uchile.cl).

J. Baselmans is with the Electrical Engineering Department of Delft University of Technology and with the Netherlands Institute for Space Research, The Netherlands.

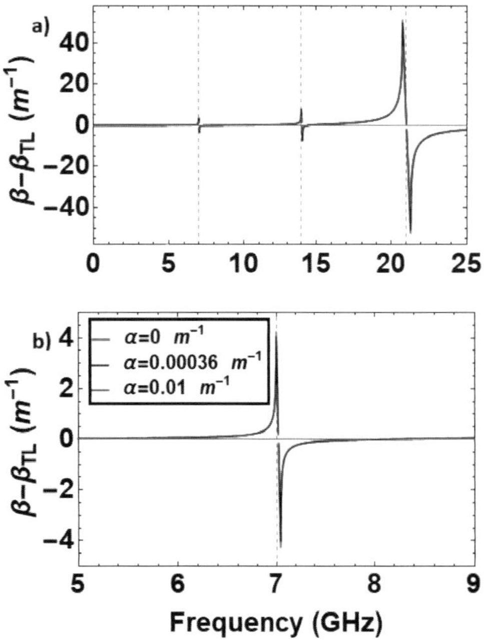


Fig. 2. Simulated dispersion relation of an engineered transmission line that uses FWM, with different losses. The difference $\Delta\beta = \beta - \beta_{TL}$ is illustrated to observe only the generated stopbands. a) The first stopband is located near the pump frequency $f_p$ at 7 GHz. An important stop-band is produced at $3f_p$ to produce FWM. b) Zoom of the first stopband.

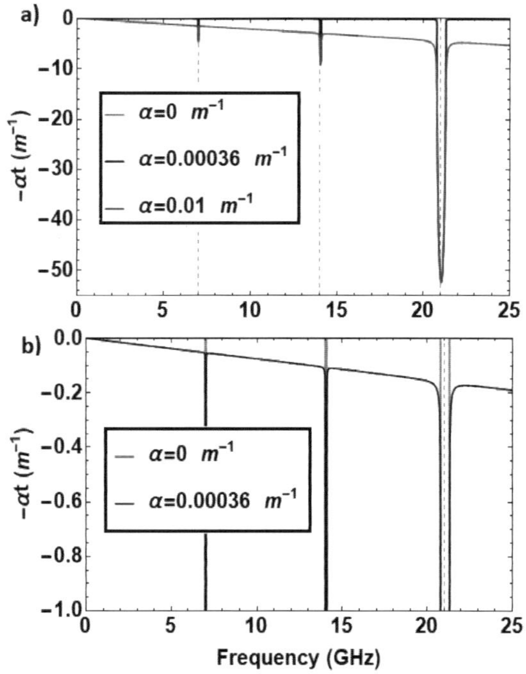


Fig. 3. Equivalent attenuation constant $\alpha_t$ of the periodic FWM transmission line. (a) Different values of losses $\alpha$ were added to the cell. It has been assumed that the unloaded and loaded elements have the same amount of losses. (b) Zoom to appreciate the effect of losses when $\alpha = 0.00036\ \mathrm{m}^{-1}$.

this work, we present simulation studies using a microstrip line. One of the main motivations is to be able to obtain a 50-Ω line suitable for readily connection with other components. Preliminary studies have shown that the necessary geometry and dimensions is achievable when using common substrates as amorphous Si [3]. However, given the necessary thickness of the substrate, the line may be prone to high dielectric losses. Therefore, we have included them in our simulations. Furthermore, we have included a DC current, $I_{DC}$, feeding the system, as a method to reduce the amplitude of the pump current. When $I_{DC}$ is included, a linear part appears in the kinetic inductance. If $I_{DC} > I_{RF}/2$, the linear part dominates over the quadratic part, and therefore, the parametric process of three-wave mixing (TWM) is produced [4]. In this case, conservation of energy also requires the creation of an idle signal according to

$$f_s + f_i = f_p.$$

In this work we compare the two parametric process, FWM and TWM.

## II. Design & Simulations

The design starts by selecting the appropriate microstrip configuration. We have selected NbTiN ($\rho_n = 150$ Ω.cm, $T_c = 15.1$ K) deposited over amorphous Si ($\epsilon_r = 10$). Given the available deposition process, a thickness of 250 $\mu$m for the substrate was selected. Furthermore, thicknesses of 60 and 300 nm were selected for the strip and the ground, respectively.

Preliminary measurements using microstrip resonators have demonstrated that microstrips with a loss of $\alpha = 0.00036\ \mathrm{m}^{-1}$ (equivalent to $\tan\delta = 1.2 \times 10^{-6}$) can be fabricated [3]. To appreciate more clearly the effect of losses, we have also used an *ad-hoc* value of $\alpha = 0.01\ \mathrm{m}^{-1}$ in the simulations described below. Furthermore, for the simulations we have used $J_C = 10^7\ \mathrm{A.m}^{-2}$ [5].

An scheme of the FWM engineered transmission line is shown in Fig. 1a. It is divided in seven transmission lines. The first four are the *unloaded* elements, that is to say, the lines that are used as a coupling to the external ports. Each of them are interleaved with three perturbations or *loaded* elements. The scheme of the TWM filter is shown in figure 1b. In this case the cell is divided in five transmission lines. The main characteristic of both filters is their propagation constant, $\gamma = \alpha_t + j\beta$, which is calculated using the ABCD matrix of the cell and assuming that the transmission line supports the propagation of a traveling-wave. We have added an attenuation constant independent of frequency in the propagation constants $k_u$ and $k_l$ during the calculation of the ABCD matrices. The dimensions of the filter were tuned to create stopbands at harmonics of the pump frequency. The parametric effect was simulated using the couple mode equations [6].

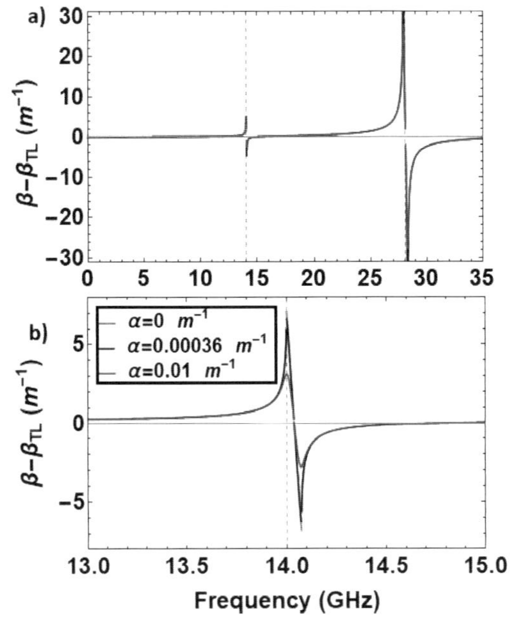


Fig. 4. Dispersion relation of the periodic transmission line, in TWM, expressed as $\Delta\beta = \beta - \beta_{TL}$. The first stopband is located at $2f_p$ to improve the amplification for TWM.

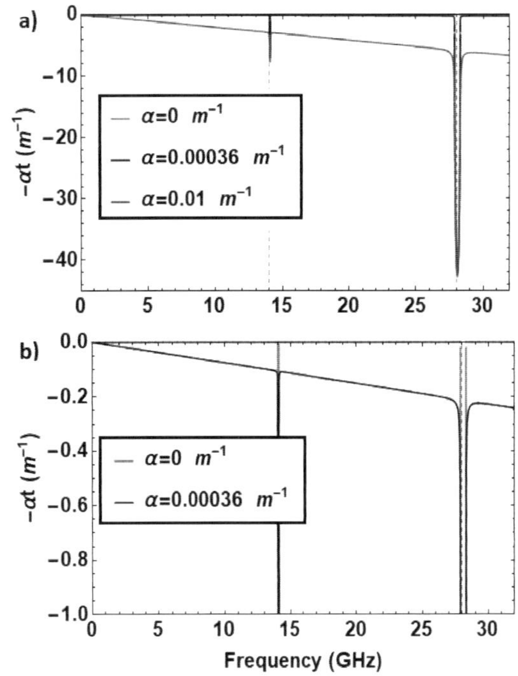


Fig. 5. Equivalent attenuation constant $\alpha_t$ of the periodic TWM transmission line. (a) Different values of losses $\alpha$ were added to the cell. It has been assumed that the unloaded and loaded elements have the same amount of losses. (b) Zoom to appreciate the effect of losses when $\alpha = 0.00036\ \mathrm{m}^{-1}$.

## III. Results

### A. Four wave mixing

Fig. 2a illustrates the propagation constant $\beta$ of the cell subtracted by the wave number of the unloaded segment $\beta_{TL}$, $\Delta\beta = \beta - \beta_{TL}$. It can be noted that the stopbands are located near or at harmonics of the pump frequency, $f_p = 7$ GHz. The first one, located near $f_p$, has a width of approximately 30 MHz. The additional dispersion around this stop band produces additional parametric amplification. The second stop-band, located near $2f_p$ with a bandwidth of 90 MHz, does not have an important impact on the amplifier. The third stop-band is located at $3f_p$ with a bandwidth of 920 MHz, which is considerably wider than the other two stop bands. This effect is expected, due to the need of removing frequencies above $3f_p$, so the power of the amplifier can be concentrated in desirable frequencies. A close up of $\Delta\beta$ around the first stop band is shown in Fig. 2b. It can be noted that an increment of the attenuation constant in the line generates a smoother curve. This behavior can be explained as a *deterioration* of the propagation constant. Losses decrease the dispersion in the transmission line, which could allow the propagation of undesirable signals. The equivalent attenuation constant $\alpha_t$ of the periodic filter is observed in Fig. 3. It can be seen that its value increases with the attenuation constant $\alpha$ for each of the sub-lines that compose the filter.

### B. Three-wave mixing

The dispersion relation of the TWM transmission line can be observed in Fig. 4a. The pump frequency has been changed to $f_p = 14$ GHz, so it can be compared with the parametric gain that uses the FWM effect. The first stop-band located near $f_p$ has a bandwidth of approximately 100 MHz, but lays outside the operation range. It can be observed that the second stop-band at $2f_p = 28$ GHz has a bandwidth of $\approx 450$ MHz, much wider when compared with the first stopband. In TWM, the deterioration of the propagation constant is also present, where a large attenuation constant generates a smoother stopband at the harmonics of the pump frequency. The equivalent attenuation constant $\alpha_t$ is shown in Fig. 5. It also presents a degradation of the stopbands when the attenuation constant of the original line increases. Furthermore, this decrement is larger at $2f_p$, where this harmonic is expected to be suppressed.

### C. Parametric Gain

Fig. 6 presents the simulated parametric gain, including dielectric loses, of two 50-Ω microstrip lines. A DC bias equal to 13% of the effective critical current was used in TWM as to obtain a similar gain to that obtained in FWM. Two important results are highlighted. First, the inclusion of a DC bias results in a parametric gain that is not affected considerably when losses are included. Secondly, TWM allows obtaining similar gain than FWM with a lower amplitude of the pump signal.

## IV. Conclusions & Future Work

In summary, we have designed and simulated two engineered transmission lines for use with three- and four-wave mixing.

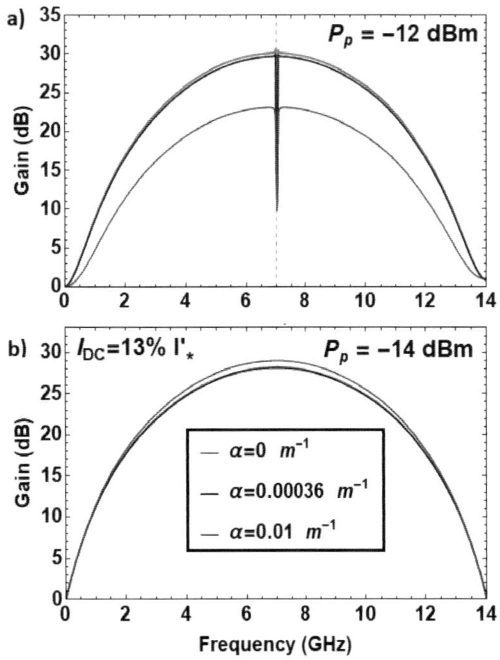


Fig. 6. Simulated parametric gain of devices, including losses of the substrate. (a) FWM and (b) TWM. The parameter $P_p$ corresponds to the power of the pump signal, and $I'_* = I_C/2\alpha$ is the effective critical current, where $\alpha$ is the fraction of kinetic inductance.

We have calculated their propagation constants and parametric gains. We observe that the design which uses three-wave mixing needs less pumping power to obtain the same gain as when four-wave mixing is used. Moreover, our simulations demonstrate that three-wave mixing is less prone to be affected by dielectric losses with the added advantage that the first stop band lays outside the operation range.

Since this work demonstrates that including losses degrades the stop bands of the engineered transmission lines, we are working on simulating the effect of having undesired tones travelling in the line. Furthermore, we are working on implementing the parametric amplifiers described here.

## References

[1] B. Eom, P. Day, H. LeDuc and J. Zmuidzinas, "A wideband, low-noise superconducting amplifier with high dynamic range," *Nature,* vol. 623, 2012.

[2] S. Aanlange and H. Snortland, "A current controlled variable delay superconducting transmission line," *in IEEE Transactions on Magnetics,* pp. 1388-1391, 1989.

[3] F. Mena, «Internal report».

[4] M. Vissers, R. Erickson, H. Ku, L. Vale, X. Wu, G. Hilton and D. Pappas, "Low-noise kinetic inductance traveling-wave amplifier using three-wave mixing," *Applied physics letter,* vol. 108, 2016.

[5] A. Klimov, W. Sysz, M. Guziewicz, V. Kolkovsky, I. Zaytseva and A. Malinowski, "Characterization of the critical current and physical properties of superconducting epitaxial NbTiN sub-micron structures," *Physica C: Superconductivity and its Applications,* vol. 536, pp. 35-38, 2017.

[6] G. Agrawal, Nonlinear fiber optics, 2001.

# Characterization of GaN-based Low Noise Amplifiers at Cryogenic Temperatures

V. Desmaris[1], D. Meledin[1], E. Sundin[1], M. Thorsell[2], N. Rorsman[2] and V. Belitsky[1] [1]

***Abstract*— In this paper, we present the first characterization of GaN-based Low-Noise Amplifiers (LNAs) at cryogenic temperatures for prospective use in radio-astronomy receivers. Both commercial and prototype LNAs fabricated in-house demonstrate a nine-fold improvement of their room-temperature noise performance when cooled to about 10 K. Very promising noise temperatures of about 8 K were measured without any specific optimization of the LNA design for cryogenic operation.**

## I. Introduction

Since its first demonstration in 1994, GaN HEMT technology has matured, but mainly focusing on power amplification for radar and telecommunication applications. GaN-based Low Noise Amplifiers (LNA) have received some attention due to their inherent added advantage in terms of robustness and enhanced linearity over InP and even GaAs pHEMT, which results in a system simplification by relaxing (and even eliminating) the requirements on limiters and filters in a radar or communication systems. For room temperature operation and cooled down to 60 K, GaN LNAs have already demonstrated noise performances similar to their GaAs counterparts [1]. Yet the performance of the GaN-based LNAs at cryogenic temperatures used for ultra-sensitive instruments, e.g., for radio astronomical applications, remains unexplored.

GaN, as a material, has recently been found to provide a good technological platform for cryogenic superconducting THz mixers and heterodyne receivers for radio-astronomy [2-3]. Also, recent works show that the behavior of GaN-based heterojunctions exhibit an enhancement of the electron mobility in the 2-dimensional electron gas, similar to GaAs and InP heterojunctions [4], when cooled down to cryogenic temperatures. Therefore, it is reasonable to assume that GaN-based LNAs would also demonstrate excellent noise performance at cryogenic temperatures and would be compatible and competitive for their use in radio-astronomy receivers.

This paper focuses on the characterization of the noise performance of GaN-based LNA at cryogenic temperatures for prospective use as a frontend at microwave frequencies and IF amplifiers in THz receiver systems. These initial measurements are carried on prototype LNAs fabricated in-house and commercially available devices designed for radar applications.

## II. Cryogenic Characterization on GaN-based LNAs

Two different low noise amplifiers were characterised, both packaged as MMIC and using SiC as substrate material: a prototype amplifier fabricated in-house [5] and a Qorvo TGA2611 amplifier. Both amplifiers featured two stages, yet the Qorvo amplifier consumed approximately half the power of our prototype LNA fabricated in-house . The LNAs were mounted into a fixture made of copper to ensure best possible cooling of the LNA chips. The fixture also included 50 Ohm transmission lines and SMA connectors as the interfaces to the RF measurement setup (Fig. 1).

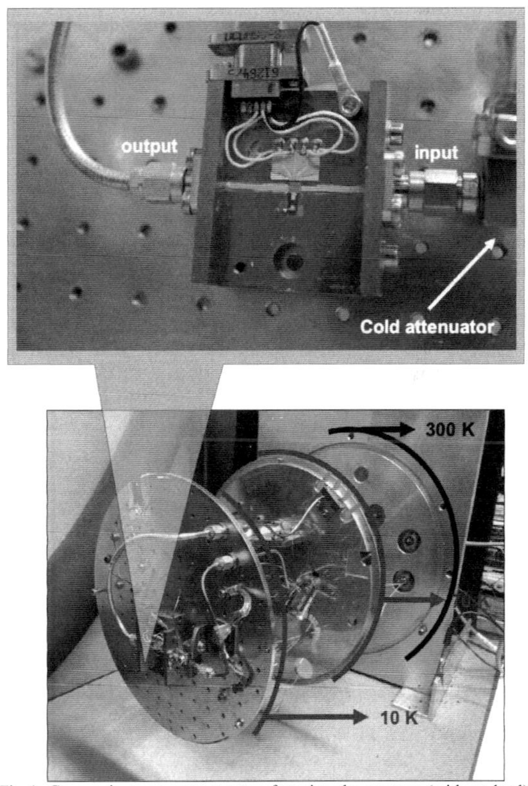


Fig 1. Cryogenic measurement setup featuring the cryostat (without lead) and LNA mounted on the fixture, attached to the 10 K plate.

[1] Group for Advanced Receiver Development, Chalmers University of Technology, SE412 96, Gothenburg, Sweden.

[2] Microwave Electronics Laboratory, Chalmers University of Technology, SE412 96, Gothenburg, Sweden.

The noise performance of the LNAs were measured in a cryogenic system comprising a cryostat equipped with a 2 stage close cycle-machine, which allows cooling down to about 10 K (Fig 1). The noise temperature measurements were performed by the standard Y-factor measurement technique and using the cold attenuator method [6] with an Agilent N4000A ENR noise diode and an Agilent MXA N9020A spectrum analyzer with Noise Figure Measurement option.

We carried out a simplified characterization of the cryostat losses and the temperature sensor to be used for the cold attenuator. The optimum noise performances of the different LNAs were determined by independently tuning the bias conditions of the different transistor stages at cryogenic temperature.

First, we recorded the average noise performance between 2 and 6 GHz when cooling down the LNAs from room temperature down to about 10 K. Fig.2 displays that both LNAs show a similar relative change in their noise temperature when cooling down from room temperature to 10 K.

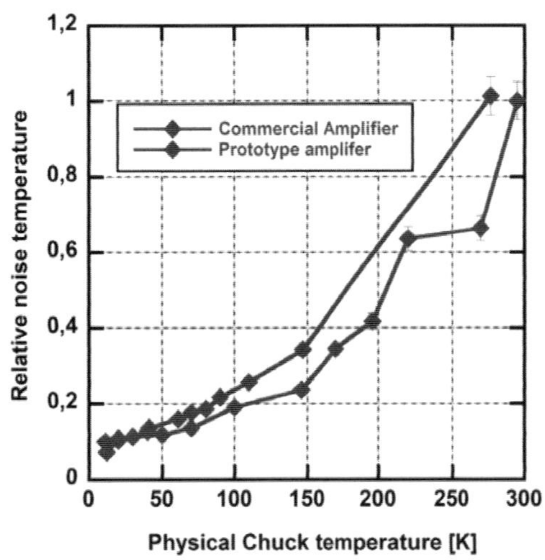


Fig.2 Relative noise temperature variation upon cooling of the LNAs

This behaviour is very similar to the one observed in InP or GaAs LNAs and can be ascribed to the enhancement of the electron mobility in the two-dimensional electron gas forming the channel of the HEMT, as shown in [4] and increasing conductivity of the imbedding circuitry.

The absolute noise performance of the Qorvo LNA over the 2-6 GHz frequency band is illustrated on Fig 3. Even though the device was designed for radar or telecom applications, it shows very decent noise performance. In fact, noise temperatures of about 8 K were measured without any specific optimization but tweaking with bias points when performing the characterization at cryogenic temperature.

This result is very encouraging and practically confirms the theoretical prediction and previous intermediate material performance measurements indication that GaN can be a compatible technology for low-noise amplification at cryogenic temperatures.

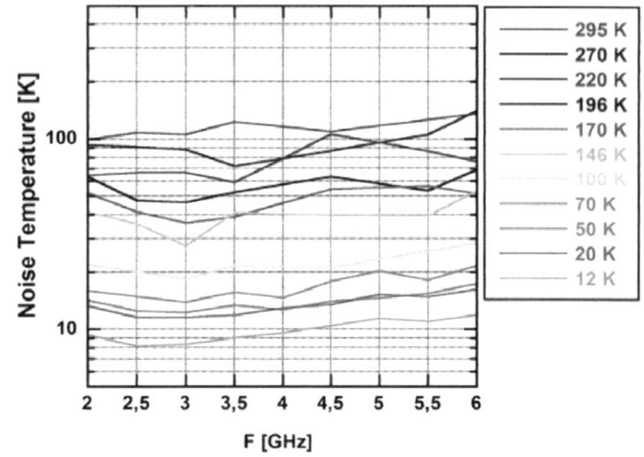


Fig. 3. 2-6 GHz noise performance at on the Qorvo LNA measured at different chuck temperature.

## III. CONCLUSIONS

In this work, we presented the first characterization of the noise performance of GaN-based LNAs at cryogenic temperature. We have shown that GaN-based LNAs demonstrate a similar enhancement of their noise performance with temperature, as the LNAs based on GaAs or InP. Also, the 8K noise temperature obtained on the GaN LNAs, not even designed for cryogenic temperature shows that the noise performance of GaN-based LNAs could potentially be competitive with other III-V technologies with some optimization while offering advantage of handling more powerful signals (higher gain could be realized in one amplifier) and higher linearity.

## REFERENCES

[1] Kobayashi *et al.*, "A Cool, Sub-0.2 dB Noise Figure GaN HEMT Power Amplifier With 2-Watt Output Power", *IEEE journal of Solid-State circuits*, vol 44 (10), p. 2648, 2009.

[2] S. Krause *et al.* "Noise and IF gain bandwidth of a balanced waveguide NbN/GaN hot electron bolometer mixer operating at 1.3 THz", *IEEE Trans. terahertz Sci. Technol*, vol 8. No 3, 2018.

[3] S. Krause *et al*, "Suspended GaN beams and membranes on Si as a platform for waveguide-based THz applications", *Journal of Micromechanics and Microengineering*, 28, 105007, 2018.

[4] A. Malmros *et al*, "Enhanced Mobility in InAlN/AlN/GaN HEMTs using a GaN interlayer", *IEEE Trans. Electron Devices, v. 66, n7, p.2910*, 2019

[5] M. Südow *et al.* "An AlGaN/GaN HEMT-based microstrip MMIC process for advanced transceiver design", *IEEE Trans. Electron Devices, v. 56, n8,* p.1827, 2008

[6] E. Sundin *et. al*, "Cryogenic IF Balanced LNAs Based on Superconducting Hybrids for Wideband 2SB THz Receivers", Proc. Of the 28th ISSTT, Cologne, Germany, 2017

# Design and Prototyping of New Flexible Stripline based Transmission Lines as Alternatives to Semi-Rigid Coaxial Cables

Marko Neric, Hamdi Mani, Thomas Mozdzen, and Chris Groppi

*Abstract*—**We present the design, assembly, and prototyping of new multi-channel flexible printed circuit board stripline based transmission lines for transmitting RF signals. Stirpline transmission lines have been used for decades and consist of a narrow center conductor that is surrounded by a dielectric substrate and then sandwitched between two grounding planes. Incorporating several striplines in the same substrate using flexible materials may be used as alternatives to industry standard stainless steel semi-rigid coaxial cables which are frequently used as RF transmission lines in a variety of devices. This is often the case for astronomy instruments that have multiple pixels working together to form a focal plane array. Larger arrays require many coax transmission line cables which can increase the footprint of a device, as well as add to the total heat load of an instrument. We have developed a single flexible circuit ribbon that can replace up to 8 individual coaxial cables. The flex circuits were designed in CST Studio Suite where transmission characteristics were simulated and subsequently optimized by adjusting the circuit dimensions. The flexible circuits make use of a novel design wherein the top and bottom stripline grounding layers are reduced to 0.015" wide strips rather than a full grounding plane. This minimization of copper grounding material increases flexibility, while decreasing heat load, all without introducing significant loss of signal. The circuits will be used in an upcoming NASA Class-D Balloon mission: GUSTO[1]. Specifically for GUSTO the circuits will need to transmit over an IF bandwidth from 0.3 - 5 GHz while one end of the ribbon is held at 20 K, and the other end is at 300 K. In this temperature configuration, a 20" length of flex circuit has a loss of 5.77 dB at 5.00 GHz which is roughly the equivalent of a stainless steel coaxial cable with an outer diameter of 0.085". In this same configuration the flex circuit loss is 9.67 dB at 10 GHz. Due to its reduced grounding layer the flex ribbon generates 75% less heat than 0.085" diameter coax. The flexible ribbon is also made of durable Kapton material which makes the circuit considerably more malleable than semi rigid coax. The full bandwidth of the flex ribbon is 0 - 10 GHz. The flex circuits have been tested at cryogenic temperatures for multiple cool-down and warm-up cycles without any signs of degradation. They are capable of delivering the comparable transmission characteristics as coaxial cables while offering improvements on versatility, and thermal conductivity.**

*Index Terms*—**RF, Cryogenics, Transmission Lines.**

## I. Introduction

DISCRETE sets of semi rigid coaxial cables are widely used in THz astronomy instrumentation. They can transmit IF signals from individual pixels in a larger focal plane array to the read out electronics. As the pixel count of future arrays increases so too does the amount of individual cables. This will make for more difficulty in routing coax cables through instruments. A new scheme for IF transmission will be needed. The motivation of this work is to replace bulky semi rigid coax cables in larger arrays.

M. Neric, H. Mani, T. Mozdzen, and C. Groppi are with the School of Earth and Space Exploration, Arizona State University, Tempe, AZ, 85287 USA e-mail: mneric@asu.edu.

One potential option for replacing coaxial cables is a flexible cryogenic printed circuit board. Multi channel flexible ribbons have been developed in the past[2]. Using closely packed striplines within a single substrate the flex ribbons could transmit up to 16 different channels in the same ribbon in a much smaller form factor than an equivalent bundle of coax[2]. Optimizing this initial design for use with the upcoming NASA Class-D Balloon Mission, GUSTO, is the focus of this work.

The design goals for this flex circuit are the same as the mission requirements for the GUSTO IF system. The IF bandwidth is 0.3 - 5.0 GHz. The circuits must operate at temperatures as low as 20 K. The circuits must be comparable in performance to UT 85 SS-SS coax while at cryogenic temperatures.

## II. Design and Simulation

Using the original design for the ASU flex circuit[2] as a starting point, we modeled the channels as stripline transmission lines. The substrate used was Dupont 8545 Kapton and it was 0.004" (0.1 mm) in thickness. The substrate was plated with 1/2 oz copper cladding. The central conductor of a stripline is buried in the substrate so a radial transition was needed to reach the top layer. Surface mount connectors would be used on both ends. Plated through hole (PTH) signal and ground vias extended from the center conductor to an anti-pad at the top and bottom of the flex circuit. The anti-pad size was determined by the size and type of surface mounted connectors. For this design we use SMP press on connector types.

A three channel flex circuit was designed using CST Studio Suite. The channels were 50 Ω stripline transmission lines. Spacing between channels directly impacts the form factor of the flex circuit. A CST simulated parameter sweep of channel spacing ranging from 0.24" (6.1 mm) to 0.05" (1.3 mm) was executed. The effects on signal transmission were only fractionally changed. Since it is preferential to use surface mount connectors, the channel spacing was set at 0.120" (3 mm) instead of a much tighter spacing. This makes it easy to hand solder connectors. At this spacing an 8 channel circuit would be 1.1" (2.8 cm) wide.

The GUSTO mission requires that the flex circuit $S_{21}$ be above -6 dB at 5 GHz over a 20 inch flex circuit. This must be achieved at cryogenic temperatures. The IF bandwidth of the circuit must be 0.3 - 5.0 GHz. The adjacent channels need to have 30 dB of isolation or better. CST simulations of a three channel circuit were done over a 5 GHz bandwidth. Figure 1 shows the simulation results.

Because GUSTO is a balloon mission it will have a limited supply of coolant for its cryogenic components. Reducing the heat load of the IF chain is a critical design goal. The heat load directly impacts the mission life expectancy. To reduce the heat load of the circuit the stripline ground planes that are traditionally solid copper layers are reduced into thin strips. CST simulations predict that a reduction in copper content of the top and bottom planes will have negligible impact on performance. As seen in Table I the heat load of a single channel of the striped flex ribbon can be reduced to 25 % of the heat load of industry standard coaxial cables.

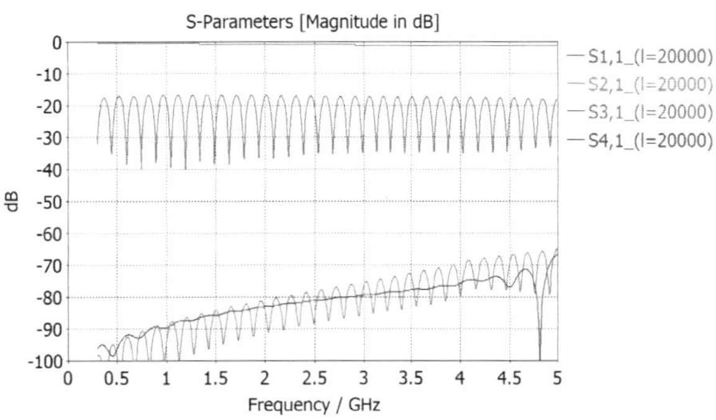


Fig. 1. CST Studio Suite Simulated transmission characteristics of a three channel flex ribbon. Input signal was from port 1 to port 2. Ports 3 and 4 were the nearest neighbors and used to simulate isolation. Design goals for insertion loss (< 6 dB), and isolation (> 30 dB) are met according to software.

UT 85 is the most commonly used coax. UT 20 is not typically used in missions since it is very fragile, but has its place when extreme space saving is necessary. Even though stainless steel has much less thermal conductivity, the striped version of the flex circuits have significantly less cross sectional conductor area. This allows them to produce a quarter of the heat per channel.

The ground strips were designed in 0.015" (0.38 mm) strips, and 0.025" (0.635 mm). These widths are chosen to protect the circuit from a misalignment between ground strips and the center conductor. A copper tab was added to the striped flex circuit for heat sinking. Fiber glass composite material FR-4 was added to strengthen the connection sites.

## III. Results

Two sets of the flex circuit were fabricated by Coast to Coast Circuits Inc. Each set of three flex circuits, much like Table I, had a circuit with 15 mil ground strips, 25 mil strips, and one that was full plane copper. One set of three ribbons was a straight flex circuit with 8-channels, and the other had a 1 inch minimum radius of curvature built in to one end.

TABLE I
THERMAL CALCULATIONS AND COMPARISON OF FLEXIBLE CIRCUIT TO SS-SS COAX

| Circuit Type | Cross Sectional Area ($m^2$) | Thermal Load ($WmK^{-1}$) | Ratio to UT 85 |
|---|---|---|---|
| Flex(15 mil) | $1.70\times10^{-8}$ | $6.83\times10^{-6}$ | 0.25 |
| Flex(25 mil) | $2.74\times10^{-8}$ | $1.10\times10^{-5}$ | 0.41 |
| Flex(260 mil) | $2.70\times10^{-7}$ | $1.08\times10^{-4}$ | 4.01 |
| Coax(UT 85) | $1.66\times10^{-6}$ | $2.70\times10^{-5}$ | 1.00 |
| Coax(UT 47) | $5.98\times10^{-7}$ | $9.75\times10^{-6}$ | 0.36 |
| Coax(UT 20) | $9.89\times10^{-8}$ | $1.61\times10^{-6}$ | 0.06 |

Thermal conductivity of stainless steal is 16.3 $WK^{-1}m^{-1}$, and for Cu: 403 $WK^{-1}m^{-1}$. Thermal analysis shows the narrowest ground stripes far surpass industry standard UT 85 coax in heat performace.

All testing was done with a Rhode & and Schwarz ZVA 24 Vector Network Analyzer (VNA) that was calibrated with a ZV-52 electronic cal-kit. The VNA was calibrated at 0 dBm test power, and averaged 16 times. Data was taken from 0.3 to 5.0 GHz at room Temperature first.

The insertion loss for a flex circuit with 15 mil ground strips is 0.3 dB higher than an equivalent length flex with full ground planes. This is a 6 % difference in power. When heat load is put into consideration such a low change in raw performance is more than acceptable. Low noise amplifiers in the GUSTO IF chain will more than make up for the difference in power regardless. Moving forward only the 15 mil ground strip flex circuits were considered.

Cryogenic testing was done in a closed cycle system capable of reaching 10 K temperatures under vacuum. The flex circuit was tested at 20 K at one end, and 300 K at the other. This operating condition will be commonplace in most cryogenic instruments. The flex circuit was bonded directly to vacuum flanges using epoxy. This is another major benefit of flex circuits over coax - the flex circuits do not need expensive and complex hermetic connectors to make vacuum tight transitions outside of a cryostat. Kapton substrate out-gassing is extremely low making it very easy to use in vacuum chambers, cryogenic systems, and space hardware.

Figure 2 is the plot of the first cryogenic test. The insertion loss of the flex circuit at 5 GHz is 5.77 dB. This is within the spec of not exceeding 6 dB. The Isolation is approximately 60 dB over the entire bandwidth. This is again better than the required spec. Another way to look at insertion loss is in terms of dB/ft in which case the flex circuit with 0.015" ground strips has 3.5 dB/ft of loss at 5 GHz. This is more than the UT 85 coaxial cable which is only 2.8 dB/ft, but again with the benefit from 75 % less heat transmission, and the fact that IF LNAs will be providing additional gain the trade off is more than worth it. Additionally the flex circuit has smaller form factor, is light-weight, and more convenient for routing makes it the more applicable choice for DC - 10 GHz transmission.

## IV. Conclusion

We have developed the prototype for a next generation of microwave flexible circuit that can be used in cryogenic instruments to replace coaxial cables over a wide IF bandwidth. The new flex circuits are ideal wherever complex routing of

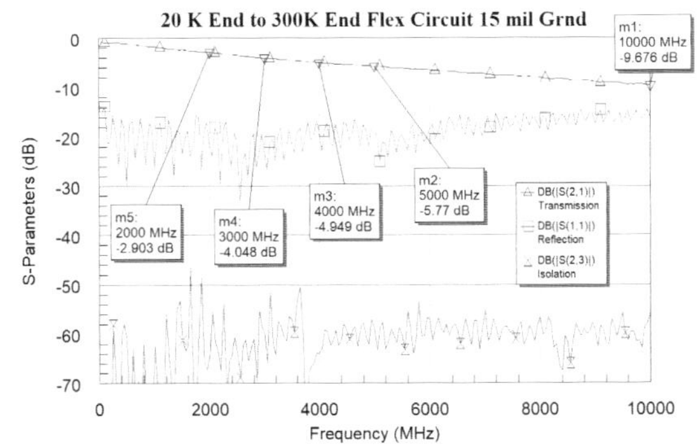


Fig. 2. S-parameters of an 8 channel flex circuit at cryogenic temperatures. The flex circuit has 3.5 dB/ft of loss at 5 GHz. The return loss is 20 dB so only 1 % of the incident power is reflected back though the port. Isolation is approximately 60 dB for nearly the entire bandwidth meaning that cross-talk between adjacent signal traces is negligible.

IF components is needed and or wherever the focal plane array has a high pixel count requiring a direct connection to their output. The improvement of heat load using a striped flex circuit is vital to increasing mission duration by reducing coolant loss.

With a 3.5 dB/ft loss at 5 GHz, stable return loss over a 5 GHz bandwidth, and outstanding isolation between channels, the prototype flex circuits meet all requirements for their role in GUSTO. The advantages of the customizable flex circuit makes it an appropriate choice for use in future missions.

## ACKNOWLEDGMENT

The authors would like to thank Bob Giambrone, and Hector Ramirez of Coast to Coast circuits for their help and support during the fabrication of the flexible test circuits.

## REFERENCES

[1] C. Walker, C. Kulesa, P. Goldsmith, C. Groppi, F. Helmich, D. Hollenbach, J. Kawamura, W. Langer, G. Melnick, D. Neufeld, J. Pineda, G. Stacey, A. Stark, A. Tielens, M. Wolre, H. Yorke, and E. Young, "GUSTO: Gal/Xgal U/LDB Spectroscopic-Stratospheric TeraHertz Observatory," Amer. ASTRO. Soc., AAS, 231, 2018

[2] P. McGarey, H. Mani, C. Wheeler, C. Groppi, "A 16-channel flex circuit for cryogenic microwave signal transmissions," SPIE, 9153, 2014

**Marko Neric** graduated from Arizona State University (ASU) in 2014 earning a B.S. in Physics. He is currently a Ph.D. candidate in the School of Earth and Space Exploration (SESE) at ASU. He is pursuing his Ph.D. in astrophysics, and an M.S. in electrical engineering. His focus is on THz astronomy, and instrument building.

**Hamdi Mani** has been an engineer at the ASU radio astronomy lab for the past 7 years. His main areas of research: design of wideband cryogenic low noise RF/ Microwave amplifiers and receivers, development of microwave and millimeter wave instrumentation, electronics packaging and cryogenic systems design . Prior to joining ASU, he worked at Caltech/JPL on developing SiGe cryogenic low noise amplifiers and wideband radio astronomy receivers.

**Thomas J. Mozdzen** received the B.S. in Physics in 1978 and the M.S. in Electrical Engineering in 1980 from the University of Illinois Urbana-Champaign, the M.S. in Physics from the University of Texas at Dallas in 1985, and the PhD in Astrophysics from Arizona State University in 2017. He has worked in the semiconductor industry for several years beginning as a reliability engineer at Mostek Corp, Dallas, Tx., and then mainly as a circuit design engineer at Siemens, Munich Germany, and at Intel, Chandler, Arizona. His PhD. Thesis focused on reducing systematic error in experiments to detect the sky-averaged redshifted 21 cm epoch of reionization signature. He holds several engineering related US patents and has authored several papers in both electrical engineering and astrophysics. He is currently employed at Arizona State University as a Postdoctoral researcher.

**Christopher E. Groppi** received the B.A. degree in astronomy (with honors) from Cornell University in Ithaca, NY, in 1997 and the Ph.D. degree in astronomy with a minor in electrical and computer engineering from the University of Arizona in Tucson, AZ, in 2003. In 2003, he joined the National Radio Astronomy Observatory as a Directors Postdoctoral Fellow. He then moved to the University of Arizona as an assistant staff astronomer in 2005. In 2006, he received an Astronomy and Astrophysics Postdoctoral Fellowship from the National Science Foundation. In 2009, he joined the Arizona State University School of Earth and Space Exploration in Tempe, AZ as an assistant professor. He became an associate professor in 2015. He is an experimental astrophysicist interested in the process of star and planet formation and the evolution and structure of the interstellar medium. His current research focuses on the design and construction of state of the art terahertz receiver systems optimized to detect the light emitted by molecules and atoms in molecular clouds, the birthplace of stars. He also applies terahertz technology developed for astrophysics to a wide range of other applications including Earth and planetary science remote sensing, hazardous materials detection and applied physics.

# A Robust 24-29 GHz Low Noise Amplifier with 1dB Noise Figure and 23 dBm P1dB

Penghui Zheng,[1,2] Xiaodong Tong,[1,2] Shiyong Zhang,[1,2] Jianxing Xu,[1,2] Rong Wang,[1,2]

A 24-29 GHz low noise amplifier (LNA) microwave monolithic integrated circuit (MMIC) based on 100 nm gallium on silicon (GaN/Si) high electron mobility transistor (HEMT) process is reported in this work. The linear gain of this LNA is 25±1dB inner the band. The LNA has an average noise figure (NF) of 1 dB over the designed band and achieves the minimum value of 0.94 dB at 27 GHz. The robustness was demonstrated by overdriving the LNA with 1 Watt continuous-wave (CW) input power for 5 minutes. Compared with the traditional GaAs and InP LNA, this GaN LNA has comparable NF, much higher robustness and linearity. Moreover, this LNA is supposed to have excellent anti-irradiation ability and wide working temperature range due to the wide band gap (3.5 eV) property of GaN material. As we know, there has been no public report of GaN LNA having NF below 1 dB at this frequency region. The LNA reported in this work has a great potential in astronomy and space detection fields [1].

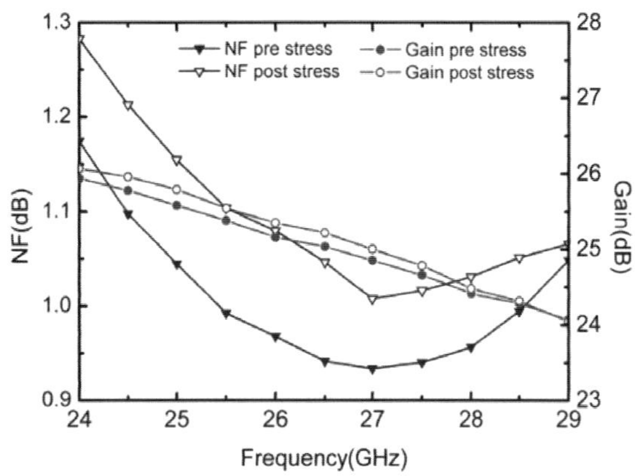


Fig. 1. Noise figure and Gain of the robust LNA measured

The LNA was designed with mixed electromagnetic and circuit simulation. The Pospiezalski model [2] based on the measured noise data was used in the circuit design. The LNA has a 3-stage cascade topology. The 4×50 um HEMT was used in the design of every state on the trade-off between robust and NF. The drain was biased at 8 Volt to achieve high linearity. A 1k Ohm resistor was used on the gate bias line of the every stage to prevent the gate finger from breakdown. The input match is critical for LNA design. To achieve the optimal noise match and input conjugate match at the same time, a source feedback was used in the design of every stage.

The measurements were conducted. The LNA has an average gain of 25±1dB in the 24-29 GHz frequency band, as the red solid line with filled circle shown in Fig.1. S11/S22 is lower than -10 dB/-18 dB inner the band. Cold source method was used as the NF measurement methodology. The LNA has an average NF of 1 dB over the designed band and achieves the minimum value of 0.94 dB at 27 GHz, as the blue solid line with filled triangle shown in Fig. 1. Additional measurements were performed for the determination of the LNA large signal performance and a preliminary evaluation of its robustness to input power. Due to the high breakdown electric field of GaN technology, a high Psat of ~28 dBm was achieved. Moreover, the 1-dB compression point output power (P1dB) of the LNA is at 23 dBm level, which indicates the high linearity of this LNA. The robustness was demonstrated by stressing the working LNA with 30 dBm CW input power at 27 GHz for 5 minutes. The NF (blue solid line with unfilled circle) and gain (red solid line with unfilled circle) after stress were measured and given in Fig. 1. It can be seen that very little change exists after stress.

GaN LNA reported in this work has low NF, high linearity and robustness. Moreover, this LNA is supposed to have excellent anti-irradiation ability and wide working temperature range. It has a great potential in the astronomy and space detection fields.

Funding information: Supported by Science Challenge Project, No. TZ2018003, China.

[1]Microsystem & Terahertz Research Center, China Academy of Engineering Physics, Chengdu,610200, China

[2] Institute of Electronic Engineering, China Academy of Engineering Physics, Mianyang,621999,China

### References

[1] S. Kaundinya and G. S. Karthikeya, "Recent advancements in design of microwave components for space applications," in *IEEE International Conference on Recent Trends in Electronics*, 2017.

[2] M. W. Pospiezalski, "Modeling of Noise Parameters of MESFETs and MODFETs and their Frequency and Temperature Dependence", *IEEE Trans. MIT*, Vol. MIT-37, pp. 1340-1350, 1989.

NOTES:

# Design of a Radio Frequency Waveguide Diplexer for Dual-band Simultaneous Observation at 210-375 GHz.

Sho Masui, Shota Ueda, Yasumasa Yamasaki, Koki Yokoyama, Nozomi Okada, Toshikazu Onishi, Hideo Ogawa, Yutaka Hasegawa, Kimihiro Kimura, Takafumi Kojima, Alvaro Gonzalez

***Abstract*—The 1.85-meter telescope has been operated at Nobeyama Radio Observatory to observe molecular clouds in nearby Garactic Plane in $^{12}$CO, $^{13}$CO, C$^{18}$O($J$ = 2-1). We are planning to relocate the telescope to the Atacama site (~2,400 m) and to newly install a dual-band simultaneous observation of CO lines at $J$ = 2-1 and $J$ = 3-2. To achieve this observation, we have designed a radio frequency waveguide diplexer to separate 211-275 GHz (ALMA band 6) and 275-373 GHz (ALMA band 7). The basic idea is to apply the waveguide frequency-separation-filter (FSF) [2], which has been successfully used for astronomical observations. However, the FSF has a narrow specific bandwidth, and we thus need to develop a new one with the wider bandwidth. A. Gonzalez et al. [3] reported the wideband diplexer with a fractional bandwidth of ~61% for the two-band local oscillator system covering the ALMA bands 7 and 8. We adopted this model and adjusted parameters for the ALMA bands 6 and 7. We obtained good performance of the FSF over the ALMA bands 6 and 7 in simulation.**

***Index Terms*—Atacama Large Millimeter/Submillimeter Array (ALMA), Frequency Separation Filter (FSF), 1.85-meter telescope, Radio astronomy, waveguide diplexer, wideband.**

## I. INTRODUCTION

RECEIVERS of radio telescope for astronomy is required to have low-noise and high-gain over wide fractional bandwidth. The receiver noise and bandwidth is critically related with observation sensitivity and efficiency. To cover wideband RF range allows to reduce the total number of receivers and simplify maintenance and operation of telescopes. In addition, wideband RF range is enable to observe new science cases. In the Atacama Large Millimeter/Submillimeter Array (ALMA), the 35-950 GHz RF range is separated by 10 bands and these fractional band width of each band is nearly 30 %. However, recent advanced technology can be produce receivers which cover much wider RF ranges.

At Osaka Prefecture University (OPU), we have operated the 1.85-meter telescope at the Nobeyama Radio Observatory to observe molecular clouds in nearby star-forming regions and along the Galactic Plane in $^{12}$CO, $^{13}$CO, and C$^{18}$O ($J$ = 2-1) (e.g., [1][2]). Now, we are planning to relocated the 1.85-m telescope to Atacama site (nearly 2,400 m) and to newly install a dual-band simultaneous observation of CO lines at $J$ = 2-1 and $J$ = 3-2. As a receiver system for dual-band simultaneous observation, we are developing radio frequency waveguide multiplexer in 210-375 GHz (the fractional bandwidth is 58.8% when its center frequency is 280.6 GHz). This multiplexer consists of ($\alpha$) wideband waveguide diplexer (Fig. 1) to separate ($\beta$) 211-275 GHz (ALMA band 6) and ($\gamma$) 275-373 GHz (ALMA band 7) and other diplexers (Fig. 1) focused on CO lines at $J$ = 2-1 and $J$ = 3-2 to achieve 2 side band observation (Fig. 1). The basic idea of waveguide diplexers is to apply the waveguide frequency-separation-filter (FSF) [3], which has been successfully used for astronomical observations. However, the FSF has a narrow specific bandwidth, and we thus need to develop a new one with the wider bandwidth. A. Gonzalez et al. [4] reported the wideband diplexer with a fractional bandwidth of ~61% for the two-band local oscillator system covering the ALMA bands 7 and 8. We adopted this model and adjusted parameters for the ALMA bands 6 and 7. In this paper, we describe the design of wideband diplexer to separate ALMA band 6 and 7.

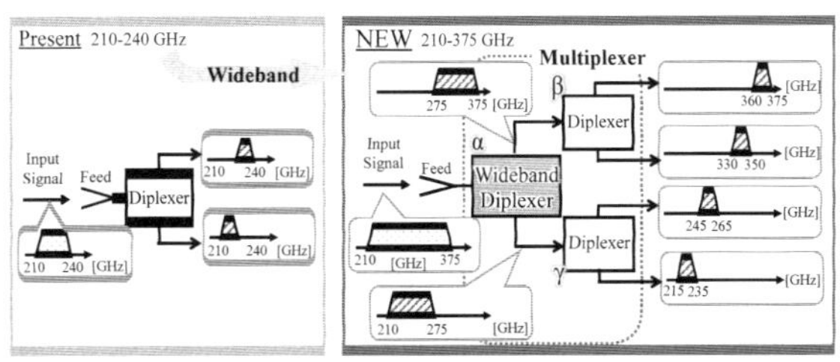


Figure 1. Schematic diagrams of present receiver and our new receiver to observe CO lines at $J$ = 2-1 and $J$ = 3-2. A diplexer installed in present receiver can separate 215-220 GHz and 230-235-GHz. Multiplexer which consists of three types of diplexer installed in new one can separate 4 bands (215-235, 245-265, 330-350, 360-375 GHz).

Munuscript received and accepted June 25, 2019. This work was supported in part by JSPS KAKENHI Grant Number 18H05440.

S. Masui, S. Ueda, Y. Yamasaki, K. Yokoyama, N. Okada, T. Onishi and H.Ogawa are with Osaka Prefecture University, 1-1 Gakuen-cho, Naka-ku, Sakai, Osaka, 599-8531, Japan (e-mail: s_s.masui@p.s.osakafu-u.ac.jp).

Y. Hasegawa is with Institute of Space and Astronautical Science, 3-1-1 Yoshinodai, Chuo-ku, Sagamihara, Kanagawa, 252-5210, Japan.

K. Kimura is with Nagoya University, Furo-cho, Chikusa-ku, Nagoya, Aichi, 464-8602, Japan

T. Kojima, A. Gonzalez are with the National Astronomical Observatory of Japan (NAOJ) and also with the Graduate University for Advanced Studies (SOKENDAI), Mitaka, Tokyo 181-8588, Japan

## II. Conceptual Design

The conceptual design of FSF [3][5] is presented in Fig. 2. The FSF consists of two 3-dB quadrature hybrid couplers and two identical high pass filters (HPF). The hybrid coupler employs a branch-line coupler (BLC) with several branches whose height and interval should be close to the quarter wavelength. An input signal to the 3 dB BLC is divided into two, and then delivered toward two output ports 3' and 4' with the almost same magnitude, having a 90° phase difference between them. When the frequency of the input signal is higher than the cutoff frequency of the internal HPFs, all the signal goes to port 4. When the input frequency is lower than the cutoff frequency, all the signal goes to port 2(Fig. 2).

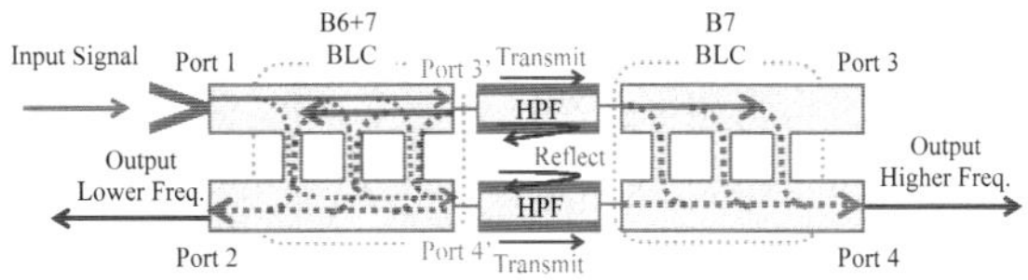


Figure 4. Schematic diagram of frequency separation filter. An input signal (purple lines) to the 3 dB BLC is divided into two, and then delivered toward two output ports 3' and 4' with the almost same magnitude, having a 90° phase difference between them. When the frequency of the input signal is higher than the cutoff frequency of the internal HPFs, all the signal (blue lines) goes to port 4. When the input frequency is lower than the cutoff frequency, all the signal (red lines) goes to port 2.

## III. Design of Building Blocks

The individual components of the diplexer have been simulated separately, and then carefully connected together. The performance and parameters of optimized components have been obtained using finite-elements software HFSS [6].

### A. Band 6 + 7 Branch Line Coupler (B6+7 BLC)

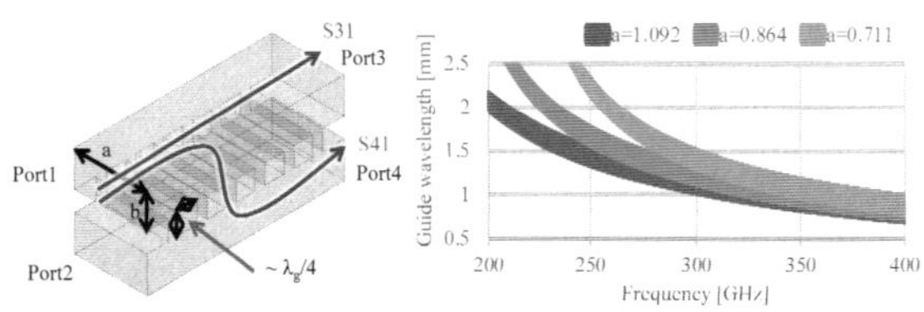


Figure 5. Left figure shows structure of BLC. The long side of waveguide is a and the short side of waveguide is b. The height and interval of several branches is set to quarter guide wavelength. Right figure shows the frequency dependence of guide wavelength. Blue line is a = 1.092, orange line is a = 0.864 and green line is a = 0.711. The larger the long side of waveguide, the smaller the amount of change of guide wavelength.

The major role of BLC is to propagate an input signal by half the intensity with phase differences of 90 degrees. It is important to set the height and interval of several branches to quarter guide wavelength. However, since guide wavelength depend on the long side (a) of waveguide and input frequency (Fig. 3), in wide frequency band it is difficult to define the height and interval of several branches. Therefore, we have tried to change the long side (a) of waveguide. Fig. 3 shows three types of the frequency dependence of guide wavelength. Orange line is fundamental size to transmit radio frequency from 170-340 GHz. Green line is smaller waveguide, and blue line is larger. As you can see, blue line is very flatten compared to other lines and it is easy to set quarter guide wavelength. However, since large waveguide cannot suppress high order mode, the performance is significantly worse. In order to solve this problem, we have tried to decrease the short side (b) of waveguide, and high order mode have been suppressed (Fig. 4). Final model and performance of Band 6 + 7 BLC is shown in Fig. 5. This waveguide size is 1.092 × 0.3 mm, and this design has 9 slots with 50 μm width. Return loss is higher than 21 dB in 210-375 GHz.

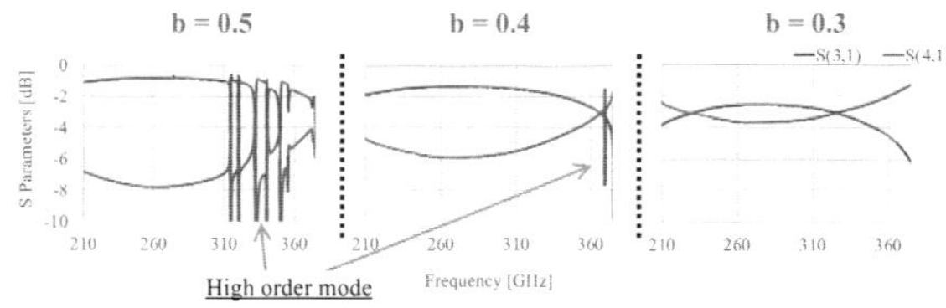


Figure 2. The short side (b) of waveguide is changed (b = 0.5, 0.4, 0.3). This figure shows that the smaller the short side (b) of waveguide, the more hi order mode is suppressed.

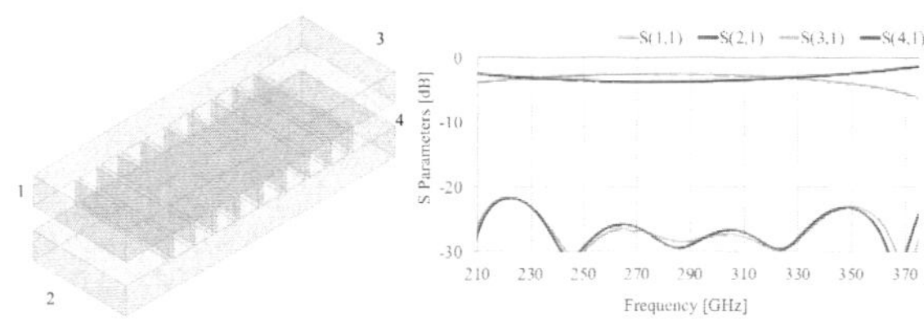


Figure 3. The model and performance of Band 6 + 7 BLC. This waveguide size is 1.092 × 0.3 mm, and this design has 9 slots with 50 μm width. Return loss is higher than 21 dB in 210-375 GHz.

### B. High Pass Filter

Conventional design of HPF is used a small size straight waveguide. However, this type HPF has high insertion loss to obtain a sharp increase of the reflection loss at the upper end of B6. And it is difficult to match characteristic impedances in wideband. Therefore, we applied a compact cavity filter [7]. This filter formed by cavity and iris resonator has wide pass band with a sharp increase and low insertion loss. Final model and performance is shown in Fig. 6. This design is very compact with thin cavities with 50 μm width and a thick cavity with 140 μm width. In 210-275 GHz, S(2,1) is higher than 27 dB and in 280-375 GHz, return loss is higher than 20 dB.

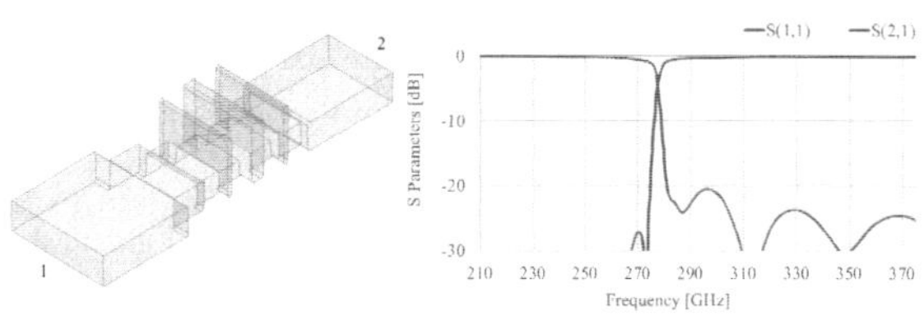


Figure 6. The model and performance of HPF. This design is very compact with thin cavities with 50 μm width and a thick cavity with 140 μm width. In 210-275 GHz, S(2,1) is higher than 27 dB and in 280-375 GHz, return loss is higher than 20 dB.

### C. *Band 7 Branch Line Coupler (B7 BLC)*

B7 BLC is only used at B7 frequencies and fractional band width is 30 %. Therefore this BLC optimized in 275-373 GHz. Standard BLC is applied to simplify the structure. Final model and performance is shown in Fig. 7. This model had 7 slots with 60 μm width and waveguide size is 0.74 × 0.29 mm. In 275-373 GHz, return loss is higher than 22 dB.

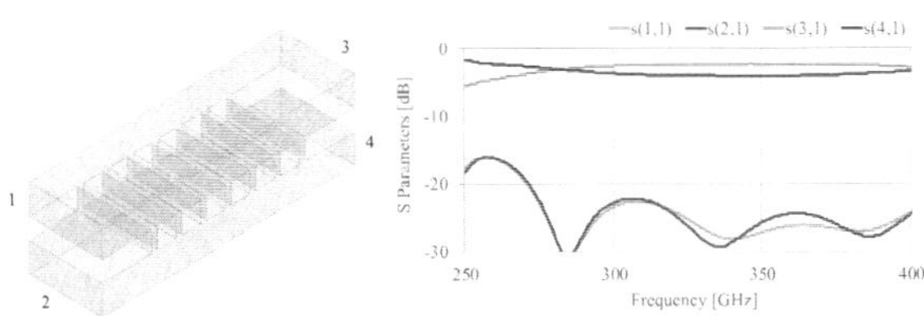


Figure 7. The model and performance of B7 BLC. This model had 7 slots with 60 μm width and waveguide size is 0.74 × 0.29 mm. In 275-373 GHz, return loss is higher than 22 dB.

## IV. Design of Waveguide Diplexer

After above components have been designed separately, these components were assembled together. We have optimized all parameters included the length between these components and obtained good performance to separate Band 6 and Band 7 with low insertion loss. If this diplexer is made of aluminum, in 210-265 GHz, return loss is higher than 20 dB and insertion loss is less than 0.15 dB at 4 K, and in 280-375 GHz, return loss is higher than 17 dB and insertion loss is less than 0.25 dB at 4 K. This design is very compact and this length from port 1 to port 3 is about 6 mm.

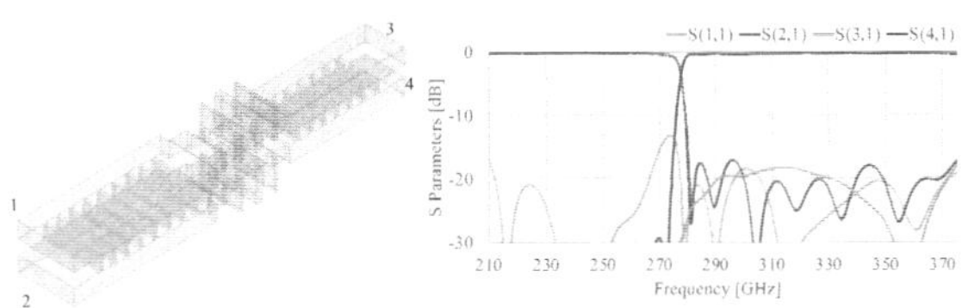


Figure 8. The model and performance of wideband waveguide diplexer. In 210-375 GHz, return loss is higher than 17 dB. If this diplexer is made of aluminum, the insertion loss is 0.25 dB in 210-375 GHz at 4 K.

## V. References


[1] T. Onishi, "A 1.85-m mm-submm Telescope for Large-Scale Molecular Gas Survey in $^{12}$CO, $^{13}$CO and C$^{18}$O (*J* = 2-1)" *PASJ*, vol. 65, no. 78, August, 2013.

[2] A. Nishimura, "REVEALING THE PHYSICAL PROPERTIES OF MOLECULAR GAS IN ORION WITH A LARGE-SCALE SURVEY IN J = 2–1 LINES OF 12CO, 13CO, AND C18O" *ApJS*, vol. 216, no. 1, December, 2014.

[3] Y. Hasegawa, "Observational demonstration of a high image rejection SIS mixer receiver using a new waveguide filter at 230 GHz" *PASJ*, vol 69, no. 91, December, 2017.

[4] A. Gonzalez, "275-500 GHz Waveguide Diplexer to Combine Local Oscillators for Different Frequency Bands" 2017, *IEEE*, vol. 7, no.6, 669--676.

[5] S. Asayama, "A Novel Compact Low Loss Waveguide Image Rejection Filter Based on a Backward Coupler with Band Pass Filters for 100 GHz Band" JIMTW vol. 36, issue 5, pp. 445–454

[6] High Frequency Structure Simulator (HFSS), "Sofware description." [On- line]. Available: http://www.ansys.com/, Accessed on: Oct 12, 2017.

[7] U. Rosenberg, S. Aman, J. Bomemann, and R. Vahldieck, "Compact pseudo-highpass filters formed by cavity and iris resonators," in *Proc. 34th Eur. Microw. Conf.*, Amsterdam, The Netherlands, 2004, pp. 985–988.

# Advanced tuning algorithms for increasing performance of high-frequency SIS mixers

T. Vos[1], R. Hesper[1], J. Barkhof[1], A. M. Baryshev[1], A. Khudchenko[1], and P. Yagoubov[2]

The NOVA Submillimeter Instrumentation Group at the University of Groningen, The Netherlands, has developed, produced and qualified the full set of 73 operational ALMA Band 9 receivers (600-720 GHz). On all key aspects, these receivers are performing within the performance specifications. Now that all receivers are in place, a potentially easy and very low-cost way to further improve performance may be by pure software-based optimization of certain tuning parameters.

Three tuning parameters are key to SIS performance: SIS bias voltage, LO pumping level and Josephson current suppression (by applying a magnetic field in the plane of the junction). The former two are relatively easy to perform automatically, although there are a couple of pitfalls. The suppression of the Josephson current, however, is much more complicated, especially in the high-current-density AlN-barrier junctions used in Band 9. The individual SIS devices can behave quite differently from each other, and often hysteretic or multiple-regime behavior is observed. In many cases there seem to be unclear tradeoffs between difference performance criteria (noise temperature and tuning stability, for instance). It is clear that in the presence of such effects, stability and repeatability issues are a serious concern. In the tuning parameters that were supplied with the receivers to the ALMA observatory, we always biased ourselves towards the safer and more repeatable regimes, for obvious operational reasons.

In this work, we present early results of the ALMA Band 9 Advanced Tuning study commissioned by ESO. The objective of this study is to investigate more "intelligent" tuning algorithms which should enable the receivers to operate in more critical regimes, which were previously avoided, but with a possibly considerable increase in performance.

In the framework of this study, we first had to develop a new software infrastructure. Our original engineering software [1], while excellently capable of qualifying production receivers (as proven by the successful ALMA Band 9 and Band 5 production campaigns), is not very suitable for adaptive "intelligent" algorithms because of the absence of conditional statements and loop constructions. The new system we developed is structured as a series of extension libraries of the Python language, entirely replacing the functionality of the old engineering code. This gives us the full power of an established high-level programming language suitable for implementation of any adaptive or interactive algorithm we can think of.

Using this infrastructure, as a first step, we have now fully automated the formerly semi-automatic and "eyeball" algorithms, arriving reliably and repeatably at very similar tuning points as with the classical methods. Working from there, we can now investigate incrementally improved and even completely different tuning methods.

As a first important result of these improvements, we found an almost linear relationship between the achievable noise temperatures and the applied magnetic field. This suggests that the focus should maybe not be so much on suppressing the Josephson current itself (as is done traditionally), but on keeping the magnetic field as low as possible while ensuring the stability of the mixer by reliably finding SIS bias voltages just outside the first Josephson feature. This approach has led to a significant increase of the sensitivity in the tested mixers. While the algorithms are primarily intended to interact with the mixers directly, in addition we can perform part of them on our archive data of all delivered ALMA Band 9 mixers. This enables us to give reasonable estimates of the expected performance increase.

## References

[1] J. Barkhof, B. D. Jackson, A. M. Baryshev, and R. Hesper, "ALMA Band 9 Cartridge Automated Test System", *Twentieth International Symposium on Space Terahertz Technology*, pp. 246-250, April 2009

[1] Kapteyn Astronomical Institute, University of Groningen, Landleven 12, 9747 AD Groningen, The Netherlands

[2] European Southern Observatory, Karl-Schwarzschild-Straße 2, 85748 Garching bei München, Germany

NOTES:

# CHAI, the CCAT-prime Heterodyne Array Instrument

Urs U. Graf[1], Netty Honingh[1], Ignacio Barrueto[1], Karl Jacobs[1], Bernd Klein[2], Rocio Molina[3], Nicolás A. Reyes[3], Pablo A. Tapia[1], and Jürgen Stutzki[1]

We present the design of the new dual color heterodyne focal plane array receiver CHAI, which is being built for the CCAT-prime telescope [1], under construction on Cerro Chajnantor in Chile.

CHAI is a 64 pixel SIS receiver operating simultaneously in the 650 μm and 350 μm atmospheric windows. Its primary scientific purpose is extended mapping of galactic sources in the important astronomical transitions of these frequency bands: CO $J$=4→3 (460 GHz), [CI] $^3P_1 \rightarrow {}^3P_0$ (492 GHz), CO $J$=7→6 (807 GHz) and [CI] $^3P_2 \rightarrow {}^3P_1$ (809 GHz).

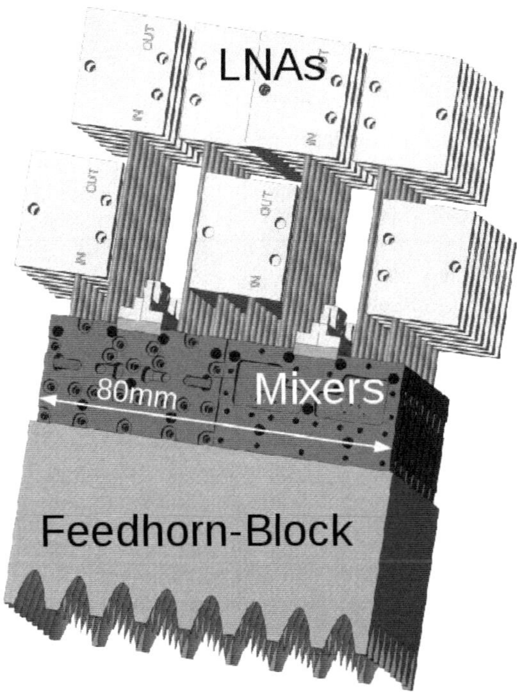


Fig. 1. Layout of CHAI's focal plane unit. It consists of a feedhorn block followed by 16 mixer blocks containing 4 mixers, each and their respective LNAs. The space between the LNAs is used for the LO signal distribution.

The two bands of the instrument are split by polarization and use two mostly identical separate cryostats. Each of them houses a subarray of 64 on-chip balanced SIS mixers, modeled after the design of [2], together with their respective low noise amplifiers (LNA). A system of waveguide splitters distributes the local oscillator (LO) signal and couples to the second waveguide port of the mixers.

The 64 mixers of each subarray are packaged in 16 identical split-block mixer units, each combining four balanced mixers with their feedhorns and the LO distribution. The LNAs are individually packaged and connected to the mixers through coaxial cables.

Cryogenic cooling of the instrument is provided by closed cycle pulse tube refrigerators. Each of the two cryostats contains two cold heads, one of which is exclusively dedicated to the mixers and the second one to cooling the LNAs and the 4K radiation shield.

The IF band extends from 4 to 8 GHz, and is processed by a straight through amplification chain without a second mixing stage. This IF band can be analyzed directly by the upcoming new generation of digital Fourier transform spectrometers [3].

The purely reflective optical system of CHAI uses mostly warm components except for a pair of cold mirrors inside each cryostat needed to funnel the array beams through the cryostat windows. The main challenge in the optical setup is to bridge the long beam paths in the telescope with a reasonably small beam cross section. In addition, part of the optics has to be retractable to provide access to the telescope focal plane for the other instruments on the observatory.

References

[1] S. C. Parshley et al., "CCAT-prime: a novel telescope for sub-millimeter astronomy", *SPIE) Conference Series,* vol. 10700, 2018.

[2] M. P. Westig et al., "A 490 GHz planar circuit balanced Nb–$Al_2O_3$–Nb quasiparticle mixer for radio astronomy: Application to quantitative local oscillator noise determination", *Journal of Applied Physics*, vol. 112, no. 9, 2012.

[3] B. Klein et al., "High-resolution wide-band fast Fourier transform spectrometers", *A&A*, vol. 542, 2012.

[1] I. Physics Institute, University of Cologne, Cologne Germany.
[2] Max-Planck-Institute for Radioastronomy, Bonn, Germany..
[3] Electrical Engineering, Universidad de Chile, Santiago, Chile

NOTES:

# 240 GHz DSB receiver performance

Kirill Rudakov, Andrey Baryshev, Ronald Hesper, Andrey Khudchenko, Daniel Montofre, Rob de Haan Stijkel, Pavel Dmitriev, and Valery Koshelets

***Abstract*— We have designed, simulated, manufactured and evaluated a double side band (DSB) receiver for the 211-275 GHz frequency range; in this design 4 – 8 GHz intermediate frequency (IF) band was realized. In order to achieve quantum limited sensitivity high quality Nb/AlOx/Nb superconductor-insulator-superconductor (SIS) tunnel junctions are employed; both a single-side and a double-side designs of the mixers elements installed in the waveguide RF hybrid block are tested. The uncorrected DSB mixer noise temperature as low as 9 K has been measured at 241 GHz in a narrow intermediate frequency (IF) band; the noise temperature is rising up to 22 K at the edges of the input frequency range. The DSB noise temperature measured at integration of the IF signal in the band 4-8 GHz do not exceed 30 K for all frequencies from 227 to 275 GHz.**

***Index Terms*— superconductor-insulator-superconductor (SIS) receivers, quantum limited sensitivity, intermediate frequency bandwidth, submillimeter waves, heterodyne terahertz receivers.**

## I. INTRODUCTION

The mixers based on superconductor – insulator – superconductor (SIS) tunnel junctions are most sensitive devices at frequencies *f* from 0.1 to about 1.2 THz. Their noise temperature is limited only by the quantum value $hf/2k_B$, where h and $k_B$ are the Plank and Boltzmann constants, respectively. The SIS mixers were successfully used both for the space missions like Hershel HIFI [1] and for the ground-based telescopes like the largest multi-element interferometer ALMA [2]).

The Russian" space observatory "Millimetron [3] with a 10-meter space telescope is presently under development. The observatory has two operational modes – the single-dish and Space-Earth interferometer modes. The second mode is aimed to observe the extremely compact objects, e.g. the immediate vicinity of black holes that require ultra-high resolution, up to tens of billionths of a second of arc. High angular resolution is provided by the orbit configuration (location near the Lagrange point L2 at 1.5 million kilometers from the Earth). For the Space-Earth interferometer the 2SB 211 - 275 GHz receivers with a noise temperature below 50 K are required.

This paper presents the results of the development and measurement of a test prototype of the DSB SIS heterodyne waveguide receiver for 211 - 275 GHz frequency band. The developed receivers can be used also for many other future ground-based radio astronomy projects.

## II. MIXER DESIGN, TECHNOLOGY AND RESULTS

To follow a successful Hybrid construction methodology of the ALMA Band 9 receiver [4] we place the mixer chip into a waveguide orthogonally to the propagation direction; the designs were developed using a Microwave Studio (CST). The Nb/AlOx/Nb SIS junction is placed into a planar Nb/$SiO_2$/Nb tuning structure made on a 125 μm thick quartz substrate. The receiving chip (width 150 μm) is located in a rectangular 1000 x 500 μm waveguide at a distance of 230 μm from the backshort in the waveguide. The mixer block consists of a few separates elements: a central part with the waveguide, a magnet block unit with two magnet pins to suppress Josephson critical current, a back piece (BP) unit where the mixer chip is installed, and an input horn. The quartz chip placed in the waveguide channel is itself a dielectric waveguide with an excitation frequency of the first mode of about 320 GHz; to prevent leakage of the RF signal through this dielectric waveguide the blocking low-pass RF filters were used. A combination of the Coplanar (CPW) and Microstrip (MSL) lines were used to tune out the intrinsic SIS capacitance and to provide the matching of the resulted SIS impedance to the waveguide at RF.

To realize a quantum-limited performance, the SIS tunnel junctions with extremely small leakage current under the gap voltage and minimal energy gap spreading $\delta Vg$ are required. This is especially important for relatively low-frequency devices (f ~ 200– 300 GHz), since the $\delta Vg$ has to be much smaller than the size of the quasiparticle step hf/e, while the leakage current at a bias voltage of about $Vg - hf/2e$ determines the noise of the mixer. The fabrication technology of the Nb–AlOx–Nb tunnel junctions is based on the fact that a very thin Al layer can completely cover the base Nb electrode [5, 6], somehow "planarizing" the column-like structure of the Nb film. This Al layer is subsequently oxidized and the top Nb electrode is deposited on the oxidized layer to form a so-called tri-layer structure.

The numerical simulations and experimental study of the receivers were supported by the Russian Science Foundation (Project No. 19-19-00618). The fabrication of the tunnel circuits was carried out at the Kotel'nikov IREE RAS within the framework of the state task.

The research leading to these results has received funding from the European Commission Seventh Framework Programme (FP/2007-2013) under grant agreement No 283393 (RadioNet3)

K. Rudakov, A. Baryshev, R. Hesper, A. Khudchenko, D. Montofre, and R. de Haan Stijkel are with the Kapteyn Astronomical Institute, University of Groningen, 9712 CP Groningen, the Netherlands (e-mail: k.rudakov@astro.rug.nl)

K. Rudakov and A. Khudchenko are also with the Kotel'nikov Institute of Radio Engineering and Electronics RAS, Moscow 125009, Russia; K. Rudakov is also with the Moscow Institute of Physics and Technology, Institutskii per., 9, 141700 Dolgoprudny, Moscow Region, Russia

P. Dmitriev and V. Koshelets are with the Kotel'nikov Institute of Radio Engineering and Electronics RAS, Moscow 125009, Russia. (e-mail: valery@hitech.cplire.ru).

The SNEAP technology was used in this work for fabrication of the SIS receiving structures based on Nb/Al-AlOx/Nb circuits, details are presented in [7–9]. To prevent etching of the quartz substrate in the process of plasma etching during the junction definition process a "monitor" layer of Nb with a thickness of about 100 nm was deposited in the substrate by DC magnetron sputtering. The SIS junctions are formed by plasma-chemical etching in CF4 by removing the top Nb layer of the tri-layer Nb/Al-AlOx/Nb structure according to the mask from the photoresist determining the junction geometry. After plasma-chemical etching, anodizing is performed up to 10 V using the same photoresist mask; then an insulating $SiO_2$ layer, typical thickness of which is 250 nm, is deposited by RF magnetron sputtering; opening of contacts to the junctions is carried out by lift-off.

The current-voltage characteristic (IVC) of a Nb/Al-$AlO_x$/Nb SIS-mixing element with an area of about 1 $\mu m^2$ is shown in Fig. 1, the IVC is measured in the voltage-bias mode, the critical current of the SIS junction is suppressed by a magnetic field. The normal resistance of the SIS junction is Rn ~ 34 Ω, the quality parameter characterized by the ratio of the resistances under and above the gap Rsg/Rn ~ 36, the gap voltage Vg ~ 2.75 mV, the energy gap spreading δV ~ 0.1 mV. It should be mentioned that the well-pronounced knee-like feature arising on the IVC at voltages slightly higher than Vg. This feature is due to the presence of a normal aluminum layer near the tunnel barrier; its presence substantially modifies the density of electron states in the superconducting electrode. A theoretical model of such a structure [10] is built on solving the quasi-classical Usadel equations with the realization of the conditions of the so-called dirty limit. Experimentally, the dependence of the effect on the parameters of the tunnel structure was investigated in [7].

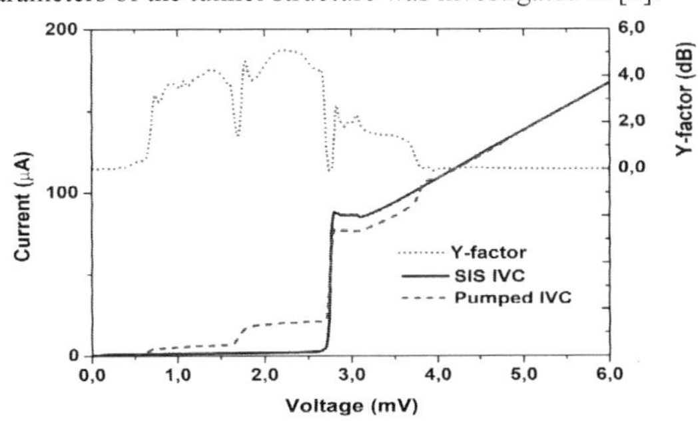


Fig. 1. The SIS IVCs (autonomous and pumped by LO at 241 GHz at was optimal LO power level); the corresponding Y-factor is presented

In order to evaluate a wideband radiation matching of the SIS mixer at RF the Michelson Fourier transform spectrometer (FTS) technique was used. A wideband GHz-THz source – glow bar - was matched with the FTS which was loaded to the SIS mixer as a detector. The mixer was voltage biased at 3mV, then a direct current response was measured versus a mirror position; these data were Fourier transformed into a mixer response on the frequency.

Noise temperature has been measured by using Y-factor method with an absorber placed in liquid nitrogen (78 K) as a cold load and a room temperature (296.4 K) absorber as a hot load (see Fig. 1) The measured DSB uncorrected noise temperature is presented in the Fig. 2 in comparison with quantum sensitivity level (Callen & Welton) [11]; the frequency range was limited to 227 – 275 GHz by parameters of the used LO source. The data were measure over the IF band 4 - 8 GHz; even lower DSB noise temperature was obtained at integration in narrow intermediate frequency (IF) band 40 MHz: the $T_r$ as low as 9 K has been measured at 241 GHz; all details will be presented elsewhere [12].

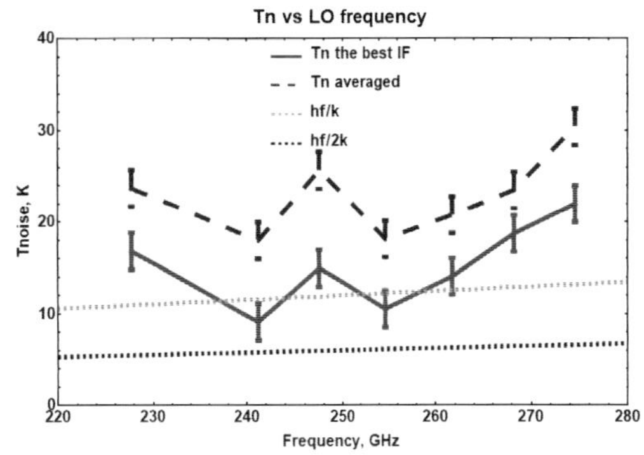


Fig. 2. Uncorrected noise temperature measured in the IF band 4-8 GHz is presented at the graph by symbols in comparison with the quantum sensitivity level $hf/k_B$ (Callen & Welton) shown by lines. The measurements uncertainty is of about 2K

## REFERENCES

[1] Th. de Graauw, et al, "The Herschel-Heterodyne Instrument for the Far-Infrared (HIFI)", Astronomy & Astrophysics, 518, L6 (2010) DOI: https://doi.org/10.1051/0004-6361/201014698

[2] https://www.almaobservatory.org/en/about-alma-at-first-glance/how-alma-works/technologies/interferometry/

[3] http://millimetron.ru/index.php/en/

[4] A.M. Baryshev, et al, "The ALMA Band 9 receiver Design, construction, characterization, and first light", Astronomy &Astrophysics, 577, A129 (2015) DOI: 10.1051/0004-6361/201425529

[5] M. Gurvitch, W. A. Washington, and H. A. Huggins, "High quality refractory Josephson tunnel junctions utilizing thin aluminum layers", *Appl. Phys. Lett.*, vol. 42, pp. 472-474, 1983.

[6] S. Morohashi, and S. Hasuo, "Experimental investigations and analysis for high-quality Nb/Al-$AlO_x$/Nb Josephson junctions", *J. Appl. Phys.*, vol. 61, pp. 4835-4849, 1987.

[7] P. N. Dmitriev, et al, "Niobium Tunnel Junctions with Multi-Layered Electrodes", *IEEE Trans. Appl. Supercond.*, vol. 9, pp. 3970-3973, 1999.

[8] L.V. Filippenko, et al, "Integrated Superconducting Receiver: fabrication and yield", *IEEE Trans. on Appl. Supercond.*, v.11, No 1, pp. 816-819, (2001).

[9] P.N. Dmitriev, et al, "High Quality Nb-based Integrated Circuits for High Frequency and Digital Applications", "*IEEE Trans. on Appl. Supercond.*", vol. 13, No 2, pp. 107-110, 2003.

[10] A.A.Golubov, E.P.Houwman, J.G.Gijsbertsen, V.M.Krasnov, J. FLokstra, H.Rogalla "Proximity effect in superconductor-insulator-superconductor Josephson tunnel junctions: theory and experiment", *Physical Review B*, vol. 51, 2 pp. 1073-1089, 1995.

[11] A.R. Kerr, M.J. Feldman and S.-K. Pan, "Receive Noise Temperature, the Quantum Noise Limit, and the Role of the Zero-Point Fluctuations", *Eighth International Symposium on space Terahertz Technology*, Harvard University, March 1997.

[12] K.I. Rudakov, et al, "240 GHz waveguide receiver with quantum limit performance", "*IEEE Trans. on THz Science and Technogy*", will be published, 2019.

# Design and Fabrication of an on-Chip Sideband Separating (2SB) Balanced SIS Mixer for 400 – 500 GHz on a 9μm Silicon Membrane

S. Widdig*, K. Jacobs, M. Schultz, C. E. Honingh, N. Wehres, and J. Stutzki
*1. Physikalisches Institut, Universität zu Köln,*
*Zülpicher Str. 77, 50937 Köln, Germany*
*Contact: swiddig@ph1.uni-koeln.de

**Abstract:** Superconductor-Insulator-Superconductor (SIS) tunnel junctions are currently used as heterodyne mixers with quantum limited sensitivity in millimeter and sub-millimeter wavelength receiver in radio astronomy. Well-engineered technology offers the opportunity to replace the traditionally used single-ended double-sideband (DSB) mixer by a balanced or sideband separating (2SB) mixer. 2SB mixers, which detect each sideband at a different output, giving the opportunity to suppress the atmospheric noise and/or the lines of the unwanted sideband, are today mostly made in waveguide technology [1][2]. Due to the large volume of the mixers in this technology, it is difficult to build many-pixel (> 32 pixels) array receivers where the footprint of each pixel must be small (e.g. (10x10) mm$^2$ for the planned CCAT-prime Heterodyne Array Instrument (CHAI) receiver [3]).

We show the development of the RF part of the first on-chip sideband separating balanced Nb-Al2O3-Nb SIS mixer for the frequency range between 400 and 500 GHz. The mixer is designed in the same technology as the existing integrated balanced mixer (IBAMI) [4]. The total size of the mixer including three 90° hybrid couplers, an RF load, four SIS junctions and an LO In-Phase power divider is (2.3 x 1.7) mm$^2$. For the first planned measurements, we designed a prototype mixer block - which includes the RF chip, the IF boards to the four G2PO connectors and the permanent magnets for suppressing the Josephson current. The prototype block allows us to amplify the IF signals of the four SIS mixer devices first and then combine them at room temperature to test predominantly the performance of the RF part of the circuit. In addition the same block will be used in an absorption cell measurement to show the sideband suppression of the mixer with the detection of molecular lines.

[1] Kamikura, M. et. al., International Journal of Infrared and Millimeter Waves (2007), **"A 385-500 GHz Sideband-Separating (2SB) SIS mixer based on a waveguide split-block coupler"**
[2] Serizawa, Y. et. al., Journal of Infrared, Millimeter and Terahz Waves (2012), **"Development of a 385–500 GHz Sideband-Separating Balanced SIS Mixer"**
[3] CCAT-prime Webpage: **http://www.ccatobservatory.org/index.cfm/page/index.htm**
[4] Westig, M. P. et. al., Journal of Applied Physics 112, 093919 (2012), **"A 490 GHz planar circuit balanced Nb-Al2O3-Nb quasiparticle mixer for radio astronomy: Application to quantitative local oscillator noise determination"**

# First Results of the Sideband Separating Mixer for 850 GHz

Andrey Khudchenko, Ronald Hesper, Andrey M. Baryshev, Jan Barkhof, Marielle E. Bekema, Rob de Haan Stijkel, Kirill Rudakov, Daniel Montofré, Duc van Nguyen, Valery P. Koshelets, Pavel Dmitriev, Michael Fominsky, Christopher Heiter, Stefan Heyminck, Rolf Güsten and Bernd Klein

*Abstract*— **We presented here the design and the first results of a new sideband separating (2SB) mixer for 800--950\,GHz, based on superconductor-insulator-superconductor (SIS) junctions. This is the first waveguide 2SB SIS mixer demonstrated at such a high frequency. The design is following the classical quadrature hybrid architecture, meanwhile additional attention was put on the reduction of reflections in the RF structure in order to minimize the RF imbalance, to achieve a high image rejection ratio (IRR). The RF waveguide block was manufactured by micro-milling and populated by single-ended SIS mixers developed earlier for upgrade of the CHAMP+ high band array on the APEX telescope. These SIS mixers have DSB noise temperatures from 210 to 400\,K. The assembled 2SB mixer yields a single-sideband noise temperature from 450 to 900\,K, with an image rejection ratio above 15dB in 95\% of the band. Comparing the DSB and SSB sensitivities, we find that the waveguide losses are as low as expected and do not exceed 0.6\,dB. The presented mixer is a prototype for use in a 2SB dual polarization receiver planned for deployment on the APEX telescope.**

*Index Terms*— **Sideband separating (2SB) mixers, image rejection ratio (IRR), submillimeter wave technology, terahertz receivers, superconductor-insulator-superconductor junctions.**

## I. Introduction

Ground based observations of astronomical objects at frequencies around 800--950\,GHz are strongly influenced by atmospheric absorption. Using sideband-separating (2SB) receivers instead of double-sideband (DSB) ones allows us to reduce the atmospheric noise contribution for spectral line sources by, ideally, a factor of two, irrespective of the actual atmospheric transparency. In practice, however, the total system noise temperature includes other contributions like mixer noise and intermediate frequency amplifier noise. These make a factor of two improvement in system noise temperature unobtainable. In addition, the actual improvement will strongly dependent on the atmospheric transparency. From historical weather conditions at the [1] and ALMA [2] sites [3], the zenith atmospheric transmission for the 800—950 GHz window can be estimated between 0.2 and 0.6. The upper limit corresponds to realistic good weather conditions, while the bottom one represents the limit at which the atmospheric opacity becomes too high for reasonable observations in this band. Within this range, the ratio of the 2SB and DSB sensitivities for spectral line observations will be on average around 1.3 for an effective atmospheric temperature of 260 K, and a state-of-the-art single sideband (SSB) mixer noise temperature of 300 K [4]. This number gives sufficient motivation to develop sideband separating receivers for this frequency range.

## II. Mixer Design

For the 2SB mixer we chose a modular design concept very similar to one for the 600—720 GHz band [5][6]. In this concept, the critical components like RF hybrid block, RF horn, LO horn and SIS holders ("back pieces") are realized as independent units, which can be easily exchanged and tested individually. This allows convenient DSB characterization of the individual SIS devices for matching purposes. Both LO and RF horns have a diagonal spline design

The quadrature hybrid is a typical five-branch coupler similar to presented in [7]. The main design goals were the reduction of the input reflection $S_{11}$ and the isolation $S_{21}$. This was done by varying the relevant dimensions (mainly slot widths and positions) while keeping the phase and amplitude balance within reasonable limits (about 0.5 deg and 0.5 dB, respectively).

## III. Test Results

The tested RF block and two horns were machined in-house at the Max Plank Institute for Radio Astronomy (MPIfR) in Bonn out of CuTeP (ASTM C14500) alloy. A liquid He cryostat was used to cool down the mixer. The noise temperature was determined with a 300/77 K hot-cold Y-

The numerical simulations and experimental study of the receivers were supported by the Russian Science Foundation (Project No. 19-19-00618). The fabrication of the tunnel circuits was carried out at the Kotel'nikov IREE RAS within the framework of the state task.

A. Khudchenko, R. Hesper, A. Baryshev, K. Rudakov, D. Montofre, J, Barkhof, M. Bekema and R. de Haan Stijkel are with the Kapteyn Astronomical Institute, University of Groningen, 9712 CP Groningen, the Netherlands (e-mail: a.khudchenko@sron.nl)

D.~Montofre is also with the Electrical Engineering Department, Universidad de Chile, Santiago 8320000, Chile.

D.van Nguyen is with Netherlands Institute for Space Research, Landleven 12, 9747 AD, Groningen, Netherlands

P. Dmitriev, V. Koshelets, M. Fominsky and also K. Rudakov and A. Khudchenko are with the Kotel'nikov Institute of Radio Engineering and Electronics RAS, Moscow 125009, Russia.

K. Rudakov is also with the Moscow Institute of Physics and Technology, Institutskii per., 9, 141700 Dolgoprudny, Moscow Region, Russia

C.~Heiter, S.~Heyminck, R.~G\"usten and B.~Klein are with the Max Planck Institute for Radio Astronomy, Auf dem Hügel 69, D-53121, Bonn, Germany

factor measurement. At the same time, the image rejection ratio was characterized according to the method described in [8] by injecting a test tone signal through a 6\,\um\ Mylar beam-splitter (6 % coupling). Both noise signal and the test tone were coupled to the mixer through a quartz window and cold reflective optics. The LO signal is applied through a separate window in the cryostat. Two LO multiplier chains were used, together covering the entire 800-950 GHz band.

The measured uncorrected single-sideband (SSB) noise temperature of the prototype mixer is shown in Fig. [1]. It varies from about 550 to 1000 K over the band. The presented USB and LSB curves can be corrected for the fraction of the 300 K noise coupled through the beam-splitter and the LO waveguide coupler (4 %; -13 dB in waveguide LO coupler minus 1 dB of additional loss in the LO path). The noise temperature corrected for these two factors will be in the range 450 to 900 K. To have an estimate of the noise penalty incurred by the waveguide structures, the sum of the DSB noise temperatures of the individual SIS mixers is presented on the same plot. It should be mentioned, that the DSB data was obtained using the same cryostat window, cold optics, IF amplifiers and isolators. For clarity, the DSB data points represent the noise temperature averaged over the 4-12 GHz IF band. From the plot, one can estimate that the corrected SSB noise temperature will be higher than the doubled DSB one by 0 to 30 % and on average about 15 %. This is in a good agreement with the waveguide losses theoretically estimated at 0.6 dB or 15 %.

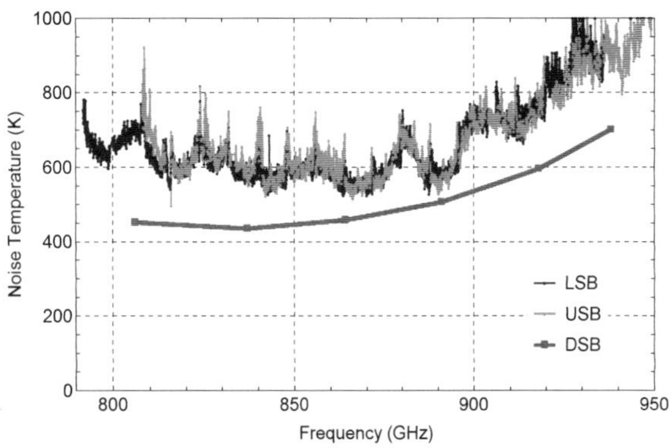


Fig. 1 Uncorrected single-sideband noise temperature of upper (USB) and lower (LSB) sidebands as function of the RF input frequency. The plot is stitched from individual 4-12 GHz IF measurements, while the LO step was 8 GHz, giving full coverage. The frequency resolution within each set is 40 MHz. For reference, the sum of the DSB noise temperatures of the two individual SIS mixer devices is plotted as well (average of two measurements). The DSB data is an averaging product over the 4-12 GHz IF band, and the points are plotted versus the LO frequency in this case.

Fig. [2] shows the image rejection ratio (IRR) obtained with the first prototype block. The IRR is above 15 dB in almost all the points, only at the end of the band it goes down to about 13 dB overall. A few points are falling down to 10 dB level, for example around 860 GHz, which is an artifact of the measurements. It is caused by phase noise and spurious harmonics in the LO signal. Nevertheless, the current results are very promising and a receiver based on this mixer has clear potential to fit ALMA-class specification of 10 dB with ample margin.

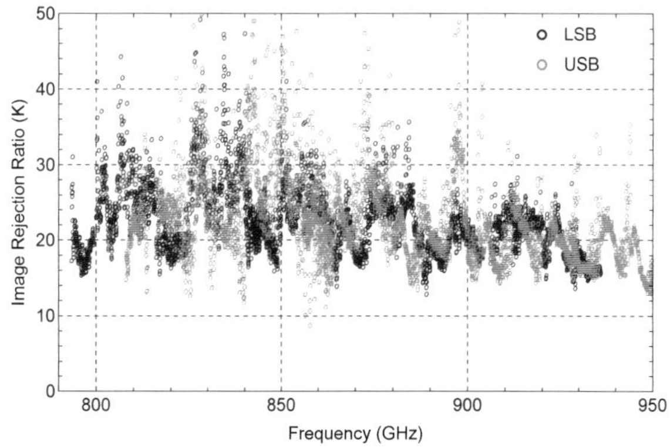


Fig. 2 Image rejection ratio with the same 2SB mixer block and SIS devices. Both LSB and USB results are presented. The data points are measured with step of 40 MHz.

REFERENCES


[1] R. Gusten, R. Booth, C. Cesarsky, K. Menten, C. Agurto, M. Anciaux, ¨ F. Azagra, V. Belitsky, A. Belloche, P. Bergman et al., "Apex: the atacama pathfinder experiment," in Ground-based and Airborne Telescopes, vol. 6267. International Society for Optics and Photonics, 2006, p. 626714.

[2] A. Wootten and A. R. Thompson, "The atacama large millimeter/submillimeter array," Proceedings of the IEEE, vol. 97, no. 8, pp. 1463–1471, 2009.

[3] A. Otarola, M. Holdaway, L. Nyman, S. Radford, and B. Butler, "Atmospheric transparency at chajnantor: 1973-2003," in ALMA Memo Series, jan 2005, p. Memo 512.

[4] M. Kroug, T. Kojima, Y. Fujii, K. Ohtawara, A. Miyachi, and Y. Uzawa, "Noise performance of alma band10 receivers employing high-jc sis mixers," in 29th IEEE International Symposium on Space THz Technology (ISSTT2018), 2018

[5] A. Khudchenko, R. Hesper, A. M. Baryshev, J. Barkhof, and F. P. Mena, "Modular 2SB SIS receiver for 600–720 GHz: Performance and characterization methods," IEEE Transactions on Terahertz Science and Technology, vol. 7, no. 1, pp. 2–9, Jan 2017.

[6] R. Hesper, G. Gerlofsma, F. P. Mena, M. C. Spaans, and A. M. Baryshev, "A sideband-separating mixer upgrade for ALMA band 9," in Twentieth International Symposium on Space Terahertz Technology, Charlottesville, apr 2009, pp. 257–260.

[7] R. Hesper, A. Khudchenko, A. M. Baryshev, J. Barkhof., and F. P. Mena, "A high-performance 650-GHz sideband-separating mixer – design and results," IEEE Transactions on Terahertz Science and Technology, vol. 7, no. 6, pp. 686–693, Nov 2017.

[8] A. R. Kerr, S.-K. Pan, and J. E. Effland, "Sideband calibration of millimeter-wave receivers," in ALMA Memo Series, mar 2001, p. Memo 357.

# Instrumentation development for the 2020 decade at the NOEMA and 30m telescopes

C. Risacher[1], A.-L. Fontana[1], S. Leclercq[1], S. Mahieu[1], D. Maier[1], J. Reverdy[1], P. Serres[1], M. Berton, Y. Bortolotti, O. Garnier, M. Parioleau, Q. Moutote, G. Perrin, B. Pissard, F. Gueth, K.-F. Schuster

IRAM operates several millimeter telescopes from two sites. The NOEMA interferometer, observing from 70-370 GHz and located in the French Alps, saw the completion of its 10th 15m antenna in September 2018 and two more antennas will be added to the array in the coming years. The 30m telescope observes in the same frequency range and is located on the Pico Veleta (3000m), Spain.

This talk will present the current instrumentation status and the plans for the near and far future. The current generation NOEMA receivers were well detailed in [1]. Without waiting for the completion of the NOEMA array, the next upgrades and projects are already moving forward and will be discussed here in detail.

Currently, 4 bands are available, 70-116 GHz, 127-180GHz, 200-276 GHz and 275-373 GHz, with dual polarization side band separation mixers having 4-12 GHz IF bandwidth for the first 3 bands and 4-8 GHz for the last band. The next step for this NOEMA receiver upgrade is to allow dual-band operation using dichroic filters for frequency separation [2], which will allow performing interferometric observations at two RF frequencies simultaneously.

On the detector side, new generation of SIS mixers are being developed, on silicon-on-insulator (SOI) substrates, with the goal of achieving extended RF and IF bandwidths.

At the same time, to allow for more efficient observations, several atmospheric monitoring projects are ongoing. A prototype of a new generation of water vapour radiometer at 22 GHz is being finalized. Those will ultimately equip all NOEMA antennas, and their improved measurement of the water line profile is used real-time to correct and improve the phase noise for each antenna. Another atmospheric monitoring project initiated and implemented with the help of the SMA observatory, is the phase-monitoring project where 2 or more satellite commercial dishes are equipped with commercial X band LNB and interferometric measurement allow retrieving the atmospheric phase variations (detail presentation in [3], this conference).

[1] IRAM, 300 rue de la Piscine, 38406, Saint Martin d' Heres, France

Fig. 1. View of the Plateau de Bure, in the French Alps, where 10 15m antennas are now operational, for the NOEMA interferometer. Two more antennas are currently under construction to be installed in the next few years.

At the 30m telescope in Spain, the NIKA2 dual-band millimeter array camera started operations a few years ago [4]. It has an instantaneous field-of-view of 6.5 arcminutes at both 1.2 and 2.0 mm with polarimetric capabilities at 1.2 mm. The 3 detector arrays are made of more than 1000 KIDs each. Recently, at the end of 2018, the polarimetric capabilities were successfully commissioned. IRAM also started the development of very large 50 pixel- 3mm and 98 pixel- 1mm multibeam receivers based on SIS mixers, which will replace the HERA 2x9 multi pixel 230 GHz array and complement the EMIR heterodyne receiver [5] spectroscopic capabilities allowing for large scale mm-wave mapping of extended objects.

## References

[1] J-Y. Chenu et al., "The NOEMA frontend", ISSTT 2015
[2] A.-L. Fontana et al., "NOEMA receivers: upgrade for simultaneous dual-band observations", ISSTT 2016
[3] S. Mahieu et al., "Atmospheric Phase Monitoring interferometer for the NOEMA observatory", ISSTT 2019
[4] A. Catalano et al., "The NIKA2 Instrument at 30-m IRAM Telescope: Performance and Results", J Low Temp Phys (2018) 193: 916
[5] M. Carter et al., "The EMIR multi-band mm-wave receiver for the IRAM 30-m telescope", Astronomy & Astrophysics, Volume 538, A89, 2012

NOTES:

# Configuring the ALMA Band 3 Cartridge into a Balanced 2SB Receiver

Doug Henke, Pat Niranjanan, and Lewis Knee

***Abstract*—Using a balanced mixer has the advantages of a reduced LO power requirement and cancellation of LO amplitude noise. A new power divider block was designed to rearrange all four mixers of an existing ALMA Band 3 test cartridge into a single-polarization, balanced and sideband-separating receiver.**

***Index Terms*—ALMA Band 3, ALMA instrumentation, balanced mixer, balanced sideband-separating receiver.**

## I. Introduction

Many low-noise receivers for radio astronomy are configured using either sideband-separating or balanced mixers, but few have been implemented that take on both aspects within the same receiver [1]–[4]. A test ALMA Band 3 cartridge was arranged so that all four mixers were used within one polarisation to configure it into a balanced and sideband-separating architecture (BAL-2SB). We wanted to test the configuration to see its impact within Band 3 and for consideration if the design could be scaled to other frequency bands.

In [5], a table of noise contributions is shown in Table 1 where the terms "LO coupled noise" and "image termination contribute ~3 K and ~5 K, respectively, towards an overall receiver noise of ~30 K. Adding a balanced architecture has the advantage of requiring less LO power, cancelling LO amplitude noise, and cancelling "image termination" noise [1],[6]—the latter being the most pertinent to the Band 3 cartridge.

## II. Implementation

Fig. 1 shows the arrangement used for the BAL-2SB receiver. The LO is fed through the front of the power divider block where it is coupled into the RF signal path. There is clear symmetry dividing the left and right sides which shows how each balanced mixer pair is combined through a 180° IF coupler. The delta port of each balanced output is then connected to the final 90° IF hybrid for image-rejection.

In the ALMA Band 3 receiver cartridge, the SIS mixers are biased through a bias-T within each LNA such that the DC bias is carried through the coaxial port of the SIS mixer block.

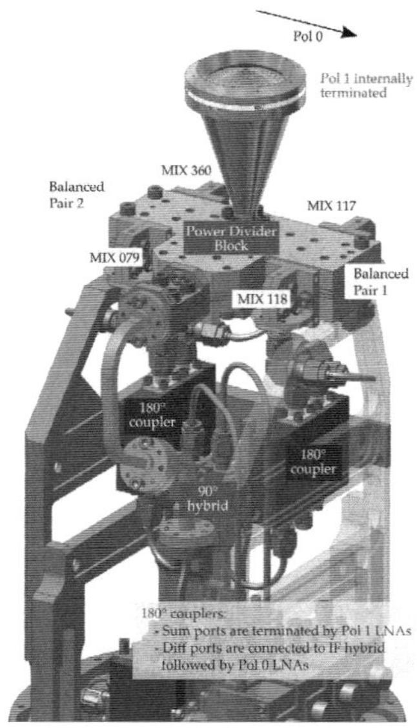


Fig. 1. Prototype cartridge for ALMA Band 3 arranged into a balanced and sideband-separating (BAL-2SB) receiver. Details of the power divider block are shown in Fig. 2.

Normally, the sum port of the 180° IF coupler would be terminated by a 50-Ohm termination, but here each sum port was instead connected to an LNA to provide the DC bias.

The power division and LO coupling was accomplished through the power divider block, as shown in Fig. 2. A turnstile has been used for both polarisation discrimination and the first power division required for sideband-separation. The LO and RF for each balanced pair are combined in-quadrature using 3 dB 90° hole couplers.

Fig. 2 (b) shows the 4-piece power divider block that used a platelet approach, instead of split-block, which enabled broad-wall hole couplers. Only one polarization was used from the turnstile and the other was internally terminated with waveguide loads. The LO was divided in-phase using a 6-port hole coupler, followed by a pair of phase shifters acting on the upper and lower LO paths to pump each balanced mixer pair.

This paragraph of the first footnote will contain the date on which you submitted your paper for review.

This material is based upon work supported by Associated Universities, Inc./National Radio Astronomy Observatory and the National Science Foundation under Cooperative Agreement No. AST-151926.

D. Henke is with NRC Herzberg Astronomy and Astrophysics Research Centre, V9E 2E7, Victoria, BC, Canada (e-mail: doug.henke@nrc-cnrc.gc.ca).

P. Niranjanan is with NRC Herzberg Astronomy and Astrophysics Research Centre, V9E 2E7, Victoria, BC, Canada (e-mail: pat.niranjanan@nrc-cnrc.gc.ca).

L. Knee is with NRC Herzberg Astronomy and Astrophysics Research Centre, V9E 2E7, Victoria, BC, Canada (e-mail: lewis.knee@nrc-cnrc.gc.ca).

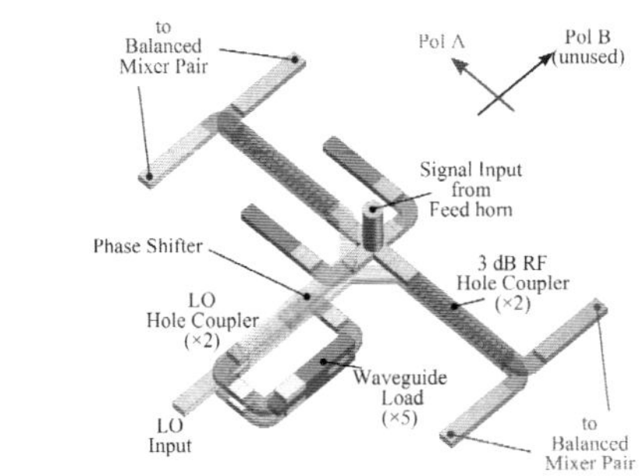


(a)

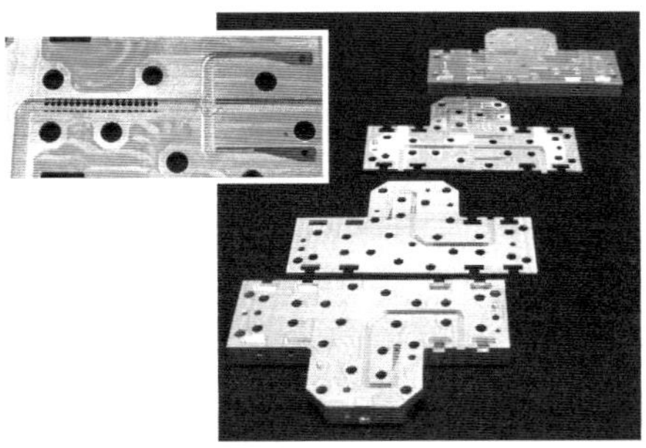

(b)

Fig. 2. Power divider block used within the balanced sideband-separating receiver (BAL-2SB). A CAD model of the waveguide channels is shown in (a) and the 4-layer machined block is shown in (b). A close-up of the turnstile and one of the 3 dB 90° hole couplers is shown inset.

Measurements of the image rejection and noise are shown in Fig. 3. Of first note, the image rejection is acceptable (i.e., better than 10 dB) and indicates that overall RF signal and LO balance has been attained. In the measured noise, there is a significant drop due to reductions in the LO coupled noise (-30 dB coupling is used within the 6-port coupler shown in Fig. 2) and image termination noise mentioned above. However, there is noticeable ripple in the middle of the band. Upon closer inspection, the noise profiles in the LSB and USB track each other indicating more of a double-sideband response rather than sideband-separated.

The apparent contradiction may be explained by considering the testing condition for sideband-separation versus that for noise. The sideband interferer was injected towards the cartridge window using an open-ended rectangular waveguide that induced a TE11 mode into the circular guide. Because of the symmetry of the power division block, higher-order modes within the circular waveguide and turnstile were not excited. However, when the noise temperature was measured using black-body calibration loads, we can assume that the polarization from each black-body load was random and energy could be transduced within the circular guide in any of the supported higher-order modes. If higher-modes are transduced they will corrupt the phase balance. This can give rise to the situation where the image-rejection appears fine from measurement, yet receiver noise measurements have degraded. We plan to test this hypothesis by implementing a linear polarization filter to see if the measured noise will change.

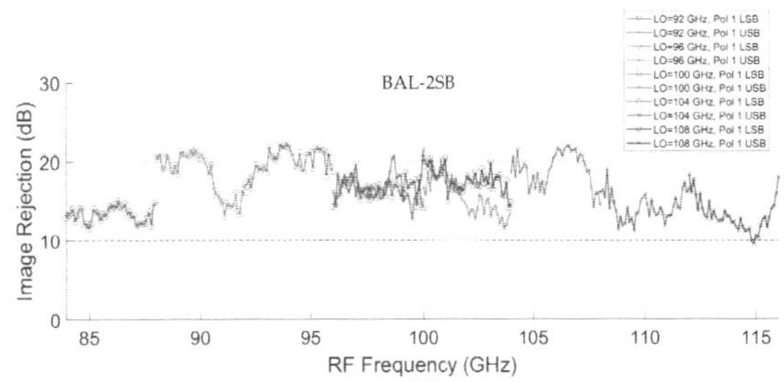


(a)

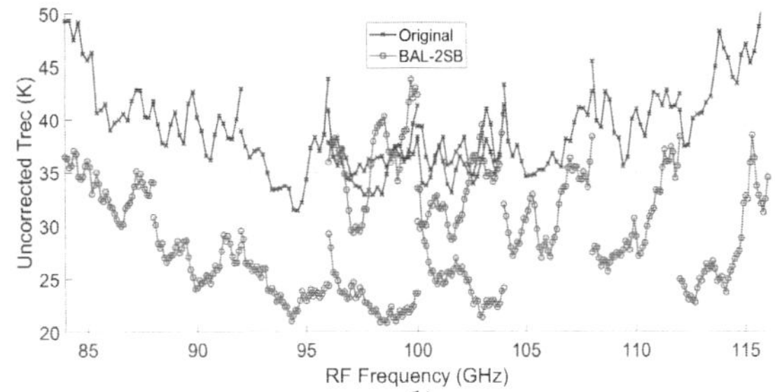


(b)

Fig. 3. Cartridge measurements of (a) image-rejection and (b) narrow-band receiver noise. In (b), the noise is compared against the original Band 3 configuration.

Integrating the turnstile as an OMT and power divider for sideband-separation was shown to be problematic and it would have been better to keep the OMT separate. Note that the same optics, mixer blocks, IF chains, and operating conditions were used throughout to allow for comparison with the original configuration.

## Acknowledgment

The authors would like to thank I. Wevers for mechanical integration and packaging; and F. Miranda and N. Kelly for precision machining. Thanks to Mike Romero of KRYTAR, Inc., who provided us with additional 180° couplers for cryogenic testing.

## References

[1] A. R. Kerr, J. E. Effland, A. W. Lichtenberger, J. G. Mangum, "Towards a Second Generation SIS Receiver for ALMA Band 6", ALMA NA Development Proposal, 2016. [Online]. Available: https://science.nrao.edu/facilities/alma/alma-develop-old-022217/2nd%20Gen%20Band%206%20Rcvr/at_download/file.

[2] A. R. Kerr and S.-K. Pan, "Design of planar image-separating and balanced SIS mixers," in *Proc. 7th Int. Symp. Space Terahertz Technol.*, Mar. 12–14, 1996, pp. 207–219.

[3] S. Claude, C. Cunningham, A. R. Kerr, and S.-K. Pan, "Design of a Sideband-Separating Balanced SIS Mixer Based on Waveguide Hybrids," ALMA Memo 316, Sept. 2000.

[4] Y. Serizawa et al., "Development of a 385–500 GHz sideband-separating balanced SIS mixer," *J. Infrared, Millimeter THz Waves*, vol. 33, pp. 999–1017, 2012.

[5] A. R. Kerr, S.-K. Pan, S. M. X. Claude, P. Dindo, A. W. Lichtenberger, J. E. Effland, and E. F. Lauria, "Development of the ALMA Band-3 and Band-6 Sideband-Separating SIS Mixers," *IEEE Trans. THz Sci. Technol.*, vol. 4, no. 2, pp. 201–212, Mar. 2014.

[6] S. A. Maas, *Microwave Mixers*, 2nd ed. Boston, MA, USA: Artech House, 1993, pp. 237–288.

# Barrier Reduction and Sub-gap Leakage in Niobium Based SIS Junctions

M. Kroug[1], T. Noguchi[2]

Superconductor-insulator-superconductor (SIS) tunnel junctions based on Nb/Al-$AlO_x$/Nb tri-layers (TL) [1] are the standard technology in superconducting electronics. This is owed to the development of a reliable fabrication process that produces junctions with current densities up to $j_c$ = 10 – 15 kA/cm$^2$ and low sub-gap leakage. Nb/Al-$AlN_x$/Nb TLs offer to make junctions with current densities as high as $j_c$ ~ 70 kA/cm$^2$ while maintaining low sub-gap leakage [2] but its use has not yet become wide spread. The deposition process for both TLs follows the same approach, with one of the merits being that a high quality insulator can be formed on the thin Al film that covers the Nb base electrode. Junctions made with either barrier type exhibit qualitatively the same behavior: Sub-gap leakage increases with current density. This is commonly seen as a signature of non-uniformity as the barrier's average thickness decreases.

Since the invention of Nb SIS junction technology in the early 1980s, it has been known that TLs with a second thin Al film near the barrier, e.g. Nb/Al-$AlO_x$/Al/Nb, thus making it a symmetric layer stack, produces junctions with higher resistance *and* lower subgap leakage [1]. Obtained with relatively low current density TLs, $j_c$ ~ 1 kA/cm$^2$, the results are attributed to the protective function of the Al layer, preventing a chemical reaction between niobium counter electrode (CE) and barrier. This poses the general question whether barrier reduction due to interface chemistry and junction quality are correlated and more specifically whether the choice of a single Nb layer as the CE limits the possibility to realize high quality junctions with current densities $j_c$ > 10 kA/cm$^2$.

A systematic study to address this issue is ongoing and to that end, we have fabricated junctions based on SIS TLs with Nb/Al base electrodes, $AlN_x$ barriers formed by nitridation, and *three different* CEs: single layer Nb, bilayer Al/Nb, and single layer NbN [3]. Plasma conditions during the nitridation process are identical and time $t$ is varied to realize a range of barrier transparencies. Preliminary results confirm observations made in [1]: A Nb CE appears to reduce the $AlN_x$ barrier resulting in substantially higher current densities than Al/Nb or NbN CEs, see Fig. 1. As for the subgap leakage levels, our results also confirm the general trend of increased sub-gap leakage towards higher current densities. However, a direct comparison between junctions with different CEs is meaningful only if their current densities are similar. To complete the set of data, the next step is to grow asymmetric Nb/Al-$AlN_x$/Nb TLs using longer nitridation times that yield current densities in the regime $j_c$ = 20 – 50 kA/cm$^2$. Samples based on asymmetric Nb/Al-$AlO_x$/Nb and symmetric Nb/Al-$AlO_x$/Al/Nb TLs are also in preparation. We believe that this study sheds light on the impact the CE material has on barrier performance, especially in high transparency TLs.

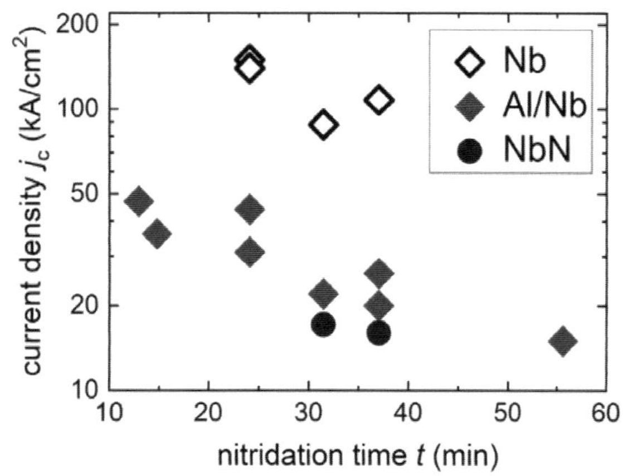


Fig. 1. Current density $j_c$ vs. nitridation time $t$ for $AlN_x$ barrier based TLs with identical base electrode Nb/Al but different counter electrodes (CE). For the same nitridation time a Nb CE (◇) yields a much higher $j_c$ compared to Al/Nb (◆) or NbN (●) CEs.

## References

[1] M. Gurvitch, M. A. Washington, and H. A. Huggins, "Nb Tunnel Junctions with Thin Layers of Al Near the Barrier," *IEDM*, vol. 81, pp. 115-117, 1981.

[2] T. Zijlstra, C. F. J. Lodewijk, N. Vercruyssen, F. D. Tichelaar, D. N. Loudkov, and T. M. Klapwijk, "Epitaxial aluminum nitride tunnel barriers grown by nitridation with a plasma source," *Applied Physics Letters*, vol. 91, no. 23, Dec 3 2007.

[3] M. Y. Torgashin, V. P. Koshelets, P. N. Dmitriev, A. B. Ermakov, L. V. Filippenko, and P. A. Yagoubov, "Superconducting integrated receiver based on Nb-AlN-NbN-Nb circuits," *IEEE Transactions on Applied Superconductivity*, vol. 17, no. 2, pp. 379-382, Jun 2007.

[1] National Astronomical Observatory of Japan, 181-8588 Mitaka
[2] Tokyo University of Electro Communications, 182-8585 Chofu

NOTES:

# Array of Multichroic Double-Slot Antennas with Cold-Electron Bolometers for the 220/240 GHz channels of the LSPE Instrument

L.S. Kuzmin, A.V. Chiginev, D.A. Pimanov, A.V. Gordeeva, S. Masi and P. de Bernardis

***Abstract*—The present paper tells about developing narrowband antennas for frequencies 220 and 240 GHz, which will be used in the LSPE balloon telescope to estimate foreground cosmic dust. For thisgoal, we used double-slot antenna with coplanar lines and Cold-Electron bolometers. The simulations show that we have achieved 5% bandwidth for 220 GHz and 5.5% bandwidth for 240 GHz for the single cell. There are also results for an array of unit cells using double-slot antennas and coplanar lines with CEBs. Besides that, the paper addresses the question of the use of SQUID readout for cold-electron bolometers.**

***Index Terms*—CMB, LSPE, B-mode, SQUID, Cold-Electron Bolometer, coplanar lines, double-slot antenna, narrowband.**

## I. Introduction

LSPE is project of a balloon-borne telescope, which is developed for observation of the circular B-mode of CMB [1]. In particular, in this project it is planned to make two frequency channels on 220 and 240 GHz, which are needed to remove the influence of the cosmic dust. The special requirement for these channels is a very narrow bandwidth, which should be equal to 5% of the operating frequency.

The whole construction of the receiving system implies the use of a back-to-back horn, which acts as an element responsible for radiation diagram of the receiver. We suggest placing a planar array of the cold-electron bolometers (CEB) [2-5] and slot antennas under the back of the horn; so that the radiation diagrams of the slot antennas themselves will not affect beam characteristics of the receiver.

The best way for accurate comparison of two neigbour narrowband signals is to develop a multichroic pixel with nanofilter on´chip. The first realization of nanofilter was invention of CEB with kinetic inductance of NbN strip [6]. The resonance was realized by capacitance of SIN tunnel junction and the kinetic inductance of NbN strip embedded into a cross slot antenna. The solution of this problem is based on the use of the kinetic inductance of the superconductor, allowing to reduce linear size of such an inductor up to 300 times compared to the geometrical inductor of the same inductance. This strip is unified with CEB and called resonant cold-electron bolometer (RCEB).

The RCEB for multichroic pixels with an internal resonance by a kinetic inductance of the NbN nanostrip and a capacitance of the SIN tunnel junctions has been realized using a single Lambda slot with two RCEBs for 75 and 105 GHz [7]. However, the fabrication of kinetic inductance meets serious technological difficulties.

In the present paper, we suggest an alternative solution for creating multichroic system using internal resonance of slot antenna and capacitance of CEB.

## II. Single cell for the double-slot antenna with folded coplanar lines and CEBs

The problem of creation of narrowband antenna is not so trivial, it requires some specific methods.

As an inductive element of a resonant circuit for matching with the capacitance of CEB, we suggest the use of the reactance of the "slot antenna + coplanar line" system. This approach implies that the imaginary part of the impedance of this reactance is positive at the operation frequency, which corresponds to the inductive character of such a system.

As the prototype of the receiving system we have chosen a COrE two-frequency single pixel at 75 and 105 GHz. This pixel is based on the two-frequency seashell antenna, which consists of pair of slot antennas connected with CEB by coplanar lines [8-9].

However, for this antenna, we chose double-slot design with central feeding, which is done by coplanar lines, they also act as waveguides. These coplanar lines are connected to a radiation detector with low resistance, which should provide the

The work is supported by the Ministry of Education and Science of Russian Federation (Project 16.2562.2017/PCh).

L.S. Kuzmin is with Chalmers University of Technology, Gothenburg SE-41296, Sweden and also with Nizhny Novgorod State Technical University n.a. R.E. Alekseev, Minin St., 24, Nizhny Novgorod 603950, Russia (e-mail: kuzmin@chalmers.se).

A.V. Chiginev is with Nizhny Novgorod State Technical University n.a. R.E. Alekseev, Minin St., 24, Nizhny Novgorod 603950, Russia and also with Institute for Physics of Microstructures, Russian Academy of Sciences, GSP-105, Nizhny Novgorod, 603950, Russia (e-mail: chig@ipmras.ru).

D.A. Pimanov is with Nizhny Novgorod State Technical University n.a. R.E. Alekseev, Minin St., 24, Nizhny Novgorod 603950, Russia (e-mail: macpimanov@gmail.com).

A.V. Gordeeva is with Nizhny Novgorod State Technical University n.a. R.E. Alekseev, Minin St., 24, Nizhny Novgorod 603950, Russia (e-mail: a.gordeeva@nntu.ru).

S. Masi is with Dipartimento di Fisica, Universit'a La Sapienza, P. le A. Moro 2, I-00185 Roma, Italy (e-mail: silvia.masi@roma1.infn.it)

P. de Bernardis is with Dipartimento di Fisica, Universit'a La Sapienza, P. le A. Moro 2, I-00185 Roma, Italy (e-mail: paolo.debernardis@roma1.infn.it)

required frequency characteristics for narrowband antenna. As a detector of microwave radiation, we suggest to use a CEB. It has many advantages compared with other types of bolometers. The most important of them are their unique sensitivity, wide dynamic range due to electron cooling, and the insensitivity to cosmic rays [5]. A schematic view of CEB and its energy diagram are shown in Fig. 1a and Fig. 1b respectively.

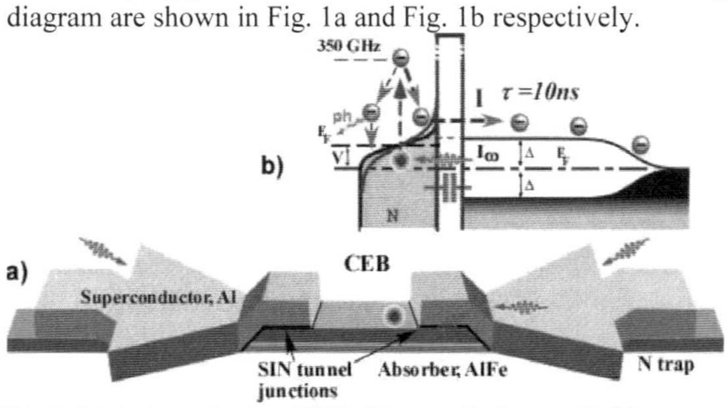


Fig. 1. (a) A schematic view of CEB; (b) energetic diagram of CEB.

Also, this antenna will be used with a back-to-back feedhorn. It makes the radiation diagrams of antenna slots not affecting the beam characteristics at the receiver. The design of feedhorn is shown in Fig. 2a.

The design for single cell of double-slot antenna is presented in Fig. 2b. The antenna consists of two resonant slots and a coplanar line, which is folded to acquire better frequency band characteristics. The coplanar lines length for each frequency was set to get the needed working frequency. The distance between antenna slots was chosen after a number of simulations in CST Microwave Studio. In total, these parameters let us obtain the required frequency characteristics. CEB is put into the gap in the central wire of the coplanar line. The position of CEB is shown as a red triangle in Fig. 2b.

For CEB, the DC biasing is provided through coplanar lines by DC connectors, which are located under the ground plane. Blue triangles in Fig. 2b show the positions of DC connectors. In numerical simulations, the influence of the DC connectors is modeled by lumped capacitances, which are placed between the ends of coplanar lines and the ground plane. These capacitances are set to 350 fF, which is the estimated value of DC connectors' capacitance.

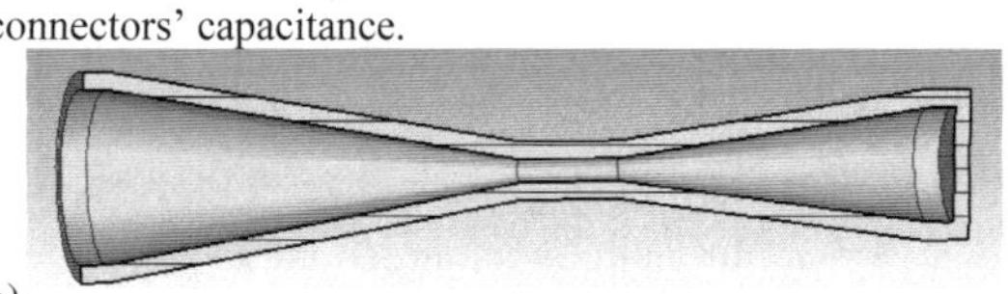

a)

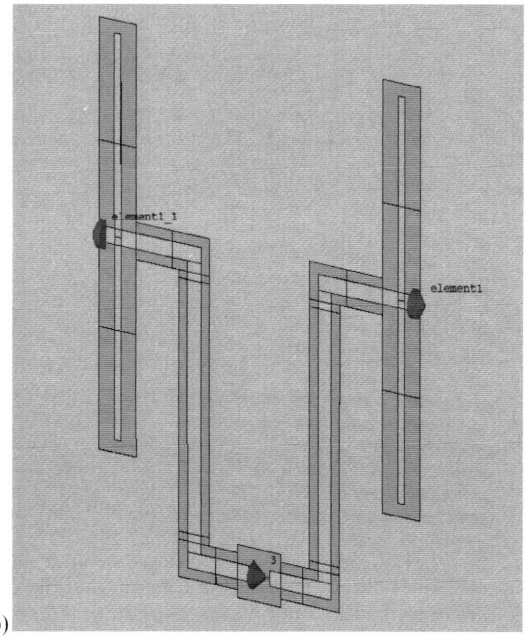


b)

Fig. 2. (a) Back-to-back feedhorn used for the antenna system; (b) single cell of double-slot antenna design overview in CST Microwave Studio.

Fig. 2b shows the single cell for antenna for 220 GHz. There is only one difference in antenna design for 240 GHz, the total length of central coplanar line; it is 60 um shorter. The calculations have been performed using frequency domain solver. In electrodynamic part of simulations, these designs have shown the following values of the impedance: ReZ (220 GHz) is 6.8 Ohm, ReZ (240 GHz) is 7.7 Ohm; ImZ (220 GHz) is 36.3 Ohm, ImZ (240 GHz) is 32.9 Ohm. We will need these values for getting the required resonances at needed operating frequencies in the schematic part of the simulations.

The equivalent circuit of the antenna connected to a CEB is presented in Fig. 3. The values of $R_{abs}$ and $C_{SIN}$ are chosen in this way: $R_{abs}=ReZ(f_0)$, $C_{SIN}=(2\pi f_0 \cdot ImZ(f_0))^{-1}$; here Z(f) is the diagonal component of the Z-matrix calculated in electrodynamic part (see above). The $C_{sin}$ values should make the total capacitances of $C_{sin}$ and $C_e$ form the series resonance with $L_e$ at frequencies 220 and 240 GHz respectively. This method of parameters choice is correct only when $ImZ(f_0)>0$ and when $|S_{mn}(f_0)|<<1$, where $S_{mn}$ are the non-diagonal components of the S-parameters matrix which is calculated in electrodynamics.

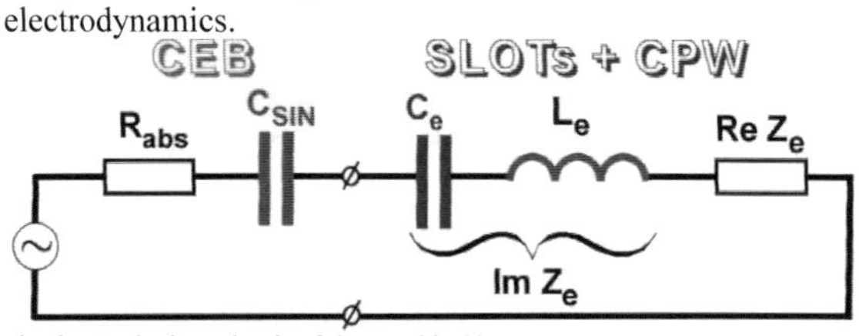


Fig. 3. Equivalent circuit of the Double-Slot antenna connected to a CEB by coplanar lines.

Fig. 4 shows the frequency characteristics of the single cell with CEB after calculations in schematics. The parameters used in schematics are following. For 220 GHz antenna, $R_{abs}$ is 6.8

Ohm, and for 240 GHz antenna, $R_{abs}$ is 7.7 Ohm. $C_{SIN}$ for 220 GHz is 19.4 fF, and for 240 GHz it is 23.4 fF. These values are set to obtain the best impedance matching at operating frequencies.

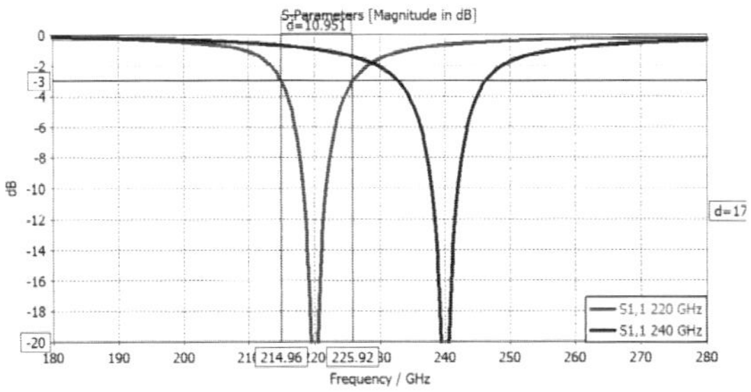


Fig. 4. Frequency characteristics of double-slot antennas with CEBs tuned to 220 GHz (red) and 240 GHz (blue) in CST MWS schematic.

As one can see in Fig. 4, the bandwidths at -3dB are even less than 11 GHz for 220 GHz and about 13.5 GHz for 240 GHz. Bandwidth for 220 GHz meets the requirements for the receiving system of LSPE well; it is 5% from operating frequency. Bandwidth for 240 GHz nearly meets the requirements for this receiving system; it is about 5.5% from operating frequency.

## III. MULTI-CELL DOUBLE-SLOT ANTENNA WITH COPLANAR LINES AND CEBS

In this paragraph, we consider an array of slot antennas, which are connected in series by coplanar waveguides (CPW). The main advantage of this model is a simple single-layer technology, which means that the DC connections between CEBs are made through the same CPWs that connect the antenna cells in AC. Thus, there is no need to use separate DC connectors to feed the cells with DC bias. This solution to a greater extent simplifies the technology, but restricts the applicability of the array to the case of the current biasing of CEBs.

The single-layer slot antenna array is shown in Fig. 5. It has 45 slots for 220 GHz channel and 45 slots for 240 GHz channel on a single 3*3 $mm^2$ substrate. At this approach, the dimensions of all parts of the unit cell for 240 GHz channel are obtained from the same ones, which are set on 220 GHz channel cell by simple scaling. On the opposite side of the substrate, there is a special waveguide port, which is used as a source of the electromagnetic radiation in the numerical modelling. In the present computations, we work with the mode $E_{01}$.

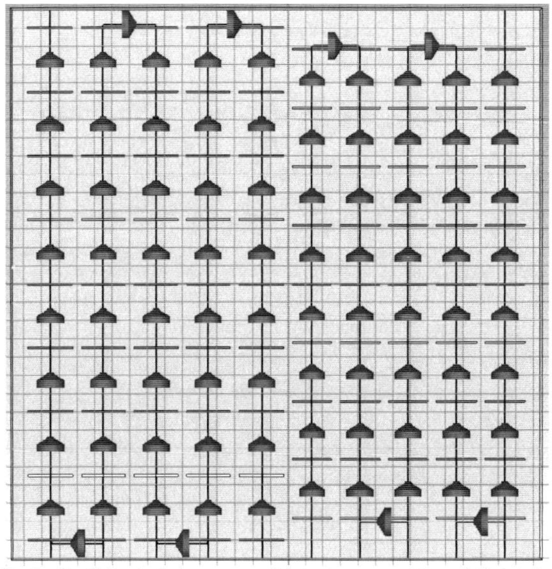

Fig. 5. The finite single-layer array of slot antennas consisting of 45 slots for 220 GHz channel (left part) and 45 slots for 240 GHz channel (right part). The blue and small red triangles denote capacitance and absorber resistance of CEB, respectively. The size of the array is 3*3 $mm^2$.

Figs. 6,7 show the power, which is accepted by all absorbers of the CEB at one frequency channel. In Fig. 6, one sees the curves for two values of absorber resistance – 10 Ohm and 17 Ohm. As we expected, the resonant curves do not shift when changing the absorber resistivity. At higher values of R the total absorption efficiency is getting higher, but the FWHM of the resonances also increases (see Table I). The decrease of R leads to corresponding decrease of the FWHM, but the absorption efficiency becomes lower.

Fig. 7 shows the accepted power for different values of SIN capacitances. We see that greater values of C lead to increase in absorption efficiency, but obviously shift the resonance frequencies towards lower values. The variations of C have only small influence on FWHM (Table I).

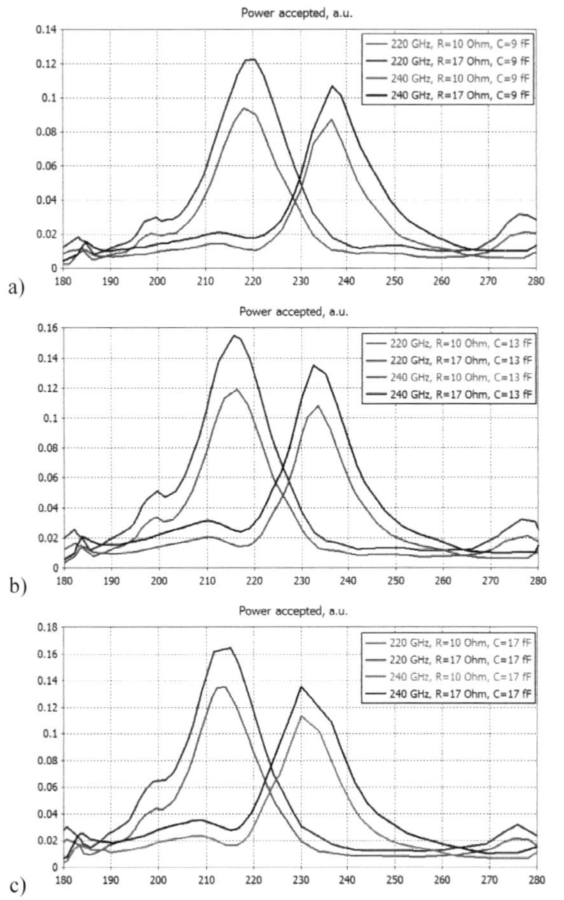


Fig. 6. Power accepted by all CEBs' absorbers, vs frequency, at a fixed CEB capacitance: (a) C=9 fF, (b) C=13 fF, (c) C=17 fF.

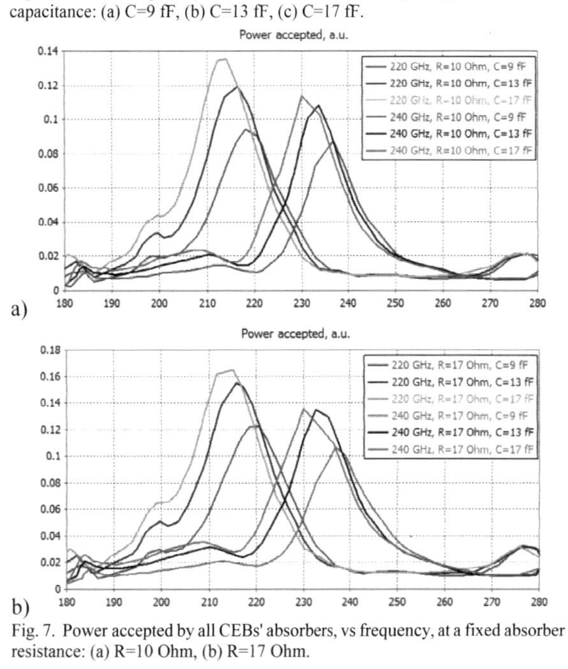


Fig. 7. Power accepted by all CEBs' absorbers, vs frequency, at a fixed absorber resistance: (a) R=10 Ohm, (b) R=17 Ohm.

TABLE I

WIDTHS OF THE RESONANT CURVES IN FIGS. 6, 7

| | | 220 GHz channel | | | 240 GHz channel | | |
|---|---|---|---|---|---|---|---|
| $R_0$, Ohm | $C_0$, fF | $F_{max}$, GHz | FWHM GHz | FWHM % | $F_{max}$, GHz | FWHM GHz | FWHM % |
| 17 | 9 | 219.59 | 18.19 | 8.28 | 236.99 | 17.54 | 7.4 |
| | 13 | 215.66 | 17.55 | 8.14 | 232.56 | 15.89 | 6.83 |
| | 17 | 215.2 | 18.3 | 8.5 | 230 | 18.4 | 8 |
| 10 | 9 | 218.14 | 16.59 | 7.61 | 236.77 | 13.97 | 5.9 |
| | 13 | 216.38 | 16.32 | 7.54 | 233.64 | 14.15 | 6.06 |
| | 17 | 213.28 | 16.4 | 7.69 | 231.67 | 16.1 | 6.95 |

Table I shows bandwidths for 220 and 240 GHz frequency channels. These bandwidth values are close to the requirements of 5% from central frequency, but the system needs further improvements.

## IV. MATCHING OF COLD-ELECTRON BOLOMETERS WITH SQUID READ-OUT

The integration of cold-electron bolometers with the multiplexing system is an important step towards the use of CEB as detectors in modern telescopes. Multiplexing can significantly simplify the readout system, which is a significant advantage for projects with a large number of detectors. Currently, the most developed systems for multiplexing are based on SQUIDs for TES bolometers, and their use also for CEB seems to be a logical development. Therefore, in this section, we use the sensitivity of SQUIDs from TES multiplexing systems to calculate the total sensitivity of the CEB with these systems.

In Fig. 8, we present the different components of noise equivalent power (NEP) of the bolometer array, composed from 45 elementary cells from Fig. 2b. There are two NEP components coming from the bolometers itself: the electron-phonon noise ($NEP_{e\text{-}ph}$) and the current and heat-flow noise through SIN junctions ($NEP_{SIN}$), one component comes from a read-out system ($NEP_{SQUID}$) and one component comes from the incoming power (photon noise $NEP_{ph}$). The sum of the first three components is defined as bolometer NEP. The sensitivity of the SQUID array is taken from [10] and equals 3.5 pA/Hz$^{1/2}$.

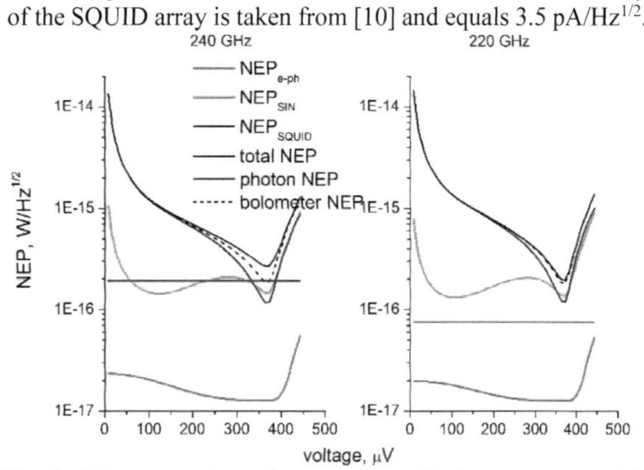


Fig. 8. NEP versus voltage of an array from 45 bolometers, connected in parallel, with SQUID read-out for 240 and 220 GHz channels.

Two channels 220 and 240 GHz have different power load 6 and 20 pW correspondingly. This power is divided between 45 bolometers in each channel.
One can see that for both channels 220 and 240 GHz there are two major NEP components: $NEP_{SIN}$ and $NEP_{SQUID}$ - noise of read-out electronics, which limit the sensitivity of the detector. The operation point in voltage bias mode for SQUID read-out is situated near the gap at 370 μV, where the resistance of the bolometer array is just 5-6 Ohms. The normal resistance of a single bolometer is 600 Ohms.
The bolometer noise is less than the photon noise for 240 GHz channel at the bias point, which means the photon noise limited operation. In order to reach this mode for 220 GHz channel one can increase the SQUID sensitivity and decrease the SIN noise by decreasing the number of bolometers in the array for this channel.
Unlike TES, CEBs can receive radiation in a wide power range without experiencing saturation problems. To illustrate this, we calculated the sensitivity depending on the received power in the range from 2 to 30 pW for the 240 GHz channel (see Fig. 9).
On this graph, one can see the level of the dark noise - the bolometer noise without power load or with small power loads. For this array it is $1.2*10^{-16}$ W/Hz$^{1/2}$. With increase of the power, the $NEP_{ph}$ increases and at 13 pW becomes higher than the bolometer NEP. At this power, the photon noise limited operation starts and continues much above 30 pW, not shown here.

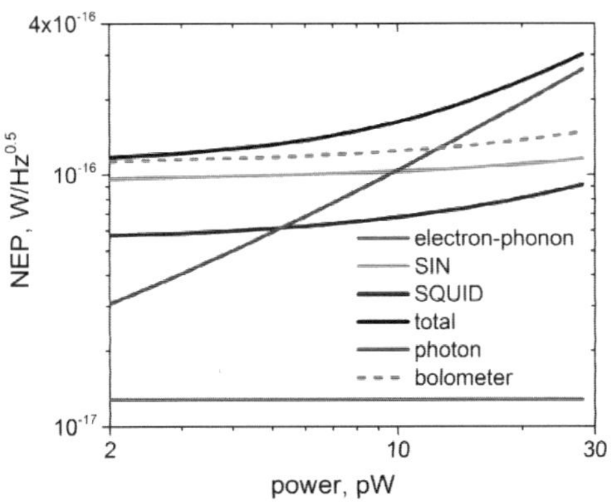


Fig. 9. NEP versus absorbed power of an array from 45 bolometers, connected in parallel, with SQUID read-out for 240 GHz channel.

Let us summarize the results of this section. The ration of the photon NEP to the bolometer NEP is $7.6*10^{-17} / 1.2*10^{-16} = 0.6$ for 220 GHz channel and $1.8*10^{-17} / 1.7*10^{-16} = 1.0$ for 240 GHz channel.

## V. Conclusions

The results, acquired from numeric modelling, show that we have developed efficient designs of single cell for a double-slot antenna with folded coplanar lines, connected to CEB, for production. The results for modeled single cell for 220 GHz meet the requirements for LSPE project well. There is already going further research to create the voltage biased array of these cells on a single substrate. The bandwidth of the single cell for 240 GHz slightly exceeds the requirements for LSPE project. So this design should be improved in terms of characteristics and used in voltage biased array too.

Also, a lot of modeling has been done for the current biased array of a various number of unit cells made of two slots with straight coplanar lines, connected to CEB. From the results, we can see that the bandwidths for this approach exceed the bandwidth requirements of 5% of operating frequencies. But they get close to the needed values. The minimal reached bandwidth at 220 GHz is 7.5% of central frequency, at 240 GHz it is 5.9%. Further research will be aimed at improving these results to meet the LSPE project requirements and to construct a single-layer antenna array with parallel connections for use with a SQUID readout.

Besides, some research was done for matching of CEBs with SQUID read-out where the possibility to use such read-out for cold electron bolometers is checked. The ration of the photon NEP to the bolometer NEP is 0.6 for 220 GHz channel and 1.0 for 240 GHz channel Also, we can say that CEBs are photon noise limited, because the bolometer noise is less than the photon noise.

## Acknowledgements

The authors would like to thank L. Lamagna, A. Sobolev and A. Pankratov for useful discussions.

## References

[1] The LSPE collaboration 2012 *Proc. SPIE* 8446, 84467A.
[2] Kuzmin L 2002 In: Pekola J., Ruggiero B., Silvestrini P. (eds) International Workshop On Superconducting Nano-Electronics Devices, Springer, Boston, MA, 145–154; Kuzmin L and Golubev D 2002 Physica C 372-376 378
[3] Kuzmin L 2008 *Journal of Physics: Conference Series* 97 012310
[4] Tarasov M A, Kuzmin L S, Edelman V S, Mahashabde S, de Bernardis P 2011 *IEEE Trans. on Appl. Supercond.* 21 3635.
[5] Salatino M, de Bernardis P, Kuzmin L S, Mahashabde S., Masi S. 2014 *J. Low Temp. Phys.* 176 323
[6] Kuzmin L S 2014, IEEE Trans. on THz Sci. Technol. 4 314-320
[7] Kuzmin L S, Mukhin A S, and Chiginev A V, 2018 IEEE Trans. on Appl. Supercond. 28 2400304
[8] Kuzmin L S, Chiginev A V, Matrozova E A, Sobolev A S, 2016 IEEE Trans. on Appl. Supercond., 26 2300206
[9] L.S. Kuzmin, A.V. Blagodatkin, A.S. Mukhin, D.A. Pimanov, V.O. Zbrozhek, A.V. Gordeeva, A.L. Pankratov, A.V. Chiginev (2019), Superconductor Science and Technology, Volume 32, 035009.
[10] Vaccaro D, Baldini A M, Cei F, Galli L, Grassi M, Nicolo D, Piendibene M, Tartari A, Spinella F, Signorelli G. "A frequency domain multiplexing system to readout the TES bolometers on the LSPE/SWIPE experiment," 2018 Nuclear Inst. and Methods in Physics Research, A936 169-171

# Specific capacitance of Nb/Al-AlN/Nb superconducting tunnel junctions

Alexey Pavolotsky[1*], Cristian D. López[1], Isabel Vrethed Tidekrans[1], Denis Meledin[1], Vincent Desmaris[1], Victor Belitsky[1]

*Abstract*— **Modern radio astronomy demands for broadband receiver systems. For SIS mixers, this translates into objective to employ superconducting tunnel junctions with a very low $R_nA$ and low specific capacitance. The traditionally used Nb/AlOx/Nb junctions have largely approached their physical limit of minimizing those parameters. It is commonly recognized that it is AlN-barrier junctions, which are needed for further progressing of the broadband SIS mixer instrumentation for radio astronomy. In this work, we present the progress in fabrication of high quality Nb/Al-AlN/Nb superconducting tunnel (SIS) junctions and their characterization in terms DC electric properties junctions' specific capacitance.**

## I. Introduction

Modern radio astronomy instrumentation projects call for twice or triple enhancement of the RF and IF bandwidths of the SIS mixers [1]. A wider RF band requires lower Q-factor and consequently, lower $R_nA$-product of the SIS junctions. The traditionally used Al-oxide material of SIS junction tunnel barrier approaches its physical limit: by using SIS junctions with higher current density and thus thinner tunnel layer that gets close to the situation when the tunnel barrier quality becomes hardly predictable. It is widely recognized by the SIS community that in general, the quality of the Al-oxide SIS junctions degrades and becomes unreliable, once $R_nA$ decreases below probably 15 Ohm·$\mu m^2$. Simultaneously, a wider IF band of SIS mixer needs the junction capacitance be lower, in turn forcing shrinking the junction size, which may cause production yield problems.

As an alternative SIS tunnel barrier material, aluminum nitride, AlN, has reduced electrical barrier height as compared to the Al-oxide tunnel barrier and thus would need a thicker tunnel barrier for the same current density as compared to the $AlO_x$ tunnel barrier junctions. Consequently, with physically thicker tunnel barrier, there is more possibility to obtain $R_nA$ well below 10.

We have earlier reported on the process development for high-quality junction fabrication based on microwave plasma nitridation [2]. We show that the Nb/Al-AlN/Nb junctions with $R_nA$ product down to ~5 Ohm·$\mu m^2$ demonstrate excellent quality. Moreover, even junctions with $R_nA$~3 Ohm·$\mu m^2$ can be fabricated having $R_j/R_n > 12$. Also, we show that the produced junctions are quite stable against the thermal annealing, at least up to 200°C, thus allowing for thermal impact during almost any possible fabricating or packaging technology processes.

In this manuscript, we present results of the Nb/Al-AlN/Nb junction specific capacitance measurements following the approach similar to reported in the paper [3]. The measurement result confirms that specific capacitance of the Nb/Al-AlN/Nb junction is noticeably lower than that reported for the Nb/$AlO_x$/Nb junctions [4], [5].

## II. Nb/Al-AlN/Nb Junctions Fabrication

The developed process [2] for fabricating of Nb/Al-AlN/Nb junctions is based on the Nb/Al-$AlO_x$/Nb process supported by GARD [6]–[10] but instead of thermal oxidation, nitridation of Al with the plasma excited by electron-cyclotron resonance (ECR) plasma source [11] is applied.

A number of batches of Nb/Al-AlN/Nb junctions had been fabricated with a demonstrated range of $R_nA$ product varying between 3 to 120 Ohm·$\mu m^2$, all with a low subgap current. The examples of the junctions' current-voltage characteristics are presented at the Fig. 1. All the measured junction in the fabricated batches showed uniform quality independently on the junction size (between 2 to 8 $\mu m^2$).

During the fabricating process and in the course of mounting/packaging, the SIS mixer chip can be exposed to the elevated temperatures. How high temperatures can be accepted during junction fabricating (baking of resists, or heating during deposition of the layers) and packaging (curing of glues or epoxies, or heating for wire bonding) is defined by the stability of the junction properties at the elevated temperatures.

To make sure that the fabricated Nb/Al-AlN/Nb junctions safely survive elevated temperatures during fabrication and handling, we have earlier carried out the study of aging and annealing behavior of Nb/AlN/Nb junctions [2]. In course of that study, the junction wafers were exposed to the aging/annealing temperature profile between room temperature and 200°C, as shown at the Fig. 2. Current-voltage characteristics at 4K temperature were recorded directly after the fabrication of the junctions and after each step of the aging/annealing temperature profile showed only minor variation of the junctions' parameters (normal resistance ($R_n$) and quality factor ($R_j/R_n$ ratio), superconducting gap ($V_g$) and its width ($\Delta V_g$)). Example of the current-voltage characteristics of the junctions experienced the whole aging and annealing sequence up to 200°C is shown on the Fig. 3.

[1] Chalmers University of Technology, Group for Advanced Receiver Development, Department of Space, Earth and Environment, Göteborg, 412 96, Sweden

[*] Corresponding author, email: Alexey.Pavolotsky@chalmers.se

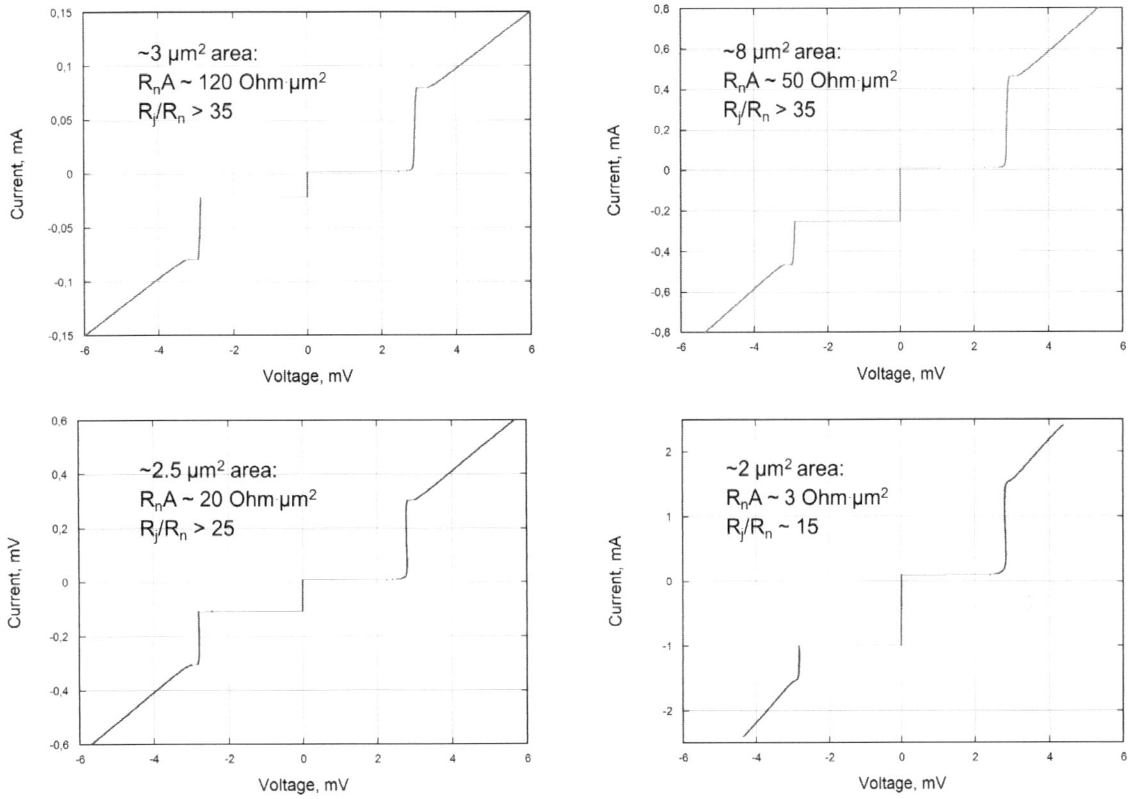


Fig. 1. Examples of Nb/Al-AlN/Nb junctions' current voltage characteristics with $R_nA$ ranging between 3 and 120 $\Omega\,\mu m^2$. The legend inside each plot panel shows junction's size, $R_nA$ and $R_j/R_n$ values.

It can be concluded that the Nb/AlN/Nb junctions are probably somewhat more temperature stable than high-quality Nb/$AlO_x$/Nb junctions (see e.g. for comparison the evolution of Nb/$AlO_x$/Nb junctions due to aging and annealing summarized in [6] vs. that for Nb/AlN/Nb junctions [2]). That is consistent also with the earlier reported results, e.g. [12].

Further post-annealing aging of the Nb/AlN/Nb junctions at room temperature for ca. 8 months after their exposure to annealing experiments shown on the Fig. 2 demonstrated only minor (<5%) drop of $R_n$ and no measurable change of $R_j/R_n$, $\Delta V_g$ and $\Delta V_g$.

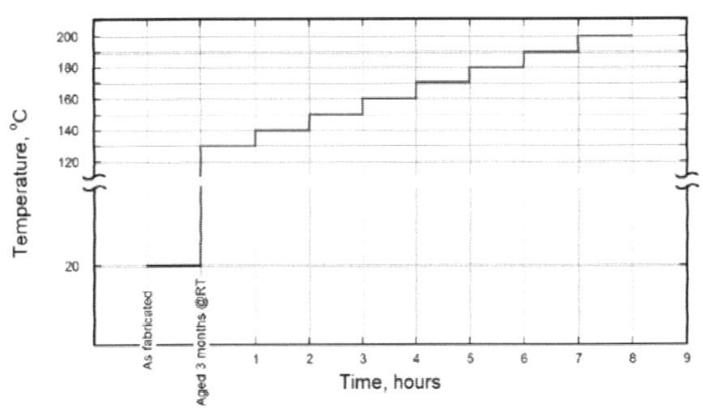


Fig. 2. Temperature profile of aging/annealing of AlN-barrier junctions [2].

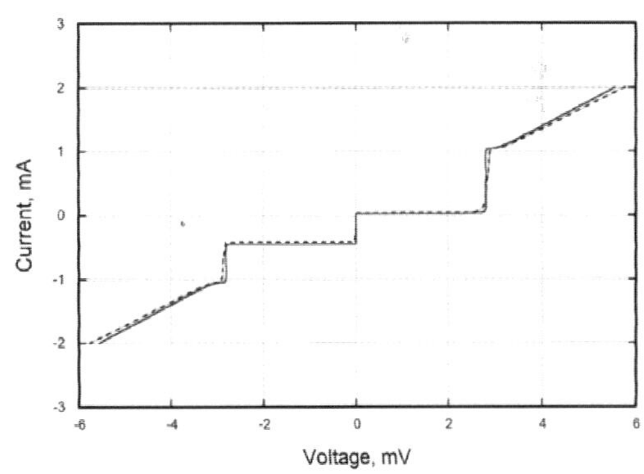


Fig. 3. Comparison of an Nb/Al-AlN/Nb junction as fabricated (red) and after annealing sequence up to 200°C (Fig. 2).

## III. Characterization of Junction Specific Capacitance

For the measurements of the Nb/Al-AlN/Nb junctions specific capacitance, we followed the approach similar to that communicated in the paper [3], using cryogenic S-parameter measurements [13].

For extracting the $C_s$ versus $R_nA$ data for Nb/Al-AlN/Nb junctions, we used the test sample layout (Fig. 4), which allowed dc-testing of the junctions with the sizes 3, 4, 5, 6 and

8 μm² on the wafer, before its dicing into the individual chips. That permitted extracting information about the junctions' $R_nA$ value. After dicing into 6 pieces, each chip contained a full set of junction sizes that allowed characterization of both $R_nA$ and $C_s$ on the single mounted sample (only re-bonding of the individual junctions was needed).

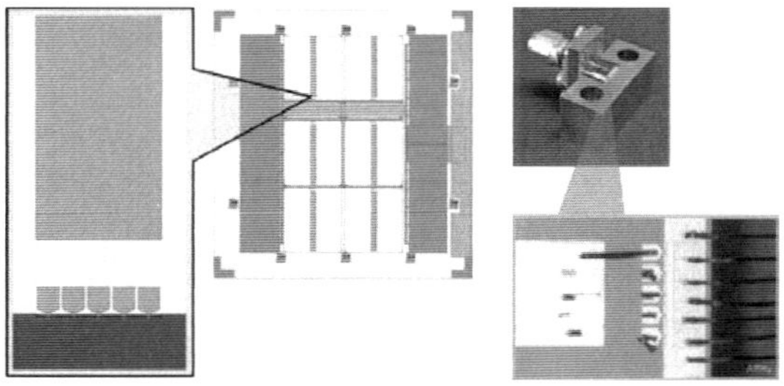

Fig. 4. The test junction wafer containing 20 single junctions of the sizes 3, 4, 5, 6 and 8 um² connected to the external contact pads. The six chips for measurement of specific capacitance each have 50Ω line and single junctions of the sizes 3, 4, 5, 6 and 8 $\mu$m² to be connected by bonding.

Specific capacitance numbers of the junctions with $R_nA \sim 20$ and 50 Ohm·μm² and nominal area of 3 – 8 μm² were measured. For lower $R_nA$ product value and bigger area junctions, the extracting of a reliable value of the junctions' specific capacitance is problematic because of domination of real conductivity due to a very low $R_n$ over the imaginary conductivity, which includes the junction capacitance contribution.

On the Fig. 5, the comparison between measured specific capacitance numbers of superconducting tunnel junctions with AlN and $AlO_x$ tunnel barriers is presented. For the so far measured, the specific capacitance of the junctions with AlN barrier is significantly (about 20%) lower than that of the junctions with $AlO_x$ barrier.

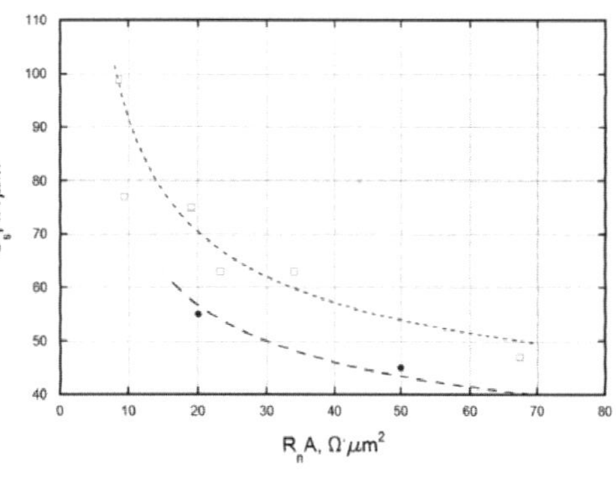


Fig. 5. Specific capacitance of Nb/Al-AlN/Nb junction (*red*) as compared with that of Nb/Al-AlOx/Nb junctions [5] (*blue*). The capacitance data for the junctions are approximated with empirical relation $C_s = a/ln(R_nA)$ [14], where $a$ is equal to 211 [5] for the Nb/Al-AlOx/Nb junctions and to 170 for the measured Nb/Al-AlN/Nb junctions.

## IV. Conclusions

In this work, we presented the progress in fabrication of high quality Nb/Al-AlN/Nb junctions. The fabricated junctions were characterized in terms of their DC electric properties and specific capacitance. The specific capacitance of the studied Nb/Al-AlN/Nb junctions is noticeably lower than that reported for the Nb/$AlO_x$/Nb junctions.

## Acknowledgement

This work was partially supported by the European Organisation for Astronomical Research in the Southern Hemisphere (ESO) in the frame of the Collaboration Agreement No. 73301/16/78225/OSZ and Radionet within the frame of the AETHRA JRA.

## References

[1] A. D. Bolatto (chair) *et al.*, "A Road Map for Developing ALMA. ASAC recommendations for ALMA 2030," 2016. [Online]. Available: https://science.nrao.edu/facilities/alma/science_sustainability/RoadmapforDevelopingALMA.pdf.

[2] A. Pavolotsky, V. Desmaris, and V. Belitsky, "Nb/Al-AlN/Nb Superconducting Tunnel Junctions: Fabrication Process and Characterization Results," in *29th IEEE International Symposium on Space THz Technology (ISSTT2018)*, 2018, pp. 107–111.

[3] P. Y. Aghdam, H. Rashid, A. Pavolotsky, V. Desmaris, D. Meledin, and V. Belitsky, "Direct Measurement of Superconducting Tunnel Junction Capacitance," *IEEE Trans. Terahertz Sci. Technol.*, 2015.

[4] P. Y. Aghdam, H. Rashid, A. Pavolotsky, V. Desmaris, and V. Belitsky, "The specific capacitance of Nb/Al-AlOx/Nb SIS junctions with extremely low RnA product," in *ISSTT 2017 - 28th International Symposium on Space Terahertz Technology*, 2017, vol. 2017-March.

[5] P. Y. Aghdam, H. Rashid, A. Pavolotsky, V. Desmaris, D. Meledin, and V. Belitsky, "Specific Capacitance Dependence on the Specific Resistance in Nb/Al-AlOx/Nb Tunnel Junctions," *IEEE Trans. Terahertz Sci. Technol.*, 2017.

[6] A. B. Pavolotsky, D. Dochev, and V. Belitsky, "Aging- and annealing-induced variations in Nb/Al- AlOx/Nb tunnel junction properties," *J. Appl. Phys.*, vol. 109, no. 2, p. 024502, 2011.

[7] P. Yadranjee Aghdam, H. Marouf Rashid, A. Pavolotsky, V. Desmaris, and V. Belitsky, "Dependence of the scatter of the electrical properties on local non-uniformities of the tunnel barrier in Nb/Al-AlO<inf>x</inf>/Nb junctions," *J. Appl. Phys.*, vol. 119, no. 5, 2016.

[8] B. Billade *et al.*, "Performance of the first ALMA band 5 production cartridge," *IEEE Trans. Terahertz Sci. Technol.*, vol. 2, no. 2, pp. 208–214, 2012.

[9] V. Belitsky *et al.*, "ALMA Band 5 receiver cartridge," *Astron. Astrophys.*, vol. 611, p. A98, Mar. 2018.

[10] V. Belitsky *et al.*, "SEPIA – a new single pixel receiver at the APEX telescope," *Astron. Astrophys.*, vol. 612, no. 31458, p. A23, Apr. 2018.

[11] L. Latrasse, M. Radoiu, J. Lo, and P. Guillot, "2.45-GHz microwave plasma sources using solid-state microwave generators. Collisional-type plasma source," *J. Microw. Power Electromagn. Energy*, vol. 51, no. 1, pp. 43–58, 2017.

[12] T. Shiota, T. Imamura, and S. Hasuo, "Nb Josephson junction with an AlN x barrier made by plasma nitridation," *Appl. Phys. Lett.*, vol. 61, no. 10, pp. 1228–1230, Sep. 1992.

[13] H. Rashid, D. Meledin, V. Desmaris, A. Pavolotsky, and V. Belitsky, "Superconducting 4-8-GHz hybrid assembly for 2SB cryogenic THz receivers," *IEEE Trans. Terahertz Sci. Technol.*, vol. 4, no. 2, pp. 193–200, 2014.

[14] V. Y. Belitsky, S. W. Jacobsson, S. A. Kovtonjuk, E. L. Kollberg, and A. B. Ermakov, "100 GHz mixer vertically integrated (stacked) SIS junction array," *Int. J. Infrared Millimeter Waves*, vol. 14, no. 5, pp. 949–957, May 1993.

# Native oxide on ultra-thin NbN films

A.V. Lubenchenko[1*], A.B. Pavolotsky[2&], S. Krause[2], O.I. Lubenchenko[1], D.A. Ivanov[1], V. Desmaris[2], V. Belitsky[2]

***Abstract*—We report study of native oxide formation over NbN ultra-thin films. With a help of XPS, chemical and phase depth profiles of NbN film of 5 nm and 10 nm thickness exposed to room air for more than a month were recorded. The surface of those films were sputtered with $Ar^+$ ions and consequently oxidized in room air for another few days. It was found that an intermediate layer of $NbN_x$ was formed between the niobium oxide layer and original NbN material.**

## I. Introduction

Ultra-thin NbN film is a material, on which the modern state-of-the-art sensitive devices like hot-electron bolometers (HEB) [1], [2] and superconducting single-photon detectors (SSPD) [3], [4] are based. The performance of such devices largely depends on the material properties of the NbN film, its chemical and phase composition as well as their depth profiles.

As-deposited films are always strained due to a mismatch of crystal lattices of the substrate and the film, as well as due to the difference of thermal expansion coefficients. The stress relaxation and reconstruction of thin films lead to structural transformation of the film itself, as well as the formation of various phases of variable composition and interfaces between them. During the process of device fabrication, the NbN film always gets exposed to room air, which causes formation of the native oxide layer over it. The native oxide layer is often left present in the final device structure between the NbN superconducting layer and the contact metallization layers. On the other hand, native oxide layer effectively withdraws part of the ultra-thin layer thickness from being a superconducting NbN material. The knowledge of the chemical and phase profiles of NbN film deposited on a certain substrate are crucial for proper understanding of the devices and their designing.

## II. Experimental Details

NbN films of 5 nm and 10 nm thickness were deposited onto a silicon substrate by reactive magnetron sputtering in the AJA Orion-5-U-D sputtering system [5], [6]. The film thickness during sputtering was controlled by the known sputtering rate (the sputtering rate was verified by HRTEM of the films).

The X-ray photoemission spectroscopy (XPS) studies of the samples surfaces were performed with the help of the electron-ion spectroscopy module based on the Nanofab 25 (NT-MDT) analytic platform. In the analysis chamber, oil-free ultrahigh vacuum was kept at the level of about $10^{-7}$ Pa. The X-ray source SPECS XR 50 without a monochromator with Mg anode as the X-ray source (1253.6 eV photon energy) was used. The spectra were recorded with a help of electrostatic hemispherical energy analyzer SPECS Phoibos 225. The energy resolution based on the full width at half maximum (FWHM) of the spectrometer at the Ag3d5/2 line (peak) was 0.78 eV for non-monochromatic X-radiation Mg Kα. The energy positions of the spectra peaks were calibrated with reference to the Cu2p3/2 (binding energy 932.62 eV), Ag3d5/2 (368.21 eV) and Au4f7/2 (83.95 eV) peaks. All survey spectra scans were recorded at a pass energy of 80 eV. The detailed scans of strong lines were in most cases recorded as wide as needed just to encompass the peak(s) of interest and were obtained with a pass energy of 20 eV. The energy analyzer was operated in Fixed Analyzer Transmission (FAT) mode.

The ion source SPECSIQE 12/38 was used for sputtering the samples. The ion source had differential pumping and was fed with 99.9995% pure Ar. The ion beam scanned an area of 2.8 mm × 4.0 mm at the incidence angle of 70° to the surface normal, the ion beam energy was of 500 eV.

## III. Chemical and Phase Profile Analysis

X-ray photo-electron spectroscopy (XPS) is one of the most efficient non-destructive methods of ultrathin films surface chemical and phase analysis. In the standard XPS realization, relative concentrations of the chemical elements are calculated with the assumption of uniform concentration of the elements across the whole depth of analysis.

Real surfaces though are always non-uniform and multi-component across the depth. Ignoring these facts cause significant inaccuracy of the analysis and often makes the extracted information questionable.

In most cases, the sample's surface not only is multilayered, but also consists of the layers of different chemical and phase composition. The extraction of the surface layer chemical and phase information from XPS spectra is a complex reverse problem with multiple unknown parameters.

In the present work, we followed the approach from [7] for XPS analysis of the oxidized surface of NbN ultrathin film. It suggests (1) the novel method for extraction of the background due to multiple inelastic electron scattering, (2) new XPS line decomposition into the component peaks accounting for

[1] National Research University MPEI, Krasnokazarmennaya, 14, Moscow 111250, Russia
[*] Corresponding author, email: LubenchenkoAV@mpei.ru

[2] Chalmers University of Technology, Group for Advanced Receiver Development, Department of Space, Earth and Environment, Göteborg, 412 96, Sweden
[&] Corresponding author, email: Alexey.Pavolotsky@chalmers.se

physical meaning of the decomposition parameters and jointly with background extraction, and (3) extracts the layer thicknesses following the simple formula.

The method assumes that the sample surface consists of a number of flat and parallel layers, each of them is uniform and can be multicomponent. Such assumption is justified by the fact that the lateral dimensions of the analyzed area are by many orders of magnitude larger than the analysis depth. Because of that, lateral non-uniformities of the surface under analysis (e.g. islands, nanoscale precipitations, elements of interface topography) are naturally averaged across the lateral dimensions. This way, structural non-uniformity will be characterized by relative concentration of chemical elements of a particular phase contained in a layer of a certain thickness.

The studied ultrathin films were deposited in vacuum. Consequent unloading the samples to the room air caused oxidation of the sample surface. As the oxidation progresses from the sample's outer surface towards its depth, the highest oxidation state will be at the immediate top surface of the film and will decrease along the sample's depth. Also, storage of the samples in the room conditions unavoidably causes hydrocarbons to deposit over the film surface.

Layer thicknesses will be calculated by the formula [7]:

$$d_i = \lambda_i \cos\theta \ln\left(\frac{I_i/(n_i\omega_i(\gamma)\lambda_i)}{\sum_{j=0}^{i-1} I_j/(n_j\omega_j(\gamma)\lambda_j)} + 1\right) \quad (1)$$

where $d_i$ is the thickness of the $i$-th layer, $n$ is the atomic concentration, $\omega_i(\gamma)$ is the differential cross-section of photoelectron production [8], $\gamma$ is the angle between the directions of incident radiation and to the energy analyzer, $\lambda$ is the IMFP (IMFP is calculated by the TPP2M formula [9]), $\theta$ is the angle between the direction to the energy analyzer and the surface normal, $I_i$ is the $i$-th peak intensity. The layers will be numbered upwards toward the surface.

## IV. Results and Discussion

In this work, we studied the influence of air oxidation onto the composition profile of NbN ultra-thin films. After deposition, the film samples were covered with an oxidation-protective layer. This layer was fully removed before the analysis in ultrasonic bath with acetone, and after that with isopropanol. The procedures of cleaning and loading samples into the analysis chamber took about 10 minutes. So the samples were air-exposed for less than 20 minutes total. The results of analysis of such films were earlier communicated in [6]. The oxidized layer thickness did not exceed 2 nm. The niobium oxide's composition was found to be $NbO_2$. In this work, we studied films that had been oxidized for more than a month after removing the protective layer. During that time the film structure changed dramatically: not only the oxide layer became thicker, up to ca. 3 nm thick, but the niobium oxide became $Nb_2O_5$.

Ion sputtering was used for removing of the surface oxide layer. Ion sputtering was done in the number of 20 minutes long steps each followed by XPS analysis. Each sputtering step removed a few monolayers. After 5th sputtering step, the highest oxide, i.e. $Nb_2O_5$, was completely removed. The mean ion

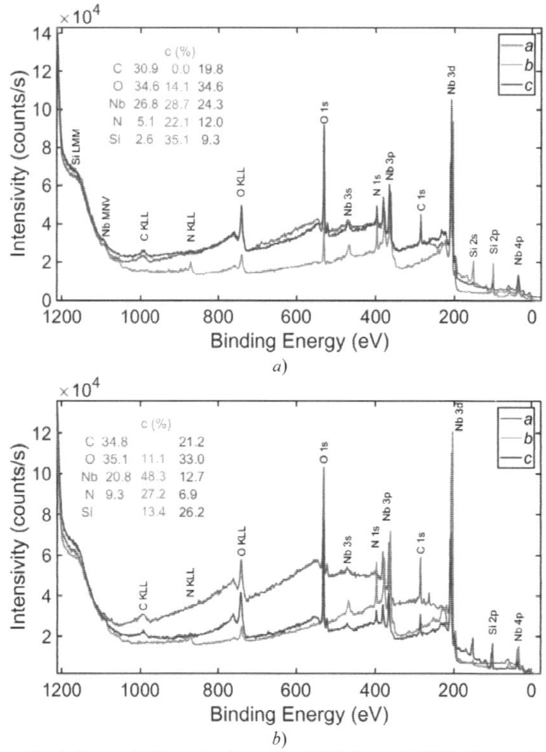


Fig. 1. Survey XPS spectra. Target: *a*) NbN, 5 nm; *b*) NbN, 10 nm. Line: (a) NbN films of 5 nm and 10 nm thickness after >1 month exposed to room air; (b) the same films after $Ar^+$ sputtering; (c) after consequent air oxidation for another few days.

penetration depth was about 1 nm, and consequently, the thickness of the modified layer was less than 2 nm. After oxide sputtering, the samples were unloaded from the analysis chamber and exposed to oxidation in air for a few days.

The following samples were analyzed by XPS: (a) NbN films of 5 nm and 10 nm thickness after more than 1 month exposure to room air; (b) the same films after $Ar^+$ sputtering; (c) after consequent air oxidation for another few days. A standard XPS analysis showed presence of C, O, Nb and Si in the samples. Fig. 1 shows the survey spectra and presents the relative atomic concentrations of the elements in the samples with 5 and 10 nm thick NbN films after each sputtering step.

To analyze composition and chemical bond, the XPS line of an element of interest needs to be decomposed into component peaks. The XPS line structure can be quite complex due to superposition of peaks of the element in its different chemical bond states and presence of satellite peaks. Moreover, shape

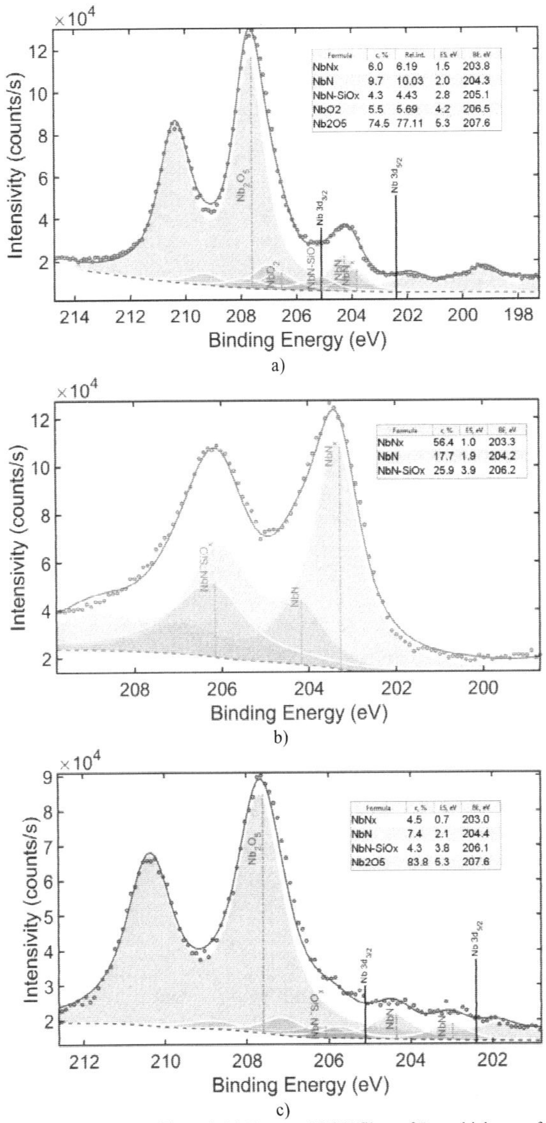


Fig. 2. XPS spectra of line Nb 3d. Target: a) NbN films of 5 nm thickness after >1 month exposed to room air; b) the same films after $Ar^+$ sputtering; c) after consequent air oxidation for another few days. Solid lines: calculation. Circles: experimental data. Area: separate calculated peaks.

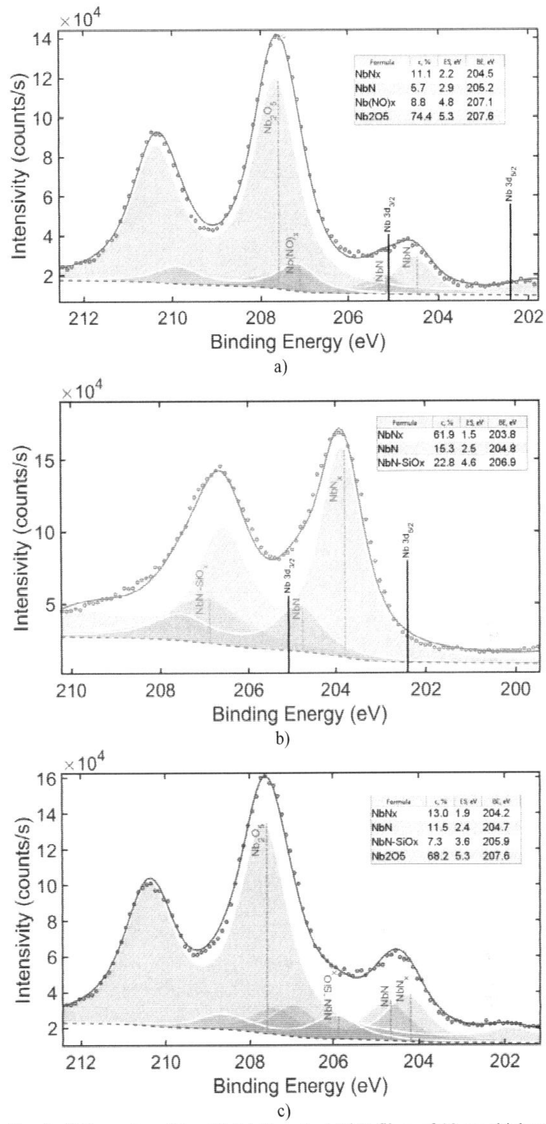


Fig. 3. XPS spectra of line Nb 3d. Target: a) NbN films of 10 nm thickness after >1 month exposed to room air; b) the same films after $Ar^+$ sputtering; c) after consequent air oxidation for another few days. Solid lines: calculation. Circles: experimental data. Area: separate calculated peaks.

and width of the peaks could be affected by a various factors. For XPS peaks deconvolution, we followed the approach presented in [7].

The spectral line shape is defined by convolution of functions describing natural shape of the line and instrumental broadening. We describe the natural shape of the line with a Doniach-Sunjic relation, while the instrumental broadening follows Gauss' function. We suggest usage of the binding energy and spin-orbit interaction energy numbers from a reference book [10]. The chemical shift energy is almost linearly proportional to the oxidation state, hence, it is sufficient to find the chemical shift energy of element with highest oxidation state, e.g. we used chemical shift energy 5.31 eV for $Nb_2O_5$. The chemical shift energy of Nb-N compounds strongly depends on their stoichiometry. The chemical shift energy dependence on the stoichiometric coefficient $x$ for the lines Nb 3d and N 1s in $NbN_x$ compounds is shown in the paper [11].

Fig. 2 and Fig. 3 present the results of Nb 3d spectral line decomposition into different phase component peaks for the samples of 5 and 10 nm thick NbN films after each analysis stage, respectively.

For the samples of NbN films, 5 nm and 10 nm thick, after more than 1 month exposure to room air, there were two niobium nitride phases identified: NbN and $NbN_x$, $x \approx 0.79$ corresponding probably to $Nb_5N_4$. The stoichiometric coefficient $x$ was extracted from the chemical shift energy with

TABLE I
CHEMICAL AND PHASE DEPTH PROFILE OF AN ULTRA-THIN NIOBIUM NITRIDE FILM 5 NM

| | Formula | *d* (nm) | | |
|---|---|---|---|---|
| | | a | b | c |
| 5 | hydrocarbons | 1.0 | - | 0.8 |
| 4 | $Nb_2O_5$ | 3.2 | - | 2.5 |
| 3 | $NbN_x$ | 1.5 | 1.8 | 0.1 |
| 2 | NbN | 3.5 | 0.5 | 0.1 |
| 1 | Nb-$SiO_x$ | 1.2 | 1.6 | 2.2 |
| | Si | substrate | | |

TABLE II
CHEMICAL AND PHASE DEPTH PROFILE OF AN ULTRA-THIN NIOBIUM NITRIDE FILM 10 NM

| | Formula | *d* (nm) | | |
|---|---|---|---|---|
| | | a | b | c |
| 5 | hydrocarbons | 0.6 | - | 0.8 |
| 4 | $Nb_2O_5$ | 5.2 | - | 2.4 |
| 3 | $NbN_x$ | 2.2 | 2.0 | 0.6 |
| 2 | NbN | not re- | 1.1 | 0.6 |
| 1 | Nb-$SiO_x$ | corded | 1.4 | 1.6 |
| | Si | substrate | | |

a use of the data from [11]. The peak position in the line Nb 3d shifts towards the lower binding energy numbers after sputtering (Fig. 2). This points to the fact that stoichiometric coefficient $x$ in niobium nitride compound $NbN_x$ in the film shifts towards lower number. Apart of the two nitride phases, the analysis identified the presence of complex oxi-nitride compound Nb-$SiN_xO_y$. This oxi-nitride could only be formed at the early stage of NbN film growth on the covered by silicon native oxide substrate.

Finally, based on the results of the XPS line decomposition and following the Eq. (1), chemical composition and phase depth profiles were extracted for each stage of the surface analysis (Tables I and II): (a) NbN film of 5 nm end 10 nm thickness after more than 1 month exposure to room air, (b) the same films after $Ar^+$ sputtering, and (c) after consequent air oxidation for another few days.

## V. CONCLUSION

In this work, we studied films that had been oxidized for more than a month after removing the protective layer. During that time the film structure changed dramatically: not only the oxide layer became thicker, ca. 3 nm thick, but the niobium oxide became $Nb_2O_5$. Moreover, a non-stoichiometric niobium nitride NbNx, ca. 1 nm thick, was found between $Nb_2O_5$ and the stoichiometric NbN layers. Further, the oxide was removed by a "delicate" Ar+ ion sputtering. After sputtering, the samples were again exposed to a room air for a few days. That caused formation of the oxide layer of $Nb_2O_5$ about 2 nm thick, and about a monolayer of non-stoichiometric $NbN_x$ phase under it.

## REFERENCES


[1] D. Meledin *et al.*, "A 1.3-THz balanced waveguide HEB mixer for the APEX telescope," *IEEE Trans. Microw. Theory Tech.*, vol. 57, no. 1, 2009.

[2] S. Krause *et al.*, "Noise and IF Gain Bandwidth of a Balanced Waveguide NbN / GaN Hot Electron Bolometer Mixer Operating at 1 . 3 THz," *IEEE Trans. Terahertz Sci. Technol.*, vol. 8, no. 3, pp. 365–371, 2018.

[3] G. N. Gol'tsman *et al.*, "Picosecond superconducting single-photon optical detector," *Appl. Phys. Lett.*, vol. 79, no. 6, pp. 705–707, Aug. 2001.

[4] A. Verevkin *et al.*, "Ultrafast superconducting single-photon detectors for near-infrared-wavelength quantum communications," *J. Mod. Opt.*, vol. 51, no. 9–10, pp. 1447–1458, Jun. 2004.

[5] S. Krause *et al.*, "Epitaxial growth of ultra-thin NbN films on $Al_x Ga_{1-x} N$ buffer-layers," *Supercond. Sci. Technol.*, vol. 27, no. 6, 2014.

[6] S. Krause *et al.*, "Ambient Temperature Growth of Mono-and Polycrystalline NbN Nanofilms and Their Surface and Composition Analysis," *IEEE Trans. Appl. Supercond.*, vol. 26, no. 3, 2016.

[7] A. V. Lubenchenko, A. A. Batrakov, A. B. Pavolotsky, O. I. Lubenchenko, and D. A. Ivanov, "XPS study of multilayer multicomponent films," *Appl. Surf. Sci.*, vol. 427, 2018.

[8] J. J. Yeh and I. Lindau, "Atomic subshell photoionization cross sections and asymmetry parameters: $1 \leqslant Z \leqslant 103$," *At. Data Nucl. Data Tables*, vol. 32, no. 1, pp. 1–155, Jan. 1985.

[9] S. Tanuma, C. J. Powell, and D. R. Penn, "Calculation of electron inelastic mean free paths (IMFPs) VII. Reliability of the TPP-2M IMFP predictive equation," *Surf. Interface Anal.*, vol. 35, no. 3, pp. 268–275, Mar. 2003.

[10] J. F. Moulder, W. F. Stickle, P. E. Sobol, and K. D. Bomben, *Handbook of X-Ray Photoelectron Spectroscopy. A Reference Book of Standard Spectra for Identification and Interpretation of XPS Data*. Eden Prairie, MN: Physical Electronics, 1995.

[11] P. Prieto, L. Galán, and J. M. Sanz, "An XPS study of NbNx prepared by ion implantation and the near-surface effects induced by Ar+ bombardment," *Surf. Sci.*, vol. 251–252, pp. 701–705, Jul. 1991.

# Design of On-chip Broadband Band Selection Filter for Multi-chroic mm/submm Camera.

Kah Wuy Chin[1,2], Tai Ohshima[2], Satoshi Ono[3], Takeshi Sakai[3], Tatsuya Takekoshi[1], Satoru Mima[4], Ryohei Kawabe[2], Masato Naruse[5], Keisuke Yoshioka[2,3], and Shinsuke Uno[1].

Astronomical observation of the broadband submillimeter continuum emission is an important approach to investigate evolution history of the large-scale structure in the universe. The Sunyaev-Zel'dovich effect (SZE), distortion of cosmic microwave background (CMB) spectrum shown between 10—1000 GHz caused by Compton scattering of CMB photons by high energy electrons in galaxy clusters[1], is an important probe of evolution history of galaxy clusters. To promote observational studies of SZE, we are developing a multichroic continuum camera system that has 2 focal planes separated at ~300 GHz by a dichroic filter[2]. We plan to realize 6-color simultaneous observations (150/220/270 GHz and 350/450/670 GHz for low and high pass focal planes, respectively) by implementation of 3 RF bandpass filters (BPFs) for each spatial pixel on detector wafers of each focal plane.

Here we report the on-chip RF filter design for the low pass band detector array. The on-chip filters for 150/220/270 GHz detectors are required to meet following requirements: (1) center frequencies are 150, 220, and 270GHz, (2) bandwidth is more than 40 GHz, (3) physical sizes of the filters are smaller than 1.0 mm x 0.5 mm. Because of the limited space, we adopted compact lumped element filters. We promoted designs of the BPFs by following steps. First, we designed the circuits of the bandpass filters (BPF) whose center frequency are 150, 220, and 270 GHz by equivalent circuit models using inductors and capacitors. Next, we designed a 150 GHz BPF as a 3rd order Chebyshev filter, and 270 GHz and 220 GHz BPFs as 5th order Chebyshev filters in order to avoid crosstalk between the filters. Then, we designed planar capacitors and inductors on a Si wafer. Finally, we integrated designed planar capacitors and inductors as same as the designed circuits of BPFs to make the on-chip filter designs. In Fig. 1, we show the on-chip 150 GHz BPF design. Its physical size was 80 μm x 414 μm. Physical sizes of 220 and 270 GHz BPFs were 96 μm x 634 μm, and 80 μm x 528 μm, respectively.

To evaluate our designs by considering the stray capacitance made by physical structure of BPFs, we calculated the S-parameters of the on-chip BPFs using electromagnetic simulation, Sonnet. As a result of calculation, 150, 220, and 270 GHz BPFs show the bandwidths of 50, 60, and 80 GHz respectively. On the other hand, we found large ripples up to 5 dB in the passband of the on-chip BPFs. We expect that these ripples can be suppressed by modification of the planar structure of the on-chip BPFs. We also plan to modify the first planar capacitor structure because the bandwidth of the 270 GHz BPF is too wide to use the 220 and 270 GHz BPFs at the same time.

In conclusion, our BPF designs are small enough to fit a spatial pixel on the focal plane wafer. We are going to fabricate our BPFs combined with MKIDs and measure the frequency response. We are also going to design on-chip BPFs for higher frequency bands (350/450/670 GHz).

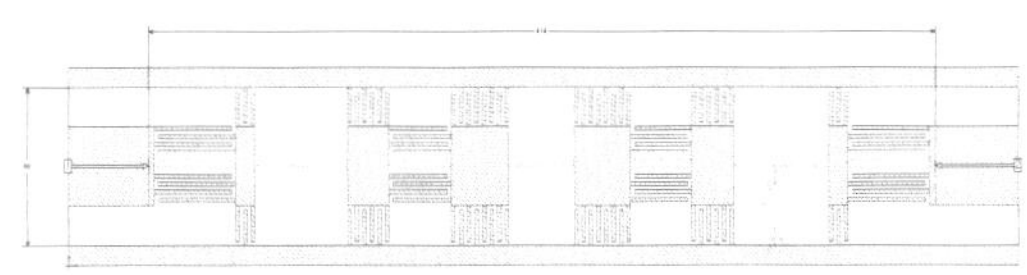

Fig. 1. Design of the 150 GHz BPF. The physical size of the BPF is 80 μm width and 414 μm length. From the Sonnet simulation result, the bandwidth was estimated to be 50 GHz.

### References

[1] Sunyaev, R. A. and Zel'dovich, Y. B.: 1980, Annual Review of Astronomy and Astrophysics 18(1), 537

[2] Oshima, T., Ohtawara, K., Takekoshi, T., Ishii, S., Izumi, N., Izumi, T., Yamaguchi, M., Suzuki, S.,Muraoka, K., Hirota, A., Saito, F., Nakatsubo, S., Kouchi, A., Ito, T., Uemizu, K., Fujii, Y., Tamura,Y., Kohno, K., and Kawabe, R.: 2018,Journal of Low Temperature Physics193

[1] The Univ. of Tokyo, Tokyo 113-0033, Japan
[2] National Astronomical Observatory of Japan, Tokyo 181-8588, Japan
[3] The University of Electro-Communications, Tokyo 182-8585, Japan
[4] RIKEN, Saitama 351-0198, Japan
[5] Saitama University, Saitama 338-8570, Japan

NOTES:

# NbN/AlN/NbN Superconducting Tunnel Junctions Fabricated for HSTDM

Jing LI[1,2], Zheng WANG[1,2], Ming YAO[1,2], Qing SHI[1,3], Bo-Liang LIU[1,3], Hong-Hu LI[1,3], Dong Liu[1,2] and Sheng-Cai SHI[1,2]

Terahertz, defined as 0.1-10THz, is an important frequency regime for astronomical observation. THz observations on ground-based telescopes are rather limited due to strong absorption by the Earth's atmosphere. Space telescopes can overcome this constraint. China is planning to launch a space station around 2020. It will be in full operation hopefully around 2022, with a 2-meter telescope flying with it in a common orbit. Onboard this telescope, there will be a high sensitivity terahertz detection module (HSTDM) based on a niobium nitride (NbN) superconductor-insulator-superconductor (SIS) receiver system operating at 0.41-0.51THz band [1].

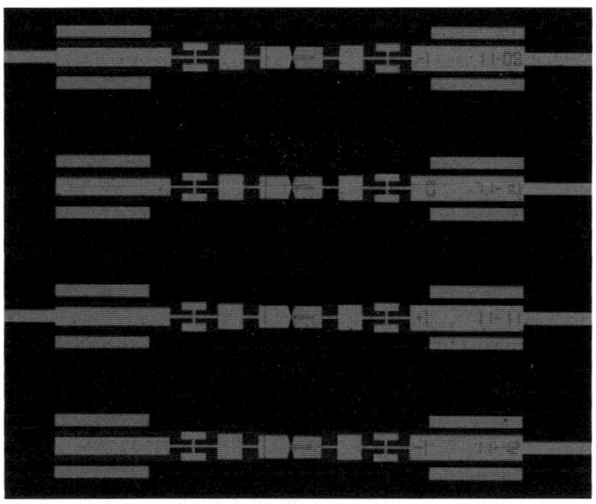

Fig. 1. NbN/AlN/NbN tunnel junctions fabricated in a cleanroom at PMO.

High sensitivity NbN SIS mixers have been proved to be of good potentials for space applications as they can work at relatively high temperatures up to 10K and have good stability [2]. Fabricating high-quality NbN SIS junctions remains challenging [3], especially for space applications. In this paper, we will mainly introduce our work on the fabrication of NbN superconducting tunnel junctions for HSTDM, including the growth of superconducting NbN films and the fabrication of high-Jc SIS junctions.

[1] Purple Mountain Observatory, Chinese Academy of Sciences, Nanjing 210034, China.
[2] Key Laboratory of Radio Astronomy, Chinese Academy of Sciences, Nanjing 210034, China.
[3] University of Science and Technology of China, Hefei 230026, China.

### References

[1] K. Zhang, S. C. Shi, Q. J. Yao, J. Li, W. Zhang, Z. H. Lin, Z. Lou, W. Miao, D. Liu and R. Q. Mao, "High Sensitivity Terahertz Detection Module," *in 2018 International Conference on Microwave and Millimeter Wave Technology (ICMMT)*, 2018.

[2] Jing Li, M. Takeda, Z. Wang, S.C. Shi, and J. Yang, "Low-Noise 0.5 THz All-NbN Superconductor-Insulator-Superconductor Mixer for Submillimeter Wave Astronomy," *Appl. Phys. Letts.* 92, 222504, June 2008.

[3] Z. Wang, A. Kawakami, Y. Uzawa, and B. Komiyama, "High critical current density NbN/AlN/NbN tunnel junctions fabricated on ambient temperature MgO substrates," *Appl. Phys. Lett.* 64, 2034, 1994.

NOTES:

# $MgB_2$ HEB Terahertz Mixers: Diffusion- or phonon- cooled?

Narendra Acharya*, Samuel Lara, and Sergey Cherednichenko

During the last few decades superconducting hot-electron bolometer (HEB) mixers have emerged as the most successful technology for extremely low-noise molecular spectroscopy observations at frequencies above 1 THz [1]. This is because, unlike SIS mixers, operation of HEBs is not limited by the superconducting gap and compared to Schottky diode mixers, their noise temperature is much lower. By far, the most extensively studied and used HEB mixers are based on phonon-cooled mechanism of non-equilibrium electrons. Despite of being state-of-the-art, the IF of NbN based phonon-cooled HEB mixers limits to 3-5 GHz due to about 12 ps of electron-phonon interaction time ($\tau_{eph}$). While $MgB_2$ offers large IF bandwidth as a result of fast $\tau_{eph}$ (1-2ps) due to its higher $Tc$ of above 30K [2], it is suspected that its low sheet resistance (one order of magnitude lower than NbN) has been problematic in optimizing the low noise temperature. In the alternative approach pioneered by Prober [3], the mixer element is a nanobridge and in this configuration, the dominant cooling mechanism for non-equilibrium electrons is outdifusion through normal metal contact pads. For such diffusion-cooled HEB mixers, material with lower sheet resistance and higher diffusion constant can be a suitable choice.

In retrospect, we investigated electron diffusion constant of $MgB_2$ ultrathin films grown in SiC substrate by hybrid physical chemical vapor deposition process and patterned in submicron bridges. Our results show that the ultrathin $MgB_2$ films has a diffusivity constant of ~ $5.0 cms^{-1}$, i.e. $\times 10$ compare to NbN ultrathin films, $\sim 0.5 cms^{-1}$. This higher electron diffusion constant in ultrathin $MgB_2$ films can have two benefits. One, compare to phonon-cooled $MgB_2$ HEB itself, the device size in diffusion cooled $MgB_2$ HEB can be reduceded to sub-micron size, thus, while ensuring an effective diffusion of non-equilibrium electrons, more uniform film surface can be utilized. On the other hand, due to larger diffusion constant of $MgB_2$, it is not necessary to reduce devices to too short i.e. nano bridges, as usually preferred in diffusion-cooled HEBs. Despite being very effective diffusion process in nanobridges, achieving a stable biasing is challenging due to the temperature distribution along the bridge. Also, too narrow bridges are likely to be susceptible to a direct detection. In this conference, we present a systematic study of superconducting properties and electron diffusion process in ultrathin $MgB_2$ films. We also discuss feasibility of utilization of low noise, high bandwidth diffusion cooled $MgB_2$ HEB mixers.

All authors are with the Department of Microtechnology and Nanoscience, Chalmers University of Technology, SE-412 96, Gothenburg, Sweden.

[2] corresponding author, narach@chalmers,se.

NOTES:

## References

[1] T. de Graauw et al., "The herschel-heterodyne instrument for the far infrared (HIFI)," *Astronomy Astrophys.*, vol. 518, 2010, Art. no. L6.

[2] E. Novoselov and S. Cherednichenko, "Gain and Noise in THz MgB2 Hot-Electron Bolometer Mixers with a 30K Critical Temperature", *IEEE Trans.THz.Sci.Technol.* vol. 7 (26) s. 704-710 (Nov. 2017)

[3] D.E. Prober, "Superconducting terahertz mixers using a transition edge microbolometer", *Appl. Phys. Lett.* 62, 17 (1993).

# Bandwidth performance of a THz normal metal TiN bolometer-mixer

N.A. Tovpeko, A.V. Trifonov, A.V. Semenov, S.V. Antipov, N.S. Kaurova, N.A. Titova, and G.N. Goltsman

***Abstract*—We report on the bandwidth performance of the normal metal TiN bolometer-mixer on top of an $Al_2O_3$ substrate, which is capable to operate in a wide range of bath temperatures from 77 K – 300 K. The choice of the combination TiN / $Al_2O_3$ is related to an advanced heat transport between the film and the substrate in this pair and the sufficient temperature coefficient of resistance.**

**The data were taken at 132.5 – 145.5 GHz with two BWOs as a signal and an LO source. Measurements were taken on TiN films of different thickness starting from 20 nm down to 5 nm coupled into a spiral Au antenna, which improves matching of incoming radiation with the thin TiN fim. Our experiments demonstrate effective heat coupling from a TiN thin film to an $Al_2O_3$ substrate (111) boosting gain bandwidth (GB) of TiN bolometer up to 6 GHz for 5 nm thin film. Current results indicate weak temperature dependence of GB on the bath temperature of the TiN bolometer. Theoretical estimations of GB performance meet with experimental data for 5 nm thin TiN films.**

***Index Terms*— Bolometer, gain bandwidth, mixer, normal metal, TiN.**

## Introduction

MODERN approaches to low signal THz observations usually based on two main technologies, namely superconducting and semiconducting one.

The superconducting technology, at the cost of cryogenic operation, is presented with such devices as SIS and HEB mixers. On the one hand, SIS usually provides ultimate noise performance of few quantum noise limits with gain bandwidth of several tens of GHz at signal frequencies up to 1 THz [1]. On the other hand, HEBs provide noise performance of several quantum limits with gain bandwidth of several GHz [2] without upper limit in terms of LO frequencies.

The semiconductor technology is associated with Schottky diodes based mixers, which are able to operate up to THz frequencies, providing good noise performance about one thousand of kelvins with gain bandwidth of several tens of GHz [3]. But their performance rapidly deteriorates, while the LO power consumption increases with rising of signal frequency above several THz [4]. Another semiconductor type of devices is mercury cadmium telluride (CMT) mixer, which is the only one available detector above 30 THz [5].

Here we report on the IF bandwidth performance of a normal metal TiN bolometer on top of an $Al_2O_3$ substrate, which has advanced heat transport between a metal film and a substrate material [6], can provide up to 6 GHz of gain bandwidth in a wide range of temperature from 77 K up to 300 K and is capable to operate in the wide LO frequency range 3-30 THz.

## Experiment

### A. Normal metal TiN bolometer

The studied bolometers were made out of a TiN film on top of an $Al_2O_3$ substrate (111). The film was deposited by means of a DC magnetron sputtering. TiN film thickness varies from 20 down to 3 nm.

### B. Experimental Setup

The experimental setup utilized two BWOs, one as a signal source with frequency tuning range from 132.5 – 145.5 GHz, while the second BWO was set to 132.5 GHz. The total output power of two sources was controlled by an RF thermistor power meter and hold at the level of 100 μWatts. Signals from two sources were further combined with the quasi-optical part and coupled through a free space and an HDTP window to the hemispherical silicon lens. The lens with a device under study was mounted of the cold plate of a cryostat. The necessity of a cryostat in our setup comes from two reasons. The first reason

This work was partially supported by the Russian Foundation for Basic Research grant No. 16-29-11779 (numerical estimation of gain bandwidth), the Russian Science Foundation grant No. 17-72-30036 (development of the experimental setup, fabrication of studied devices and experimental studies of gain bandwidth performance of fabricated devices).

N.A. Tovpeko is with the Department of Physics, Moscow Pedagogical State University, M. Pirogovskaya St. 29, Moscow, 119435, Russia (e-mail: mouseninka@mail.ru).

A.V. Trifonov is with the Department of Physics, Moscow Pedagogical State University, M. Pirogovskaya St. 29, Moscow, 119435, Russia (e-mail: andrey89trifonov@gmail.com).

A.V. Semenov is with the Department of Physics, Moscow Pedagogical State University, M. Pirogovskaya St. 29, Moscow, 119435, Russia (e-mail: av.semyonov@mpgu.edu).

S.V. Antipov is with the Department of Physics, Moscow Pedagogical State University, M. Pirogovskaya St. 29, Moscow, 119435, Russia (e-mail: sv_antipov@mail.ru).

N.S. Kaurova is with the Department of Physics, Moscow Pedagogical State University, M. Pirogovskaya St. 29, Moscow, 119435, Russia (e-mail: nkaurova@yandex.ru).

N.A. Titova is with the Department of Physics, Moscow Pedagogical State University, M. Pirogovskaya St. 29, Moscow, 119435, Russia (e-mail: titovana@mail.ru).

G.N. Goltsman is with the Department of Physics, Moscow Pedagogical State University, M. Pirogovskaya St. 29, Moscow, 119435, Russia (e-mail: goltsman10@mail.ru).

is the temperature range from 300 K down to 77 K. The other reason in the bias voltage of our bolometer, which varies in the range of 1 - 10 mV, and can cause degradation of TiN film, due to its' oxidization at room temperature. The gain bandwidth data were recorded by a spectrum analyzer (Rohde&Schawarz FSV 9 kHz – 13.5 GHz) as the difference spike between to signal and LO frequencies at the RF input of the bolometer.

## Results and Discussion

The measurements were taken on several different devices, in term of the TiN film thickness, which was presented by three different sizes: 20 nm, 10 nm and 5 nm. If the phonon cooling mechanism is considered for the type of films, then the cut-off frequency of such films can be derived as follows: $f_{3dB,ph} = 1/(2\pi\tau) = G_{int}/(2\pi Cd)$. Where $\tau$ - the response time of the bolometer, $G_{int}$ – the interface heat conductance, C – the film heat capacitance, d – the thickness of the film, $\kappa$ - the heat conductivity. The above equation is considered under the condition of $G_{film} = \kappa d >> G_{int}$ , which is automatically met if d < 100 nm. Considering numerical values of above specified film parameter ($\kappa \sim 20\ Wm^{-1}K^{-1}$; $C \sim 3\times10^{6}\ Jm^{-3}K^{-1}$; $G_{int} \sim 5\times10^{8}\ Wm^{-2}K^{-1}$) 3 GHz of gain bandwidth for d = 10 nm is expected.

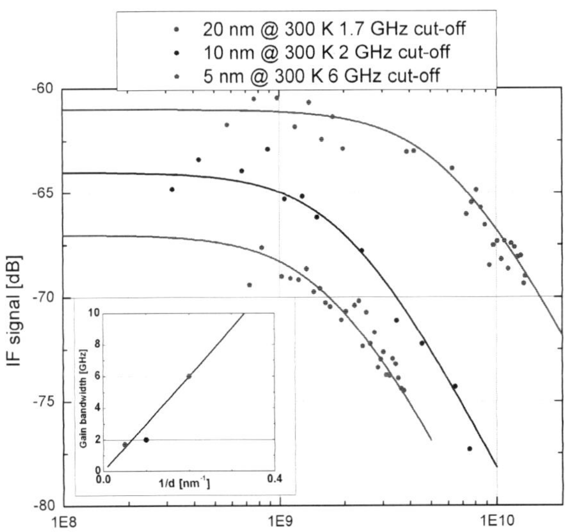


Fig. 1. Gain bandwidth as the function of the film thickness. The data were fitted with the second order Lorenz. The black line on the insert stands for the theoretical expectations of the gain bandwidth performance. Colored dots indicate measured results.

As can be seen on the insert in Fig.1, there is some mismatching between the theoretical expectations and the measured result for 10 nm thin film. We believe, that the origin of this difference is related to the film thickness, which probably is ~ 15 nm instead of 10 nm. Further studies of the impact of an operation temperature on the gain bandwidth indicate very weak correlation between the two parameters, as shown in Fig.2.

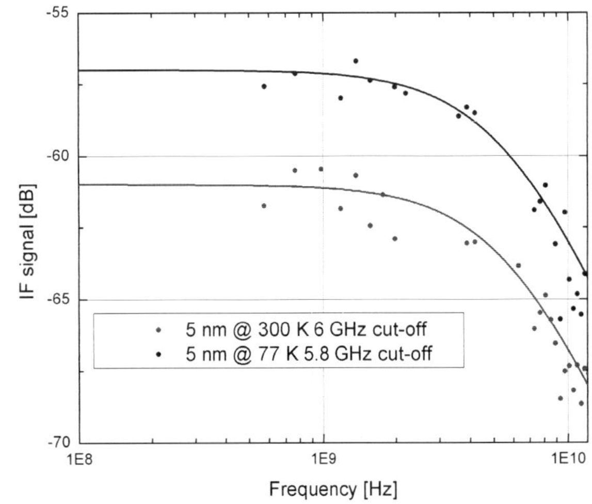


Fig. 2. The gain bandwidth as the function of operation temperature.

Lack of visible dependence of the gain bandwidth (the response time) of the TiN bolometer on its' temperature in the range 77-300 K, is related to the fact, that C - the heat capacitance and $G_{int}$ - the interface heat conductance (as well as $\kappa$ - the heat conductivity of the film) have nearly the same (nearly linear) dependence on the temperature.

## Conclusion

The normal metal TiN THz bolometer-mixer on a sapphire (111) substrate can provide an advanced gain bandwidth performance of 6 GHz at 5 nm thin TiN film. Such gain bandwidth performance can place TiN bolometers in one row with the mature Schottky technology, but for higher operation frequencies.

## References

[1] C. E. Tong, L. Zeng, P. Grimes, W.-C. Lu, T.-J. Chen, Y.-P. Chang, M.-J. Wang, "Development of SIS Receivers with Ultra-wide Instantaneous Bandwidth for wSMA". Proceeding 29th Internation. Symposium on Space THz Technology, 2018.

[2] S. Krause, D. Meledin, V. Desmaris, A. Pavolotsky, V. Belitsky , "Noise performance of a balanced waveguide NbN HEB mixer utilizing a GaN buffer-layer at 1.3 THz ". Proceeding 29th Internation. Symposium on Space THz Technology, 2018.

[3] J. Treuttel, L. Gatilova, A. Maestrini, D. Moro-Melgar, F. Yang, F. Tamazouzt, T. Vacelet, Y. Jin, J. Mateos, A. Feret, C. Chaumont, C. Goldstein, "A 520-620 GHz Schottky receiver front-end for Planetary Science and Remote Sensing with 1070K-1500K DSB noise temperature at Room Temperature". *IEEE Trans.Terahertz Science and Tech.*, vol. PP, no 99, p.1-8 , 2015.

[4] I. Mehdi, J. V. Siles, C. Lee and E. Schlecht, "THz Diode Technology: Status, Prospects, and Applications," in *Proceedings of the IEEE*, vol. 105, no. 6, pp. 990-1007, June 2017.

[5] W. Lei, J. Antoszewski, and L. Faraone, "Progress, challenges, and opportunities for HgCdTe infrared materials and detectors", Appl. Phys. Rev. 2, 041303 (2015).

[6] R. M. Costescu, M.A. Wall and D.G. Cahill "Thermal conductance of epitaxial interfaces", *Phys. Rev. B*, vol. 67, pp. 054302, 2003.

J. Böhm[1], N. Honingh[1], and K. Jacobs[1]

# Development of a HEB mixer for the observation of molecular hydrogen on SOFIA

Based on the successful development of superconducting THz Hot Electron Bolometer (HEB) mixers for the Stratospheric Observatory for Infrared Astronomy (SOFIA) [1], we are currently developing a HEB mixer for the high resolution observation of molecular hydrogen at 10.7 THz.

As Local Oscillator we will use a Quantum Cascade Laser [2], similar to that at 4.7 THz that has been implemented successfully by our group in the upGREAT receiver [1].

Contrary to our 4.7 THz mixer which is a waveguide mixer, we use an open structure mixer including a planar antenna and a dielectric lens. This type of mixer is successfully used up to a frequency of 5.3 THz [3].

The RF design of the mixer cannot completely be done with simulation solvers using spatial discretization into cells, as the cell dimensions are directly related to the wavelength of interest. As such, our computational resources prohibit the use of our commercial software CST design studio for the spatial discretization of the dielectric lens of the mixer design at 10.7 THz. We therefore use a hybrid approach calculating the planar antenna field in CST design studio and using an additional, in house developed, software to simulate the propagation of this field through a lens of arbitrary form to the focus. The software combines two techniques. It uses a plane wave decomposition of the initial field, which is propagated as rays based on the spectral theory of diffraction (STD).

We will present the current status of the RF design in detail, including our newly developed software.

## References


[1] Risacher, C., et al. "The upGREAT dual frequency heterodyne arrays for SOFIA." , *Journal of Astronomical Instrumentation* Vol. 07, No. 04, p 1840014 (2018)

[2] Bosco, Lorenzo, et al. "A patch-array antenna single-mode low electrical dissipation continuous wave terahertz quantum cascade laser." *Applied Physics Letters* 109.20 (2016): 201103.

[3] Zhang, W., et al. "NbN hot electron bolometer mixer at 5.3 THz." Millimeter, Submillimeter, and Far-Infrared Detectors and Instrumentation for Astronomy V. Vol. 7741. International Society for Optics and Photonics, 2010.


[1] I. Physics Institute, University of Cologne, Zülpicher Straße 77, 50937, Cologne, Germany

NOTES:

# $MgB_2$ HEB Mixers with Nanopatterned Surfaces: Effect on the Noise Temperature and the LO Power.

Narendra Acharya and Sergey Cherednichenko*

Heterodyne receivers are capable of detecting and spectrally resolve fine structures of terahertz wave emission coming from interstellar medium, stars, and planets. For frequencies above 1THz, superconducting Hot-Electron Bolometer (HEB) mixers enable such heterodyne receivers on ground, air- and space based platforms.[1] For HEB mixers, ultra thin superconducting films are required with a ps-rate electron phonon interaction $\tau_{eph}$, and a fast phonon-to-substrate escape time $\tau_{esc}$ The resulting electron temperature relaxation time sets the upper limit for the utilizable Intermediate Frequency (IF). So far, such combination of properties has been observed in two materials. In NbN, with a critical temperature of 8-11K in thin films, $\tau_{eph}$ is about 12ps, and the maximum IF at 4-6GHz. In $MgB_2$, with a critical temperature of 30K in thin films, $\tau_{eph}$ is about 1-2ps, and the maximum IF is at 11-13GHz. Besides a wider bandwidth, $MgB_2$ HEB mixers are much less critical to the cooling, allowing for operation >10K (20K low noise operation has been reported [3])

Despite of obvious advantages offered by $MgB_2$ HEB mixers, there are two issues, which have to be resolved for efficient applications. Both of this issues are related to a rather low resistivity (hence, sheet resistance) of $MgB_2$ films, about a factor of 10 lower compared to NbN films. Even for 5nm thick $MgB_2$ films, the sheet resistance is in the range of 50-70 Ω/□, which requires the HEB bridges to have an aspect ratio w/l≤1 in order to keep the bridge resistance matched to the impedance of THz antennas (30-100Ω). On the other side, due to contact resistance (remember, that HEB mixers operate at a few THz), the width of the bridges has to be kept not too small in order to keep contact losses are low as possible. Earlier, we have observed that reducing the HEB width from 1μm down to 300nm the noise temperature is increasing. E.g. NbN HEB mixers have a width 2-4μm. With the first and the second constrains in mind, the optimized for the lowest noise $MgB_2$ HEB mixers would be e.g. 2μm×2μm. The LO power for HEB mixers is known to be proportional to the HEB area. Therefore, optimized for low noise, $MgB_2$ HEB mixers would require too high LO power.

Both authors are with the Department of Microtechnology and Nanoscience, Chalmers University of Technology, SE-412 96, Gothenburg, Sweden.

[2] corresponding author, serguei@chalmers,se.

In this study, we discuss artificial surfaces for $MgB_2$ films, such that the effective sheet resistance could be increased, and the aspect ratio w/l could be decreased. We achieve this effect by nano patterning $MgB_2$ films with a set of nano holes (see Fig.1). We vary the hole size and the patters (from 200nm to 20nm, and from an arranged order to a quasi-random).

We observe that both the critical temperature and the critical current are not affected by such patterning (comparing to reference non-patterned devices), whereas the sheet resistance can be tuned in a wide range (determined by the filling factor). On the conference, we will present results for both dc and THz characterization.

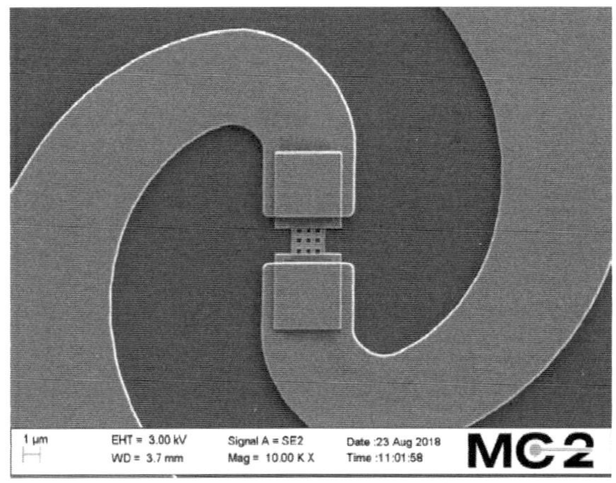


Fig. 1. Scanning Electron Microscope (SEM) image of a nano patterned $MgB_2$ microbridge, integrated with a gold spiral antenna. The microbridge size is 1.5μm×1.5μm.

REFERENCES

[1] T. de Graauw et al., "The herschel-heterodyne instrument for the far infrared (HIFI)," *Astronomy Astrophys.*, vol. 518, 2010, Art. no. L6.

[2] E. Novoselov and S. Cherednichenko, "Low noise terahertz MgB2 hot-electron bolometer mixers with an 11 GHz bandwidth", *Appl. Phys. Lett.* 110, 032601 (2017).

[3] E. Novoselov and S. Cherednichenko, "Gain and Noise in THz MgB2 Hot-Electron Bolometer Mixers with a 30K Critical Temperature", *IEEE Trans.THz.Sci.Technol.* vol. 7 (26) s. 704-710 (Nov. 2017)

NOTES:

# Development of a Ti hot electron bolometer based on Johnson noise thermometry

W. Miao, H. Gao, Z. Wang, W. Zhang, Y. Ren, K.M. Zhou, J.Q. Zhong, and S.C. Shi

Johnson noise thermometry is an important measurement technique used to probe the thermodynamic properties of hot electrons in conductors. Based on this technique, a normal metal hot electron bolometer with high sensitivity and high saturation power has been demonstrated [1]. In this paper, we present the development of a titanium (Ti) hot electron bolometer (HEB) based on Johnson noise thermometry. The HEB consists of a micro-size Ti microbridge and a log spiral antenna. The antenna is made of niobium (Nb) and used as a superconducting Andreev reflector. We measure the power responsivity, thermal conductance and noise equivalent power (NEP) of the Ti HEB at different bath temperatures between 3 K and 12 K. To understand the thermal transport inside the Ti microbridge, we also measure the bolometers with different microbridge lengths at different temperatures. Detailed experimental results and analysis will be presented.

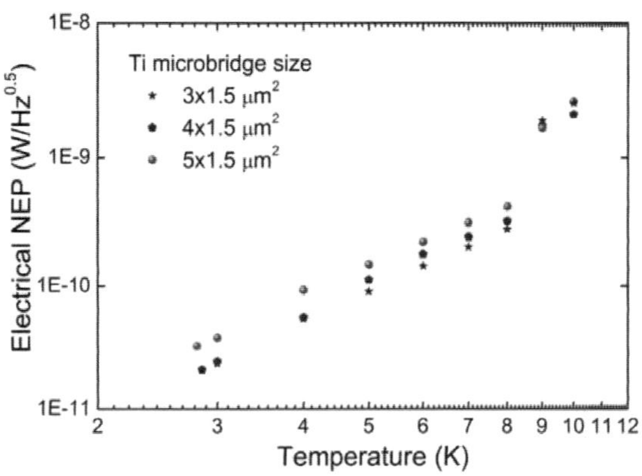


Fig. 1. Measured electrical NEP of the Ti HEBs of different microbridge lengths.

References

[1] B.S. Karasik, C.B. McKitterick, T.J. Reck, and D.E. Prober, "Normal-Metal Hot-Electron Nanobolometer With Johnson Noise Thermometry Readout," *IEEE Transactions on Terahertz Science and Technology*, vol. 5, no. 1, pp.16 –20, Jan. 2015.

Purple Mountain Observatory, CAS, Nanjing 210033, China.

NOTES:

# Measurements of Receiver Noise Temperature of an Ni-NbN HEBM at 2 THz

Yoshihisa Irimajiri* and Akira Kawakami

*National Institute of Information and Communications Technology (NICT), Koganei, Tokyo, 184-0015 Japan*
*Contact: irimaji@nict.go.jp

***Abstract*— We are developing an HEBM at the 2 THz for SMILES-2 for the measurement of Oxygen atom at 2.06 THz and OH at 1.83 THz. We have reported a wide IF bandwidth of ~6.9 GHz for an Ni-NbN HEBM of which a magnetic thin film was deposited between an electrode and a superconducting strip to suppress superconductor under an electrode. In this work, we have measured receiver noise temperature of the same device with a length of 0.1 μm. In order to measure the receiver noise temperature using a thin beam splitter, a LO power was increased by cooling two triplers of an AMC source to 50 K. As a result, more than 100 μW was achieved at 1.85~1.97 THz. The uncorrected receiver noise temperature of an Ni-HEBM was measured to be $T_{rx}$(DSB) ~1,220 K. After the correction of the loss of a band pass filter at the 2 THz which was used to avoid direct detection effect, $T_{rx\_BPFloss\_corr}$(DSB) was ~810 K.**

## I. Introduction

We are developing an HEBM at the 2 THz band for SMILES-2 [1] for the measurement of emission line spectra of Oxygen atom at 2.06 THz and OH at 1.83 THz. We have reported a wide IF bandwidth of ~6.9 GHz for an Ni-NbN HEBM of which a magnetic thin film was deposited between an electrode and a superconducting strip to suppress superconductor under an electrode [2]. We measured receiver noise temperature of the Ni-NbN HEBM to research whether the same device also shows good noise performance.

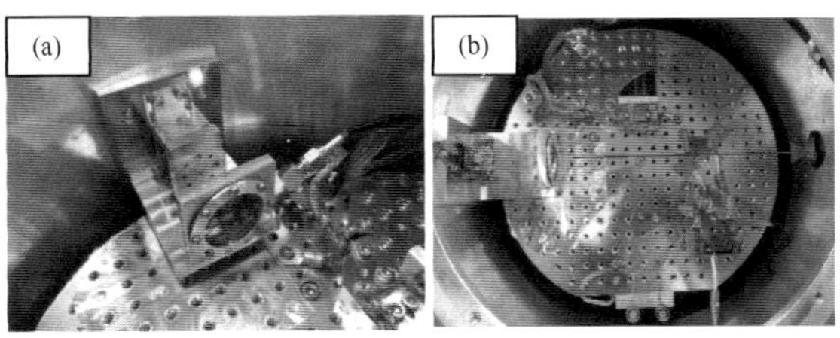


Fig.1. (a, b) Photographs of the two triplers installed at 50 K shield of the cryostat. The output of the THz source is collimated using an AR-coated@150 μm Si lens with f25. The LO signal is taken out from the window at the opposite side.

## II. Experiment and Results

We measured the Ni-NbN HEBM with a length of 0.1 μm and a width of 0.5 μm using conventional Y-factor method. LO output power from an amplifier multiplier chain (AMC) was not enough to pump the HEBM if we use a thin beam splitter film to reduce contribution from RF optics. Therefore, to increase the power, two triplers of the AMC were cooled to 50 K attaching to a 50 K radiation shield of a 4 K mechanical cooler [3]. Figure 1(a, b) show photographs of two triplers installed in the cryostat. The output of the THz source is collimated using an AR-coated@150 μm Si lens with f25. The LO signal is taken out from the window at the opposite side.

Two triplers inside the cryostat and a 140-220 GHz band source outside the cryostat were connected by a WR5.1 waveguide made by stainless with a length of 73 mm (Fig. 2). The insertion loss of the waveguide was measured by a network analyzer to be -1.2~-1.8 dB at 140~220 GHz without a vacuum window of a 6-μm thick Mylar film. The measured waveguide consists of a copper-plated waveguide and a gold-plated thin waveguide inside.

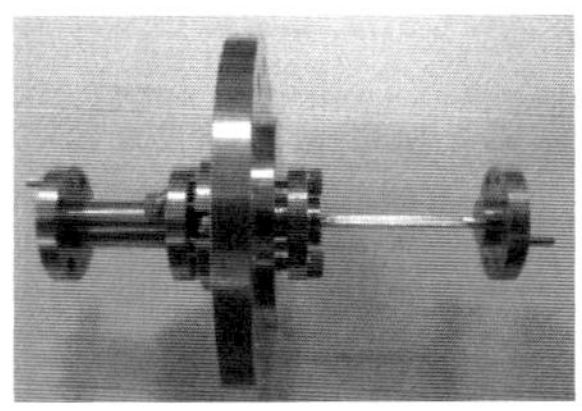

Fig.2. Waveguide (WR5.1) made by stainless to connect a 140-220 GHz band source at 300 K and two triplers at 50 K inside of the cryostat. The left part of the waveguide is plated by copper and right part (thin waveguide) is plated by gold. The measured losses are -1.2~-1.8 dB at 140~220 GHz. The waveguide was made in Japanese company (Oshima). The waveguide installed into a cryostat, however, is copper-plated inside in all parts.

The output power of the LO at the 2 THz was measured by a pyroelectric detector. As a vacuum window, a 6-μm thick Mylar film was used. In this measurement, the waveguides plated by copper inside in all parts were used. The measurement setup and the result are shown in Fig. 3(a, b). The power increased ~4 times compared to the original one also thanks for VDI giving a new tripler. More than 100 μW was achieved at 1.85~1.97 THz. We also measured the output power without water vapor absorption using a pyroelectric detector which is set in a nitrogen gas filled box. The measured result indicates the deep depression at 1.87 THz, 1.92 THz, and 2.04 THz shown come from water vapor absorption.

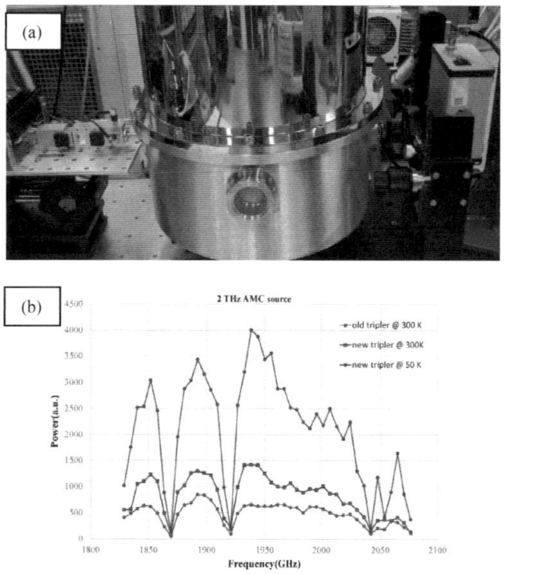


Fig.3. (a) Measurement set up of an output power of a cryogenic LO using a pyroelectric detector. (b) Measured LO power of original triplers at 300 K (blue line), new triplers at 300 K (brown line), and cryogenic triplers at 50 K (red line). The LO power increased ~4 times compared to the original one. More than 100 µW was achieved at 1.85~1.97 THz. The deep depression at 1.87 THz, 1.92 THz, and 2.04 THz come from water vapor absorption.

In order to measure receiver noise temperature, the LO signal was taken out from the cryostat and reflected by a polyester beam splitter (BS) with a thickness of 4 µm of which the reflection coefficient is ~5 %, and was fed into an HEBM cooled by an another LHe dewar (Fig. 4). As a first step experiment, the BS and HEBM were put outside the cryostat because alignment is difficult inside the cryostat at the high frequency.

By pumping the HEBM at an optimal level, the uncorrected receiver noise temperature ($T_{rx}$) was measured to be ~1,220 K(DSB). After the correction of the loss of a band pass filter at 2 THz which was used to avoid direct detection effect, $T_{rx\_BPFloss\_corr}$(DSB) was ~810 K. We measured the same HEBM device which shows a wide IF bandwidth.

Fig.4. Photograph of measurement set up of receiver noise temperature of the Ni-NbN HEBM. The LO signal was taken out from the cryostat and reflected by a polyester beam splitter (BS) with a thickness of 4 µm of which the reflection coefficient is ~5 %, and was fed into an HEBM cooled by an another LHe dewar. As a first step experiment, the BS and HEBM were put outside the cryostat because alignment is difficult inside the cryostat at the high frequency.

## III. Summary

We measured receiver noise temperature of the Ni-NbN HEBM to be ~810 K ($T_{rx\_BPFloss\_corr}$) using a cryogenic LO of which two triplers are cooled to 50 K. We confirmed the Ni-NbN HEBM which shows a wide IF bandwidths of ~6.9 GHz also shows a good performance in noise temperature. We plan to develop a waveguide-type HEBM at 2 THz with a corrugated or a spline-profile horn for the applications of atmospheric observations. We expect the LO coupling could be better for the waveguide mixer.

## References

[1] S. Ochiai, P. Baron, T. Nishibori, Y. Irimajiri, Y. Uzawa, T. Manabe, H. Maezawa, A. Mizuno, T. Nagahama, H. Sagawa, M. Suzuki, and M. Shiotani, "SMILES-2 Mission for Temperature, Wind, and Composition in the Whole Atmosphere," *SOLA*, vol. 13A, pp. 13–18, 2017. DOI: 10.2151/sola.13A-003

[2] A. Kawakami, Y. Irimajiri, T. Yamashita, S. Ochiai, Y. Uzawa, "Broadening the IF band of a THz hot electron bolometer mixer by using a magnetic thin film," *IEEE Trans. on THz Sci. and Technol.*, vol. 8, no. 6, pp. 647–653, 2018. DOI: 10.1109/TTHZ.2018.2874355

[3] N. Reyes, R. Güsten, U. Graf, C. Risacher, O. Ricken, "A cryogenic solid state LO source at 1.9THz", in *Proc. 28th Int. Symp. Space THz Techn.*, Cologne, Germany, Mar. 2017.

# Bandwidth of a 4.7 THz asymmetric Fourier grating

Y. Gan, B. Mirzaei, J. R. G. Silva, W. Laauwen, F.F.S. van der Tak, and J. R. Gao

*Abstract*—**We present an analysis of the bandwidth of a preliminary designed asymmetric 8-pixel Fourier grating as the beam multiplexer for the 4.7 THz local oscillator of the GUSTO mission. We take the GUSTO grating as an example to address the bandwidth question although GUSTO itself does not need to operate over a wide frequency range. By illuminating single beams with different frequencies from 4.445 THz to 5.045 THz to the grating, we simulated the changes in the grating's performance in three aspects using COMSOL Multiphysics: diffraction efficiency, power distribution, and the angular distribution of the output beams. These parameters can reduce the coupling efficiency between the output beams of the grating and the beams of the mixer array of GUSTO. The grating's bandwidth is calculated to be 250 GHz, which is sufficient for many applications.**

*Index Terms*— **Fourier grating, Bandwidth, Mixer array, Coupling efficiency.**

## I. Introduction

HETERODYNE detection is widely used to detect atomic fine structure lines and molecular rotational lines in the terahertz (THz) frequency region from the interstellar medium (ISM). This technique provides very high spectral resolution, R > $10^6$. Heterodyne receivers convert a sky signal in THz down to gigahertz frequency by mixing the weak celestial signal with a signal from a local oscillator (LO). In the supra-THz region ( > 1 THz ), quantum cascade lasers (QCLs) provide considerably higher output powers compared to LO sources based on multipliers [1]. Therefore, the 4.745 THz band in the Galactic/ Extragalactic ULDB Spectroscopic Terahertz Observatory (GUSTO) [2], aiming for detection of [OI] line emission from the Milky Way and nearby galaxies, especially the Magellanic Clouds, combines a QCL and a Fourier grating as the LO for the 8-pixel hot electron bolometer (HEB) lens-antenna coupled mixer array. The array is crucial to enhance observation speed. In order to detect the Doppler shift caused by the linear velocity of the ISM beyond our galaxy, a sizable tuning range of LO will be necessary since the IF bandwidth of a HEB mixer available in this frequency range is limited. So, an interesting question is how large the bandwidth of a Fourier grating can be.

A Fourier grating as a multiplexer to diffract a single beam from the QCL to multiple beams is designed to work across a finite frequency range. For a given Fourier grating, the change of the source frequency causes a loss in the coupling efficiency between the image beams of the grating and the beams of the mixers [3]. So the bandwidth of a Fourier grating is characterized by the interplay between the grating and an array. Thus, the bandwidth analysis should take the specific array used into account. Fig. 1 shows a conceptual diagram of the 4.745 THz array receiver . A single QCL beam is first diffracted to 8 beams by a reflective Fourier grating, which are collimated by a parabolic mirror, becoming parallel to each other. These 8 parallel beams are then coupled to the lens-antennas of the mixer array. We analyze the bandwidth of an 8-pixel Fourier grating designed for the 4.745 THz band of GUSTO, as an example. We notice that GUSTO itself does not require a large bandwidth since it aims to detect [OI] lines from the Milky Way. Furthermore, we took a GUSTO grating design available at the time when we performed the analysis, which is not the final one. We do believe that the approach we present here should be applicable for any combinations of a grating LO with an array.

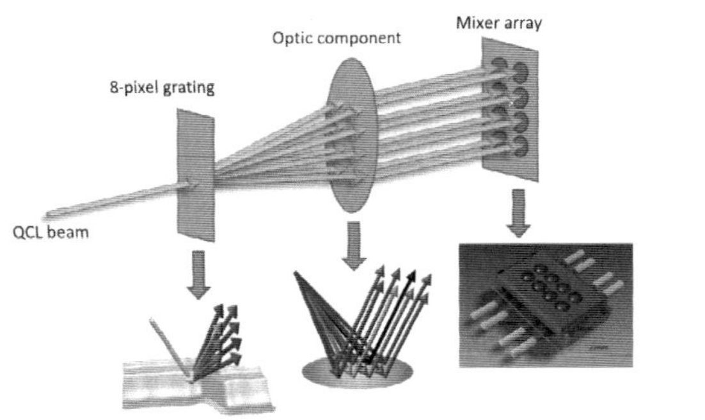


Fig.1. Conceptual diagram of the 4.745 THz 8 beam local oscillator for the GUSTO mission. The QCL beam is first diffracted by a

This work was supported in part by EU RadioNet Project under Grant 730562 and in part by TU Delft Space Institute. Y. Gan is supported by China Scholarship Council Studentship (201706030153) and by University of Groningen.

Y. Gan, B. Mirzaei, J. R. Silva, W. Laauwen, F.F.S. van der Tak and J. R. Gao all work in SRON Netherlands Institute for Space Research, Groningen / Utrecht, the Netherlands. (emails: y.n.gan@sron.nl, b.mirzaei@sron.nl, j.r.g.d.Silva@sron.nl, w.m.laauwen@sron.nl, F.F.S.van.der.Tak@sron.nl and j.r.gao@sron.nl).

Y. Gan, J. R. Silva and Floris van der Tak have second affiliation in Kapteyn Astronomical Institute, University of Groningen, 9747 AD, Groningen, The Netherlands.

B. Mirzaei and J. R. Gao have second affiliation in faculty of applied sciences of Delft University of Technology, Delft, the Netherlands.

reflective phase grating, and is then collimated by a parabolic mirror to make the 8 parallel beams. These 8 beams are coupled to a quasi-optical mixer array.

## II. Simulated 4.745 THz Fourier Grating

A phase grating consists of a periodic structure to diffract a single beam to multi-beams in different directions through phase modulation. According to diffraction theory, the diffracted far field distribution from a grating can be expressed as the Fourier transform of the grating's transmission/reflection function [4]. Gratings using Fourier synthesis technique to achieve continuous phase-only groove shapes are called Fourier gratings [5]. To design a Fourier grating, we expanded the phase modulation function of the grating to Fourier series with a set of Fourier coefficients $a_n$. Using the Fast Fourier Transform and the Standard multidimensional minimization algorithm in Matlab, we found a set of $a_n$ for a one-dimensional grating with the desired number of pixels. A two dimensional grating is generated by superimposing two 1D profiles orthogonally. In our case, one direction is a 2-pixel 1D grating, and the other direction is a 4-pixel 1D grating. According to the grating equation $D(sin\theta_m - sin\theta_i) = m \cdot \lambda$, the direction of the output beam $\theta_m$ in the diffraction order m (integer) is determined by the grating period $D$, the incident angle with respect to the normal of the grating $\theta_i$ and working wavelength $\lambda$. An asymmetric grating is designed to accommodate the requirements of the GUSTO optical system by employing consecutive diffraction orders [6]. For the 4-pixel 1D grating, we chose the (-2,-1,0,+1) diffraction orders, and for the 2-pixel 1D grating, we chose the (-1,0) diffraction orders. The surface topology and specifications (as simulated with COMSOL Multiphysics [7]) of the design are shown in Fig. 2 and Table I, respectively.

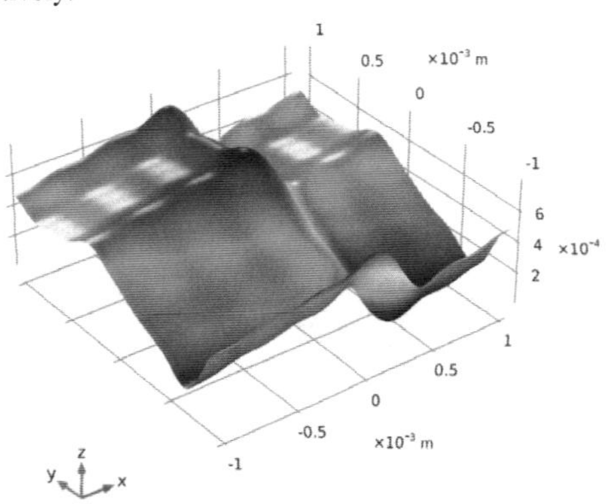


Fig. 2. The surface topology of one unit cell of the grating [6]. The input QCL beam illuminates the grating surface with an incident angle 15° in the x direction and 0° in the y direction.

TABLE I
Grating Specifications

| | |
|---|---|
| Working frequency | 4.745THz |
| Material | Aluminum |
| Angular distribution | 1.83° |
| Incident angle | 15° |
| Unit cell size | 2.04mm×1.979mm |
| Diffraction orders | (-2,-1,0,1)(-1,0) |
| Diffraction efficiency | 70.2% |
| Uniformity deviation = $(I_{max}-I_{min})/I_{average}$ | 12.5% |

If the incident angle increases, the power variation among the output beams becomes larger [8]. The different unit cell size in two directions is to make the angular distribution in these two directions the same. The diffraction efficiency, defined as the ratio of the total diffracted beam power to the power of the incoming beam. The power variation is simulated by using COMSOL Multiphysics and by importing the surface topology of the grating. We apply the periodic port with periodic boundary condition in the RF module, and extract the S-parameters of the port. The power distribution of the output beams is plotted in Fig. 3(a), where we find the largest variation to be 12.5% (between the maximum power and the minimum power). By importing the surface profile of the designed unit cell of the grating and by repeating it in both orthogonal directions, while taking the input as a Gaussian beam, we simulated the far field beam pattern of the grating. The outcome is shown in Fig. 3(b), where the $m$ and $n$ are the diffraction orders in both directions. From the results in Fig. 3, we conclude that the grating achieves a good power uniformity among the output beams.

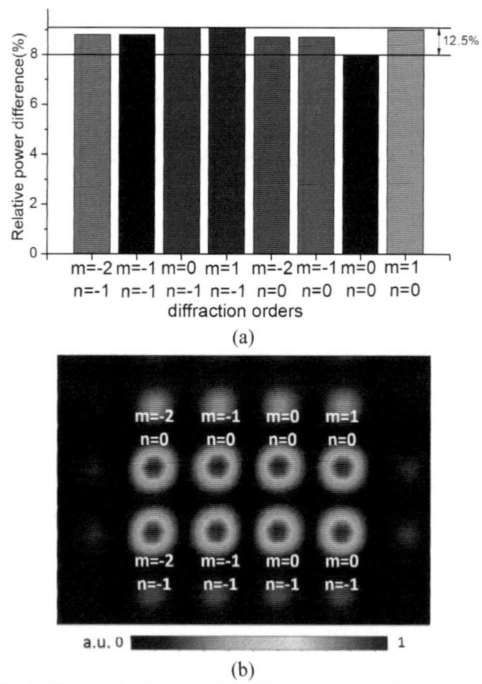


Fig. 3. (a) Power distribution of the 8 output beams from the grating. The largest variation is 12.5% (between the maximum power and the minimum power). (b) Far field beam pattern of the grating. The m and n are the diffraction orders in two directions.

## III. Calculation of the Grating Bandwidth

The 8 output beams from the grating are used to pump an 8-pixel mixer array in the 4.745 THz band of GUSTO. The power variation among the LO beams can degrade the sensitivity of the mixers in the array since it depends on LO power. We assume that the latter should be within 5% of the optimal value, and we also assume all HEBs in the array require the exact same LO power. The corresponding LO power variation is estimated to be ~ 21% using the isothermal technique [9,10,11]. *In this paper, we use this criterion to define the bandwidth of the grating, namely, as the frequency range changes such that the LO power variation among the array mixers is within 21%.*

The change of the frequency affects the performance of the grating in three aspects: (a) frequency change causes a change in the diffraction efficiency $\eta$; (b) Frequency change causes a change in the power distribution of the outcome beams; (c) According to the grating equation, the frequency change affects also the angular distribution of the output beams, which reduces optical coupling to the lens-antennas of the mixer array.

Based on these three aspects, we defined Gaussian beams with different frequencies from 4.445 THz to 5.045 THz to illuminate the grating used in COMSOL Multiphysics. The simulation results, in which the changes in diffraction efficiency when the frequency of the beams is changed, are shown in Table II.

TABLE II

The diffraction efficiencies $\eta$ corresponding to different frequencies $f$ from 4.445THz to 5.045 THz

| $f(THz)$ | 4.445 | 4.545 | 4.645 | 4.745 |
|---|---|---|---|---|
| $\eta$ | 70.6% | 70.5% | 70.4% | 70.2% |
| $f(THz)$ | 4.845 | 4.845 | 5.045 | |
| $\eta$ | 70.0% | 69.7% | 69.4% | |

Table II suggests that the maximal change in diffraction efficiency by varying the frequency is 0.8%, which is negligible compared to other effects (the power distribution and the angular distribution).

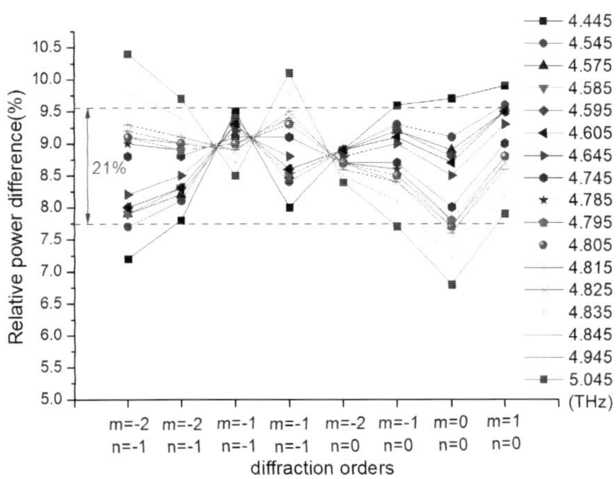


Fig. 4. Power distribution of the output beams when the working frequency changes from 4.445 THz to 5.045 THz. Red dashed lines indicate the LO power boundaries, within which the powers of the output beams vary within 21% around their average value.

When the frequency changes, the power distribution among the output beams varies. When the grating works at the nominal (or designed) frequency of 4.745 THz, its maximal power variation of the output beams is 12.5% (between the maximum power and the minimum power). When the frequency changes, this number becomes larger. Fig. 4 plots the power distribution of the output beams of the grating operated at different frequencies. The red dashed lines indicate the LO power boundaries, within which the power of the output beams varies within 21% around their average value. From Fig. 4, when the frequency changes between 4.575 THz and 4.825 THz, the relative powers of all the output beams from the grating vary within 21% around their average value (8.6%). Base on this we derive that the bandwidth of the grating is 250 GHz.

Now we examine the angular distribution of the output beams as a function of the frequency. Since the pixel spacing of the mixer array is fixed, the change in the spacing of the beams can lead to offsets with respect to those of the mixer array. According to the relation between the offset and Gaussian beam coupling [12], we can calculate the coupling loss caused by the offset at different frequencies. The results are shown in Fig. 5.

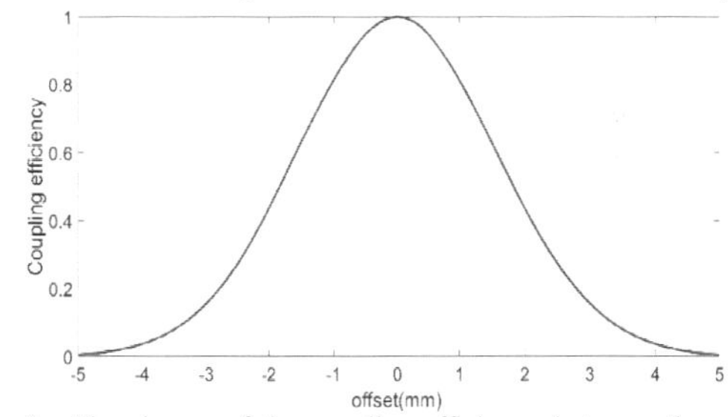


Fig. 5. The change of the coupling efficiency between the output beams and the mixer array caused by the spatial offset between them (or called misaligned).

Based on the grating equation, we calculated the angular distribution of the output beams and the offset between the output beams and the lens-antennas at different frequencies. Then, taking advantage of the offset dependence of coupling efficiency in Fig. 5, we obtained the coupling loss caused by the offset. The next step is to calculate the distribution of powers that are effectively coupled to the mixer array. The results are shown in Fig. 6.

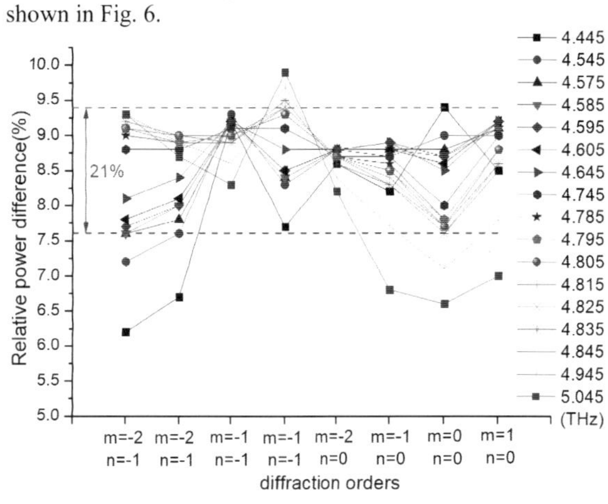

Fig. 6. Power distribution of the output beams, taking the angular distribution into consideration when the working frequency changes from 4.445 THz to 5.045 THz. Red dashed lines indicate the LO power boundaries, within which the powers of the output beams vary within 21% around their average value.

In Fig.6 the red dashed lines indicate the LO power boundaries, within which the powers of the output beams vary within 21%, where the average value is 8.5 %. We find that the coupled powers are still within 21% with frequency changing between 4.575 THz and 4.825 THz. Therefore, from this aspect, the bandwidth of the grating is also 250 GHz.

## IV. Conclusion

We analysed the bandwidth of an asymmetric grating preliminary designed for the 4.745 THz local oscillator for GUSTO in three aspects; diffraction efficiency, power distribution and angular distribution. We found that for a 4.745 THz asymmetric grating, its output beam power variation remains within 21% when the operating frequency changes from 4.575 THz to 4.825THz, which gives a 250 GHz frequency bandwidth. The effect is dominated by the power distribution and angular distribution, while the diffraction efficiency remains nearly unchanged. The 250 GHz bandwidth corresponds to 15,000 km/s linear velocity, which is much more than the line widths of the galactic (~100 km/s) and extragalactic (~1000 km/s) objects.

## References


[1] J. V. Siles, R. H. Lin, C. Lee, E. Schlecht, A. Maestrini, P. Bruneau, A. Peralta, J. Kloosterman, J. Kawamura and I. Mehdi, "Development of high-power multi-pixel LO sources at 1.47 THz and 1.9 THz for astrophysics: present and future," 26th International Symposium on Space Terahertz Technology (ISSTT 2015), pp. 40–42.

[2] Walker, C. K. et al. GUSTO: Gal/Xgal U/LDB Spectroscopic-Stratospheric Terahertz Observatory. In Am. Astron. Soc. Meeting Abstracts Vol. 231 (2018).

[3] Robert May, J. Anthony Murphy, Creidhe O'Sullivan, Marcin Gradziel, Neil Trappe, "Gaussian beam mode analysis of phase gratings", Proc. SPIE 6893, Terahertz Technology and Applications, 68930G (2008).

[4] Goodman, J.W., "Introduction to Fourier Optics," McGraw-Hill, 1968.

[5] P. BliekR. DeleuilM. BreidneD. Maystre, "Microwave verification of a numerical optimization of fourier gratings," Appl. Phys. 24, 147-150 (1981).

[6] B. Mirzaei, Y. Gan, J. R. G. Silva, S. Yates, W. Laauwen, and J. R. Gao, " Asymmetric phase grating as 4.7 THz beam multiplexer for GUSTO", 30th International Symposium on Space Terahertz Technology (ISSTT 2019).

[7] COMSOL Multiphysics, https://www.comsol.com/rf-module.

[8] B. Mirzaei, J. R. G. Silva, Y. C. Luo, X. X. Liu, L. Wei, D. J. Hayton, J. R. Gao and C. Groppi, "Efficiency of multi-beam Fourier phase gratings at 1.4 THz," Opt. Express 25, 6581-6588 (2017).

[9] H. Ekström, . S. Karasik, E. Koll erg and K. S. Yngvesson, "Conversion gain and noise of niobium superconducting hot-electron-mixers," IEEE Trans Microw. Theory Tech. 43, 938–947 (1995).

[10] P. Khosropanah, J. R. Gao, W. M. Laauwen, and M. Hajeniusb, "Low noise NbN hot electron bolometer mixer at 4.3 THz," Appl. Phys. Lett. 91, 221111, (2007).

[11] The isothermal technique is based on the assumption that the mixer has the same response to LO power and DC power, we can estimate the absorbed LO power in a mixer from the pumped current-voltage curves. In [9] P. Khosropanah, et al measured a set of current-voltage curves of a similar HEB mixer pumped with different LO power operated at 4.3 THz, and from the data when the LO power varies within ~21% the change in HEB mixer's noise temperature is ~5%.

[12] Paul F Goldsmith, Institute of Electrical, Electronics Engineers, Microwave Theory, and Techniques Society, "Quasioptical systems: Gaussian beam quasioptical propagation and applications," IEEE press New York (1998).

# A planar silicon metamaterial lens with integrated anti-reflection coating for frequencies around 150 GHz

J. Jauk[1,*], G. Wang[1,*], M. Parioleau[1], A.L. Fontana[1], S. Leclercq[1], and E.F.C. Driessen[1]

For quasi-optical elements in the millimeter and sub-millimeter range, silicon is an interesting material. Its high refractive index facilitates the production of compact and lightweight elements. Moreover, its thermal conductivity allows better thermalization at cryogenic temperatures, and the loss tangent of bulk high-resistivity silicon ($\tan \delta < 10$-4) is without competition.

Silicon is however very difficult to machine, and the high refractive index necessitates the use of anti-reflection coatings. Micromachined anti-reflection coatings have been developed for planar substrates but become increasingly more difficult for curved surfaces of e.g. lenses.

In this work, we follow a different approach. We use the fact that it is possible to modulate the refractive index of a material by inserting sub-wavelength voids and changing the fill factor of the voids. This way, a silicon metamaterial with a dielectric constant between 3.3 and 11.7 can be generated [1].

From (visible) optics it is well known that a curved surface, such as a lens, can be mimicked by a planar element that has the appropriate refractive index gradient. Thus, we designed a planar silicon element, with an expected focal length of 180 mm. The lens has a diameter of 50 mm, and the effective dielectric constant varies continuously from 11.2 in the center to 3.3 at the edges of the lens, by the means of an hexagonal array of holes with a period of 104 μm and a hole size varying from 13.4 μm in the center to 81 μm at the edges. The total thickness of the lens is 1 mm, which is about an order of magnitude thinner than an equivalent (curved) polyethylene lens with a similar focal length.

The element was fabricated out of four 250 μm thick, high-resistivity silicon wafers that were micromachined using a Bosch process in an inductively coupled plasma etcher. The wafers were aligned using dowell pins, and pressed together in a dedicated holder. The same process was used to fabricate two anti-reflection coating layers, using 250 μm wafers with an adapted dielectric constant profile, such that the dielectric constant is given by $\varepsilon AR(x) = \sqrt{\varepsilon lens(x)}$. The thickness of the AR coating is chosen such that the averaged reflection over the surface of the lens is minimized, since the optimal quarter-wave adaptation is impossible using a planar design with varying dielectric constant. Note that the modular design of our devices would easily allow for more intricate AR coatings consisting of multiple layers of varying thickness and dielectric constant.

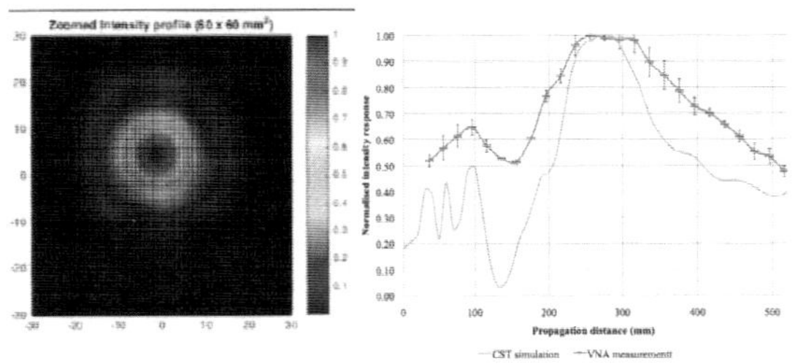


Fig. 1. (left) Measured beam profile of a microwave feed horn imaged by the silicon metamaterial lens. (right) On-axis intensity as a function of distance, measured (red) and as simulated with CST (green). Both measurements and simulations are made at a frrequency of 150 GHz.

The fabricated lens was subsequently characterized in an antenna range. We find that the imaging properties of the lens are excellent, but that the effective focal length is approximately 30% smaller than designed. The measured value is confirmed by CST simulations of the structure (see Fig. 1). This difference is attributed to the fact that the lens is of comparable thickness to the wavelength, privileging near-field effects, and rendering ray-optical approaches such as used for the design insufficient. Simple transmission measurements indicate the effectiveness of the AR coating.

The presented technology offers great perspective in terms of compact, planar, low-loss optics. Moreover, the technology can be easily integrated with silicon detector wafers, and future developments that involve more elaborate anti-reflection coatings, integrated filtering, or microlens arrays, are just part of the possibilities.

### References

[1] K.F. Schuster et al., "Micro-machined quasi-optical elements for THz applications", *16th International Symposium of Space Terahertz Technology*, p. 524-528 (2005).

[1] Institut de Radioastronomie Millimétrique, Grenoble, France
[*] These authors contributed equally to this work.

NOTES:

# Asymmetric phase grating as 4.7 THz beam multiplexer for GUSTO

B. Mirzaei, Y. Gan, J. R. Silva, M. Finkel, S. Yates , W. Laauwen, A. Young , C. Walker, and J. R. Gao

***Abstract*—The grating design for the 4.7 THz channel of GUSTO (Galactic/ Extragalactic ULDB Spectroscopic Terahertz Observatory) has been reported in this paper, which acts as a beam multiplexer for coupling a single QCL (Quantum Cascade Laser) beam to an 8-pixel mixer array. The design and analysis are based on modeling and simulations showing a successful accommodation of the requirements from the designed optical system. The asymmetric feature is used and to be applied for the first time at such a frequency range.**

***Index Terms*— Array receivers, GUSTO, Heterodyne, Phase grating, Terahertz.**

## I. Introduction

GUSTO is a super-THz heterodyne instrument planned to be launched at 2021 from Antarctica to mainly study the star formation and the life cycle of the interstellar clouds in our galaxy and beyond. Molecular and atomic fine structure lines at three scientifically valuable frequencies of 1.4, 1.9 and 4.7 THz will be measured by GUSTO, each with an 8-pixel heterodyne receiver, continuously for around 100 days, the most efficient super-THz observation ever.

A single heterodyne receiver with extremely high spectral resolving power of $10^7$ consists mainly of a mixer and a local oscillator (LO). The former mixes the celestial signal with LO in order to down-convert it to the gigahertz range. A multi-pixel heterodyne receiver requires multiple parallel receivers; each with its own LO beam. In general, advanced fabrication technology allows making relatively uniform array of mixers, so that within a certain variation of the LO power they still deliver receiver sensitivities with negligible differences.

While the mixer technology (superconducting hot electron bolometer mixers) is the same for all three frequency bands of GUSTO, the local oscillators of the two lower bands are solid-state frequency multiplied sources and for the 4.7 THz channel, a QCL [1]. Waveguide splitters are used to generate 8 LO beams for lower channels and a phase grating is applied for multiplexing a QCL beam into 8 for the highest channel. The phase grating is the only applicable multiplexer technology at such a high frequency. It is a periodic arrangement of a unit cell with a specific surface morphology for phase manipulation of the incident coherent radiation in order to make multiple image beams in the far-field. Here we report the design and analysis of a phase grating fulfilling the requirements of the optical system design for the highest frequency receiver of GUSTO.

## II. Requirements

Although we have recently demonstrated and published the base technology of THz phase gratings [2,3], geometrical limits of the higher frequency channel of GUSTO demands an advanced design to accommodate the tough requirements on the beam distribution scheme. The force is originated from the fixed distance of the mixer array to the cryostat window, which together with the strict volume constraint of the warm coupling optics determine the beam size on the phase grating to be 2.95 mm of radius. This consequently defines the unit cell size upper limit since for grating to function, at least two unit cells should be covered by the beam.

Since symmetry makes the application and analysis of an optical component easier, previously demonstrated THz gratings [2,3] provide a symmetric spatial distribution among the diffracted beams. 8-pixel gratings for example have symmetric employment of the diffraction orders (±1, ±3), which leave a gap of one beam in between the adjacent pixels. The tight required angular distance of 1.83° in combination with the incident angle of 15° put a lower limit on the unit cell size too, since the latter has an inverse proportion to the beams separation. Such a confined range for the unit cell size does not leave any room to miss the intermediate diffraction orders in the symmetric structures. In other words, all the consecutive orders have to be employed to realize the small distances between the beams, which immediately breaks the symmetry since the array has a 2x4 arrangement. We have created an alternative design based on asymmetry to accommodate the abovementioned requirement being elaborated in the next section.

## III. Design

We use the method given in references [2,3] with a

B. Mirzaei, J. R. Silva, M. Finkel, S. Yates, W. Laauwn, and J. R. Gao all work in SRON (Netherlands institute for Space Research) with the following corresponding emails: b.mirzaei@sron.nl, j.r.g.d.Silva@sron.nl, matveyf@sron.nl, s.yates@sron.nl, w.m.laauwen@sron.nl, and j.r.gao@sron.nl. B. Mirzaei and J. R. Gao have second affiliation in faculty of applied sciences of Delft University of Technology.

Y. Gan is a PhD student in Kapteyn Astronomical Institute of University of Groningen and SRON with the email: y.n.gan@sron.nl

A. Young and C. Walker are both in University of Arizona with the corresponding emails: agyoung@email.arizona.edu, and cwalker@as.arizona.edu.

modification for generating asymmetric feature for designing the grating. Since the surface profile in the mentioned references is symmetric and so an even function, the Fourier series include only cosine terms. For the asymmetric design however, we applied both sine and cosine terms in the Fourier series. Doing so and performing an optimization process we could come up with a surface profile fulfilling the requirements mentioned in the previous section. The 3D plot of such a profile together with the cross sections are shown in Figure 1. Each orthogonal cross section is responsible for multiplexing the beam either to 2 or to 4 so that a rectangular 8-beam pattern is generated.

The computational characterization is performed using 3D modeling in COMSOL Multiphysics, while the design is done

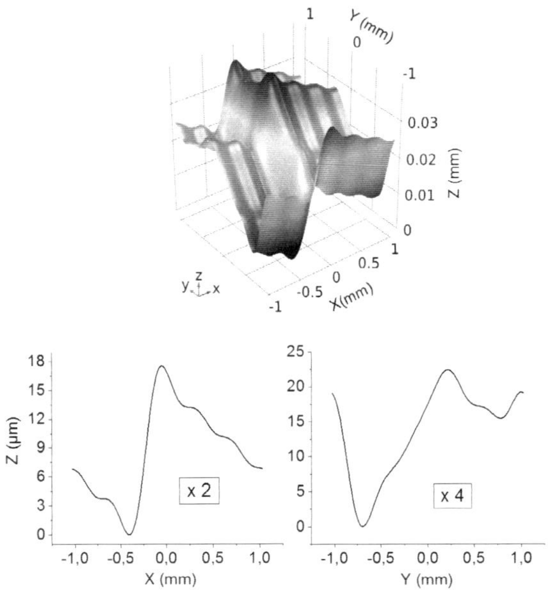


Fig. 1. 3D plot of the designed surface profile (top). At the bottom, orthogonal cross sections are plotted, each multiplexing the beam to 2 (left) or 4 (right).

in MATLAB. The performance of the grating can be assessed in 3 main aspects. The first is the efficiency defined as the ratio between the total power in the desired modes and the power of the incoming beam. We expect an efficiency of about 70 % for this design.

The second aspect is the uniformity of the power distribution among the diffracted beams. Assuming an array of similar mixers optimized for a certain LO power, variation among the diffracted beams degrades the average sensitivity of the array. In this sense more uniformity is desirable. For this design we expect a power variation of about 13 %, which is defined as $(P_{max}-P_{min})/P_{average}$ . This level of variation lowers the array's average sensitivity by less than 2 %.

The third important factor is the spatial distribution of the beams, which should comply with the requirement of the optical system, in which the grating will be implemented. The angular distance between adjacent beams is expected to vary between 1.83° and 1.849°. Such a small deviation from the requirement of 1.83° has a negligible effect on the coupling of the beams to the mixer array.

## IV. Influence of the Machining Accuracy

The designed surface of the grating will be transferred to an aluminum plate using a CNC (Computer numerical control) micro-milling machine. Figure 2 shows an SEM (Scanning Electron Microscope) photo of the 4.7 THz grating reported in [2] giving a good impression of the milling capability. Having a large minimum radius of curvature of 1.4 mm makes it easier to manufacture in the sense that the structures are not extremely fine. However, reaching the required machining accuracy might be another issue, which is discussed here in this section.

In order to effectively model the effect of possible machining inaccuracies on the grating performance a Monte Carlo study is desirable. However, since our simulation tool

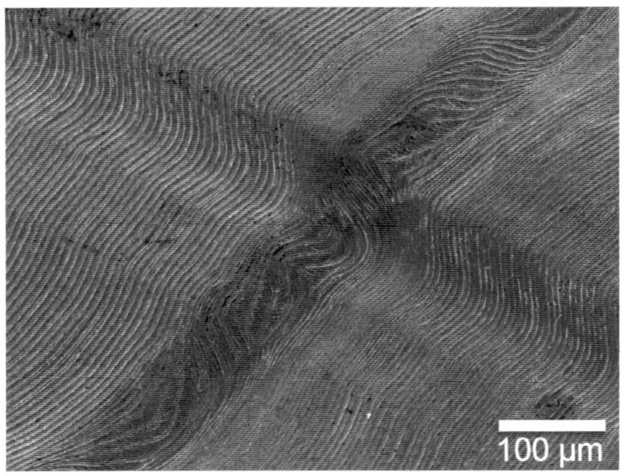


Fig. 2. An SEM photo of the machined 4.7 THz grating reported in [2].

uses finite element modeling with a long processing time, it is not feasible to have such a study. Therefore we include surface deviations to our model to make the closest similarity to the potential machining errors. For that, a sine variation is added on top of the ideal grating profile with 1 mm of period and different amplitudes. This variation is applied in two orthogonal phases to cover the possible effect of the phase in order to derive the largest effect on the performance. Such variations with ± 1 µm are represented in figure 3, where all the surfaces are in the same scale let the reader have a clearer comparison.

Since the beams traveling direction is solely dependent on the unit cell size, this factor does not change with surface deviation. This simulation shows that such surface deviations cause negligible effect on the grating efficiency, while the main influence is on the power uniformity among the beams. Table 1 summarizes the latter for the phase with larger effect for different extents of deviation from ± 0.2 to ± 1 µm. For GUSTO we choose a ± 0.5 µm allowable tolerance in machining, which together with power variations caused by

Fig. 1. Applied sine variations with 1 mm of period and 1 μm of amplitude on top of the grating surface profile.

other elements in the optical system degrades the average sensitivity of the array by 3 %. This is seen to be tolerable for the mission science goals.

It should be mentioned that GUSTO would have a finely adjustable mount for grating which can correct any possible errors caused during mounting i.e. tilt, rotation and translation. We also evaluate the effect of possible temperature variation during the mission on the unit cell size through thermal expansion/contraction. We find a negligible influence on the grating performance.

TABLE I
SIMULATION RESULTS OF THE EFFECT OF THE SURFACE VARIATION ON THE POWER UNIFORMITY OF THE GRATING

| Variation (μm) | Uniformity Deviation |
|---|---|
| ± 1 | 25 % |
| ± 0.8 | 23 % |
| ± 0.7 | 21 % |
| ± 0.6 | 19 % |
| ± 0.5 | 17 % |
| ± 0.4 | 15 % |
| ± 0.2 | 14 % |
| Ideal | 13 % |

## V. Conclusion

The design and expected performance of an asymmetric phase grating being implemented as a 4.7 THz beam multiplexer has been discussed in this paper. This grating is capable to fulfill the requirements by the designed optical system of the higher frequency channel of GUSTO. An analysis over the effect of the machining accuracy is also given in this paper for the first time.

## References

[1] A. Khalatpour, J. Reno, and Q. Hu, "Phase-locked photonic wire lasers by π coupling," Nat. Photonics 13, 47–53 (2019).

[2] B. Mirzaei, J. R. G. Silva, D. Hayton, C. Groppi, T. Y. Kao, Q. Hu, J. L. Reno, and J. R. Gao, "8-beam local oscillator array at 4.7 THz generated by a phase grating and a quantum cascade laser," Opt. Express 25, 29587-29596 (2017)

[3] B. Mirzaei, J. R. G. Silva, Y. C. Luo, X. X. Liu, L. Wei, D. J. Hayton, J. R. Gao and C. Groppi, "Efficiency of multi-beam Fourier phase gratings at 1.4 THz," Opt. Express 25, 6581-6588 (2017).

# Development of mm/submm Frequency Selective Filters made with FPC Fabrication Technology

Shinsuke Uno[1,2], Keisuke Yoshioka[3,4], Tai Oshima[3], Tatsuya Takekoshi[1], Kotaro Kohno[1], and Kah Wuy Chin[2,3]

We report the initial evaluation results of the prototyped mm-wave metal mesh filters, which are designed as a band-pass filter for a mm/submm broadband multichroic camera covering 130–295 GHz.

Band-pass filters are one of the key components for mm/submm wave imaging and spectroscopy instruments, but it typically takes a very long time for its procurement because of the very limited availability of the design and fabrication company. Here we propose a new method to produce a broad-band mm/submm-wave band-pass filters, which exploits recent rapid advancement of commercially available flexible printed electronic circuit (FPC) fabrication technology.

We have simulated transmittance of mesh patterns using the multiphysics simulation software COMSOL RF module. and have prototyped metal mesh filters with the same method for printed circuits. Our pattern design is based on the idea that optical properties of mesh patterns can be simply represented as transmission line equivalent circuits[1]. From electromagnetic field simulations, we concluded that the best mesh pattern for broad band-pass and steep cutoff is a stack of combination layers of hexagonal grid mesh and hexagon with optimal spacing. Our goal here is to fabricate a band-pass filter covering 130 - 295 GHz, but considering the availability of transmittance measurement, we also produced scaled models of the filters covering 260-590 GHz, which is measurable using THz-TDS.

After microscopic inspection of the fabricated patterns, we measured the transmittance using THz-TDS. An example of measured and simulated transmission curves is shown in Fig. 1. We find that the pass-bands were broader than designed because of narrower line widths presumably by over-etching during the fabrication process. This line width narrowing will mitigated by the preceding calibration of the etching process. Measurements also show higher transmittance than that of simulations, and unexpected resonances in bands. By simulations these resonances are qualitatively reproduced in the presence of small (<15°) but non-negligible incident angle, which was assumed to be zero in the original design.

Based on these evaluations, we concluded that the proposed method, which exploits the recent advancement of the flexible printed circuit fabrication technology, is promising to produce mm/submm-wave band-pass filters with a reasonably short turn-around time.

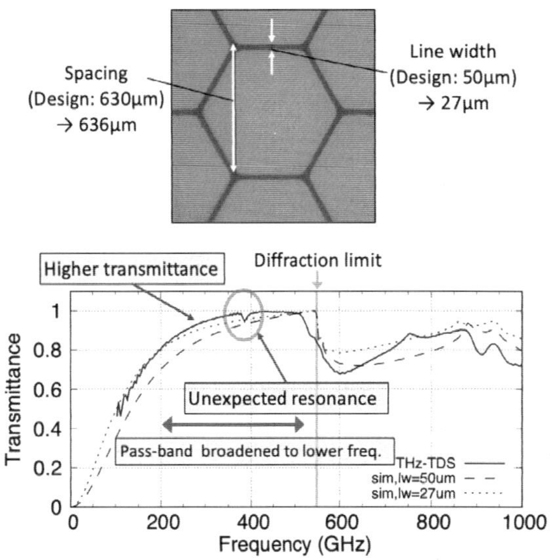


Fig. 1. Top panel: microscopic image of FPC filter prototype. Bottom panel: the transmission curve of a prototype of FPC hexagonal grid filter (solid line) and of simulations (dot lines).

REFERENCES

[1] Ulrich R., Infr. Phys., Vol 7, pp. 37-55, 1967.

[1] Institute of Astronomy, The University of Tokyo, Tokyo 181-0015, Japan.
[2] Department of Astronomy, School of Science, The University of Tokyo, Tokyo 113-0033, Japan.
[3] National Astronomical Observatory of Japan, Tokyo 181-8588, Japan.
[4] The University of Electro-Communications, Tokyo 182-8585, Japan.

NOTES:

# Development of mm/submm broadband anti-reflection coating exploiting the various expanded PTFEs measured with THz-TDS

Tai Oshima[1], Keisuke Yoshioka[1,2], Tatsuya Takekoshi[3], Kah Wuy Chin[1,4], Shinsuke Uno[3,4], and Takeshi Sakai[2]

Large scale sky survey in the millimeter/submillimeter bands with broadband multi-color continuum camera utilizing the highly sensitive low temperature detectors is indispensable for promoting the sciences such as efficiently estimating the redshift of distant star-forming galaxies, studying the internal structure of hot plasma in clusters of galaxies using the Sunyaev-Zel'dovich effect, and constraining physical properties of the dust in star--forming regions. Especially, recent increase in the number of observing colors and expansion to higher observing frequencies has pushed the optical components of the camera to fully cover the multiple frequency bands up to ~ 1 THz. Furthermore, rapid growth in the aperture size of the low temperature detectors requires vacuum windows of higher mechanical strength which results in thick windows. Thus the broadband antireflection (AR) of the vacuum window is essential. Therefore, we are developing a broadband AR covering the frequency bands of aforementioned astronomical interest.

Widely used conventional ARs are composed of mono- or multi-layer of expanded PTFE (polytetrafluoroethylene) such as Zitex and Porex glued on polyethylene substrate. However, lack in the variety of refractive indices of these expanded PTFE has limited the flexibility in optimizing the transmissivity of frequency bands of interest. Therefore, we have conducted a survey and evaluated the expanded PTFEs such as Poreflon and C-porous of various porosity and thickness with THz-TDS. The range of porosity and thickness were 40 to 80 % and 60 to 400 $\mu$ m, respectively.

In order to precisely quantify the optical properties of highly transparent and thin materials, uncertainty due to the low frequency power level fluctuation in the source and detector must be circumvented. For this purpose, a sample holder was built that automatically moves the sample in and out of the beam every 30 second and measures background and sample spectra in an interleaved manner. The achieved repeatability of the transmissivity and refractive index were better than 0.5 % and 0.004, respectively, at 0.1 to 3 THz.

As a result, the obtained refractive indices of expanded PTFEs varied from 1.06 to 1.27 with the fine spacing of ~ 0.03. Exploiting the improved flexibility in choice of the refractive index, AR capable of covering the observing bands of interest was designed. In Fig. 1, we show an example of simulated transmissivity of tri-layer AR with the atmospheric transmissivity at Atacama, one of the best mm/submm observing site.

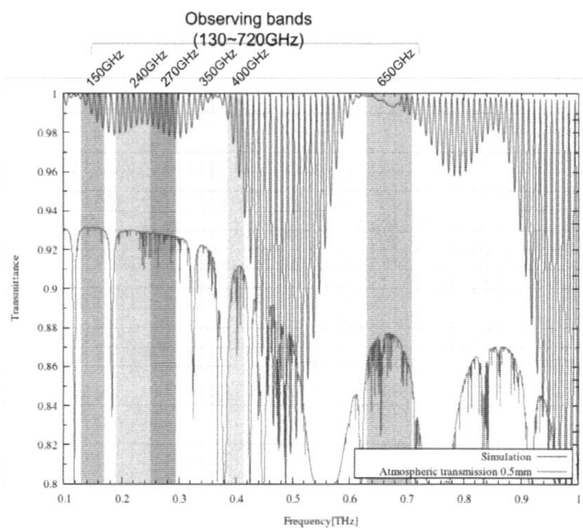


Fig. 1. (cyan line) An example of transmissivity curve of a vacuum window with AR composed of tri-layer expanded PTFEs. (green line) Typical atmospheric transmissivity at Atacama. (Colored boxes) The observing frequency bands of interest.

[1] National Astronomical Observatory of Japan, Tokyo 181-8588, Japan.
[2] The University of Electro-Communications, Tokyo 182-8585, Japan.
[3] Institute of Astronomy, The University of Tokyo, Tokyo 181-0015, Japan.
[4] Department of Astronomy, School of Science, The University of Tokyo, Tokyo 113-0033, Japan.

NOTES:

# Measurements of a Prototype 20 GHz Metamaterial Flat Lens

Cassandra Whitton[1], Christopher Groppi[2], Philip Mauskopf[3], and Paul Goldsmith[4]

*Abstract*— **In this paper, we present measurements of a prototype metamaterial flat lens. Flat, lenses with short focal lengths are of particular interest due to their potential use in quasi-optical observing in space-based cubesat applications. Our metamaterial flat lens was manufactured by using 11 layers of RO3003 circuit board laminate with etched sub-wavelength-sized copper patterning. The copper patterning is designed in such a way as to maximize the transmittance of the lens while applying the correct phase shift across the lens plane to give the lens its focusing properties. The lens was measured by scanning a receiver horn through one axis of the image plane of a transmitting horn. This measurement demonstrated that the waist of the focused gaussian beam is 30% wider than ideal. It is suspected that this non-ideality is caused by phase error in the design process, though simulations would be necessary to confirm this. Further measurements will be useful to fully characterize the lens's focal properties and determine how much loss it incurs.**

## I. Introduction

CubeSats may be an attractive prospective for those wishing to perform terahertz observations, due to the high atmospheric attenuation at these frequencies which makes ground-based observing difficult or impossible [6], and due to the often-prohibitive costs of other mission types which are able to observe from above the atmosphere. However, CubeSat missions come with their own set of design challenges, which particularly includes requirements for low weight and small form factor [2]. In particular, we would like to advocate for the use of metamaterial lenses as primary observing apertures in such systems, due to a number of advantages which help ameliorate CubeSats' particular challenges.

Metamaterials, which involve the structured embedding of metal elements within dielectric substrates, and metamaterial lenses in particular, have recently seen much advancement and development into the millimeter wavelength regime [3], [4]. The lenses which have been created so far are both thin and lightweight compared to a conventional lens of equivalent $f$-number, freeing up weight budget and making it easier to place and stow the lens, if a deployable design is necessary or desirable. Furthermore, these design techniques by Ref. [4] ensure that no anti-reflection coating is necessary to minimize reflection losses. Finally, such lenses have been found to theoretically have less than half a dB of loss, which is significantly better than that of a Fresnel zone plate lens, which, while flat and light, can exhibit on the order of 3 to 4dB or more of loss [5].

Here we have created and tested a metamaterial flat lens which operates at 20 GHz. The lens we present here is intended to act as a low-frequency prototype to test our design procedure. A successful design procedure should allow us to experiment with more expensive high-frequency designs, operating at 600 GHz or even above 1 THz.

Submitted for review 25 June 2019.

This work was supported in part by the JPL SURP program. RO3003 material used in this work was provided gratis by Rogers Corporation.

[1]Cassandra Whitton is with the School of Earth and Space Exploration, Arizona State University, Tempe, AZ 85287, USA (e-mail: cassandra.whitton@gmail.com)

[2]Christopher Groppi is with the School of Earth and Space Exploration, Arizona State University, Tempe, AZ 85287, USA (e-mail: cgroppi@asu.edu)

[3]Philip Mauskopf is with the School of Earth and Space Exploration, Arizona State University, Tempe, AZ 85287, USA (e-mail: philip.mauskopf@asu.edu)

[4]Paul Goldsmith is with the Jet Propulsion Laboratory, Pasadena, CA 91109, USA (e-mail: paul.f.goldsmith@jpl.nasa.gov)

## II. Lens Design

The lens is designed in a narrow bandwidth around 20 GHz, and is designed in such a way as to transform an incoming plane wave to a gaussian beam. This phase transformation is given by Refs. [4] and [1] as

$$\phi(r) = -\frac{\pi r^2}{\lambda R} \tag{1}$$

where $\lambda$ is the operational wavelength, $r$ is the distance on the lens plane from the lens center, $R$ is the radius of curvature of the phase front, given by

$$R(f) = f + \frac{\left(\frac{\pi w_0^2}{\lambda}\right)^2}{f} \tag{2}$$

and $w_0$ is the waist of the focused beam at the focal plane. The focal length of the lens is given by $f$.

In the case of our lens, the diameter of the active area is 254 mm, and the focal length is 105 mm, making our lens an $f/0.41$ lens. These parameters in combination give the phase transformation shown in the top plot of Fig. 1.

Given this phase transformation, we subdivide the surface of the lens vertically and horizontally into pixels, each of square dimensions $\lambda/10$, which is, in this case, about 1.50 mm. Each of these pixels is assigned a single phase transformation based on the above equations. In our case, each pixel has 10 metal layers, with 10 copper squares of metal, stacked on top of each other and separated by 11 surrounding dielectric layers. The dimensions of each square may be picked freely. Then, using techniques as described in Ref. [4], each pixel is optimized to give the desired phase transformation and maximum transmittance. These optimizations work by automatically tweaking the dimensions of each metal square until the desired conditions for that pixel are met. Because each pixel is treated independently from each other pixel, this is relatively computationally simple, as it

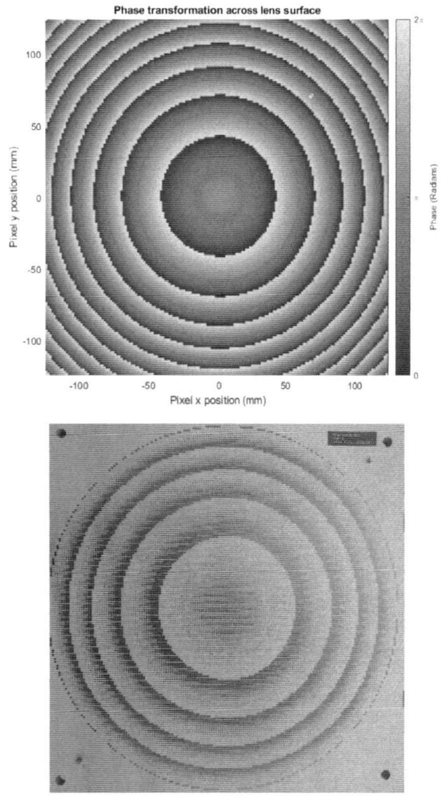


Fig. 1: The designed lens phase transformation (above) is compared with one layer of the manufactured lens (below). The layer's pattern is made up of thousands of copper squares.

only requires the optimization of around 5 free parameters (one for each layer of the lens, divided by 2 due to symmetry across its center) per pixel.

The lens was manufactured on RO3003 circuit board laminate with 760 mm thickness, with 1 ounce copper cladding. Due to manufacturing tolerances, the metal squares were constrained to be no smaller than 200 um in dimension, with at least 200 um between adjacent squares. A single layer of the manufactured lens is shown in the bottom half of Fig. 1. The layers were then stacked together, as shown in Fig. 2, using alignment holes that were drilled into the laminate layers during manufacturing. The completed lens is roughly 0.59 cm thick.

Though not directly relevant to this experiment, the lens is designed in such a way that it may be scaled from 20 GHz to 600 GHz. In doing so, the layer thicknesses and metal square sizes would be reduced by a factor of 30. Though this places much stricter tolerances on the manufacturing process, we have independently confirmed that there exist manufacturers with these capabilities.

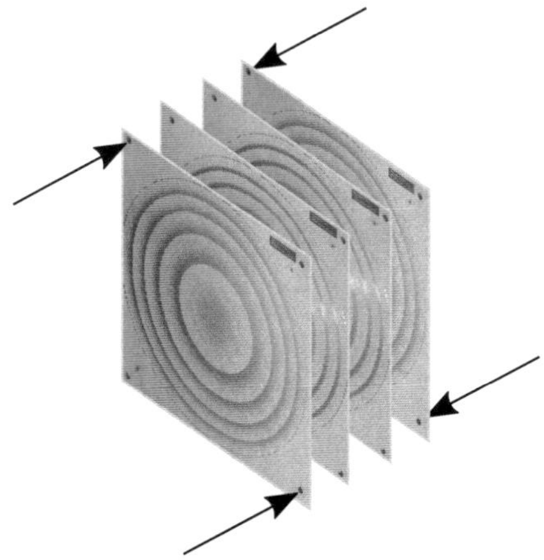

Fig. 2: Multiple layers of the lens are stacked together, aligned with guide-pins, and held together through boltholes at the corners. While this figure demonstrates only 4 layers, our lens has 11.

## III. Experimental Setup

In order to test that the lens operates correctly, we performed a simple image-plane measurement. The physical setup for the experiment is shown in Fig. 3. To accomplish this, the transmitter (Tx) and receiver (Rx) were each placed 2 focal lengths away from the lens. We do this because placing the Tx $2f$ away from the lens causes its image to be $2f$ from the lens on the other side, as shown by the well-known lens equation:

$$\frac{1}{f} = \frac{1}{d_o} + \frac{1}{d_i} \tag{3}$$

Here, $d_o$ is the distance of the object from the lens and $d_i$ is the distance of the image from the lens. Thus, the receiver directly measures the image of the transmitter. In addition, the resultant absolute magnification of the image is 1.

The transmitter and receiver each consist of a K-band pyramidal horn antenna with nominal 14 dBi gain coupled to a WR-42 waveguide. The transmitter is fed by a signal generator emitting a 20 GHz tone at -20 dBm. The receiver was connected to a power meter. The power meter only nominally operates up to 18 GHz; however, we tested that the power meter responded linearly to power input at 20 GHz. Therefore, while the absolute measurements of the power meter were likely incorrect, we are confident in the relative power measurements that it provided.

Once the transmitter's position was set, the receiver was scanned manually through the image plane until the point of maximum power reception was found. This was used as the zero-point for the measurement. The receiver was then manually moved up and down through the image plane in increments of a couple of millimeters. At each stopping point, the height of the receiver relative to its zero-point was recorded, along with the power measurement from the power meter.

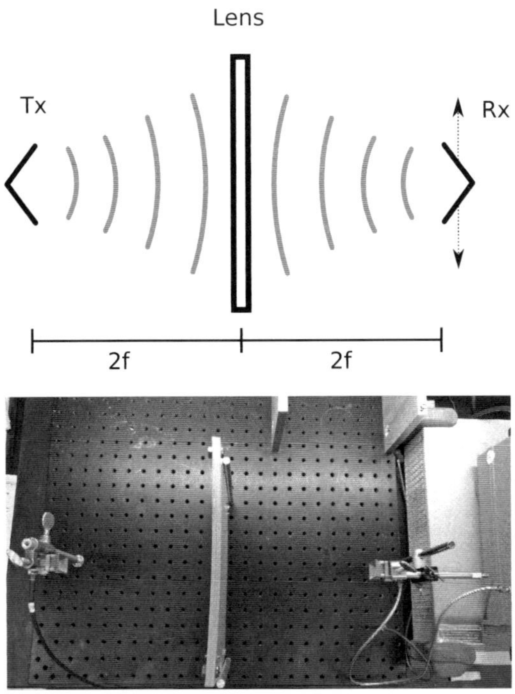


Fig. 3: A diagram of the test setup is display, with the realized setup displayed below it. The receiver and transmitter are both placed two focal lengths from the lens, such that the receiver sees an un-magnified image of the transmitter.

## IV. Results

Plotted with red circles in Fig. 4 is the result of the image plane measurement as described above. The magenta line is the best gaussian beam fit to the measurement.

Plotted in blue is what we would expect to measure if the lens were acting as an ideal lens. This was calculated using the properties of the horn antennas, which we believe have beam waists of 8.4 mm at 20 GHz. By convolving the gaussian beam shapes of the two horns together, as described in Ref. [1], we expect the measured image to have an effective beam waist of approximately 11.9 mm.

The effective beam waist of the gaussian beam fit was 15.4 mm, which was about 30% wider than ideal. While we have not demonstrated that our lens focuses perfectly, we have demonstrated that it is functional as designed.

## V. Conclusion

We have successfully demonstrated the focusing abilities of our 20 GHz metamaterial flat lens. However, the gaussian beam focus is 30% wider than ideal.

Further testing would be necessary to determine whether this is due to manufacturer tolerance error, phase error in the initial design, or an inaccurate estimate of the beam waists of the transmitter and receiver horns which were used. We suggest that a simulation of the designed lens, with its phase error included, could reveal whether or not the phase

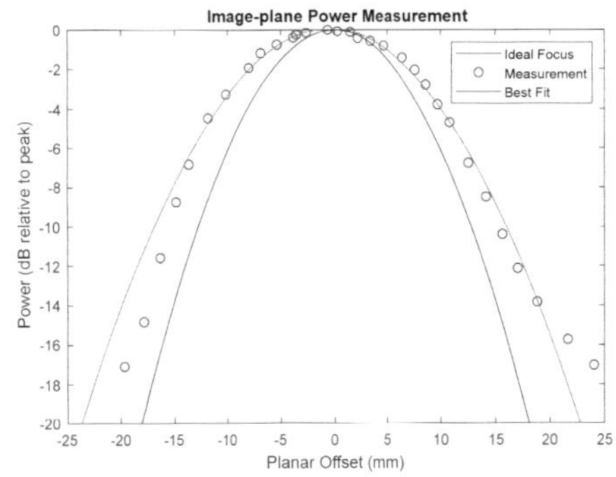


Fig. 4: Here is the plotted measurement of the image-plane scan compared to the expected beam measurement for an ideally focusing lens. The measured beam-width is about 30% wider than ideal.

error was the cause. We also believe that a full near-field measurement of the lens would help to quantify the lens's performance in more meaningful ways.

In any case, we have demonstrated that our design process works well enough to accomplish an acceptable focus with our lens. We believe that scaling the lens to higher frequencies in the future is very achievable, as well as relevant. Doing so would allow their eventual application in CubeSat-based terahertz observations.

## Acknowledgements

We would like to thank Rogers Corporation for providing us with RO3003 circuit board material free of charge, and the JPL SURP program for their financial support.

## References

[1] P.F. Goldsmith. *Quasioptical Systems: Gaussian Beam Quasioptical Propogation and Applications*. IEEE Press Series on RF and Microwave Technology. Wiley, 1998.

[2] Eva Peral, Eastwood Im, Lauren Wye, Simon Lee, Simone Tanelli, Yahya Rahmat-Samii, Stephen Horst, Jim Hoffman, Sang-Ho Yun, Travis Imken, et al. Radar technologies for earth remote sensing from cubesat platforms. *Proceedings of the IEEE*, 106(3):404–418, 2018.

[3] G Pisano, C Tucker, PA R Ade, P Moseley, and MW Ng. Metal mesh based metamaterials for millimetre wave and thz astronomy applications. In *Millimeter Waves and THz Technology Workshop (UCMMT), 2015 8th UK, Europe, China*, pages 1–4. IEEE, 2015.

[4] Giampaolo Pisano, Ming Wah Ng, Fahri Ozturk, Bruno Maffei, and Vic Haynes. Dielectrically embedded flat mesh lens for millimeter waves applications. *Applied optics*, 52(11):2218–2225, 2013.

[5] JM Rodriguez, Hristo D Hristov, and Walter Grote. Fresnel zone plate and ordinary lens antennas: Comparative study at microwave and terahertz frequencies. In *Microwave Conference (EuMC), 2011 41st European*, pages 894–897. IEEE, 2011.

[6] David M Slocum, Thomas M Goyette, and Robert H Giles. High-resolution terahertz atmospheric water vapor continuum measurements. In *Terahertz Physics, Devices, and Systems VIII: Advanced Applications in Industry and Defense*, volume 9102, page 91020E. International Society for Optics and Photonics, 2014.

# Low-loss Silicon MEMS Phase Shifter at 550 GHz

S. Rahiminejad[1], M. Alonso del Pino[1], C. Jung-Kubiak[1], T. Reck[2] and G. Chattopadhyay[1]

Submillimeter-wave spectrometer and radiometer instruments provide essential information for remotely studying atmospheric composition, measuring the surface properties of cold bodies. Submillimeter-wave radars are currently being developed to measure a variety of new science objectives such as cold ice and comet particle density and velocities. Unfortunately, all of these instruments require mechanical scanning or even re-orientation of the spacecraft to map a scene due to a lack of low-loss phase shifters at these frequencies. A phased array antenna with integrated phase shifters can steer the main beam electronically. Integrated circuit technologies such as SiGe and CMOS are most commonly used to fabricate phase shifters, however, they cannot operate above 300 GHz without considerable loss. This work presents a low loss MEMS phase shifter operating at 550 GHz with up to 180° phase shift, Fig. 1.

The phase shifter consist of a silicon MEMS motor that moves an electrically thin slab into the E-Plane of a hollow waveguide. When the dielectric slab is inserted into a hollow waveguide, the phase velocity of the incoming wave is decreased, thus it results in a phase shift. The slab is designed to have 180 degree phase shift when it is fully inserted into the waveguide, and it can be tuned from 0 to 180 degrees when the slab stands in an intermediate position. The permittivity of the silicon slab is lowered in order to improve the reflection coefficient. This permittivity is synthesized by etching subwavelength hexagonal holes inside the slab [1]. Depending on the silicon-to-air ratio in the slab, the permittivity can be anything between 1 (air) to 11.9 (Si), however, the fabrication constrains the minimum permittivity of 1.6. Simulations shows a 180° phase shift at 550 GHz, with a reflection below -25 dB and a transmission above -0.14 dB throughout the frequency band, 500-600 GHz.

The MEMS motor consists on a large deflection electrostatic actuator previously developed for a compact submillimeter-wave waveguide switch [2]. The comb-drives and beams have been designed to have a deflection of +/-100 um without reaching the onset electrostatic instability. The comb teeth have a uniform length to ensure an almost linear response on the movement of the slab as a function of the voltage. The motor also contains an extra pair of combs that move together with the slab and allow monitoring the position of the slab by measuring the interdigitated capacitance.

The MEMS motor and dielectric slab are fabricated using Deep Reactive Ion Etching on a SOI wafer. The SOI wafer has a device layer of 30 μm corresponding to the thickness of the slab and motor and a handle wafer of 350 μm that provides mechanical support. The waveguide synthesized on an E-Plane split-block that is CNC machined. This architecture allows a seamless integration of the receiver with the phase shifter on the same metal block.

The phase shifter is being fabricated and measurement results are intended to be presented at the conference.

The research described herein was carried out at the Jet Propulsion Laboratory, California Institute of Technology, Pasadena, California, USA, under contract with National Aeronautics and Space Administration.

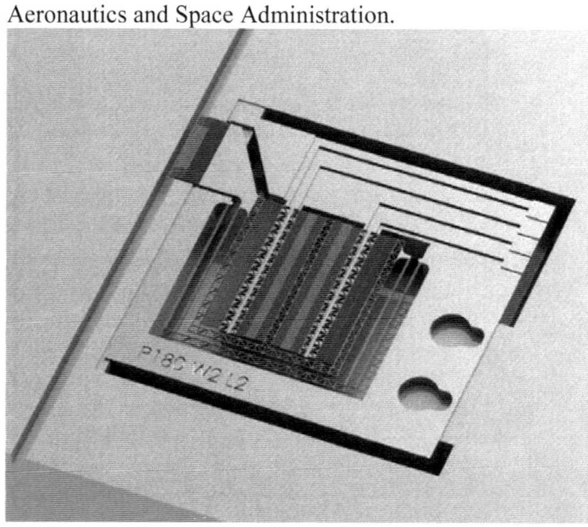


Fig. 1. The Silicon MEMS phase shifter placed in a metal block. The perforated Silicon slab is placed so that it can move freely in and out of the milled waveguide. The MEMS motor is actuated by applying a Voltage over the squared pads.


REFERENCES

[1] Menzel C, Rockstuhl C, Paul T, Lederer F, Pertsch T. Retrieving effective parameters for metamaterials at oblique incidence. Physical Review B. 2008 May 29;77(19):195328.

[2] T. Reck, C. Jung-Kubiak and G. Chattopadhyay, "A 460 GHz MEMS-Based Single-Pole Double-Throw Waveguide Switch," *2018 IEEE/MTT-S International Microwave Symposium - IMS*, Philadelphia, PA, 2018, pp. 773-775.




[1] Jet Propulsion Laboratory, Pasadena, CA 91109, USA.

[2] Was at JPL, now at Virginia Diodes Inc., Charlottesville, VA 22902, USA

NOTES:

# Multilayer dielectric diagonal horn for reshaping THz QCL beam pattern

Haotian Zhu[1], Francois Joint[1], Etienne Herth[2], Dominique Decanini[2], Yan Delorme[1],

Raffaele Colombelli[2] and Quan Xue[3].

We present the development of a diagonal dielectric horn in order to improve the beam shape of THz quantum cascade lasers (QCL). A high resistivity (HR) silicon based diagonal horn which consists of 7 pyramidal grades has been designed and the working frequency is 2.7 THz. The HR-silicon based substrate integrated image guide (SIIG) [1] (cross-sectional size: 80 μm x 14 μm) is used to guide the EM energy from a metal-metal 1st-order QCL (cross-sectional size: 50 μm x 14 μm) to the seven-layer horn. The mode in SIIG and QCL is $E^{y}_{11}$ mode and $TM_{00}$ mode, respectively. The simulated insertion loss of the SIIG is ~ 0.10 dB/mm at 2.7 THz. The loss induced by the SIIG is then relatively low considering its small dimension (length ~ 240 μm). Full wave EM simulations of the dielectric horn have been performed and we have obtained the beam diagram with a reduced FWHM and a near Gaussian shape. To ease handle and transfer, the handling structures around the horn are applied, as shown in Fig. 1.

The fabrication process is novel and challenging. Since it is impossible to pattern the complex 7-layer 3D mask using traditional photo-lithography or electron-beam lithography, two-photon lithography technology is used to form the mask for the 3D pyramid-shape diagonal horn. First, we built a perfect 1:1 photo resist (PR) mask, as shown in Fig.1. The deep reactive ion etching (DRIE) is then performed to transfer the mask to HR-silicon wafer. The next fabrication plan is transferring the photoresist pattern to a ~900 nm-thickness silica, and using the silica as the etching mask for silicon etching. The fabrication process is similar to [2]. The fabrication is in process now.

Compared to the traditional silicon lens or the micro-transverse-electromagnetic-horn antenna [3], the proposed dielectric diagonal horn is much more compatible with planar circuit integration and with the GaAs based QCL fabrication process. Moreover, the micromachined silicon based multilayer horn can be easily scaled up to higher frequency QCL applications.

Fig. 1. Fabricated 1:1 PR mask of the pyramid-shape diagonal horn.

### References

[1] A. Patrovsky and K. Wu, "Substrate integrated image guide (SIIG)- a planar dielectric waveguide technology for millimeter-wave applications," *IEEE Trans. Microw. Theory Tech.*, vol. 54, no. 6, pp. 2872–2879, Jun. 2006.

[2] C. Jung-Kubiak, *et al.*, "A multistep DRIE process for complex terahertz waveguide componets," *IEEE Trans. THz Sci. Technol.*, vol. 6, no. 5, pp. 690–695, Jun. 2016.

[3] W. Maineult, "Metal-metal terahertz quantum cascade laser with micro-transverse-electromagnetic-horn antenna," Appl. *Phys. Lett.*, 93, 183508. 2008.

1. LERMA, Observatorie de Paris, PSL Research University, CNRS, Sorbonne Universités, UPMC Univ. Paris 06, 75014, Paris, France.
2. Centre de Nanosciences et de Nanotechnologies, CNRS, Universités Paris-Sud, CNRS, Universités Paris-Saclay, 91120, Palaiseau, France.
3. School of Electronic and Information Engineering, South China University of Technology, Guangzhou, 510640, China.

NOTES:

# Broadband Antireflective Silicon Optics for Terahertz instruments

C. Jung-Kubiak[1], F. DeFrance[2], S. Rahiminejad[1], J. Sayers[2], J. Connors[3], T. Macioce[2], G. Chattopadhyay[1], and S. R. Golwala[2]

Submillimeter wave and THz instruments are central to contemporary astronomy, astrophysics, and cosmology research. Many applications from tens of GHz to THz frequencies, in ground and in space, would greatly benefit from large optics with broadband antireflective (AR) treatments. Silicon is a very interesting material at millimeter and submillimeter wavelengths. It has very low losses, even at ambient temperature, which makes it ideal for windows and optics, but its high refractive index ($n_{Si}$ = 3.42) makes AR treatment essential. Microfabrication of surface textures is an attractive technique for varying the effective refractive index of a material, thus, creating the AR structures [1]. Etching allows considerable flexibility in designing the texture geometry, which determines the effective index of refraction. Multilayers structures, where each layer is used to achieve a different index, can also be designed, providing a wider bandwidth compared to single-layer structures. Silicon Deep Reactive Ion Etching (DRIE) is used to individually pattern the wafers with posts or holes, of varying depth and width, calculated depending on the desired n index. The high resistivity wafers are then bonded together to produce the complete optic. Single-layer and double-layer designs have been simulated, fabricated and tested in the 190-330 GHz atmospheric window [2] with results showing less than -20 dB of reflectance over the full spectral band.

Building upon this work, we have simulated and fabricated a four-layer structure, with a 4:1 bandwidth, that is sufficient to cover the atmospheric windows at 125-170 GHz, 190-310 GHz and 335-355 GHz. As shown in Fig. 1, the two top layers T1 and T2 are posts with n index of 1.18 and 1.56 respectively. The two bottom layers T3 and T4 are holes with n index of 2.19 and 2.9 respectively. Using the DRIE process developed in [3], the two higher index layers (T3 and T4) are fabricated onto the same high resistivity wafer (W2), on the front and on the back. The second hole layer (T2) is fabricated on the backside of two additional wafers (W1 and W3). The 3 wafers (W1, W2 and W3) are then bonded together. After bonding, the lower index layer T1 is etched on the frontside and the backside of the stack. Fig. 1 presents Scanning Electron Microscope (SEM) images of the fabricated structures.



[1] Jet Propulsion Laboratory (JPL), California Institute of Technology, Pasadena, CA 91109

[2] Division of Physics, Mathematics, and Astronomy, California Institute of Technology, Pasadena, CA 91125

[3] now at NASA Goddard Space Flight Center, Greenbelt, MD, 20771

Simulations show -20 dB of reflectance over all 3 atmospheric windows, measurement results are expected in the next few weeks and will be presented at the conference.

The research described herein was carried out at the Jet Propulsion Laboratory, California Institute of Technology, Pasadena, California, USA, under contract with National Aeronautics and Space Administration.

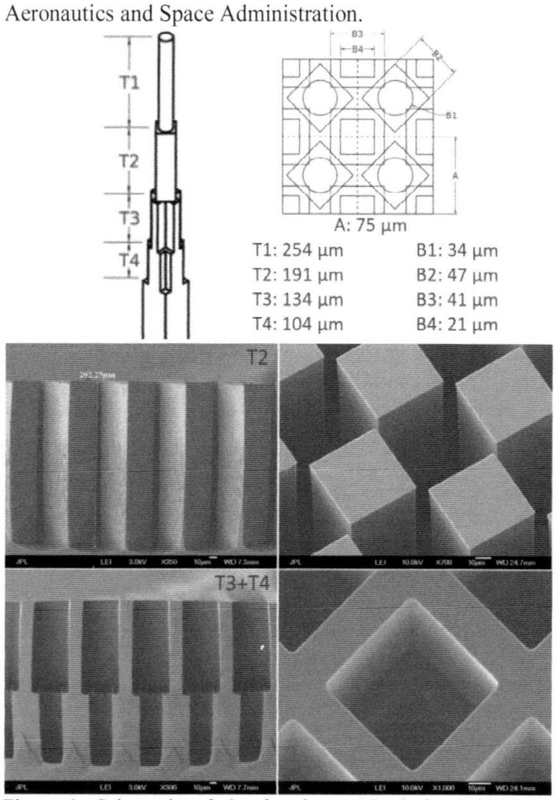


Figure 1: Schematic of the four-layer AR design with SEM images of the fabricated structures T2 and T3+T4, before bonding and etching of T1.

[1] A. J. Gatesman, et al., "An anti-reflection coating for silicon optics at terahertz frequencies," *IEEE Microwave and Guided Wave Letters*, vol. 10, no. 7, pp. 264–266, Jul. 2000

[2] F. Defrance, et al., "A 1.6:1 Bandwidth Two-Layer Antireflection Structure for Silicon Matched to the 190-310 GHz Atmospheric Window," *Applied Optics*, vol. 57, no. 18, pp. 5196-5209, Jun. 2018

[3] C. Jung-Kubiak, et al., "A Multi-Step DRIE Process for Complex Terahertz Waveguide Components," *IEEE Transactions on Terahertz Science and Technology*, vol. 6, no. 5, pp. 690-695, Sep. 2016

NOTES:

# Optical performance of laser-patterned high-resistivity silicon wafer in the frequency range of 0.1 - 4.7 THz

Evaldas Svirplys, Simonas Indrišiūnas, Heiko Richter, Andrzej Urbanowicz, Gediminas Račiukaitis, Till Hagelschuer, Heinz-Wilhelm Hübers, and Irmantas Kašalynas

***Abstract*— Optical performances of the high-resistivity silicon with laser-ablated surface were studied in the transmission mode in the frequency range of 0.1 -4.7 THz. It was demonstrated that surface irregularities causes THz waves to scatter significantly. This effect can be estimated using a THz wave scattering theory and the effective refractive index model.**

***Index Terms*—terahertz transmittance, silicon optics, diffractive lenses, waves scattering, laser-ablation**

## I. Introduction

DIRECT laser ablation (DLA) is a mask-less technology used for the research and development of optical components of various materials [1]. The relevance of the DLA technology has been verified demonstrating functional optical components including the multilevel phase Fresnel lenses on a high-resistivity (HR) silicon and the Soret zone plates developed on a free standing metal-foil [2-3]. In order to reduce reflection losses, the anti-reflection structures on the back side of a silicon wafer can be patterned by the same DLA technology as this has been proposed recently [4]. The effect of surface roughness on the performance of silicon diffractive components after laser ablation has been recently investigated [5]. In this work we extended the studies of optical transmission of laser ablated HR silicon in the frequency range from 0.1 THz up to 4.7 THz.

## II. Samples and Experimental Setup

The samples were prepared on a 500 μm thick, both-sides polished silicon wafer by varying the surrounding environment as well as the DLA parameters in order to modify the composition and roughness of the laser-ablated surface. Fig. 1 shows a photograph of the set samples. In total 38 samples were prepared changing the process parameters such as the pulse repetition rate, the pulse energy, the impulse overlap and the processing atmosphere. Most of the samples were fabricated in ambient air, while others were developed in an argon-rich atmosphere at pressures of 1 atm and 2 atm.

Fig. 1. Photo of four samples with four craters of 5x5 $mm^2$ area each. They wereabricated on 500 μm thick, both-sides polished silicon wafer using DLA technology.

A stylus profiler and a scanning electron microscope were employed to characterize the sample morphology. Optical transmittance was measured with a Golay cell detector employing the THz beam of a quantum cascade lasers (QCL) operating at 2.5 THz, 3.1 THz and 4.7 THz. The dielectric constants dispersion for each sample was obtained by a THz time domain spectroscopy (TDS).

## III. Results

The dependence of the transmittance on the surface roughness was measured at different frequencies. The results for 4.7 THz are presented in Fig. 2. The data allowed us to identify the critical value $R_{acrit}$ at which the transmittance dropped by 20%. For example it was found that the critical $R_{acrit}$ value at 4.7 THz was about 1.9 μm. Also $R_{acrit}$ values were measured for different frequencies of the THz QCL and the THz TDS system. The results are summarized in Table I. And $R_{acrit}$ decreases with increase of the frequency, i.e. smaller roughness on the surface changes the THz waves scattering at

The work was supported by the European Regional Development Fund according to the supported activity 'Research Projects Implemented by World-class Researcher Groups' under Measure no. 01.2.2-LMT-K-718, supporting the project KOTERA-PLAZA (Grant no. DOTSUT-247).

E. Svirplys, S. Indrišiūnas, A. Urbanowicz, G. Račiukaitis and I. Kašalynas were with the Center for Physical Sciences and Technology, Saulėtekio 3, LT-10257 Vilnius, Lithuania. (e-mail: Simonas.Indrisiunas@ftmc.lt; Andzej.Urbanovic@ftmc.lt;Gediminas.Raciukaitis@ftmc.lt;Irmantas.Kasalynas@ftmc.lt)

H. Richter, T. Hagelschuer, H.-W. Hübers were with German Aerospace Center (DLR), Institute of Optical Sensor Systems, Rutherfordstr. 2, 12489 Berlin, Germany. (e-mail: Heiko.Richter@dlr.de; Till.Hagelschuer@dlr.de; Heinz-Wilhelm.Huebers@dlr.de)

higher frequencies. The results are also shown in Fig. 3. Experimental data were fitted with an empirical expression that relates the critical surface roughness $R_{acrit}$ and the radiation frequency $f$ as following:

$$R_{acrit} = a + b \cdot \exp\left(-\frac{f}{c}\right), \quad (1)$$

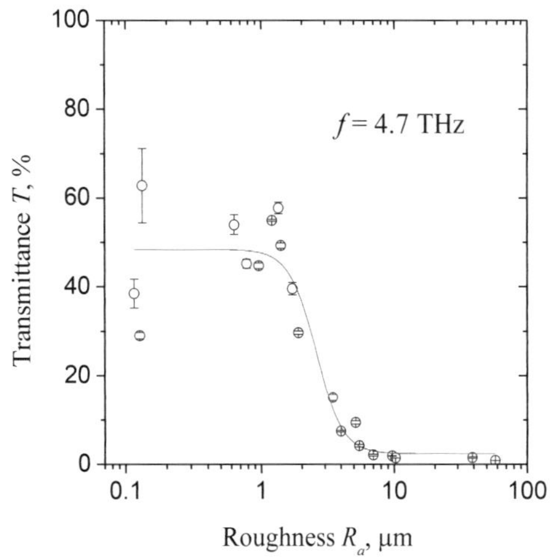


Fig. 2. Transmittance of the silicon wafers with different surface roughness (Ra) at frequency of 4.7 THz.

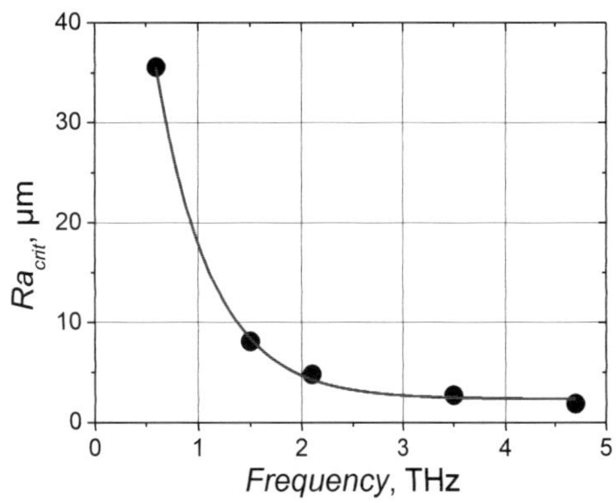


**Fig. 3.** The critical surface roughness $R_{acrit}$ (at which the transmittance of laser ablated silicon droped by 20%) dependence on the radiation frequency $f$. Solid line shows data calculated using equation (1).

TABLE I

Critical $R_a$ values, at which the transmittance dropped by 20%, obtained from experiment and modeling by using equation (1).

| Frequency, THz | Experimental *Ra* value, µm | Calculated *Ra* value, µm |
|---|---|---|
| 4.7 | 1.9 | 3.0 |
| 3.1 | 2.7 | 4.5 |
| 2.5 | 4.8 | 5.5 |
| 1.5 | 8.1 | 9.2 |
| 0.6 | 35.6 | 23.1 |

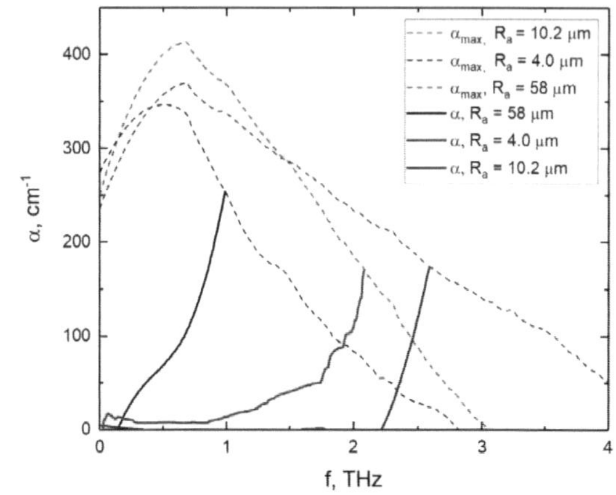


**Fig. 4.** The absorption coefficient spectra (solid lines) of laser-ablated silicon samples with a surface roughness ($R_a$) of about 4.0 µm, 10.2 µm and 10.2 µm. The maximum absorption coefficient values that can be measured with the THz TDS are indicated as dashed lines.

The data were fitted using parameter values of a = 2.4, b = 103.5, and c = 5.3. The measurement results allowed us to estimate the effective refractive index $n_{eff}$ of the silicon with laser-ablated surface. Itit was found out to be $n_{eff} \approx 2$ [5].

The sample transmittance was measured with a THz TDS system. Results are shown in Fig. 4. It demonstrates that higher surface roughness leads to smaller critical frequency above which the value of the absorption coefficient can increase above 100 cm$^{-1}$. Moreover, no significant difference between the optical properties of samples fabricated either in ambient air or in argon enriched environment were found in the THz regime [5].

## IV. CONCLUSION

To conclude, we have demonstrated that surface irregularities cause THz waves to scatter significantly. This effect can be estimated using a THz wave scattering theory and the effective refractive index model.

## REFERENCES


[1] B. Voisiat, S. Indrišiūnas, R. Šniaukas, L. Minkevičius, I. Kašalynas, and G. Račiukaitis, "Laser processing for precise fabrication of the THz optics", Proc. SPIE vol. 10091, LAMOM XXII; 100910F (2017); Event: SPIE LASE, 2017, San Francisco, CA, USA

[2] S. Indrišiūnas, H. Richter, I. Grigelionis, V. Janonis, L. Minkevičius, G. Valušis, G. Račiukaitis, T. Hagelschuer, H.-W. Hübers, and I. Kašalynas, "Laser-processed diffractive lenses for the frequency range of 4.7 THz," Opt. Lett. vol. 44, pp. 1210-1213 (2019).

[3] L. Minkevičius, S. Indrišiūnas, R. Šniaukas, B. Voisiat, V. Janonis, V. Tamošiūnas, I. Kašalynas, G. Račiukaitis, and G. Valušis, "Terahertz multilevel phase Fresnel lenses fabricated by laser patterning of silicon," Optics Letters Vol. 42, pp. 1875-1878 (2017).

[4] M. Tamosiunaite, S. Indrisiunas, V. Tamosiunas, L. Minkevicius, A. Urbanowicz, G. Raciukaitis, I. Kasalynas, and G. Valusis, "Focusing of terahertz radiation with laser-

ablated antireflective structures", IEEE Trans. Terahertz Sci. Technol. 8, 541–548 (2018).
[5] S. Indrišiūnas, E. Svirplys, H. Richter, A. Urbanowicz, G. Račiukaitis, T. Hagelschuer, et all., "Laser Ablated Silicon In The Frequency Range of 0.1 – 4.7 THz," submitted for publication in IEEE Transactions on Terahertz Science and Technology.

# Spectral measurements of THz radiation from intrinsic Josephson junction BSCCO stacks; phase locking of the BSCCO oscillators

Valery P. Koshelets, Nickolay V. Kinev, Andrey B. Ermakov, Raphael Wieland, Eric Dorsch, Olcay Kizilaslan, Dieter Koelle, Reinhold Kleiner, and Huabing Wang

*Abstract*—**Coherent THz emission from stacks of intrinsic Josephson junctions (IJJs), created naturally in the $Bi_2Sr_2CaCu_2O_{8+x}$ unit cell, was measured by a superconducting integrated receiver (SIR). The noise temperature of the SIR is as low as 120 K and its spectral resolution is better than 0.1 MHz, which exceeds the resolution of modern terahertz-range Fourier spectrometers by several orders of magnitude. In this report, the results of the spectral measurements of THz radiation emitted from intrinsic Josephson junction stacks are summarized. The phase-locked SIR has been used also for the locking of the BSCCO oscillator under the test. About 10 % of the power emitted by the BSCCO oscillator operating at 563 GHz with free-running linewidth of 13.5 MHz has been phase locked. The possibility of mutual locking of two BSCCO oscillators fabricated on one substrate has been investigated by direct measurements of emitted radiation spectra by the SIR.**

***Index Terms*—oscillators and spectrometers, phase locking, stacks of intrinsic Josephson junctions, superconducting integrated circuits, terahertz receivers.**

## I. Introduction

In recent years, coherent THz emission has been obtained from stacks of intrinsic Josephson junctions (IJJs), created naturally in the BSCCO unit cell with the CuO layers forming the superconducting electrodes and the BiO and SrO layers forming the barrier layer [1, 2]; a 1- µm-thick crystal consists of about 670 IJJs. Terahertz emission from BSCCO mesa has been obtained both at a low-bias (where the temperature distribution in the stack is almost homogeneous) and a high-bias regime (where an over-heated part and a cold part of the sample coexist) [3, 4].

Coherent emission above 1 THz by intrinsic Josephson BSCCO junction stacks with improved cooling has been demonstrated [5, 6]. Due to the variable size of the hot spot and the temperature rise caused by the self-heating, the emission frequency can be tuned over a wide range of up to 700 GHz [5]. So far, emitted by one device a maximum power of up to 30 µW was obtained [7, 8]. That is already enough for practical implementations, although for most applications good spectral properties are required.

The spectral characteristics of the oscillator were studied using the low-noise SIR with superconducting local oscillator, which was developed at Kotel'nikov IREE [9 - 12] to perform spectral studies of the electromagnetic radiation in the frequency range 450–700 GHz. The SIR was successfully implemented for measuring the profiles of the spectral lines of the gas-molecule radiation and absorption on board of high-altitude balloon [10-12] and can be used for the spectral study of any external terahertz oscillator radiating in the operation frequency range of the receiver. The best noise temperature of the SIR is 120 K and its spectral resolution is better than 0.1 MHz, which exceeds the resolution of modern terahertz-range Fourier spectrometers by several orders of magnitude.

Two configurations for the oscillator and receiver location were used: in the first case the oscillator was located in a cryostat of the SIR in the vicinity of the mixing unit [4, 13]; in the second case the oscillator and the receiver were located in independent cryostats with Mylar quasioptical windows. The SIR operates at temperature of about 4.5 K, whereas the optimal BSCCO-oscillator temperature is 20–50 K. The spectral lines of the oscillator radiation are recorded by the SIR and displayed on the spectrum-analyzer screen in the intermediate-frequency range 4 – 8 GHz. The spectrum analyzer allows one to average the signal, read it by a computer, and perform other necessary digital operations for the spectrum analysis and processing.

Application of the SIR has allowed to measure radiation emitted from intrinsic Josephson junction stacks both at a low-bias and a high-bias regime with spectral resolution better than 1 MHz [4]. While at low bias we found that the linewidth (LW) is not smaller than 500 MHz, at high bias, the emission LW turned out to be in the range 10–100 MHz. We attribute this to the hot spot acting as a synchronizing element; a LW as narrow as 7 MHz has been recorded at high bias [14].

The work was carried out at IREE RAS within the framework of the state task and partially supported by the Russian Foundation for Basic Research, grant No. 17-52-12051, and by the Deutsche Forschungsgemeinschaft, Project KL-930/13-2.

V. P. Koshelets, N V. Kinev, A B. Ermakov are with the Kotel'nikov Institute of Radio Engineering and Electronics RAS, Moscow 125009, Russia. (e-mail: valery@hitech.cplire.ru).

R. Wieland, E. Dorsch, O. Kizilaslan, D. Koelle, R. Kleiner are with Physikalisches Institut and Center for Collective Quantum Phenomena in LISA+, Universität Tübingen, D-72076 Tübingen, Germany (e-mail: kleiner@uni-tuebingen.de).

H. B. Wang is with the Research Institute of Superconductor Electronics, Nanjing University, Nanjing 210093, China (e-mail: hbwang@nju.edu.cn).

It is important to note that the tuning of the BSCCO oscillator frequency is continuous over the range; that was confirmed by fine-tuning of the SIR local oscillator (LO) frequency. Actually for such measurements the lowest frequency is about 450 GHz due to the design of the SIR mixer, while losses in the Nb interconnection lines of the SIR restrict the measurements at frequencies higher than 730 GHz. A combination of the BSCCO mesa and the SIR was used to accurately measure the terahertz absorption spectra of ammonia and water vapor [15, 16]. In this experiment, the bias current through the BSCCO emitter is kept at a constant value, tuned to the respective gas-line frequency, and intermediate-frequency (IF) spectra are taken using the SIR. These are quite encouraging results, although for most practical applications phase-locking of the cryogenic oscillator to a stable reference is required.

## II. Phase-Locking of a BSCCO Oscillator

To check a principle possibility of such locking we used the phase-locked SIR not only for detection of the BSCCO oscillator emission, but also for further locking of the oscillator under test [13, 14]. A simplified block diagram of the experimental setup for phase locking of the BSCCO oscillator is presented in Fig. 1. The BSCCO oscillator signal initially down-converted by the SIR to the IF band 4-8 GHz was then down-converted one more time to a secondary IF band 0.1 – 0.9 GHz. The obtained IF signal is actually a convolution of the BSCCO oscillator signal and stable phase-locked SIR LO. This signal is applied to the room-temperature phase-locking loop (PLL) system, where the phase of the signal is compared with the phase of the stable reference ($f_{ref}$ = 400 MHz). The PLL was equipped by additional Frequency Discriminator (FD), which compares the signal with an internal 400 MHz resonance tank; the FD error signal was applied to the oscillator in parallel with PLL signal and can be adjusted separately. Note that all reference sources used in the experiment (400 MHz, 6 GHz, and tunable 19-21 GHz LO used for the FFO phase-locking) were internally synchronized to the common 10 MHz reference. The error signal is returned back to the BSCCO oscillator to control its phase via an additional 5 Ohm resistor placed in the bias line of the IJJs stack; the resistor is mounted directly on the oscillator holder.

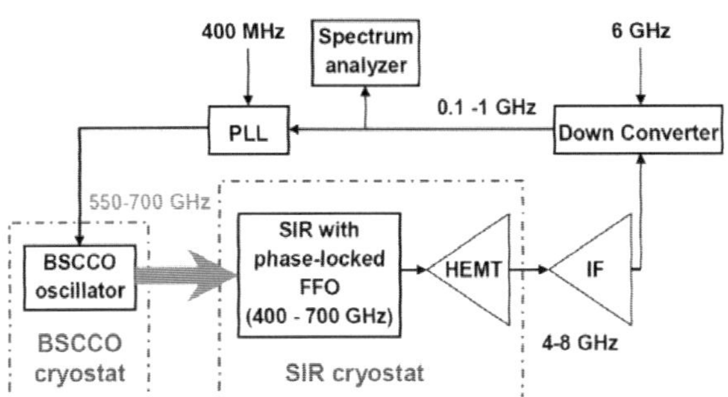


Fig. 1. Simplified block-diagram of the experimental set-up for the phase locking of a BSCCO oscillator by using the SIR with phase-locked FFO.

Results of the BSCCO oscillator frequency and phase locking are presented in Fig 2. The LW of the BSCCO oscillator frequency locked at 563 GHz is 13.5 MHz (Fig. 2a); about 10 % of the oscillator power has been phase locked. The ratio of the phase-locked power to the total power emitted by the oscillator is called a "spectral ratio" (SR); the obtained SR value of about 10 % is reasonably good result taking into account the wide linewidth of the BSCCO oscillator and the long length of the PL loop. At decreasing of the spectrum analyzer resolution bandwidth (RBW) the signal power in the phase-locked peak remains almost unchanged while the unlocked power in the wings is lowering proportionally to the RBW (Fig. 2b, 2c). A signal-to-noise ratio (SNR) of about 47 dB in a bandwidth of 9.1 Hz has been recorded (see Fig. 2c). An even better SNR of about 57 dB was measured for the RBW of 1 Hz [14].

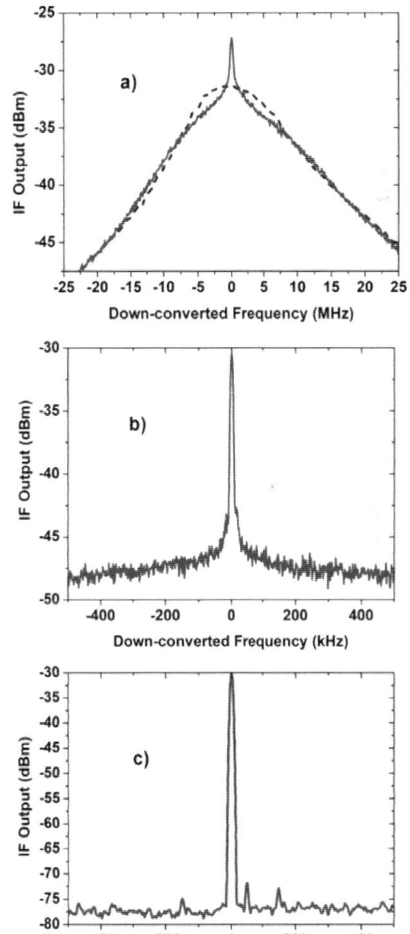


Fig. 2. Spectra of the BSCCO oscillator measured by the SIR relative to the phase-locked at 563 GHz FFO: a) Frequency (dashed) and phase locked (solid), Span 50 MHz, Resolution Bandwidth (RBW) = 470 kHzinewidth = 13.5 MHz; b) Phase locked, Span 1 MHz, RBW = 9.1 kHz; c) Phase locked, Span = 1 kHz, RBW = 9.1 Hz; the signal-to-noise ratio is 47 dB as measured in a bandwidth of 9.1 Hz.

It should be mentioned that such phase-locking was well possible only for some BSCCO oscillators demonstrating a free-running LW of about 10 - 15 MHz (that is comparable to regulations bandwidth of the room-temperature PLL, limited by the length of the cable from the PLL to the BSCCO oscillator). Another important issue is time constants in the PLL regulation loop. To realize a reasonably wide locking range, quite fast variations of the oscillator frequency (voltage on the IJJs stack) are required; the voltage variations should follow the PLL control signal with a delay smaller than 0.1 µs. Such a small delay is quite problematic in the high-bias regime with a large and "inertial" hot-spot region; note that reasonable linewidth values below 20 MHz can be achieved up to now only in the high-bias regime [4, 14].

## III. Mutual Locking of Two BSCCO Oscillators

In order to overcome the drawbacks of single JJs and to produce a significant off-chip radiation 1D or 2D arrays of Josephson Junctions (JJs) can be used; the development of such arrays has a long history [17, 18]. To significantly advance the performance of the THz radiation sources one has provide conditions to mutually phase-lock the junctions in the array. For 1D distributed JJ arrays of resistively shunted Nb/AlOx/Nb tunnel junctions, a power exceeding 10 µW was detected on-chip at frequencies from 300 to 500 GHz, the minimum inferred linewidth near 400 GHz, was about 100 kHz [19]. To this end mutual interaction of the IJJs stacks is a rather interesting issue [20]. As a first step in this direction we study the interaction of two BSCCO oscillators fabricated on one substrate.

We perform our measurements on BSCCO stacks embedded between gold electrodes, so-called gold- BSCCO-gold (GBG) structures. These GBG structures were fabricated on a common gold electrode; a sketch of the sample geometry is shown in Fig. 3a. The preparation of the sample is described in detail in Refs. [5, 21]. In brief, a BSCCO single crystal is glued onto a sapphire substrate with epoxy resin. A 100-nm-thick gold film is deposited on the crystal immediately after cleaving. As the third step, the stacks plus contact pads are pre-formed on top of the crystal as the "bone"-shaped structures in Fig. 3a with a total length of 630 µm and a thickness of about 1 µm. The sample is then glued face-down to a second sapphire substrate using epoxy. The base crystal is cleaved away by removing the first sapphire substrate, leaving approximately 0.7-$\mu$m-thick BSCCO structures contacted by gold and surrounded by epoxy. The fresh BSCCO surface is immediately covered with a 100-nm-thick gold layer. Photoresist is patterned in a rectangular 200 × 1450 $\mu m^2$-wide area using photolithography, and then the whole structure is etched down to the gold layer facing the second substrate by ion milling, resulting in five GBG structures with lateral dimensions of 300 × 50 $\mu m^2$ connected by the common gold layer (Fig. 3a). The nominal thickness of the stack corresponds to about 450 IJJs. The sample is fabricated from an as-grown BSCCO single crystal near optimal doping with $T_c \approx 89$ K. Finally, the sapphire substrate is glued onto a hemispheric sapphire lens. Current-voltage characteristics (IVCs) of two IJJs stacks "A" and "B" measured at bath temperature 4.2 K are shown in Fig. 3b and 3c, respectively.

To control the temperature of the sample it was mounted on the lens holder thermally connected to the cryostat bottom only by a copper heat link (cross-section of about 1 mm$^2$). Both stacks "A" and "B" were biased simultaneously by two independent computer-controlled current sources; the currents were swept from 0 up to 45 mA and then slowly decreased to a value that provides emission at frequencies of about 650 GHz. The SIR LO was phase-locked on this frequency; by tuning currents $I_A$ and $I_B$ it was possible to record emission lines both in low and upper sidebands. All data presented below were measured in the low sideband for frequencies of the oscillators from 642 to 646 GHz. The sample-holder temperature was about 14 K due to self-heating of the stacks; it should be mentioned that sweeping of one current (e.g. $I_A$) results in a change of the sample temperature and a shift of the stack "B" frequency, which is much smaller than the shift for the stack "A", but still quite considerable (see Fig. 4 and 5).

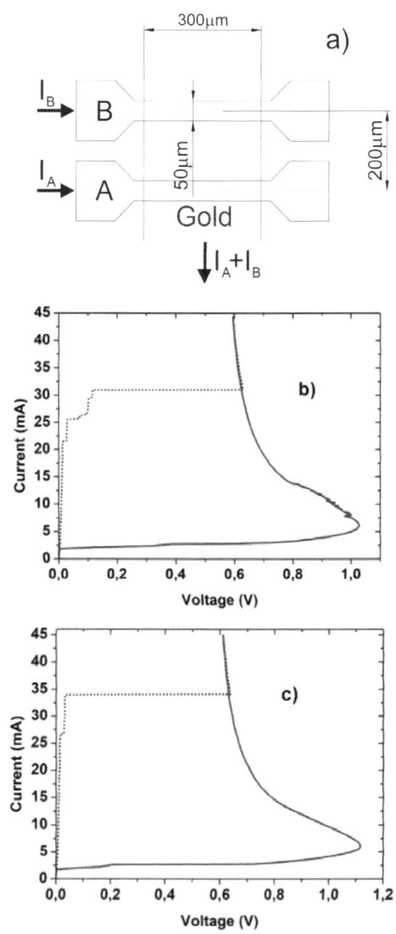


Fig. 3. a) Layout of the BSCCO oscillator circuit based on stacks of intrinsic Josephson junctions (only two stacks used in the experiment are shown); b) and c) Current-voltage characteristics of the IJJs stacks A and B, respectively.

Down-converted spectra emitted by stacks "A" and "B" at a frequency of about 644 GHz are presented in Fig. 4 (log scale for IF output power) and in Fig. 5 (linear scale). The spectra were measured by the SIR with phase-locked LO at frequency 650 GHz at small variation of the bias current for stack "B", while the current $I_A$ was fixed. When the current $I_B$ increases the voltage (and the frequency) for stack "B" decreases (partially due to extra heating), as a result the down-converted frequency increases (low sideband). The frequency of stack "A" also decreases due to extra heating; that corresponds to a slow motion of the peak "A" on the graphs from left to right. It should be mentioned that amplitudes of all peaks at down-converted frequencies of about 5.9 GHz are suppressed due to standing waves in the IF chain; this unevenness was not corrected for presented data.

When the frequencies of the two stacks coincide (solid curves in Fig. 4) the power in the resulting peak increases; that is even more pronounced in linear scale (Fig. 5). It is important to note that the power emitted by the two stacks exceeds the power of the singe stack more than two times; moreover, the linewidth of the locked stacks is 24 MHz (diamonds and solid line in Fig. 5b) compared to 35 MHz for a single stack (see asterisks and dotted line). Such mutual "locking" of two distantly spaced oscillators might be explained by resonant conditions provided by electromagnetic surrounding.

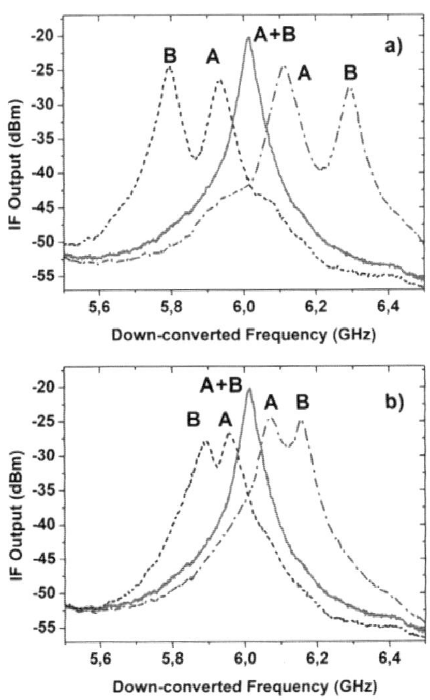


Fig. 4. Down-converted spectra emitted by stacks "A" and "B" at frequency of about 644 GHz for small variation of the bias current for stack "B"; the spectra were measured by the SIR with phase-locked local oscillator at frequency 650 GHz: a) $I_A$ = 24.571 mA, $I_B$ = 32.313 mA (dashed), 32.363 mA (solid) and 32.438 mA (dash-dotted); b) $I_A$ = 24.571 mA, $I_B$ = 32.338 mA (dashed), 32.363 mA (solid) and 32.400 mA (dash-dotted).

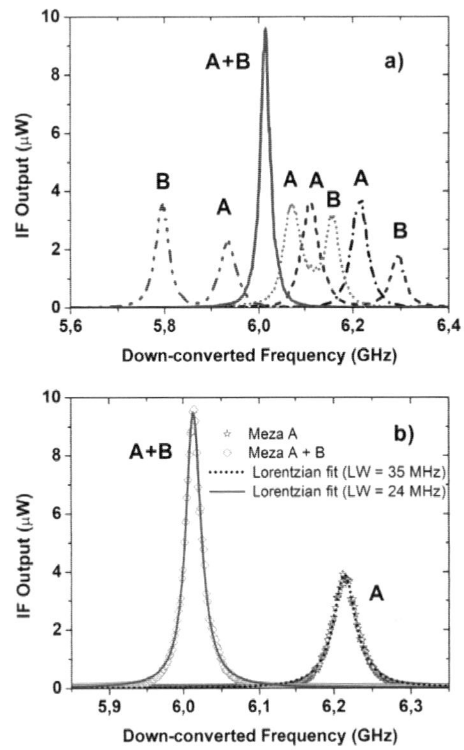


Fig. 5 Down-converted spectra emitted by stacks "A" and "B" (linear scale) at frequency of about 644 GHz for small variation of the bias current for stack "B"; the spectra were measured by the SIR with phase-locked local oscillator at frequency 650 GHz: a) $I_A$ = 24.571 mA, $I_B$ = 32.313 mA (dash-double dotted), 32.363 mA (solid), 32.400 mA (dotted), 32.400 mA (dashed), and 32.438 mA (dash-dotted); b) $I_A$ = 24.571 mA, $I_B$ = 32.363 mA (diamonds), 32.363 mA (asterisks); Lorentz fits to experimental spectra are shown by lines. Note that the power from the locked stacks (A+B) exceeds the value for the most "powerful" stack "A" by 2.6 times (about 4.3 dB); the linewidth of the locked stacks is 24 MHz compared to 35 MHz for a single one.

## IV. Conclusion

In this report, the results of the spectral measurements of THz radiation emitted from intrinsic Josephson junction stacks are summarized. The SIR was successfully implemented for the spectral measurements of THz radiation emitted from intrinsic BSCCO Josephson junction stacks; a linewidth as narrow as 13.5 MHz has been recorded in the high-bias regime at 563 GHz; about 10 % of the oscillator power has been phase locked by using the phase-locked SIR. That is a very important step towards the development of fully high $T_c$ phase-locked local oscillator, which opens prospects for various practical applications. The possibility of mutual locking of two BSCCO oscillators fabricated on one substrate has been demonstrated by direct measurements of emitted radiation spectra. The resulting linewidth of two IJJs stacks of 24 MHz is noticeably less than 35 MHz measured for each single BSCCO oscillator.

## REFERENCES


[1] R. Kleiner, P. Muller, G. Kunkell, F. Steinmeyer, "Intrinsic Josephson effects in $Bi_2Sr_2CaCu_2O_8$ single crystals", *Phys. Rev. Lett.*, vol. 68, pp. 2394-2397, April 1992.

[2] L. Ozyuzer, A. E. Koshelev, C. Kurter, N. Gopalsami, Q. Li, M. Tachiki, K. Kadowaki, T. Yamamoto, "Emission of coherent THz radiation from superconductors", *Science,* vol. 318, no. 5854, pp. 1291-1293, Nov. 2007.

[3] H. B. Wang, S. Guénon, B. Gross, J. Yuan, Z. G. Jiang, Y. Y. Zhong, M. Grünzweig, A. Iishi, P. H. Wu, T. Hatano, D. Koelle, and R. Kleiner, "Coherent Terahertz Emission of Intrinsic Josephson Junction Stacks in the Hot Spot Regime," *Phys. Rev. Lett.*, vol. 105, no. 5, pp. 057002-1-057002-4, July 2010

[4] M. Y. Li, J. Yuan, N. Kinev, J. Li, B. Gross, S. Guénon, A. Ishii, K. Hirata, T. Hatano, D. Koelle, R. Kleiner, V. Koshelets, H. Wang, and P. Wu, "Linewidth dependence of coherent terahertz emission from $Bi_2Sr_2CaCu_2O_8$ intrinsic Josephson junction stacks in the hot-spot regime," *Phys. Rev. B.*, vol. 86, pp. 060505-1-060505-5, August 2012.

[5] M. Ji, J. Yuan, B. Gross, F. Rudau, D. Y. An, M. Y. Li, X. J. Zhou, Y. Huang, H. C. Sun, Q. Zhu, J. Li, N. Kinev, T. Hatano, V. P. Koshelets, D. Koelle, R. Kleiner, W. W. Xu, B. B. Jin, H. B. Wang, and P. H. Wu, "$Bi_2Sr_2CaCu_2O_8$ intrinsic Josephson junction stacks with improved cooling: Coherent emission above 1 THz", *Appl. Phys. Lett.*, vol. 105, pp. 122602-1-122602-5, September 2014.

[6] T. Kashiwagi, T. Yamamoto, T. Kitamura, K. Asanuma, C. Watanabe, T. Nakade, T. Yasui, Y. Saiwai, Y. Shibano, H. Kubo, K. Sakamoto, T. Katsuragawa, M. Tsujimoto, K. Delfanazari, R. Yoshizaki, H. Minami, R. A. Klemm, and K. Kadowaki, "Generation of electromagnetic waves from 0.3 to 1.6 terahertz with a high-Tc superconducting $Bi_2Sr_2CaCu_2O_{8+\delta}$ intrinsic Josephson junction emitter", *Appl. Phys. Lett.*, vol. 106, no. 9, pp. 092601-1-092601-5, March 2015.

[7] D. Y. An, J. Yuan, N. Kinev, M. Y. Li, Y. Huang, M. Ji, H. Zhang, Z. L. Sun, L. Kang, B. B. Jin, J. Chen, J. Li, B. Gross, A. Ishii, K. Hirata, T. Hatano, V.P. Koshelets, D. Koelle, R. Kleiner, H.B. Wang, W. W. Xu, and P. H. Wu, Terahertz emission and detection both based on high-Tc superconductors: Towards an integrated receiver, *Appl. Phys. Lett.*, vol. 102, pp. 092601-1-092601-4, March 2013.

[8] S. Sekimoto, C. Watanabe, H. Minami, T. Yamamoto, T. Kashiwagi, R.A. Klemm, and K. Kadowaki, "Continuous 30 μW terahertz source by a high-Tc superconductor mesa structure," *Appl. Phys. Lett.*, vol. 103, pp. 182601-1-282601-5, October 2013.

[9] V. P. Koshelets and S. V. Shitov, "Integrated Superconducting Receivers," *Supercond. Sci. Technol.*, vol. 13, no. 5, pp. R53-R69, May 2000.

[10] V. P. Koshelets, A. B. Ermakov, L. V. Filippenko, N. V. Kinev, O. S. Kiselev, M. Y. Torgashin, A. de Lange, G. de Lange, S. I. Pripolzin and V. L. Vaks, "Superconducting integrated THz receivers: development and applications," *Proc. SPIE,* 2010, pp. 78540J-1-78540J-13, 2010.

[11] G. de Lange, M. Birk, D. Boersma, J. Dercksen, P. Dmitriev, A. Ermakov, L. Filippenko, H. Golstein, R. Hoogeveen, L. de Jong, A. Khudchenko. N. Kinev, O. Kiselev, B. van Kuik, A. de Lange, J. van Rantwijk, A. Selig, A. Sobolev, M. Torgashin, E. de Vries, G. Wagner, P. Yagoubov and V. Koshelets, "Development and Characterization of the SIR Channel of the TELIS Atmospheric Sounder," *Supercond. Sci. Technol.*, vol. 23, no. 4, pp. 045016-045024, March 2010.

[12] O. Kiselev, M. Birk, A. Ermakov, L. Filippenko, H. Golstein, R. Hoogeveen, N. Kinev, B. van Kuik, A. de Lange, G. de Lange, P. Yagoubov, and V. Koshelets, "Balloon-Borne Superconducting Integrated Receiver for Atmospheric Research," *IEEE Trans. on Appl. Supercond.*, vol. 21, no. 3, pp. 612-615, June 2011.

[13] N. V. Kinev, L. V. Filippenko, M. Y. Li, J. Yuan, H. B.Wang, and V. P. Koshelets, "Spectral properties of a terahertz oscillator based on the $Bi_2Sr_2CaCu_2O_{8+\delta}$ mesastructure", *Radiophysics and Quantum Electronics*, vol. 56, No. 8–9, pp. 582-590, January 2014.

[14] V.P. Koshelets, P.N. Dmitriev, M.I. Faley, L.V. Filippenko, K.V. Kalashnikov, N.V. Kinev, O.S. Kiselev, A.A. Artanov, K.I. Rudakov, A. de Lange, G. de Lange, V.L. Vaks, M.Y. Li, H.B. Wang, "Superconducting Integrated Terahertz Spectrometers", *IEEE Transactions on Terahertz Science and Technology*, vol. 5, pp. 687- 694, July 2015; DOI: 10.1109/TTHZ.2015.2443500.

[15] H. Sun, Z. Yang, N. V. Kinev, O. S. Kiselev, Y. Lv, Y. Huang, L. Hao, X. Zhou, M. Ji, X. Tu, C. Zhang, J. Li, F. Rudau, R. Wieland, J. S. Hampp, O. Kizilaslan, D. Koelle, B. Jin, J. Chen, L. Kang, W. Xu, R. Kleiner, V. P. Koshelets, H. Wang, and P. Wu, "Terahertz spectroscopy of dilute gases using $Bi_2Sr_2CaCu_2O_8$ intrinsic Josephson junction stacks", *Phys. Rev. Applied*; vol. 8, p. 054005, 2017; DOI: 10.1103/PhysRevApplied.8.054005.

[16] E. Sobakinskaya, V.L. Vaks, N. Kinev, M. Ji, M. Y. Li, H.B. Wang, V.P. Koshelets, «High resolution terahertz spectroscopy with a noise radiation source based on high-$T_c$ superconductors», *Journal of Physics D: Applied Physics*, vol. 50, no 3, paper#035305, 7 pages, 2017; DOI: 10.1088/1361-6463/50/3/035305

[17] A.K. Jain, K.K. Likharev, J.E. Lukens, and J.E. Sauvageau, "Mutual phase-locking in Josephson junction arrays" Phys. Reports, vol. 109, No. 6, pp. 309 – 426, 1984.

[18] M. Darula, T. Doderer and S. Beuven, "Millimetre and sub-mm wavelength radiation sources based on discrete Josephson junction arrays", *Supercond. Sci. Technol*., vol. 12, p. R1-R25, 1999.

[19] S. Han, B. Bi, W. Zhang, and J. E. Lukens "Demonstration of Josephson effect submillimeter wave sources with increased power", *Appl. Phys. Lett.*, vol. 64, p. 1424, 1994; doi: 10.1063/1.111904.

[20] T. M. Benseman, K. E. Gray, A. E. Koshelev, W. K. Kwok, U. Welp, H. Minami, K. Kadowaki, and T. Yamamoto, "Powerful terahertz emission from $Bi_2Sr_2CaCu_2O_{8+\delta}$ mesa arrays," *Appl. Phys. Lett.*, vol. 103, pp. 022602-1-022602-4, July 2013.

[21] Huili Zhang, Raphael Wieland, Wei Chen, Olcay Kizilaslan, Shigeyuki Ishida, Chao Han, Wanghao Tian, Zuyu Xu, Zaidong Qi, Tong Qing, Yangyang Lv, Xianjing Zhou, Nickolay Kinev, Andrey B. Ermakov, Eric Dorsch, Marc Ziegele, Dieter Koelle, Hiroshi Eisaki, Yoshiyuki Yoshida, Valery P. Koshelets, Reinhold Kleiner, Huabing Wang, and Peiheng Wu, "Resonant cavity modes in $Bi_2Sr_2CaCu_2O_{8+x}$ intrinsic Josephson junction stacks", *Phys. Rev. Applied,* vol. 11, p. 044004, 2019; DOI: 10.1103/PhysRevApplied.11.044004

# 4.7 THz GaAs Schottky Diode Receiver Components

Peter J. Sobis[1,2], Vladimir Drakinskiy[1], Divya Jayasankar[1], Arvid Hammar[2], Martin Anderberg[2], Anders Emrich[2], Elena Saenz[3] and Jan Stake[1].

We present the preliminary design of membrane integrated Schottky diode mixer and multiplier components for a 4.7 THz heterodyne room temperature receiver system employing a frequency stabilized Quantum Cascade Laser QCL local oscillator. The development includes a broadband 600 GHz LO based on a cascaded membrane integrated X2X2X2 diode multiplier chain with more than 5 mW of demonstrated peak output power using a single chip doubler as a last stage. The 600 GHz LO source drives an 8$^{th}$ harmonic GaAs Schottky mixer at 4.7 THz, that is used for the phase locking of a 4.7 THz QCL. The main output of the QCL's is in turn optically coupled to a fundamental Schottky diode mixer operating at 4.7 THz.

The work includes the design, modeling, and manufacturing of GaAs Schottky diode structures with sub-micron anodes [1-3] in different balanced configurations, optimized substrate-less THz circuit networks with integrated MIM-capacitors, and practical low loss circuit housings employing optimized smooth walled spline horn type structures.

This research has in part been carried out in the GigaHertz Centre in a project financed by VINNOVA, Chalmers, Omnisys Instruments, Low Noise Factory, Wasa Millimeter Wave, and RISE. The development is done in connection to the European Space Agency tender No. "ITT 1-9512/18/NL/AF – Frequency Stabilization of a Quantum Cascade Laser for Supra-THz Applications".

[1] Terahertz and Millimetre Wave Laboratory, Chalmers University of Technology, SE 412 96, Gothenburg, Sweden.
[2] Omnisys Instruments AB, SE-421 32, Västra Frölunda, Gothenburg, Sweden.
[3] European Space Agency – ESA/ESTEC
Noordwijk, PO Box 299, NL-2200 AG, The Netherlands.

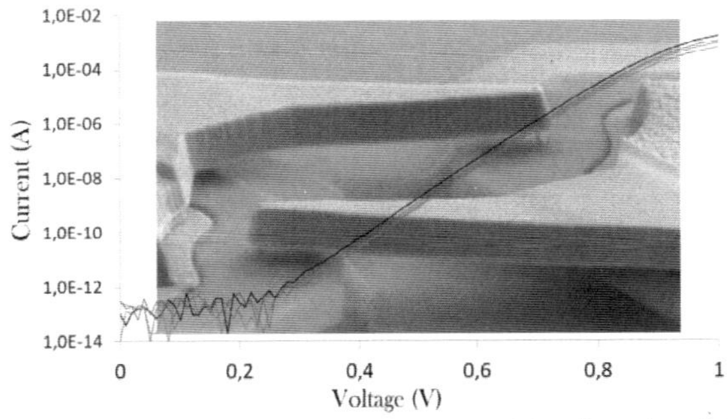


Fig. 1. Fabricated sub-micron (area < 0.1µm$^2$) antiparallel diode structure together with measured current-voltage DC characteristics of devices with different size ranging from 0.1µm$^2$ and below.

## References

[1] B. T. Bulcha, J. L. Hesler, V. Drakinskiy, J. Stake, A. Valavanis, P. Dean, L. H. Li, and N. S. Barker, "Design and characterization of 1.8–3.2 THz Schottky-based harmonic mixers," IEEE Trans. Terahertz Sci. Technol., vol. 6, no. 5, pp. 737–746, Aug. 2016. http://dx.doi.org/10.1109/TTHZ.2016.2576686

[2] A.-Y. Tang and J. Stake, "Impact of Eddy Currents and Crowding Effects on High-Frequency Losses in Planar Schottky Diodes," IEEE Trans. Electron Devices, vol. 58, no. 10, pp. 3260–3269, Aug. 2011. http://dx.doi.org/10.1109/TED.2011.2160724.

[3] A.-Y. Tang, T. Bryllert, and J. Stake, "Geometry optimization of THz sub-harmonic Schottky mixer diodes," IEEE 37th International Conference on Infrared, Millimeter, and Terahertz Waves (IRMMW-THz), September 2012. http://dx.doi.org/10.1109/IRMMW-THz.2012.6379506

NOTES:

# Reliability assessment of GaAs and InP THz mixers and frequency multipliers fabricated on 3" wafers

Josip Vukusic*, Vladimir Drakinskiy*, Daniel Heinerås+, Divya Jayasankar*, Peter Sobis†,*, Tomas Bryllert+, Vaclav Valenta‡, Marie Genevieve‡, Fernando Martinez Martin‡ and Jan Stake*

*Abstract*—**We report on the developments in this two-year European Space Agency funded project that aims at performing a preliminary reliability study of 300 GHz InP heterostructure barrier varactor diode multipliers and 1.2 THz GaAs Schottky diode mixers. Fabrication of the monolithically integrated circuits will be done on 3" wafers using established III-V processing. The reliability tests that will be performed include thermal and electrical step-stress studies, as well as shock, humidity and accelerated lifetime tests. We will present results and analysis of these experiments.**

## Summary

THz sources and detectors have in the recent decade found their way into applications in fields such as security imaging, telecommunications and bio sensing. But radio-astronomy has been the traditional application for mm-wave and THz devices. Several space bound exploratory missions have carried instrumentation operating in this frequency range. Whether deployed to orbit earth or to explore outer space, it is vital for the mission success that the instrumentation is highly reliable, i.e can perform as expected throughout its planned lifetime. It is therefore essential to conduct reliability tests on device/component level to ensure projected performance in future missions.

This work describes the preliminary reliability testing of 300 GHz heterostructure barrier varactor diode frequency multipliers [1] and 1200 GHz Schottky diode mixers [2] fabricated on 3" wafers.

Initial electrical and thermal step-stress tests will help us outline the maximum possible temperature, current, voltage or RF bias that a component can endure. This is executed by raising the applied voltage bias for instance on a device, at 24-hour steps, and then measuring the devices electrical characteristic (i.e current-voltage, capacitance-voltage, S-parameter). The results from the step-stress tests will be used as boundary conditions for the accelerated lifetime tests.

Accelerated lifetime tests will be used to estimate the longevity of the THz circuits. Because of the exponential dependence on temperature of the physical and chemical changes in the device material and structure it is possible to stress the devices/circuits by operating at elevated temperatures. In this way, the components can be driven to failure in much shorter time (<< 1 year) than their expected lifetime at room temperature. By doing this at several elevated temperatures we can extrapolate the device lifetime at corresponding realistic circumstances. Within this project we will study storage accelerated lifetime, DC bias accelerated lifetime, and RF accelerated lifetime. In addition, we will investigate the influence of humidity, thermal cycling and mechanical shock on the devices/circuits and module packaging.

The complexity, functionality and integration level of today's THz circuits render them of millimeter size, thus requiring large wafer fabrication to reduce cost and open up for their use in various low- to mid- volume applications.

By scaling THz monolithically integrated circuits fabrication to 3" wafers and reliability testing these, we aim at priming THz sources and detectors for future applications and high volume supply.

A 3" GaAs wafer divided into 10x9 $mm^2$ chips intended for reliability testing.


[1] J. Vukusic, T. Bryllert, Ø. Olsen, J. Hanning, and J. Stake, "Monolithic HBV-Based 282-GHz Tripler With 31-mW Output Power," *IEEE El Dev Lett*, vol. 33, no. 6, pp. 800–802, Jun. 2012

[2] P. Sobis, V. Drakinskiy, N. Wadefalk, P. Å. Nilsson, A. Hammar, D. Nyberg, A. Emrich, H. Zhao, T. Bryllert, A. Y. Tang, J. Schleeh, J. Grahn, and J. Stake, "SWI 1200/600 GHz Highly Integrated Receiver Front-Ends", Proceedings of the 36th ESA Antenna Workshop on Antennas and RF Systems for Space Science, ESA/ESTEC, Noordwijk, The Netherlands, October 2015


*Chalmers University of Technology, Sweden
+Wasa Millimeter Wave AB, Sweden
†Omnisys Instruments AB, Sweden
‡ESA/ESTEC, Netherlands

# YBaCuO Josephson generators as THz sources for bolometer characterization

L.S. Revin[1,2], A.L. Pankratov[1,2,3], D.V. Masterov[1], A.E. Parafin[1], S.A. Pavlov[1], A.V. Chiginev[1,2], I.V. Rakut'[2,3], A.V. Gordeeva[1,2], V.O. Zbrozhek[2], A.V. Blagodatkin[1,2], L.S. Kuzmin[2,4]

In a long junction the mode of fluxon motion may occur under the action of external magnetic field in which solitons are created at one edge of the junction, move along the junction, and are radiated at the other edge. Such regimes can be useful in creation of THz band oscillators [1], heterodyne and Hilbert spectrometers.

New technology of preliminary mask (PM) with $CeO_2$ buffer layer, which determines the necessary topology of the structure directly during the growth of the YBCO film has been developed [2]. The PM method has been successfully used to create Josephson junctions on a bicrystal YSZ substrate with a sublayer of epitaxial cerium dioxide $CeO_2$.

To register the radiation from long bicrystal junction the sample of Cold Electron Bolometer (CEB) [3] was used. CEBs represent SINIS junctions with nanoabsorber made of aluminum with suppressed superconductivity. CEBs are connected in series and parallel at DC for optimal matching with the amplifier. A pair of half lambda dipoles with wide electrodes is connected to another pair by high-inductive 1 μm wide lines. The design is optimized for the frequency band of 240-280 GHz.

Changing the current through the Josephson junction the bolometer response was measured as well as the voltage on YBCO oscillator (Josephson frequency $f = 2eV_{YBCO}/\hbar$). Figure 1 shows the bolometer response for each radiation frequency. Comparing the result with the CEB amplitude-frequency characteristic based on the calibrated backward-wave source (enlarged part of Fig. 1), it can be concluded that the signal peaks are determined by the characteristic of the receiving system rather than the oscillating one.

[1] Institute for Physics of Microstructures of Russian Academy of Sciences, Nizhny Novgorod, Russia, rls@ipmras.ru

[2] Nizhny Novgorod State Technical University n.a. R.E. Alexeev, Nizhny Novgorod, Russia

[3] Lobachevsky Nizhny Novgorod State University, Nizhny Novgorod, Russia

[4] Chalmers University of Technology, Gothenburg, Sweden

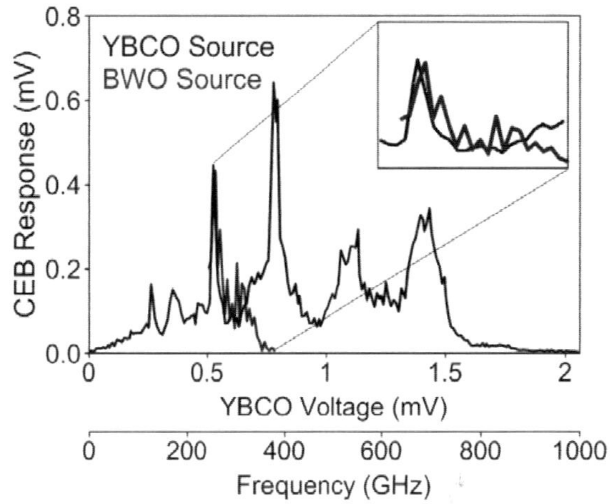


Fig. 1. Maximum registered response of the bolometer depending on the frequency of radiation. Comparison with experiments using BWO.

In conclusion, the samples of YBCO long Josephson junctions on YSZ bicrystal substrate have been fabricated using original technology of preliminary topology mask and good characteristics of the samples have been achieved. Using the cold-electron bolometer, the subTHz emission up to 900 GHz was registered. It can also be used as a THz band cryogenic network analyzer in combination with a cryogenic bolometer.

The work is supported by RSF (projects 16-19-10478 and 16-19-10468).

## References

[1] Stepantsov, E., Tarasov, M., Kalabukhov, A., Kuzmin, L., Claeson T., "THz Josephson properties of grain boundary YBaCuO junctions on symmetric, tilted bicrystal sapphire substrates", *Journal of Appl. Phys.* 2004. V. 96, No. 6. P. 3357.

[2] Masterov, D.V., Parafin, A.E., Revin, L.S., Chiginev, A.V., Skorokhodov, E.V., Yunin, P.A., Pankratov, A.L., "YBCO long Josephson junctions on bicrystal $Zr_{1-x}Y_xO_2$ substrates fabricated by preliminary topology masks", *Supercon. Sci. Technol.*, vol. 30, no. 2, pp. 025007, 2017.

[3] A.V. Gordeeva, V.O. Zbrozhek, A.L. Pankratov, L.S. Revin, V.A. Shamporov, A.A. Gunbina, L.S. Kuzmin, *Appl. Phys. Lett.*, 110, 162603 (2017).

NOTES:

# A 4.7 THz QCL phase locking experiment

S. Mahdizadeh[1], M. Justen[1], and U.U. Graf[1]

We present the last status of a 4.7 THz QCL phase locking experiment, similar to [1]. We use a QCL with a four quantum well bound-to-continuum structure and a patch antenna array as beam forming element [2]. The experiment consists of a multiplier chain followed by a super lattice harmonic mixer [3] at the 24th harmonic to produce the mixing signal with the QCL which is coupled into the super lattice via an ellipsoidal mirror and a diagonal horn. The mixing product from the room temperature super lattice, amplified and filtered in the IF, is fed to the phase locking circuit. The mixing product is 12dB over noise floor at 3MHz RBW. We compare the noise properties of two different AMC multiplier chains and its impact of the SNR in the IF.

References

[1] A. V. Khudchenko, "Phase locking a 4.7 THz quantum cascade laser using a super-lattice diode as harmonic mixer", 39th International Conference on Infrared, Millimeter, and Terahertz waves (IRMMW-THz), Sept. 2014

[2] L. Bosco, "A patch-array antenna single-mode low electrical dissipation continuous wave terahertz quantum cascade laser" *Applied Physics Letters* 109, 201103, Nov. 2016.

[3] D. G. Paveliev, "Experimental Study of Frequency Multipliers Based on a GaAs/AlAs Semiconductor Superlattices in the Terahertz Frequency Range," *Physics of Sem. Dev., Vol. 46, No. 1, pp. 121–125*, 2012.

[1] 1. Physikalisches Institut, University of Cologne, Cologne, Germany.

NOTES:

# A 900GHz Broadband Balanced Frequency Quadrupler

F. Yang[1]

Terahertz sources have attracted recent interest for both high resolution radar imaging and radio astronomy applications. For such applications, it is important to have compact sources that produce enough power to pump the front end of the transceiver.

For terahertz multiplier sources, cascading multipliers is a pragmatic approach. It typically consists of a chain of doublers and triplers[1][2] which are selected to yield the desired output frequency. The final efficiency of a multi-stage chain, as a result, is often on the order of a few percent or less [3]. Moreover, inter-stage mismatches in the chain can readily influence adjacent stages by pulling them from their optimum operating status and reducing efficiencies. Direct multiplication to a high-order harmonic greater than the third is desirable for higher frequency band, even with the challenges including proper termination of all intermediate harmonics (idlers).

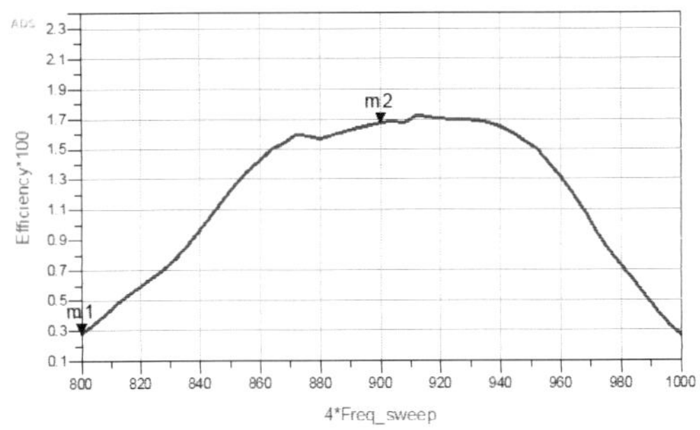


Fig.1. Power efficiency simulation of the 900 GHz GaAs membrane frequency Quadrupler

The work presented here applies the GaAs membrane substrate technology to implement an integrated frequency quadrupler. This Quadruplers features 2 anodes in series balanced configuration and is connected to a split waveguide-block by metallic beam-leads. The simulation results gives about 0.8%-1.7% multipling efficiency from 830 GHz to 980 GHz and 0.4 mW of maximum output power with 24 mW of input power at 225 GHz.

[1] State Key Lab of Millimeter Waves, Southeast University, Nanjing, 210096 , Jiangsu, China.
Email: yangfei@seu.edu.cn

### References

[1] Maestrini A, Gatilova L, Treuttel J, et al. 1200GHz and 600GHz Schottky receivers for JUICE-SWI[C], International Symposium on Space Terahertz Technology. 2016.

[2] J. Treuttel; L. Gatilova ; F. Yang ; et al., A 330 GHz frequency doubler using European MMIC Schottky process based on e-beam lithography, 2014 XXXIth URSI General Assembly and Scientific Symposium (URSI GASS), Beijing, 2014, pp. 1-4.

[3] Mehdi I, Siles J V, Lee C, et al. THz Diode Technology: Status, Prospects, and Applications[J]. Proceedings of the IEEE, 2017, 105(6):990-1007.

NOTES:

# A 410-510GHz Local Oscillation Source for SIS Mixers

Peng Chen[1], Xian-Jin Deng[1], Li Li[1], Kun Huang[1],Sheng-Cai Shi[2] and Kun Zhang[2],
[1]Institute of Electronic Engineering, Microsystem and Terahertz Research Center,
China Academy of Engineering Physics, Mianyang, Sichuan, 621900, P. R. China
[2]Purple Mountain Observatory, National Astronomical Observatories, Nanjing, Jiangsu, 210008, P. R. China

***Abstract*—A 410-510GHz local oscillation source for SIS mixers has been designed, fabricated, and tested. The source is a × 24 frequency multiplier working at room temperature. It converts the input low band microwave signals to submillimeter signals. It is comprised of an E band quadrupler, a power amplifier, a D band doubler and a final trippler. Both of the D band doubler and the final trippler use plannar Schottky diodes and quartz based circuits to realize frequency multiplication. Over the 410-510GHz band, the output power of the source is above 50uW, and the maximum is 122uW in 461.2GHz. A horn antenna is connected with the source to radiate the power to SIS receivers which are used for submillimeter radio telescope in Purple Mountain Observatory in China.**

## I. INTRODUCTION

THE submillimeter regime is the last window to be fully explored in astronomical observations, and scientifical studies in this area are becoming more and more important. Recent radio astronomical projects such as SMA, ALMA, and Herschel all operated in the submillimeter region [1].

The superconducting SIS (Superconductor-Insulator-Superconductor) mixers have excellent low-noise performance, and they are the most sensitive devices for the coherent detection of submillimeter signals. Heterodyne receivers based on a SIS mixer are attractive for astronomical observations, because the noise temperature of the receivers are approaching to the quantum limit and the requirements of local oscillator power are only at microwatt or submicrowatt [2]. A high quality local oscillator is of great importance to such a receiver.

At present, many kinds of LOs can be used for a SIS mixer, such as Gunn diodes followed by frequency multipliers, multiplier chains driven by frequency synthesizers in microwave region [3] [4], quantum cascade lasers, photonic local oscillators, and etc. In all of these LOs, multiplier chains driven by frequency synthesizers are the best choice for SIS mixers in the submillimeter region, since they have small size and weight comparatively, and it is very convenient to make electronic sweep to change the frequency and power of the local oscillator.

In this paper, a 410-510GHz local oscillation source for SIS mixers is presented. The SIS mixers are used to build the high sensitivity receiver of a radio telescope in Purple Mountain Observatory in China. The 410-510GHz source is a ×24 frequency multiplier working at room temperature. It mainly consists of four modules, and two modules of them work in E band: a quadrupler and a power amplifier. The other two modules are a D band doubler and a final trippler. Both of the quadrupler and the power amplifier are designed with GaAs MMICs. The D band doubler and the final trippler use plannar Schottky diodes and quartz based circuits to realize frequency multiplication. A horn antenna is connected with the source to radiate the power to SIS mixers.

## II. RESULTS

The 410-510GHz local oscillation source has been designed, fabricated, and tested. The initial product is shown in Fig.1. Input RF power is provided by an Agilent signal generator, and the power supply is +12Vdc.

**Fig. 1.** Picture of the fabricated 410-510GHz Local Oscillation source

The test results are shown in Fig.2. Over the 410-510GHz band, the output power of the source is above 50uW, and the maximum is 122uW in 461.2GHz.

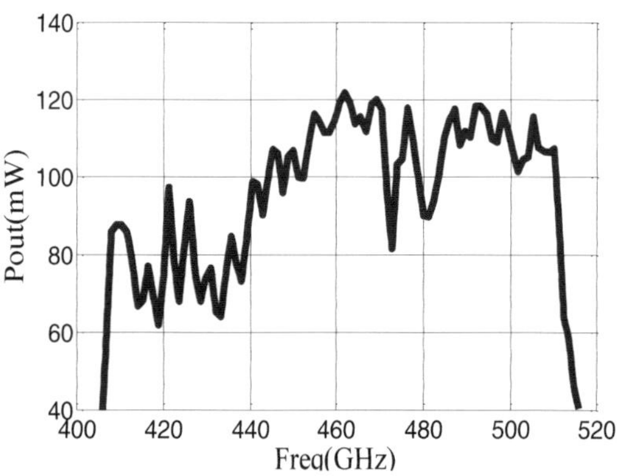


**Fig. 2.** Measured output power versus output frequency of the source

## REFERENCES

[1]. Shi Sheng-Cai, "Development of submillimeter superconducting SIS mixers for radio astronomy," *Publications of Purple Mountain Observatory*, vol. 22, pp. 21-38,2003.

[2]. Satoshi Kohjiro, Kenichi Kikuchi, Masaaki Maezawa, and etc, "SIS mixers as a noise detector for optimization of photonic local oscillators in terahertz range," *IEEE Transactions on Applied Superconductivity*, vol. 19, pp. 389-394, 2009.

[3]. A. V. Raisanen, "Frequency multipliers for millimeter and submillimeter wavelengths," *Proc. IEEE*, vol. 80, pp. 1842-1852, 1992.

[4]. I. Mehdi, "THz local oscillator technology," *Proc. SPIE*, vol. 5498, pp. 103-112, 2004.

# Characterization of Digital Real-Time Spectrometers for Radio Astronomy and Atmospheric Remote Sensing

Axel Murk, Mikko Kotiranta

*Abstract*— **High resolution real-time spectrometers are widely used for radio astronomy and atmospheric remote sensing. Today these are mostly realized by high-speed digital signal analyzers. In this contribution we compare the radiometric performance of different commercially available digital Fast Fourier Transform (FFT) and Polyphase Filter-Bank (PFB) spectrometers. It is found that the radiometric noise of the individual channels corresponds for both types very well to the theoretical value, but that the FFT and PFB spectrometers have a different behavior when channels are binned. We also report on spectroscopic linearity issues in one of the older FFT spectrometers.**

*Index Terms*—**Digital signal processing, microwave radiometry, radio astronomy, remote sensing, real-time spectrometer.**

## I. Introduction

High resolution real-time spectrometers are needed for both radio astronomy and atmospheric remote sensing. In the past discrete filter banks, acousto-optical spectrometers (AOS) or chirp-transform spectrometers (CTS) were used for this purpose, but with the advances of fast analog-to-digital converter (ADC) and field programmable gated array (FPGA) circuits it became possible to process sufficiently wide bandwidths in real-time using digital signal analyzers. One of the first commercially available models was based on the AC240 digitizer from the company Acqiris. It uses an 8-Bit ADC and calculates a Fast-Fourier Transformation (FFT) with 16384 channels over a bandwidth of 1 GHz [1][2]. More recent digital spectrometer models have ADCs with higher sampling rates to process bandwidths of 4 GHz or above. They can be also equipped with ADCs with more bits of resolution, which improves the spurious free dynamic range of the instrument.

The first digital spectrometers used a standard FFT analysis, which results in a channel response of $|\sin(x)/x|^2$. This relatively high spectral leakage can be reduced by tapering the time domain signals with an optional window function. But since this also leads to a coarser frequency resolution and a significant loss of sensitivity it is in most cases preferred to operate the FFT spectrometer without a window function.

Most modern digital spectrometers process the time domain data with a digital polyphase filter bank (PFB) algorithm instead of an FFT. This results in a channel response with a much faster sidelobe roll-off and almost ideal channel separation compared to an unwindowed FFT. Equally important, it does not lead to a significant loss of sensitivity [3].

The Institute of Applied Physics (IAP) at the University of Bern operates a variety of microwave radiometers for remote sensing of the Earth's atmosphere. This includes observations of ozone, water vapor, wind and temperature in the stratosphere where the pressure broadening of the emission lines allows to retrieve vertical profiles of these quantities. A high linearity and a well-known channel response of the spectrometer are essential for these observations. Over the years we replaced all filter banks, AOS and CTS of our radiometers with digital FFT and PFB spectrometers. In this paper we compare the noise performance and linearity of these commercially available digital back-ends.

## II. Spectrometer Comparison

Table I gives an overview of the spectrometers which were included in the comparison.

TABLE I
SPECTROMETER PROPERTIES

| Model | Bandwidth [MHz] | Channels | ADC Bits | Type |
|---|---|---|---|---|
| *AC240 Acqiris* | 1000 | 16384 | 8 | FFT |
| *U5303 Acqiris* | 1600 | 16384 | 12 | PFB |
| *AFFTS 1500 RPG* | 1500 | 8129 | 8 | PFB |
| *AFFTS 100 RPG* | 100 | 16384 | 8 | PFB |
| *XFFTS 500 RPG* | 500 | 32489 | 10 | PFB |
| *USRP X310 Ettus* | 200 | 16384 | 14 | FFT |

The AC240 was introduced in 2005 and resulted from a collaboration between the company Acqiris and the Swiss universities ETHZ and FHNW [1]. The available processing resources which were available on the Virtex-2 FPGA at that time did not allow to calculate the FFT over this bandwidth and resolution without any truncation errors. These lead to small numerical artifacts and nonlinearities on the accumulated data which are only noticeable after long integration times. Nevertheless, the AC240 has been successfully used by different groups for atmospheric remote sensing and radio astronomy. This outdated hardware is no longer available and

A. Murk and M. Kotiranta are with the Institute of Applied Physics at the University of Bern, Bern, Switzerland. (e-mail: axel.murk@iap.unibe.ch).

has been replaced by the Acqiris U5303. For this powerful dual channel signal acquisition board ETHZ, FHNW and IAP developed with Acqiris and advanced PFB spectrometer firmware. It has a much better dynamic range and does not produce any numerical truncation errors. It also includes several advanced features such as I/Q signal processing with amplitude and phase corrections, cross correlation, and a Kurtosis analysis for RFI mitigation [4]. The capabilities of this new spectrometer will be reported elsewhere in an upcoming publication.

The AFFTS is the first generation of PFB spectrometers developed at the Max-Planck Institute for Radio Astronomy (MPIfR). Its successor XFFTS from MPIfR is capable to process a wider bandwidth with a higher resolution and a better dynamic [3]. It is distributed with firmware for different bandwidths by the company Radiometer Physics (RPG) [5].

The USRP X310 is a software defined radio (SDR) platform developed by Ettus Research [6]. Its on-board FPGA can be used for signal processing through the open source GNU Radio software framework, but it is also possible to develop customized firmware for it with the LabVIEW Real-Time environment from National Instruments. We have used this option to develop a FFT spectrometer firmware with 200 MHz processed bandwidth. As an option the SDR can zoom into a spectral region with higher frequency resolution, or the bandwidth can be extended by frequency switching at the cost of integration time in each spectrum. This spectrometer is currently used in one of our operational radiometers for the observation of stratospheric winds [7].

### *A. Channel Response*

Fig. 1 compares the measured channel response of the PFB spectrometers AFFTS and U5303 with the FFT spectrometer AC240 without a window function. It shows the significant sidelobes of the FFT, which are above -15 dB. According to [3] the AFFTS and XFFTS should have a similar sharp roll-off to less than -50 dB as the U5303. It is currently not clear whether the observed shoulders starting at the -30 dB level are an artefact of phase noise in this measurement, or a firmware bug in this specific early AFFTS model which was available at IAP for this measurement.

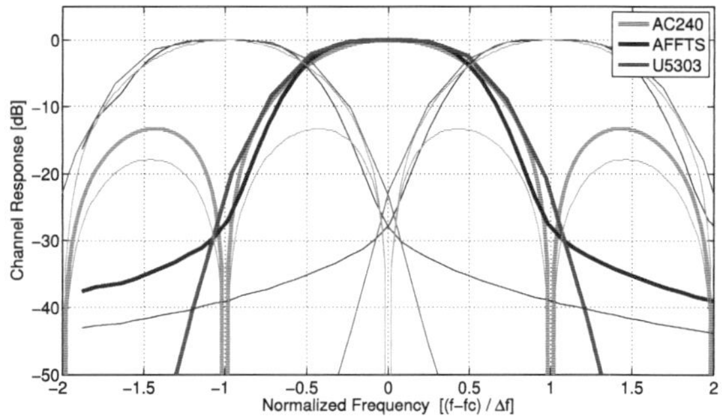


Fig. 1. Channel response of the FFT spectrometer AC240 compared to two PFB spectrometers with significantly lower sidelobes. The frequency axis is normalized with the channel spacing, and the response of neighboring channels is shown in thinner lines of the same color.

### *B. Radiometer Setup*

The two Acqiris spectrometers AC240 and U5303 and two USRP X310 were connected to a single radiometer frontend to allow simultaneous observations of hot and cold blackbody calibration targets or atmospheric emission lines. The radiometer is a single sideband heterodyne receiver with an uncooled WR-10 low noise amplifier (LNA), and it is tuned to the 110.8 GHz ozone line. It also includes two noise diodes which can be injected before the LNA for calibration purposes and to verify the linearity of the system. One of the two USRP receivers was tuned to a weaker CO line at 115.3 GHz. This SDR channel USRP-B was operated with 20 instead of 200 MHz bandwidth and had thus a ten times higher frequency resolution than USRP-A. Figure 2 shows the schematic block diagram of the radiometer which is described in more detail in [8]. The overall system noise temperature of the setup was around 550 K in the ozone band and 650 K around the CO band which is beyond the nominal frequency band of the LNA.
The AFFTS and XFFTS spectrometers were not integrated in this 110 GHz radiometer setup.

### *C. Spectral Line Observations*

Figure 2 compares the calibrated ozone spectra which were measured simultaneously by the AC24, U5303 and USRP spectrometers. For this figure the channels of the different spectrometers were binned to a comparable resolution.

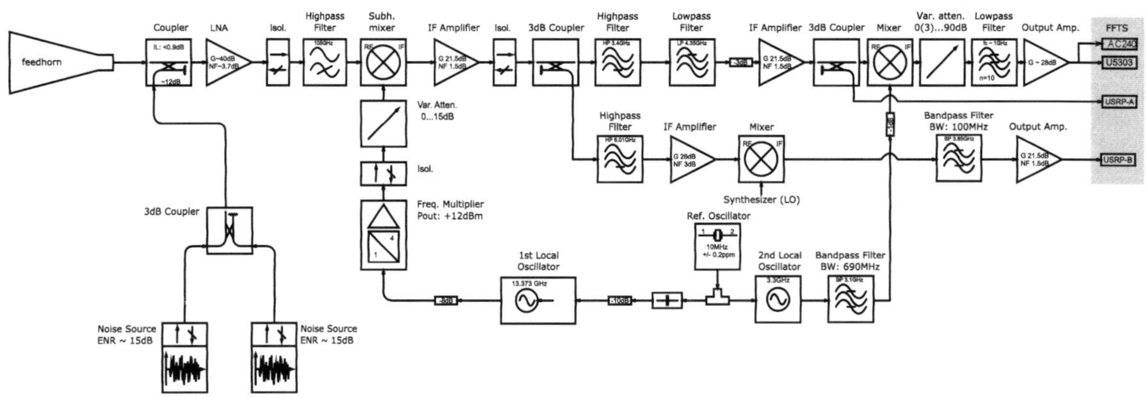


Fig. 2. Block diagram of the 110 GHz radiometer which was used for the spectrometer comparison.

The spectra of the USRP and U5303 coincide very well with each other. The AC240 spectrum, however, has a systematically smaller line amplitude than the two other spectra. Also the slope on the AC240 spectrum, which is caused by the line wing of a strong oxygen line at 118 GHz, has a different inclination than with the two other spectrometers.

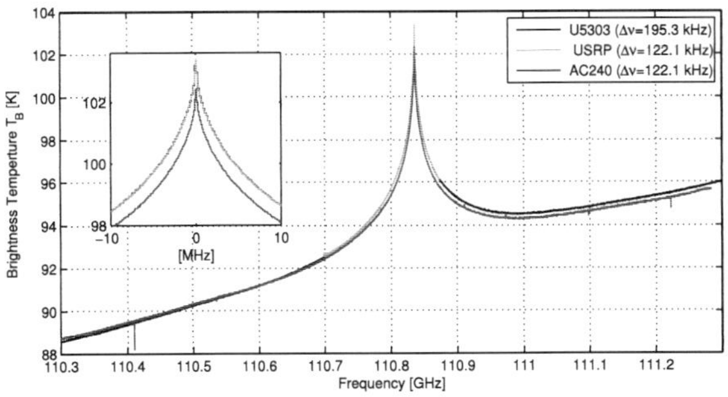


Fig. 3. Ozone line observed simultaneously with three different FFT spectrometers. The zoom in the inset highlights the systematic bias of the AC240 spectrometer.

The observed behavior cannot be explained by a simple analog gain compression effect in the AC240 input circuit. Such non-linearities would lead to an overall offset over the full bandwidth, including the line wings. This was also verified by comparing the signal levels for different combinations of hot and cold load measurements in which either one, two or none of the noise diodes is switched on. The excess noise ratio (ENR) of the noise diodes remains the same for all spectrometers independent whether they are turned on over the hot or cold input signal, which indicates that the bias of the AC240 only occurs for spectral line signals. It is also not possible to explain the bias with the $|\sin(x)/x|^2$ channel response of the FFT. For that reason we suspect that this systematic error is caused by the numerical rounding errors, which might lead to a higher spectral leakage which is not observed in the channel response measurements with the CW signal. We are planning to investigate this in the future by an analysis of the bias between the ozone spectra under different weather conditions, i.e. for different line amplitudes and distances from the two reference temperatures, as well as by additional laboratory measurements.

### *D. Radiometric Noise*

According to the radiometer equation a noise signal with an expectation value of its power $\langle P\rangle$ can be observed with a standard deviation $\sigma_P$ depending on the the integration time $\tau$ and the fluctuation bandwidth $\Delta f_{\mathrm{neq}}$:

$$\sigma_P \,/\, \langle P\rangle \;=\; 1/\sqrt{\Delta f_{neq}\;\tau}$$

For an ideal FFT spectrometer without a window function the theoretical fluctuation bandwidth $\Delta f_{\mathrm{neq}}$ is identical with the channel spacing $\Delta f$. In order to investigate whether this criterium is met with the spectrometers under test we recorded time series of several thousand spectra with integration times of either 50 ms or 1 s. In order to distinguish between radiometric noise and instrumental drift we first subtract consecutive spectra from each other before we calculate the standard deviation. Because of this differentiation the result needs to be scaled by a factor of $\sqrt{2}$ to obtain the correct value of $\sigma_P$. This test is very similar to a standard Allan variance measurement [9], but here we are focusing only on the radiometric noise at short integration times to assess the efficiency of the spectrometers. The Allan time $\tau_A$, where the instrumental drift dominates over the radiometric noise, will be in most cases determined by the radiometer front-end.

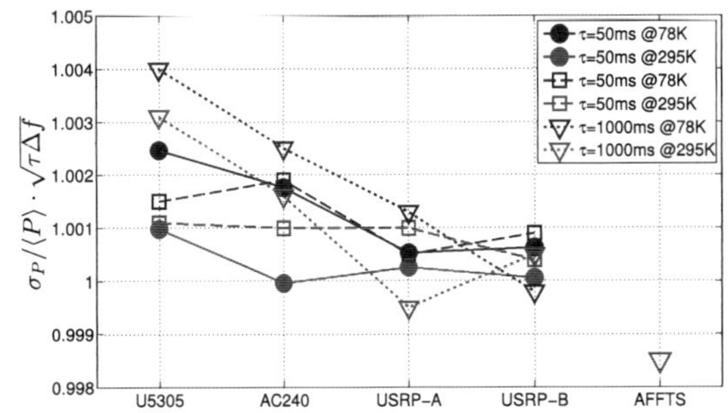


Fig. 4. Normalized radiometric noise of the different spectrometers measured with different integration times while observing a hot or a cold target.

The Allan variance is usually plotted on a double logarithmic scale, which does not allow to identify small deviations from the nominal noise bandwidth. Fig. 4 shows the normalized radiometric noise for a single spectrometer channel. To improve the statistics of the measurement it was calculated as an average of the normalized noise over all spectrometer channels. Anomalous channels at the edge of the spectrometer bandpass or which are affected by spurious signals were excluded from the average since they would bias the result. The remaining thousands of channels did not indicate any systematic variations of the normalized noise over the bandwidth, even if the total powers vary over several dB of magnitude over the spectrometer bandwidth.

The values in Fig. 4 deviate only by a few ppm or less from unity. This indicates that the noise performance of the FFT spectrometers is almost identical with the theoretical value. For the PFB spectrometers this behavior is not self-evident since the channel response and noise performance will depend on the selected window functions. With the AFFTS a value of about 0.9985 is observed, which means that the fluctuation bandwidth is by a factor 1.003 wider than the channel spacing. According to [3] the PFB has been designed for a nominal noise equivalent bandwidth of 1.16 of the channel spacing, i.e. we should observe an even smaller level of radiometric noise. However, since this measurement was obtained only with a 1s integration time we cannot rule out that this measurement was not affected by instrumental drift. Also for the U5303 the measurements with 1 s integration time have a noticeably higher normalized variability than with 50 ms, which indicates that the Allan variance curve deviates already at these integration times from the expected behavior of $1/\sqrt{\tau\,\Delta f}$.

Another interesting feature in Fig. 4 is the fact that the normalized radiometric noise on the cold load seems to be systematically higher than on the ambient load. This could probably be explained by quantization noise of the received thermal radiation, or by short term fluctuations due to standing waves from the instable LN2 surface of the cold load.

### E. Channel Binning

In most applications several spectrometer channels will be binned together to reduce the noise and the number of channels. It is thus of interest to know whether the radiometric noise is reduced according to the square root of the number of binned channels.

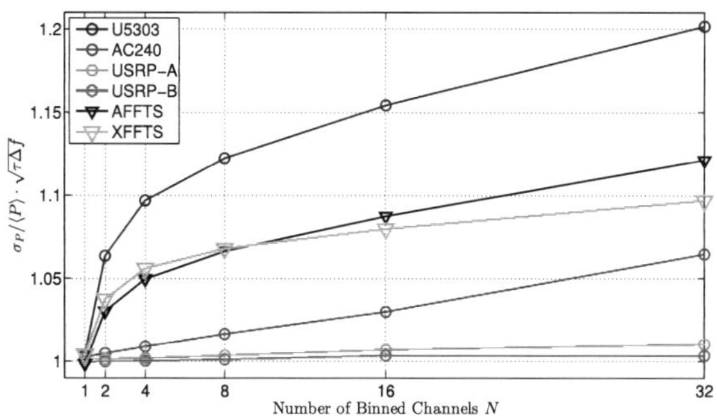


Fig. 5. Normalized radiometric noise of the different spectrometers depending on the number of binned channels.

A similar analysis was performed in [9] for the AOS of the HIFI instrument on the Herschel spacecraft. Since the AOS channels have a significant overlap the noise between adjacent channels is correlated. As a result, binning of $N$ channels does not reduce the noise by a factor of $\sqrt{N}$.

With an ideal digital FFT spectrometer the noise in adjacent channels should be uncorrelated. Figure 5 shows the normalized radiometric noise of the different digital spectrometers depending on the number of binned channels. The two FFT spectrometers USRP-A and USRP-B with an analyzed bandwidth of 200 MHz and 20 MHz, respectively, behave under channel binning very similar to the ideal case. With the AC240 the noise seems to be reduced less efficiently by the binning and a linear increase of the normalized noise with the number of binned channels is observed. Part of this could be probably explained by the increasing influence of gain variations similar as in an Alan variance plot, but his effect should be rather small since these measurements were conducted with a very short integration time of 50 ms. The largest deviations are observed for all PFB spectrometers, in particular for the first two channels that are binned. This indicates a significant correlation of the noise in adjacent channels, which is surprising since the channel response of these spectrometers is very close to an ideal rectangular filter shape.

## III. Conclusions

Digital spectrometers with high frequency resolution have become a key technology for radio astronomy and microwave remote sensing. In this paper we characterized and compared the radiometric noise performance, channel response, binning artifacts and linearity of a different FFT and PFB spectrometers. For the outdated FFT spectrometer AC240 we observed a significant spectroscopic nonlinearity, which leads to a scaling error of the observed emission lines. It will be important to understand the systematics of this effect in more detail and to find a way to correct this systematic error in the already existing long-term atmospheric data series which were collected with this type of spectrometer.

More recent FFT and PFB spectrometers do not suffer from this effect. The PFB models provide an excellent channel response and low spectral leakage, which is especially important if the observations are disturbed by narrow band radio interferences. However, our tests also indicate that the different PFB models suffer from correlated noise between adjacent channels. As a result the noise reduction by channel binning is less efficient than for a standard FFT spectrometer. The reasons for this effect are currently not understood and will need further investigations.

## References


[1] A. O. Benz, P. Grigis, V. Hungerbühler, H. Meyer, C. Monstein, B. Stuber, and D. Zardet, "A broadband FFT spectrometer for radio and millimeter astronomy", Astronomy and Astrophysics 442, 767–773 (2005).

[2] S. C. Müller, A. Murk, C. Monstein, and N. Kämpfer, "Inter-comparison of Digital Fast Fourier Transform and Acoustooptical Spectrometers for Microwave Radiometry of the Atmosphere", IEEE Transactions on Geoscience and Remote Sensing 47/7 (2009).

[3] B. Klein, S. Hochgürtel, I. Krämer, A. Bell, K. Meyer, and R. Güsten, "High-resolution wide-band fast Fourier transform spectrometers", Astronomy and Astrophysics 542, L3 (2012).

[4] Acqiris, "Enhanced Fast Fourier Transform Application on 12-bit High-Speed Data Acquisition PCIe Card", https://acqiris.com

[5] Radiometer Physics GmbH, https://www.radiometer-physics.de

[6] Ettus Research, a National Instruments Brand, https://www.ettus.com/

[7] J. Hagen, A. Murk, R. Rüfenacht, S. Khaykin, A. Hauchecorne, and N. Kämpfer, "WIRA-C: a compact 142-GHz-radiometer for continuous middle-atmospheric wind measurements", Atmos. Meas. Tech., 11, 5007-5024 (2018)

[8] M. Kotiranta, R. M. Gomez, G. Nedoluha, N. Kämpfer, A. Murk, "Receiver Development for the Microwave Ozone Profiling Instrument MOPI 5", Proc. of IGARSS (2019).

[9] V. Ossenkopf, "The stability of spectroscopic instruments: a unified Allan variance computation scheme", Astronomy and Astrophysics, 479(3), 915–926, (2008).

# Astroclimatic studies of the sites for forthcoming radio astronomical observatories

Grigoriy M. Bubnov, *Member, IEEE,* Valeriy F. Grigorev, Vyacheslav F. Vdovin, *Member, IEEE,* Peter M. Zemlyanukha and Igor I. Zinchenko

***Abstract*— We are presenting results of the observations our research group has been conducting for about 7 years. Over this time, we have gathered statistical data from a number of sites which can be used for radio astronomical observations in the millimetre and sub-millimetre bandwidths. We have just finished a year-long cycle of tests in Svalbard and are now able to compare its atmospheric transparency with other sites. The astroclimatic analysis is a prerequisite for selecting a site for radio astronomical observations in the millimetre bandwidth.**

***Index Terms*—Astroclimate, radio astronomy, millimeter wave propagation, atmospheric measurements.**

## I. Introduction

THE ATMOSPHERIC propagation of terahertz waves strongly depends on the atmospheric conditions, and one of the topical problems of today is to investigate this dependence and develop techniques, instrumentation and mathematical modelling of the atmospheric transparency investigations. We have developed instrumentation [1] and improved the methods [2] for investigation of atmospheric propagation of terahertz waves. Since 2012 we have gone on 8 expeditions and explored atmospheric absorption at over 11 sites [3 and References]. Our goal is to find the most appropriate place for a radio telescope operating in the millimetre and sub-millimetre wavelength. For estimating the atmospheric absorption (also referred to as optical depth or tau, and measured in Nepers) we use the atmospheric-dip method in automatic remote measurements mode with the help of the radiometric system MIAP-2 operating in 84-99 GHz (3 mm) and 132-148 GHz (2 mm) bands.

Over the last 6 years we have collected a broad array of statistical data on atmospheric absorption, which allows us to compare different places in terms of suitability for radio astronomy. We ourselves believe that the Suffa plateau deserves being most promoted as a site of a prospective radio astronomy project. Therefore, we are presenting the Suffa plateau statistical data sets in comparison with Svalbard, Northern Caucasus, Badary and other locations.

## II. Sites Description

Among all sites where we have been collecting astroclimatic data, are only four where we were able to collect enough data to plot statistical diagrams and make conclusions as regards averaged seasonal variations of the atmospheric transparency. These sites (either functioning observatories or research stations), unlike the rest of them, are also equipped with service staff and supplied with electricity and the Internet connection.

### *A. Svalbard*

Geographic coordinates: 78° 5'42.22“N 14°12'35.98“E;

Observation period: June 2018 – June 2019;

Elevation: 100 m;

Climatic conditions: Arctic, but weather often warm due to the impact of the Gulf Stream;

Svalbard is in the Arctic being the northernmost habitable land featuring permanent residents and provided with such utilities as electricity and the Internet. The archipelago has two residential towns and several seasonal settlements. The high latitudes of the location make it reasonable to suppose that the optical depth will not be large here. Its highest point is 1,713 m, but available sites of sufficient areas for mounting a telescope are at the elevation of about 1,500 m. Among drawbacks of Svalbard intended for an observational radioastronomy site we can mention the Gulf Stream flowing by the southern part of the archipelago bringing in warm and humid air.

We did our research at the research station of the Polar Geophysical Institute found 3 km north of Barentsburg. Since we set up our equipment on June 9, 2018 the automatic recording of the atmospheric depth has been carried on with 10-minute intervals. We have already quoted some of the results of these observations in our previous publications [3,4,5]. Below are statistical diagrams including histograms and cumulative distribution plots of the optical depth observed. All data are grouped according to similarity of weather conditions: December, January, February and March are considered the winter season; October, November, April and May the transitional period; and June, July, August and September the summer season.

This work was supported in part by the Scoolarship issued by RF President under Project СП-5103.2018.3 and by Institute of Applied Physics RAS state program 0035-2019-0005. The authors acknowledge the support from Presidium RAS Program No. 5: Photonic technologies in probing inhomogeneous media and biological objects.

Grigoriy M. Bubnov and Vyacheslav F. Vdovin are with Institute of Applied Physics RAS, and with Nizhny Novgorod State Technical University n.a. R.E. Alekseev, N.Novgorod, Russia. (e-mails: bubnov@ieee.org; vdovin_iap@mail.ru).

Valeriy F. Grigorev is with Polar Geophysical Institute, Murmansk, Russia. (e-mail: valgri@pgi.ru)

Peter M. Zemlyanukha and Igor I. Zinchenko are with Institute of Applied Physics RAS, N.Novgorod, Russia. (e-mails: g.pzemlyan@gmail.com, igorzin50@gmail.com)

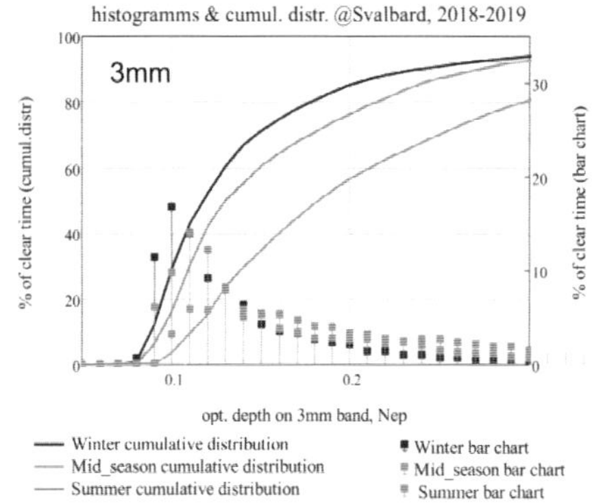


Fig. 1. Histograms and cumulative distribution plots of optical depth measured on Svalbard at 3 mm wavelength in 2018-2019.

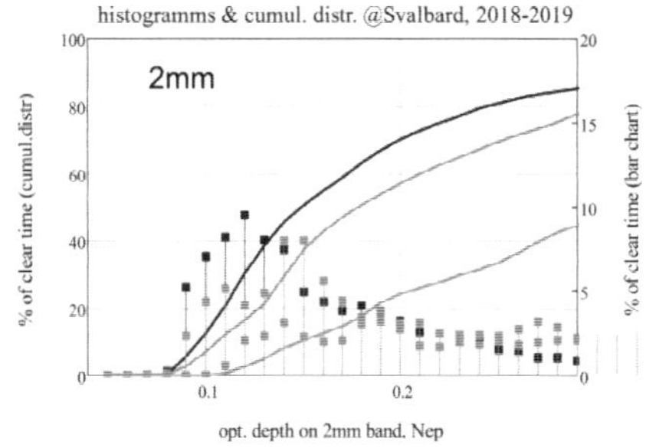


Fig. 2. Histograms and cumulative distribution plots of optical depth measured on Svalbard at 2 mm wavelength in 2018-2019.

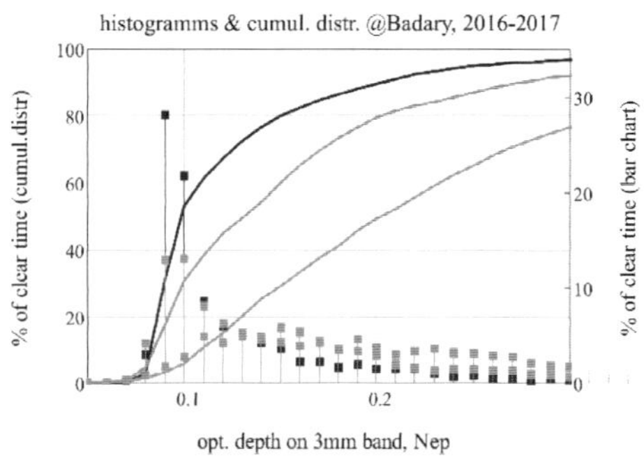


Fig. 3. Histograms and cumulative distribution plots of optical depth measured at the Badary observatory at 3 mm wavelength in 2016-2017.

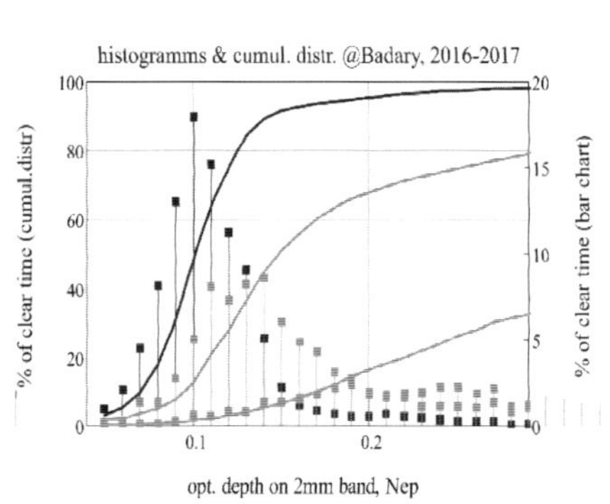


Fig. 4. Histograms and cumulative distribution plots of optical depth measured at the Badary observatory at 2 mm wavelength in 2016-2017.

*B. Badary observatory*

Geographic coordinates: 51°46'10.75"N 102°14'6.13"E;

Observation period: June 2016 – June 2017;

Elevation: 700 m;

Climate zone: Sharply continental, but it's often humid due to the valley's terrain;

Climatic conditions: Sharply continental, but often humid due to the peculiarities of the Tunkin Depression;

The Badary Observatory of the Institute of Applied Astronomy is a part of the Kvazar-KVO radiointerferometric network. It is located in the Tunkin Depression 300 km west of Lake Baikal. The extreme continental climate is characteristic of this entire region with its dry frosty winters and relatively short summers. The averaged precipitable water vapour (PWV) is relatively low, which is seen also by the weather satellites data (we are not quoting these data here). However, the Tunkin Depression facilitates collection and accumulation of water vapour thus nullifying the advantages of the sharply continental climate. To the west of the Badary observatory, the Observatory of the Institute for Solar and Terrestrial Physics is found near the village of Mondy; the elevation of this site is 2,100 m. Our observations have shown that the astroclimatic conditions in this latter place are far better than in the depression, but we were unable to conduct a full-scale year-long observation cycle. [6]

Below are histograms and cumulative distribution plots of the optical depth observed. All data are grouped by seasons similarly to the previous example.

*C. Special astrophysical observatory (N. Caucasus)*

Geographic coordinates: 43°38'51.45"N41°26'31.00"E;

Observation period: 2014 – 2015;

Elevation: 2100 m;

Climatic conditions: Predominantly temperate climate, but the site borders the subtropical climate zone;

The Special Astrophysical Observatory of the Russian Academy of Sciences (SAO RAS) is one of the largest optical observatories of the world. It is located at 2,100 m above the sea level near the village of Arkhyz in the Northern Caucasus. Among the reasons for the choice of the place for the observatory were its good astroclimatic conditions: the stability of atmosphere in the optical band, high percentage of occurrence of the clear sky, and relatively dark nights. Located at the height of more than 2 km above the sea level, the SAO RAS site enjoys low oxygen absorption; however, the Black Sea being only 90 km away enhances the absorption of non-condensed water vapor (the sky remains cloudless) up to quite high values.

We carried out our measurements through one of the windows of the BTA (Large Altazimuth Telescope) tower in different months from 2014 through 2015. The observation time covers about 10 months, which seems enough for reconstruction of statistics of the whole year. At the time the 2-millimetre channel of the radiometer was not operable, and that is why the statistics has been gathered for the 3-mm channel only. In detail, the results of these observations were presented in [7].

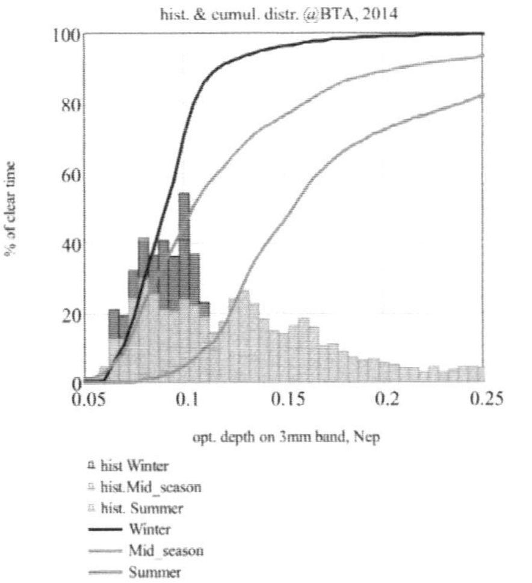


Fig. 5. Histograms and cumulative distribution plots of optical depth measured at the SAO RAS observatory at 3 mm wavelength in 2014-2015.

*D. Suffa plateau*

Geographic coordinates: 39°37'28.46"N 68°26'54.16"E;

Observation period: uninterrupted since 2012;

Elevation: 2400 m;

Climatic conditions: Dry sharply continental climate;

The Suffa plateau was chosen for the setup of the 70-metre radio telescope RT-70 similar to the Yevpatoria RT-70 telescope in the Soviet times. The construction started with the foundation of the telescope and some structures. Then, in the 1990's, the construction work stopped to be renewed only in 2018 and 2019 when some international intergovernmental contracts were drawn. Located at the height of 2,400 m above the sea level, the Suffa plateau enjoys a high percentage of occurrence of clear sky both in the optical and millimetre bandwidth. Observations on astroclimate started here as early as the end of the 1980's and included launches of weather balloons, standard meteorological records and research of the convectional stability of the atmosphere in the optical bandwidth. In 2012 first test measurements were taken, and since 2014 up to now full-scale regular observations have been carried out of the millimetre bandwidth astroclimate. Statistical data regarding the Suffa plateau itself is abundant; see the statistical diagrams on figures 6 and 7.

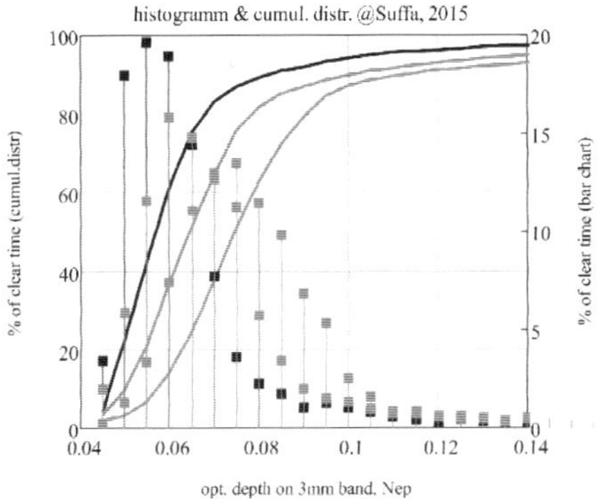


Fig. 6. Histograms and cumulative distribution plots of optical depth measured on the Suffa plateau at 3 mm wavelength in 2015.

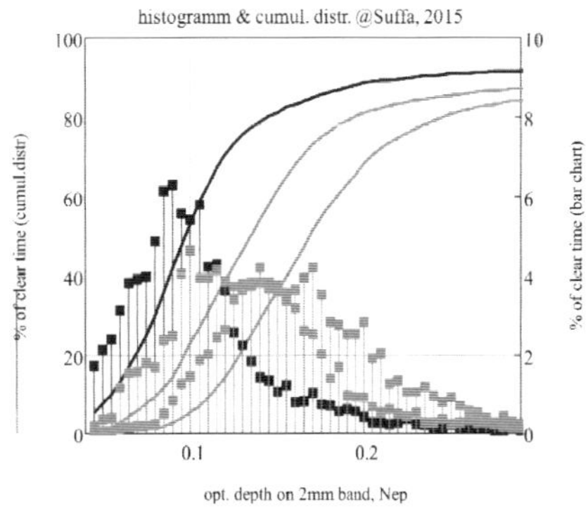


Fig. 7. Histograms and cumulative distribution plots of optical depth measured on the Suffa plateau at 2 mm wavelength in 2015.

In one of our expeditions we climbed a mountain 3,300 m high located in the vicinity of the plateau. We could see dense low clouds at the height of 100 to 200 m above the plateau, but at 3,300 m the sky was clear. The local staff told us, and their words are indirectly corroborated by the radiometer MIAP-2 data, this was a normal situation in that place (see Fig. 8).

It is difficult to evaluate quantitatively the advantages of the 3,300-m site in comparison with the 2,400-m site without taking direct measurements; however, at this stage of research we would describe this advantage as "substantial" for the millimetre and especially the sub-millimetre bandwidth. We are planning to carry out measurements in an expedition in 2019 or 2020.

Fig. 8. This photo was taken from the mountain 3,300 m high located near the plateau. The low-altitude clouds covering the Suffa plateau can be seen.

## III. Sites Comparison

We have gathered enough data on the sites described in the previous section for statistical comparison. As for the remaining sites mentioned in this paper, we were able to dedicate to them only short-time observations from a couple of hours to several months. One should bear this in mind when studying the plots and diagrams with the median values of optical depth indicated because the averaging period is different for each particular site. The median values have been calculated for a month-long uninterrupted record or shorter, depending on a particular case. Initially one of the MIAP-2 tools was operating in the single-channel mode only, this is why we do not have 2-millimetre transparency window measurements for all sites.

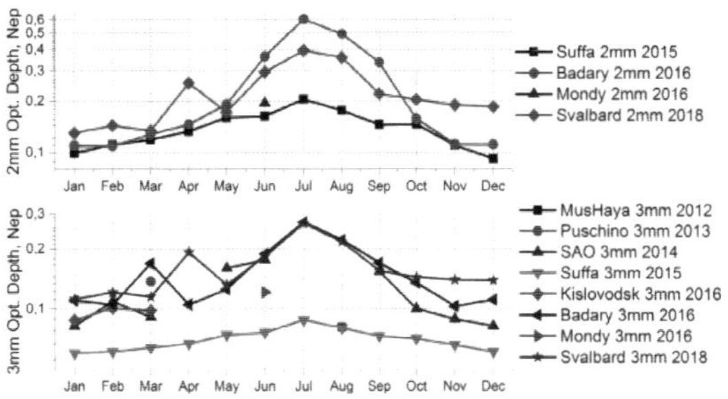


Fig. 9. Monthly median values of optical depth data obtained by our group on different sites starting from 2012.

Among all the sites described the Suffa plateau has most advantages. The median optical depth values are not higher than 0.2 Nepers for the 2-millimetre band and 0.1 Nepers for 3-millimetre. This is to be expected as the Suffa plateau is located high in the mountains in dry sharply continental climate. The site on Mus-Khaya Mountain in the south-east of Yakutia which is also located in conditions of sharply-continental climate near the Pole of Cold (Oymyakon) can rival with that on the Suffa plateau. However, the nearest habitation to Mus-Khaya Mountain is 100 km away, and one will have to pass the dense taiga on the way. On the Suffa plateau, on the other hand, there are basics of conveniences: there is a road, and electric power and running water supply.

The worst transparency values for the millimetre bandwidth are those sites that are closest to the sea: Kara-Dag (Crimea), Svalbard, and the SAO RAS. The Badary Observatory stands apart in this list as its climate is characterized by low atmospheric humidity, but the local impact of the Tunkin Depression collecting vapor cancels the climatic advantages of the location.

The optical depth in the millimetre bandwidth is defined by water vapor and oxygen absorption. The oxygen absorption depends mostly on the elevation of the site, while the water vapour part varies depending on the weather. Therefore, we can numerically describe the astroclimate of a site through precipitable water vapour. The method of calculation of PWV was discussed in our previously published papers [6,3]. The comparative analysis of sites' PWV is shown below.

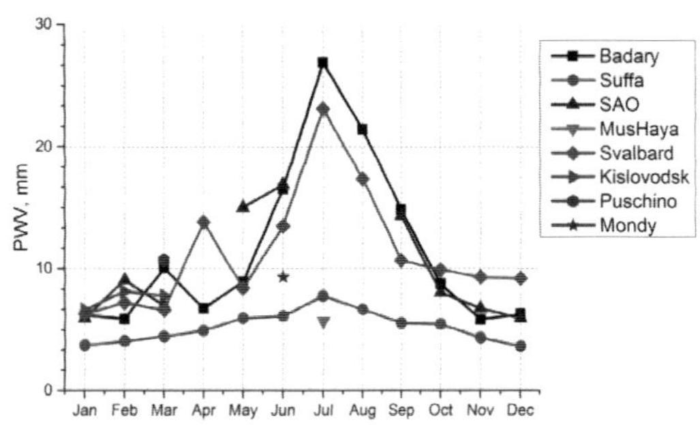


Fig. 10. Monthly median values of the PWV data obtained by our group on different sites starting from 2012.

The Suffa plateau and Mus-Khaya Mountain are also the best as far as PWV is concerned. The site of the Institute for Solar and Terrestrial Physics near the village of Mondy has also shown good results in PWV by the measurements taken over two days. At this stage of research, we can say that Mondy is the best place for the millimetre-bandwidth radioastronomy in Russia as it is equipped with all necessary utilities being a functioning optical observatory. Adding the millimetre bandwidth to its toolkit would be the most rational decision in view of both the astroclimate and the construction expenses.

For most sites our observations did not last longer than a year. Climatic conditions can change from year to year, and statistics of transparency may vary. The radiometric system MIAP-2 on the Suffa plateau is operating continuously, and its data can be used for tracking yearly changes (see Fig. 11).

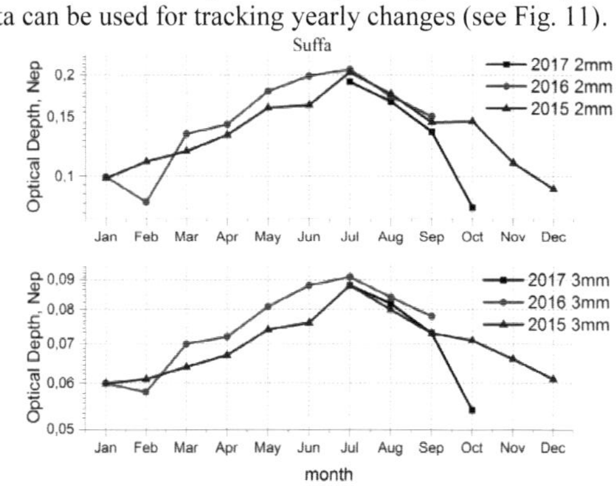


Fig. 11. Monthly median values of optical depth data obtained in different years on the Suffa plateau.

To draw the final line, we are offering this table of comparison of all the sites for which we have gathered enough data to present seasonal statistics. The table shows the share of the year when the optical depth observed is lower than 0.1 nepers. Two values are indicated for the Suffa plateau, the first value (2,400 m) is for the plateau itself following the immediate measurements, while the second one (3,300 m) is an extrapolation of the Suffa data for the 3,300-metre elevation made with the use of the standard atmospheric profile. In the observatory of the Institute for Solar and Terrestrial Physic near Mondy village we were not able to conduct long-term observations, but the results of our short-time measurements permit to extrapolate the Badary observatory data and complete the statistics for a year with some degree of accuracy.

TABLE I

CLEAR SKY TIME PERCENTAGE COVERING OPTICAL DEPTH BELOW 0.1 NEP

| Site | Altitude, m | 3mm | 2mm |
|---|---|---|---|
| Svalbard | 100 | 16% | 7% |
| Badary | 700 | 30% | 21% |
| Mondy* | 2100 | 47 % | 45% |
| SAO | 2100 | 14% | - |
| Suffa | 2400 | 90% | 27% |
| Suffa* | 3300 | 94% | 51% |

* Extrapolation of the data obtained at the closest point.

## Conclusion

In this paper we are presenting a comparison of different sites which can be suggested for construction of a radio astronomical observatory operating in the millimetre bandwidth. Among all sites having ever been studied with our radiometer MIAP-2, the Suffa plateau is the best choice for astroclimatic reasons. However, our observations have shown that the mountain peak close by might prove a better choice. We are planning to go on expedition to study that mountain in 2019. At present there are plans to cut some of the bandwidths of the RT-70 radio telescope in order to lower the expenses and accelerate the construction process. We, on the other hand, are proposing to add to the Suffa observatory another tool operating in the sub-millimetre bandwidth at the elevation of 3,300 m. Assembling a commercial compact instrument will be easier and faster than constructing the RT-70 which has not gone further than the designing stage so far. Our instrument will permit to collect data in the Suffa observatory in near future without waiting for the construction completion of the RT-70.

## Acknowledgment

This work was work was supported in part by the IEEE Antennas and Propagation Society Eugene F. Knott Memorial Doctoral Research Grant. The authors would like to thank Barentsburg stations of Polar Geophysical Institute and Arctic and Antarctic Research Institute for technical support. Special thanks for station's staff Yu.A. Kvashnin, V.G. Kuleshov.

The authors would like to thank Dmitri Sofronov for the help in editorship.

## References

[1] V. I. Nosov et al., "A dual-wave atmosphere transparency radiometer of the millimeter wave range", Instruments and Experimental Techniques, Vol. 59, Issue 3, pp 374-380, May 2016. 10.1134/S0020441216020111

[2] Grigoriy M. Bubnov et al., "A refined method of the data processing for astroclimate measurements in mm-waves", Proc. of 29th ISSTT, Pasadena, California, USA, March 26-28.

[3] Grigoriy Bubnov et al., "Svalbard astroclimate research: expedition and first results", Terahertz and Microwave Radiation: Generation, Detection and Applications (TERA-2018), N.Novgorod, October 22 - 25, 2018

[4] G.M. Bubnov, V.F. Vdovin, V.I. Nosov, and P.M. Zemlyanukha, "On the issue of optical depth measurements by using a wide band radiometer," Institute of Applied Physics RAS, N.Novgorod, Russia. Feb. 25-28. [Online] Available: http://msmw.iapras.ru/files/abstracts2019.pdf

[5] G.M. Bubnov, V.F. Vdovin, V.F. Grigorev, and P.M. Zemlyanukha, "Svalbard astroclimate investigations in mm waveband: the first results and summer period features", Kotelnikov Institute of Radioengineering and Electronics (IRE), Moscow, Russia. Nov 26-30, 2018. p.285-289.

[6] Grigoriy M. Bubnov et al. "Millimeter-wave astroclimate investigations on Badary observatory near Baikal lake" 32nd URSI GASS, Paper F6-1(2225), Montreal, 19-26 August 2017. [Online] Available: http://www.ursi.org/proceedings/procGA17/GASS2017Program.pdf

[7] Grigoriy M. Bubnov et al., "Searching for New Sites for THz Observations in Eurasia", IEEE Tr. on TST, Volume: 5, Issue: 1, pp.64-73, 2015.

[8] Yury N. Artemenko et al. "New stage of the Suffa Submm Observatory in Uzbekistan Project", *to be published in* Proc. of 30th IEEE International Symposium on Space Terahertz Technology, Gothenburg, Sweden, Apr 15-17, 2019 p.124

**Grigoriy M. Bubnov** (M'2016) was born on 29 October 1989 in Nizhny Novgorod, Russia (then Gorky, USSR). Graduated from the Nizhny Novgorod State Technical University (NNSTU) in 2013 with a master's degree in radiotechnological engineering. The title of his master's thesis is "MM-wave radiometer for atmospheric absorption measurements". Finished the postgraduate school of NNSTU in 2017. His scientific interests include methods and techniques of millimetre and sub-millimetre wave spectroscopy and atmospheric radiometry, and laboratory and field investigation of atmospheric absorption. Took part in and organized several expeditions on astroclimatic research. Is heading two projects supported by respective grants for astroclimatic research. Since 2011 to the present time has worked at the Institute of Applied Physics of the RAS. Current position is Junior Researcher.

**Valeriy F. Grigorev** was born in Kotovsk, Odessa Region, USSR, in 1946. He received his training as a radio engineer at the Odessa Electrotechnical Institute of Communications in 1970. At various times he worked either as a technician or in a managerial post in the armed forces of the USSR and the Russian Federation. From 1997 to the present time he has been holding the position of Deputy Director in the Polar Geophysical Institute, Murmansk, Russia.

His main research interests are propagation of radio waves in various environments, remote electromagnetic sounding of the earth's ionosphere and lithosphere, and long-term observations of geophysical processes specifically in the Arctic zone. He is the author of a number of scientific publications and patents.

Mr. Grigorev is awarded the titles of Honourable Worker of Communications of the Russian Federation, Honorary Radio Operator and Honorary Polar Explorer.

**Vyacheslav F. Vdovin** (M'94) received his master's degree in Radiotechnology from the Gorky Polytechnic Institute (nowadays the Nizhny Novgorod State Technical University) in 1978, Ph.D. degree in Radiophysics from the Lobachevsky Nizhny Novgorod State University in 1993, and Doctor of Sci. in Devices and Methods of Experimental Physics from the Radiophysical Research Institute in 2006.

From 1978 up to now he has been working for the Institute of Applied Physics of the Russian Academy of Sciences, and since 2008 for the Astro Space Centre at the P.N. Lebedev Physical

Institute of the Russian Academy of Sciences. In 2010 he became a Professor of the Nizhny Novgorod State Technical University.
Member of the Cryogenic Society of America (CSA). Vice-President of the World Federation of Scientific Workers (WFSW).

**Petr M. Zmlyanukha** was born in Nizhny Novgorod, Russia, in 1991. He received his BS and MS degrees in radiophysics at the Lobachevsky University of Nizny Novgorod.
From 2012 he has been a Junior Researcher at the Department of Microwave Receivers and Astronomy of the Institute of Applied Physics of the RAS. Author of 8 articles in refereed journals. His main interests are microwave astronomy, star formation and interstellar medium, spectroscopy and statistical approaches to data processing. Peter is the associate reviewer of *Radiophysics and Quantum Electronics*.

**Igor I. Zinchenko** graduated from the Radio Physics Department at the Lobachevsky Nizhny Novgorod State University in 1972. He received his Candidate's of Science (more or less equivalent to the Ph.D.) degree in the field of astrophysics and radio astronomy from the Radiophysical Research Institute in Nizhny Novgorod in 1982, and the Doctor's of Science degree in the same field from the same institution in 1998.
From 1972 to 1977 he held the position of a Junior Researcher at the Radiophysical Research Institute in Nizhny Novgorod. From 1977 up to now he has been with the Institute of Applied Physics of the Russian Academy of Sciences in Nizhny Novgorod. Since 1987 he has been Head of the Department of Receiving Radio Equipment and Millimetre-wave Radioastronomy at this institute. Since 2002 he also has been working as Professor for the Faculty of Radio Physics (Department of radiowaves propagation and radioastronomy) at the Lobachevsky Nizhny Novgorod State University. His major fields of research are millimetre and submillimetre wave radioastronomy, astrophysics, and the properties of the atmosphere in the millimetre and submillimetre wavebands. He has published about 200 papers dedicated to various topics in these spheres.

Prof. Zinchenko is a member of the International Astronomical Union, International Union of Radio Science (URSI), European Astronomical Society, Euro-Asian Astronomical Society, and Astronomical Council of the Russian Academy of Sciences.

# Atmospheric Phase Monitoring Interferometer for the NOEMA Observatory

S. Mahieu[1], B. Pissard[1], M. Bremer[1], C. Risacher[1], R. Blundell[2], R. Kimberk[2], J. Test[2]

*Abstract*—**IRAM is currently adapting, in collaboration with the Smithsonian Astrophysical Observatory (SAO), an atmospheric phase monitoring system as was developed for the Submillimeter Array (SMA) to be possibly used for the NOEMA interferometer.**

## I. Introduction

In this paper, we present the proof-of-concept of the atmospheric phase monitoring system and show the first results obtained so far. The final goal of this project is to provide a permanent monitoring system of the observing conditions for the NOEMA interferometer, located on the plateau de Bure, at 2500m above sea level in the French Alps. Indeed, this system would make observations more efficient, by being able to choose the correct observing band right away and/or anticipate on whether to start or stop observing.

After initial tests carried out on the roof of the IRAM Grenoble headquarters, the system was moved to the plateau to make first onsite data acquisition at the end of October 2018.

## II. System Description

The atmospheric phase monitoring system demonstrator is based on a dual off- axis aluminium satellite dishes (Fuba DDA 110: 1090 x 991 $mm^2$) interferometer, that receive a broadband white noise-like Ku Band (~11.85 GHz) signal from a geostationary satellite and focus it to the center of the feed signal source.

The Low Noise Blocks (LNB) that down convert this signal to the Intermediate Frequency (IF) one (~1.2 GHz) have been modified to be fed with a common Local Oscillator.

The IF signals are then amplified, filtered and transported through optical fibers down to the building where they are processed by a commercial analog correlator (IQ demodulation card) that produces the phase delays between pairs (only one so far) of antennas from the I & Q signals.

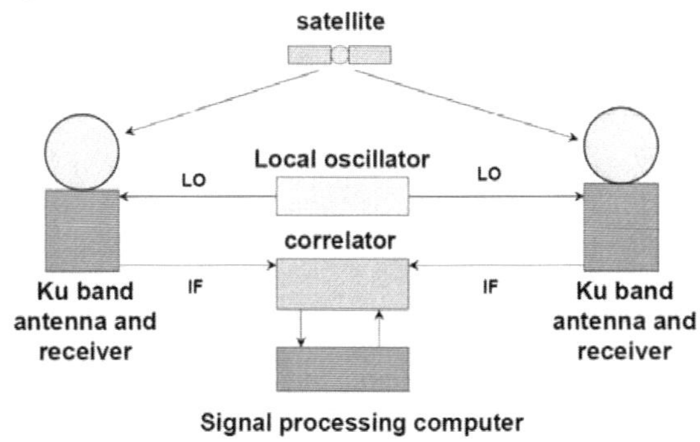


Fig. 1. Phase Monitoring System Principle.

The data are then further processed by a LINUX pc that runs software, which amongst other parameters, works out the atmospheric RMS phase over different time intervals. This real-time statistical data measured in the direction of the satellite provides an estimate of the phase front distortion experienced at the same time by the NOEMA interferometer.

## III. System Implementation

During the development phase at the IRAM headquarters, some optimizations were implemented in order to achieve the best possible results. Despite these efforts, we were still facing, at the end, electromagnetic pollution and environmental limitation and decided to move the atmospheric phase interferometer to the plateau de Bure for a short data collection campaign. The first trials on the plateau, in parallel with the NOEMA interferometer are encouraging as there is good agreement between the two systems.

## IV. Conclusion

This paper has briefly described the phase monitoring system work that has been ongoing for the past two years at IRAM. More detailed description of the system will be provided during the conference, together with test results.

## References

[1] Robert S. Kimberk & All "A Multi-Baseline 12 GHz Atmospheric Phase Interferometer with One Micron Path Length Sensitivity", May. 2012.

[1] Institut de Radio Astronomie Millimétrique, IRAM, 300 rue de la piscine, 38400 Saint Martin d' Hères, France.

[2] Harvard-Smithsonian Center for Astrophysics, 60 Garden Street, Cambridge, MA 02138, USA.

NOTES:

# New Optics for SEPIA- Heterodyne Facility Instrument for the APEX Telescope

I. Lapkin[1,2], S.-E. Ferm[1], M. Fredrixon[1], D. Meledin[1], M. Strandberg[1], V. Desmaris[1], A. Ermakov[1,3], A. Pavolotsky[1], E. Sundin[1] and V. Belitsky[1]

[1]Group for Advanced Receiver Development, Chalmers University of Technology, Gothenburg, Sweden
[2]Institute of Applied Physics of the Russian Academy of Sciences (IAP RAS), N. Novgorod, Russia
[3]Institute of Radio Engineering and Electronics, Russian Academy of Sciences, Moscow, Russia

***Abstract* — The design of SEPIA (Swedish ESO PI Instrument for APEX) was driven by the idea of using ALMA receiver cartridges on the APEX telescope. SEPIA was installed at the guest position of the Naismith cabin A, APEX telescope in early 2015. The SEPIA cryostat and optics was designed to accommodate up to 3 ALMA cartridges. In 2017, the APEX facility instrument SHeFI was decommissioned and SEPIA was accepted as its successor. Moving SEPIA from its PI into Facility Instrument position brought additional constrains due to the severe limitations of the available space. That had led to the necessity of complete redesigning of the SEPIA tertiary optics. During February-March 2019, the new tertiary optics was installed in the APEX Cabin A and SEPIA was placed at its final Facility Instrument position. Here, we present the details of the optical design, layout of the optical component placement, the beam alignment technique, the results of the alignment and SEPIA technical commissioning results at the APEX telescope.**

***Index Terms*—Radio astronomy, Radio-wave propagation, Mixers, Receivers, ALMA, APEX, SEPIA Instrument, Gaussian optics.**

## I. Introduction

THE initial idea with the implementation of ALMA cartridge receivers at the APEX telescope was inspired by the interest of the ESO astronomical community to the future observations with Band 5 [1] at ALMA (started in 2018). This idea transformed into the SEPIA [2] (Swedish-ESO PI Instrument for APEX), a mutual project of the GARD, OSO and ESO. SEPIA project was started in the beginning of 2014, and during February 2015 SEPIA was installed and commissioned at the APEX telescope with modified ALMA Band 5 pre-production receiver and updated with ALMA Band 9 [3] in February, 2016. For SEPIA Band 9, GARD provided receiver infrastructure and optics.

SEPIA receiver was designed initially as a PI instrument, dedicated for the installation at guest position of the APEX Nasmyth cabin A and accommodate up to 3 receiving channels, compatible with the ALMA receiving cartridges Bands 5-10. In 2017, the APEX facility instrument SHeFI [4] was decommissioned and SEPIA was accepted as its successor.

This work was supported by the Swedish Research Counsel via its grant to Onsala Space Observatory in part of the Swedish support of the Atacama Pathfinder Experiment (APEX) telescope.

Moving SEPIA from its PI into Facility Instrument position brought additional and severe space limitations constrained by the support structure, the facility calibration load unit and requirements for the optical path clearance for the PI instruments, located at the left and right sides of the facility instrument position. That had led to the necessity of the complete redesigning of the SEPIA tertiary optics. The new optics was designed and parts were manufactured in 2018 and SEPIA was installed onto its new position in February-March 2019

## II. Optics design

The SEPIA optics design implements frequency-independent illumination of the secondary for all receiving channels with edge taper aiming around -12 dB. To fulfill this condition within the entire working frequency range (159-722GHz) [2], the complete optical path works as a re-imaging system, transferring the image of the secondary onto the aperture of the corrugated horn for each receiver channel. Table I and Fig 1 represent the entire SEPIA illumination path through the Nasmyth and Cassegrain cabins.

The new SEPIA optical design based on the system of equations describing a 2-mirror refocusing optical system [5], derived from the ray transfer matrices for Gaussian beams. The design was performed with the assumption of the mirrors rims sizes of 5 beam waists. The actual optics layout (Fig.2) employs individual active channel mirrors (NMA2-1, NMA2-2, NMA2-3) together with the common active mirror (NMA1). Along with the existing Nasmyth Cabin A tertiary optics, the new optics accommodate the input beams of the SEPIA receiver cartridges to the APEX antenna. Compared to the previous variants of the optical layout, the first common active mirror (NMA1) is located in the vicinity to the Instrument Selection Mirror (NMF1), which allowed to accommodate all elements of the receiver, including the channels' optics in the space inside Flexlink support frame in the middle of the APEX Cabin A, dedicated for the Facility Instrument (Fig.2, Fig.5).

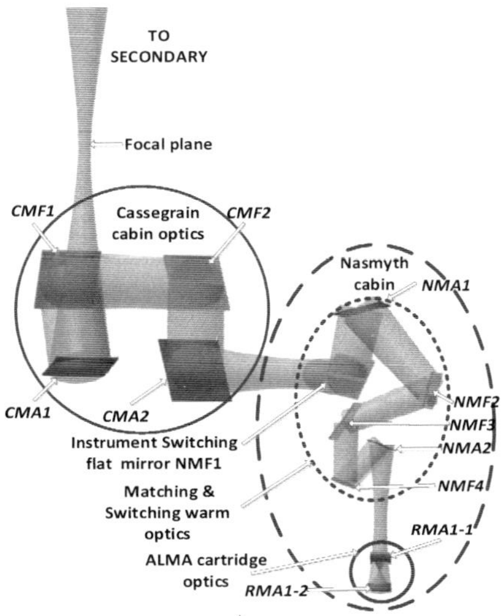


Fig. 1. The 3D CAD model of the total SEPIA optics path.

TABLE I
SEPIA TOTAL OPTICS LAYOUT

| | Element's Tag | Distance (mm) | Focal distance (mm) | |
|---|---|---|---|---|
| | Cassegran cabin optics | | | Common optical path |
| | Focal plane | 5882.86* | - | |
| | CMA1 | 750 (6632.86)** | 700 | |
| | CMF1 | 361.86 | Fold Instrument Selection Mirror | |
| | CMF2 | 1838.14 | Fold Mirror | |
| | CMA2 | 300 (2500)** | 1800 | |
| | | 1764.2 | Fold Mirror | |
| | Nasmyth cabin warm optics | | | |
| | NMF1 | 400 | Fold Mirror | |
| Modified optics layout | NMA1 | 234.30 (2398.5)** | 760 | |
| | NMF2 | 369.23 | Fold Mirror | |
| | NMF3 | 304.34 | Fold Mirror | |
| | NMF4 | 205.18 | Fold Channel Selection Mirror | |
| | NMA2-1 NMA2-2 NMA2-3 | 234.85 (1055.77)** | 295.66 | |
| | Receiver position 1 Cold optics ALMA Band5 | | | Individual channel optical path |
| | RMA1-1 | 500 | 67.192 | |
| | RMA2-1 | 140 | 32.756 | |
| | Horn1 | 60.05 | (60.17/9.0)*** | |
| | Receiver position 2 Cold optics ALMA Band7**** | | | |
| | RMA1-2 | 498 | 60 | |
| | RMA2-2 | 133.78 | 30.41 | |
| | Horn2 | 52.04 | (20.46/6.9)*** | |
| | Receiver position 3 Cold optics ALMA Band9 | | | |
| | RMA1-3 | 446.2 | 39.41 | |
| | RMA2-3 | 95.9 | 24.86 | |
| | Horn3 | 44.48 | (15.72/5.07)*** | |

*- distance from the telescope secondary mirror
**- unfolded distance between active mirrors
***- horn's parameters (slant length/ aperture diameter)
****- cartridge optics designed in GARD, compatible with the existing Band 7 ALMA front-end optics

The flat Channel Selection Mirror (NMF4) is used to switch between the bands. Each frequency band has individual adjustable active mirror (NMA2-1, NMA2-2, NMA2-3) and a set of mirrors that form common optical path. As a consequence of the space constraint, we had to use several additional flat mirrors (NMF2, NMF3, NMF4) to fold the beam inside the allocated space. The picture (Fig.5) shows actual placement of the SEPIA in the APEX Cabin A.

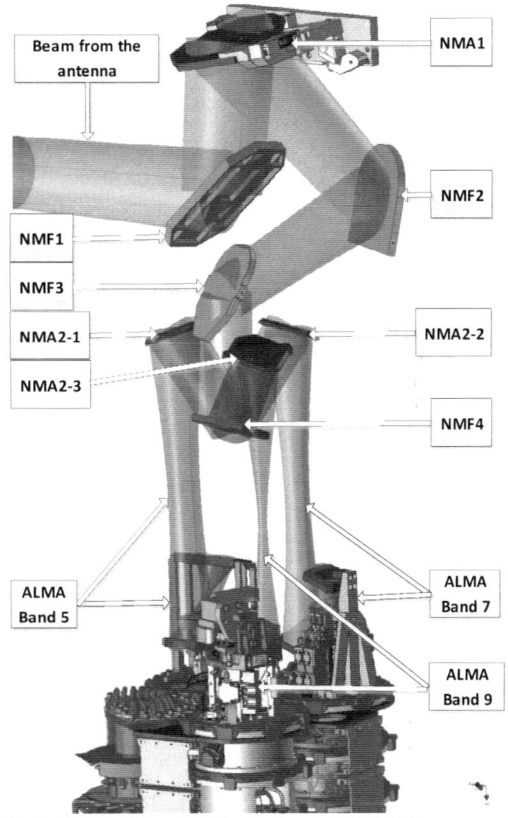


Fig. 2. The 3D CAD model of the new layout of the SEPIA instrument at the Facility Instrument position. NMF1-Swicting mirror between PI and SEPIA instruments; NMA1 – active adjustable refocusing mirror, NMF4 - Channel switching mirror; NMA2-1, NMA2-2, NMA2-3- Individual channel refocusing mirrors NMA2; ("ALMA Band 5"; "ALMA Band 7"; "ALMA Band 9")- ALMA receiving cartridges.

Verification of the optical system using the GRASP program [6], confirmed the required parameters of the illumination system, (Fig3,4) As a feed, Gaussian beam, matched with the horn parameters was implemented. Special attention was paid to the level of cross-polarization losses and optical distortions in the optical system, generated by reflections from off-axis mirrors. Compare with the initial layout, modified optical system employed mirrors with smaller F/D parameters, and large incidence angles; it was a compromise, allowed to compact optical layout. Estimation was made with the Murphy formulas [7] (1), (2), based on the parameters of the beam radii at the mirror ($w_{beam}$), the focal length ($f$), and incidence angle ($i$).

$$P_{cross-pol} = {}^{1}\!/_{8} \cdot \left({}^{w_{beam}}\!/_{f}\right)^2 \cdot \tan^2 i \qquad (1)$$

$$P_{distortion} = {}^{1}\!/_{4} \cdot \left({}^{w_{beam}}\!/_{f}\right)^2 \cdot \tan^2 i \qquad (2)$$

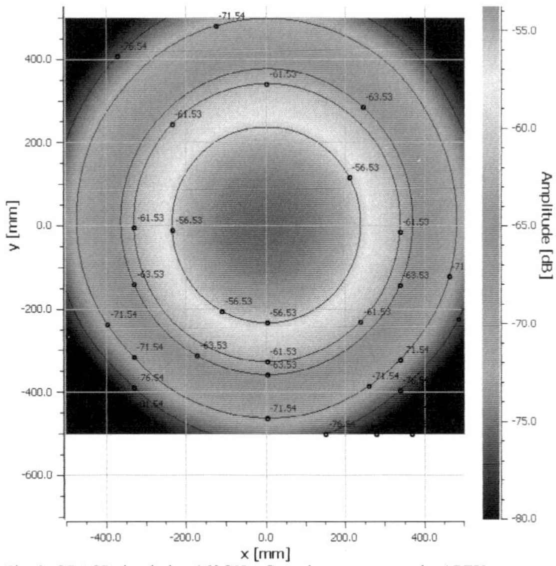

Fig. 3. GRASP simulation 163GHz, Co-pol component at the APEX secondary mirror.

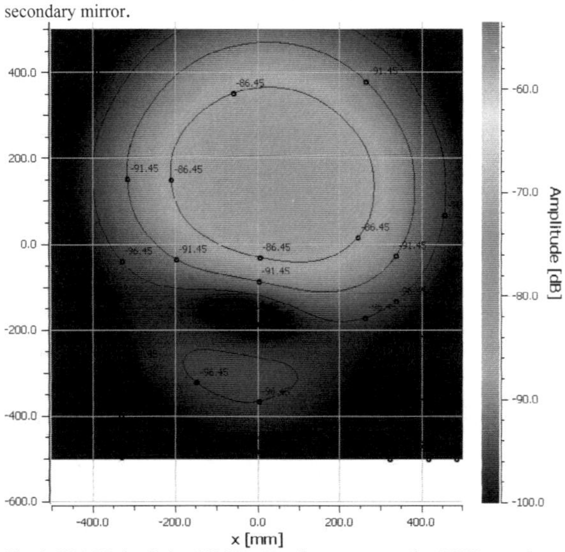

Fig. 4. GRASP simulation 163GHz, X-pol component at the APEX secondary mirror.

Our simulations using GRASP physical optics software package (Fig.4, 5) confirm that the integral level of the cross-polarization losses along the entire optical system, from the horn to the secondary mirror, which included 6 active mirrors and 4 flat mirrors was less than -23 dB.

## III. Optics Fabrication, Alignment and Installation

The entire SEPIA optics unit was built at GARD workshop in house. Before shipping the optics to the telescope, all optical components were pre-aligned in the lab using a laser placed at the optical central axis. To facilitate the laser-assisted alignment the mirror's surfaces was additionally polished at their central area to provide sufficient reflectivity. During the alignment procedure, the laser was placed in the position of the axis of the input beam. for each respective frequency channel. The pre-aligned optics unit then was shipped and installed at the telescope.

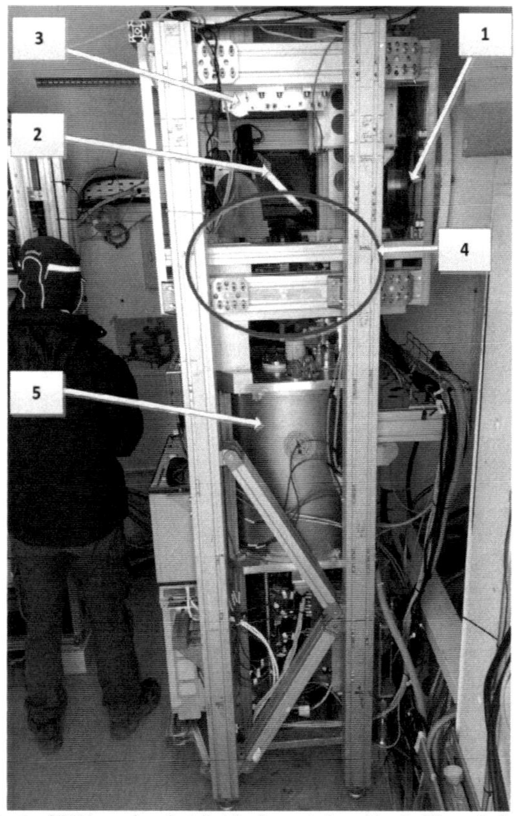

Fig. 5. SEPIA receiver installed in the central position;1- Nasmyth flange, Cabin A; 2- Instrument Selection Mirror (PI or Facility Instrument);3- SEPIA Active Top mirror support bracket; 4- SEPIA New Optics; 5- SEPIA Receiver.

Total alignment and installation procedure at the telescope included 3 steps. At the first step (Fig. 6) the laser was installed in the Cabin A Nasmyth tube (the position of the input beam axis). A position sensitive device (PSB) was attached to the SEPIA optics unit frame using temporary mechanical centering and alignment structures and the entire optics frame was aligned using the laser beam as a reference while placing the PSD in the three firmly arranged positions along the optical axis of the input beam inside the optics unit as depicted in the Fig. 6. Using the PSD sensor installed in these different positions and iterative adjustment of the entire optics frame, we achieved alignment accuracy at the level of a fraction of millimeter.

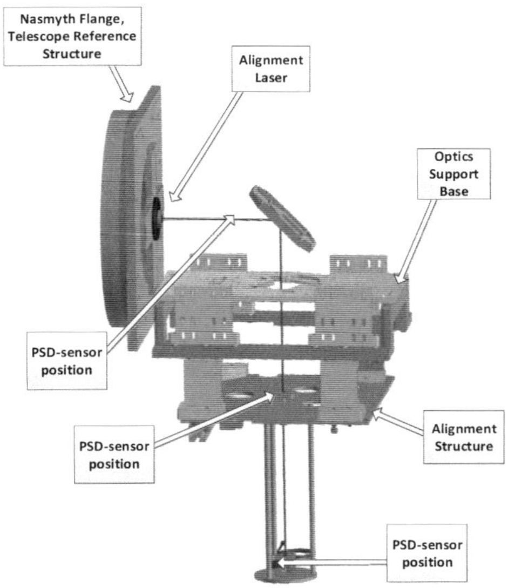


Fig. 6. Laser alignment during the optics installation at the telescope. Step1.

At the second step (Fig. 7), after the Optics unit frame was installed and aligned, the optics components that was temporary removed could be integrated back into the Optics unit, and its alignment verified, and if necessary, corrected using the laser with procedure similar to what we used in the lab and as shown in the Fig. 7.

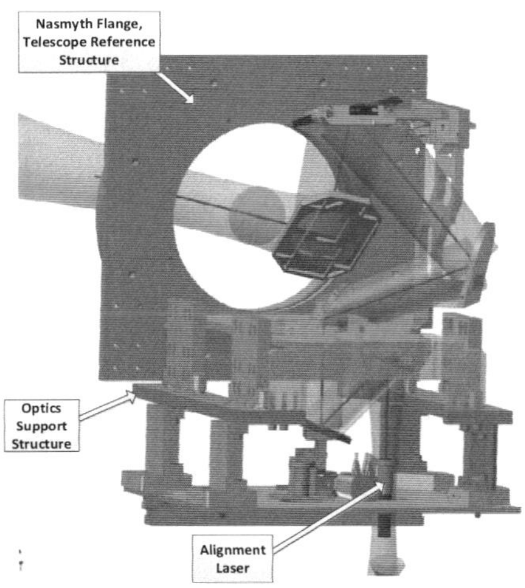


Fig. 7. Installation procedure at telescope. Step2

At the third step, the SEPIA receiver was attached to the new Optics unit. Using a cold load (77K) and a warm load (300K), the RF beams of the Band 5 (SEPIA180) and the Band 9 (SEPIA660) was then checked and fine-tuned with the individual channel adjustable mirrors NMA2 for best pointing.

## IV. Commissioning

In the spring of 2019, the installation of the new optics and the SEPIA receiver were completed and commissioning was conducted via set of verification observations with the SEPIA receiver installed into the Facility Instrument position.

The *SiO* maser observation of *R Dor* at 171 GHz was used for the SEPIA180 verification procedure [8]. The *SiO* maser emission is essentially a point source and thus the map depicts the beam shape. Results of observation (Fig. 8, Fig. 9) demonstrated good quality beam shape down to 2% (below 2% the noise starts to show up). The contour levels start at 2% and end at 20% of peak value at the center. The (Fig. 9) show the radial cross-section of the beam pattern for the same data with the Bessel beam fit for illumination tapering of -14 dB. The half power beam width (FHPBW) at 171 GHz estimated as 36.5 ± 0.5 arcsec, which is close to expected value.

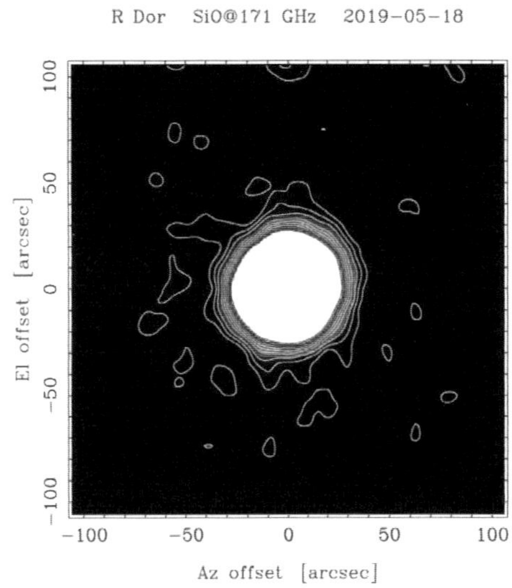


Fig. 8. SEPIA facility instrument position, SEPIA180 tests observation [8]. Beam map at 171 GHz (*SiO* J=4-3 v=2 maser) toward the star R Dor.

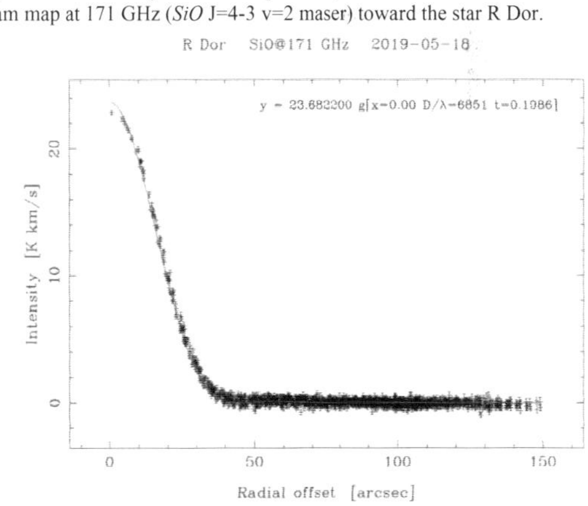


Fig.8. SEPIA facility instrument position, SEPIA180 tests observation [8]. Beam pattern (Fig.7) radial cross-section. Solid line-Bessel fit for the -14dB edge taper illumination.

In addition, the SEPIA180 Neptune measurements at 208 GHz confirmed the value of the antenna efficiency, and found to be consistent and close to the results of previous years. The SEPIA660 passed similar verification procedure [8], based on the source *VX Sgr* water line observation as depicted in Fig. 10,

Fig. 11.

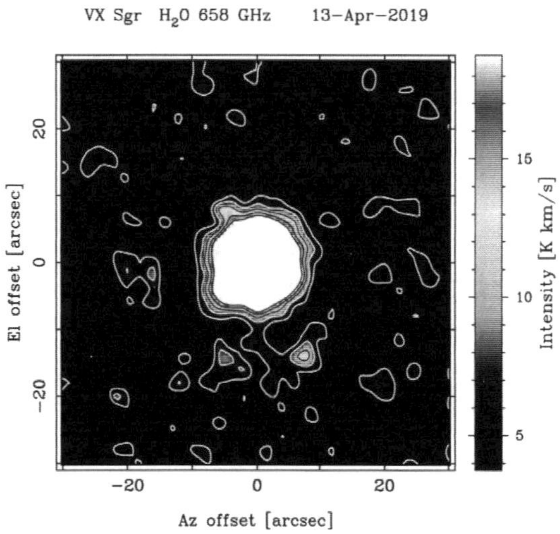

Fig. 9. SEPIA facility instrument position, SEPIA660 tests observation [8].

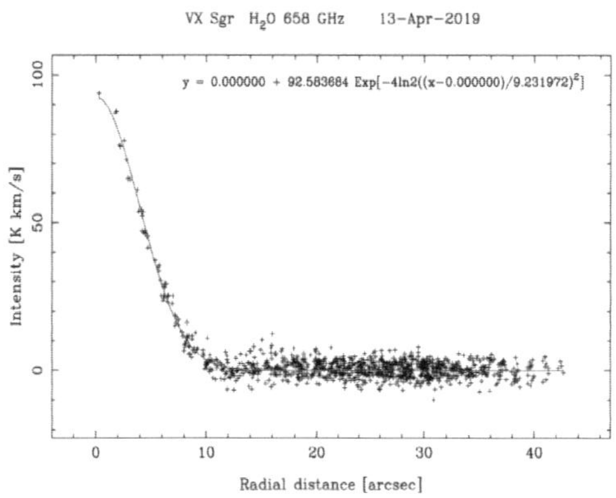

Fig. 10. SEPIA facility instrument position, SEPIA660 tests observation [8]. Beam pattern (Fig.9) radial cross-section. Solid line-Bessel fit for the -14dB edge taper illumination

## V. Conclusion

At the time of writing this manuscript, the SEPIA receiver is fully operational, commissioned and actively used for observations with its both bands, the SEPIA180 and SEPIA660.

## References


[1] V. Belitsky, M. Bylund, V. Desmaris, A. Ermakov, et.al. "ALMA Band 5 receiver cartridge Design, performance, and commissioning", *Astronomy & Astrophysics*, Volume 611, id. A98, (2018).

[2] V. Belitsky, I. Lapkin, M. Fredrixon et. al., "SEPIA - a new single pixel receiver at the APEX telescope," *Astronomy & Astrophysics*, Volume 612, id.A23, (2018).

[3] A. Baryshev, R. Hesper, F. P. Mena et. al., "The ALMA Band 9 receiver Design, construction, characterization, and first light" *Astronomy & Astrophysics*, Volume 577, id.A129, 12 pp, May 2015.

[4] V. Belitsky, I. Lapkin, V. Vassilev, R. Monje, A. Pavolotsky, D. Meledin, D. Henke, O. Nyström, V. Desmaris, C. Risacher, M. Svensson, M. Olberg, E. Sundin, M. Fredrixon, D. Dochev, S.-E. Ferm and H. Olofsson, "Facility Heterodyne Receiver for the Atacama Pathfinder Experiment Telescope", *Proceedings of joint 32nd International Conference on Infrared Millimeter Waves and 15th International Conference on Terahertz Electronics* .

[5] O. Nyström, I. Lapkin, V. Desmaris, et.al "Optics Design and Verification for the APEX Swedish Heterodyne Facility Instrument (SHeFI)", *Journal of Infrared, Millimeter, and Terahertz Waves,* July 2009, Volume 30, Issue 7, pp 746–761, https://doi.org/10.1007/s10762-009-9493-7

[6] GRASP https://www.ticra.com/software/grasp

[7] J. A. Murphy et. Al. "Distortion of Simple Gaussian beam on Reflection from Off-Axis Ellipsoidal Mirrors", *[8] Int. Jour. of Infrared and Millimeter Waves,* , Volume 8, Issue 9, pp 1165–1187, 1987

[8] Private communication, P. Bergman (Onsala Space Observatory, per.bergman@chalmers.se).

# Far-sidelobe Measurements of LiteBIRD Low Frequency Telescope 1/4-Scaled Model

Hayato Takakura, Yutaro Sekimoto, Junji Inatani, Shingo Kashima, Hiroaki Imada, Takashi Hasebe, Toru Kaga, Yoichi Takeda, and Norio Okada

***Abstract*—LiteBIRD is a satellite for polarization observation of the Cosmic Microwave Background. We have carried out near-field antenna pattern measurement of one of its telescopes, the Low Frequency Telescope, in 1/4 scale. We have investigated the far-sidelobes at the center and at the edges of the 20-degree field of view. The measured far-sidelobe patterns are consistent with the simulated one at −50 dB level, and the patterns for two orthogonal polarization directions are consistent with each other down to −40 dB level.**

***Index Terms*—Antenna pattern measurement, Cosmic Microwave Background, Far-sidelobes, LiteBIRD.**

## I. Introduction

The footprints of the inflationary universe are expected to be observable as a unique polarization pattern of the Cosmic Microwave Background (CMB). To detect faint polarization signals at large angular scales, space observation in broad frequency bands with a wide Field of View (FoV) is demanded. LiteBIRD is the only space mission for the CMB observation in 2020s [1], [2], and the Low Frequency Telescope (LFT) is being developed as one of its telescopes. The frequency range of the LFT is 34–161 GHz and the aperture diameter is 400 mm. A crossed-Dragone design for the LFT has been investigated in the former studies [3], [4].

This study aims to examine the optical performance of the LFT and especially focuses on the measurement of its far-sidelobes, which cause contamination of the CMB signals with strong radiations from the Galactic plane.

## II. Measurement Setup

We have conducted antenna pattern measurement of the LFT in 1/4-scale, so that quick iteration of the optical design can be easily made. The scaled LFT was designed and built at the machine shop of Institute of Space and Astronautical Science, Japan Aerospace Exploration Agency. It has a focal plane area of 100 mm × 50 mm to cover its FoV of 20° × 10°.

To measure the far-sidelobes over the wide FoV, we developed a near-field antenna pattern measurement system [5] based on a vector network analyzer (Fig. 1). Antenna patterns

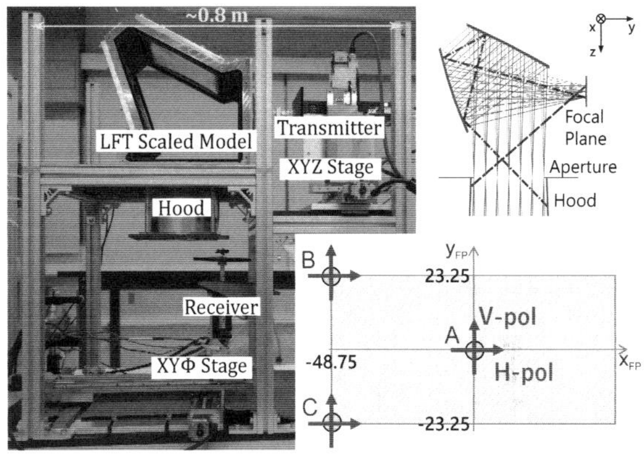


Fig. 1. (*left*) The 1/4-scaled model of LFT and its near-field antenna pattern measurement system. A conical horn with the transmitter of a vector network analyzer is placed at three positions on the focal plane of the scaled LFT with the XYZ stage, as shown in the *lower right* panel. A probe horn with the receiver measures both the amplitude and phase distributions near the aperture with the XYΦ stage. Note that this arrangement is a time-reversed configuration of a real CMB observation. (*upper right*) The ray diagram of the scaled LFT. The dashed and dot-dashed lines represent two types of stray light, direct paths and three-time reflected paths, both of which are reduced at the hood.

were measured at three positions of the focal plane. For each position, the measurement was conducted for two orthogonal polarization directions named H-pol and V-pol. The measurement wavelength were also scaled to 1/4 size, and 140, 160, 180, 200 and 220 GHz have been chosen as the measurement frequencies. This frequency range roughly corresponds to the lowest frequencies of the LFT bands, 35–55 GHz, at which larger far-sidelobes are expected due to diffraction.

## III. Results and Discussion

One of the causes of the far-sidelobes is stray light, such as the rays coming from the sky to the focal plane through direct paths and three-time reflected paths. Such stray light is designed to be reduced at the hood, as shown in Fig. 1. To confirm the effects of the hood, we measured far-sidelobe patterns of the scaled LFT without and with the hood, and compared them with a simulated pattern that predicts the far-sidelobes without any stray light. The results are shown in Fig. 2. Note that $\theta_x$ and $\theta_y$ in Fig. 2 are the arcsines of the spatial frequencies in $x$ and $y$ directions, respectively. The simulation calculates the propagation of the electromagnetic waves based on [6, Eq. 1.22].

H. Takakura is with Department of Astronomy, School of Science, University of Tokyo, Tokyo Japan e-mail: takakura@astro.isas.jaxa.jp

H. Takakura, Y. Sekimoto, J. Inatani, T. Hasebe, T. Kaga, Y. Takeda and N. Okada are with Institute of Space and Astronautical Science (ISAS), Japan Aerospace Exploration Agency (JAXA).

S. Kashima and N. Okada are with National Astronomical Observatory of Japan.

H. Imada are with Laboratoire de l'Acclrateur Linaire, Universit Paris-Sud, CNRS/IN2P3, Universit Paris-Saclay.

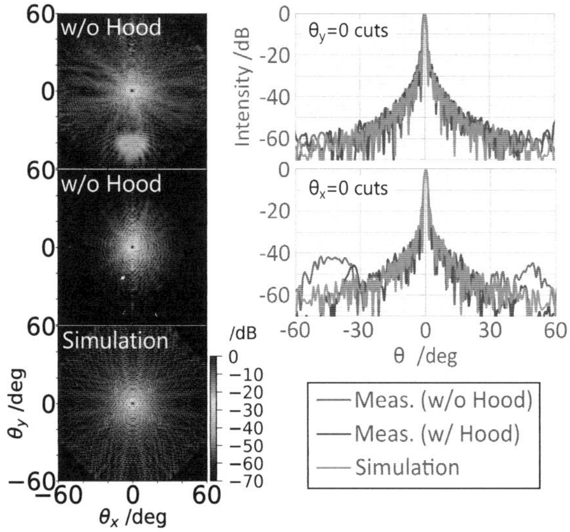


Fig. 2. The far-sidelobe patterns measured without and with the hood, and a simulated beam pattern at Position A. The simulation predicts the far-sidelobe pattern without the hood nor any stray light. The measurements and simulation are at 140 GHz, which corresponds to 35 GHz of the full-scale LFT. For without-hood case, sidelobe components are observed around $\theta_y = -45°$ and $\theta_y = 50°$. For with-hood case, these far-sidelobe features are drastically reduced, and the measured far-sidelobe pattern agrees with the calculated one at down to $-50$ dB level.

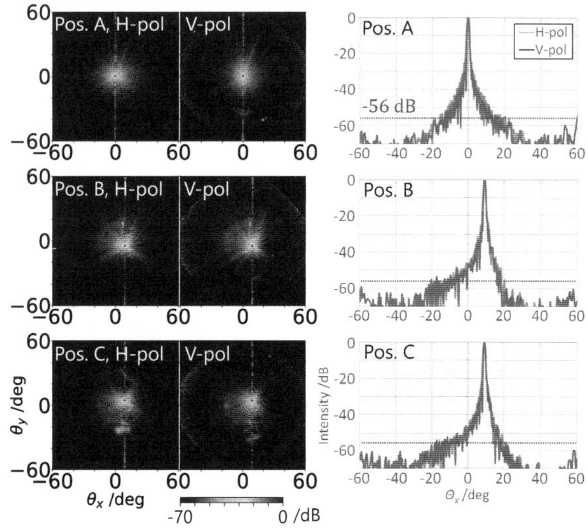


Fig. 3. H-pol and V-pol far-sidelobe patterns at each feed horn position of A, B and C, measured at 220 GHz (corresponding to 55 GHz), and their cuts in $\theta_x$ direction.

For without-hood case, two sidelobe components with the peak level of nearly $-40$ dB are observed around $\theta_y = -45°$ and $\theta_y = 50°$. These two far-sidelobe features correspond to the predicted direct paths and three-time reflected paths, respectively, and they are reduced by around 20 dB when measured with the hood. The measured far-sidelobe pattern with the hood agrees with the simulated far-sidelobe pattern without stray light at down to $-50$ dB level.

Far-sidelobe patterns measured at the three positions on the focal plane are shown in Fig. 3. The peak angles of each beam are at $(\theta_x, \theta_y) = (0°, 0°)$, $(9°, -5°)$ and $(9°, 5°)$ for the Positions of A, B and C, respectively. The measurements show that far-sidelobe patterns of two orthogonal polarization directions, H-pol and V-pol, are consistent with each other at down to $-40$ dB level, even at the edges of the focal plane. For both polarization directions, most of the far-sidelobes outside of $\sim \pm 25°$ are suppressed below $-56$ dB level. Some small-scale far-sidelobe features in $\theta_x \sim 0°$ direction are considered to be caused by the remaining stray light. Also, in each map, a vertical feature with a level of around $-50$ dB is found at $\theta_x \sim 0°$. This feature has been identified as near-field scanning noise caused by phase variation.

## IV. Conclusion

We have conducted far-sidelobe measurement of the LiteBIRD LFT in 1/4 scale. We have confirmed that some far-sidelobe features are due to stray light and are drastically reduced by the hood, and that other far-sidelobe components are mostly less than $-56$ dB. We have also found that the far-sidelobe patterns for H-pol and V-pol agree with each other down to at least $-40$ dB level.

Minute description on the measurement setup as well as more results and analysis, including the measurements of the cross-polarization of the LFT, will be presented in the full paper [7].

## Acknowledgment

This work was supported by the acceleration program of JAXA research and development directorate, NAOJ support program, and JSPS/MEXT KAKENHI Grant Numbers JP17H01115, JP15H05891, JP15H05743.

## References

[1] M. Hazumi et al., "LiteBIRD: a small satellite for the study of B-mode polarization and inflation from cosmic background radiation detection," in *SPIE Astronomical Telescopes + Instrumentation*, M. C. Clampin, G. G. Fazio, H. A. MacEwen, and J. M. Oschmann, Eds., vol. 844219, no. September, aug 2012, p. 844219.

[2] Y. Sekimoto and LiteBIRD Pre-phase A2 members, "Concept design of the LiteBIRD satellite for CMB B-mode polarization," *Proceedings of SPIE - The International Society for Optical Engineering*, 2018.

[3] S. Kashima et al., "Wide field-of-view crossed Dragone optical system using anamorphic aspherical surfaces," *Appl. Opt.*, vol. 57, no. 15, pp. 4171–4179, 2018.

[4] H. Imada et al., "The optical design and physical optics analysis of a cross-Dragonian telescope for LiteBIRD," in *Space Telescopes and Instrumentation 2018: Optical, Infrared, and Millimeter Wave*, H. A. MacEwen, M. Lystrup, G. G. Fazio, N. Batalha, E. C. Tong, and N. Siegler, Eds. SPIE, jul 2018, p. 157.

[5] A. Yaghjian, "An overview of near-field antenna measurements," *IEEE Transactions on Antennas and Propagation*, vol. 34, no. 1, pp. 30–45, jan 1986. [Online]. Available: http://ieeexplore.ieee.org/document/1143727/

[6] T. Kitsuregawa, *Advanced Technology in Satellite Communication Antennas: Electrical and Mechanical Design*, ser. Artech House antenna library. Artech House, 1990.

[7] H. Takakura et al., "Far-sidelobe Antenna Pattern Measurement of LiteBIRD Low Frequency Telescope in 1/4 Scale," *IEEE Transactions on Terahertz Science and Technology*, submitted for publication.

# Holographic Measurement System for the CCAT-prime Telescope – System Design and Novel Software Approach

Xiaodong Ren[1], Pablo Astudillo[2], Urs Graf [1], Richard Hills[3], Bojan Nikolic[3], Patricio Mena[2], Stephen Parshley[4] and Nicolás Reyes[2].

We here describe a plan for measuring and setting the reflector surfaces of the CCAT-prime telescope [1]. The most effective technique for measuring large mm-wave telescopes is "holography" [2], where the errors in the surfaces are inferred from measurements of the amplitude and phase of the beam pattern. For CCAT-prime, however, the very high surface accuracy (goal 7um), together with the crossed-Dragone optical design [3], which consists of two large off-axis reflectors, presents significant new challenges.

To reduce systematic errors, due to e.g. phase deviations in the feed horns, we use a higher frequency (~300GHz) than in previous systems. A coherent source will be securely located about 300m up-slope of the telescope to minimize atmospheric propagation errors. The signal will be collected by a receiver slightly behind the nominal focal plane and a reference receiver will be located on the telescope yoke. We will scan the antenna rapidly to provide frequent calibration of the instrumental phase and amplitude.

The holography technique exploits the Fourier transform relationship between the beam pattern and the aperture fields. The standard approach is to perform and inverse FT on the data. This method is however not well suited to the CCAT-prime case, because it cannot easily discriminate between the errors contributed by the two surfaces. To avoid this difficulty, we chose to treat it as an inference problem.

We can easily do the 'forward' calculation (simulation of the far or near-field beam patterns based on any given set of errors of the two reflectors.) Therefore, by least square fitting, we can find the set of errors which produces the simulated beam map that best fits the measurement data. In order to break the degeneracy between the two surfaces and to get the unique solution of the errors in both mirrors, we will do the measurements with the holography receiver at several different positions in the telescope focal plane.

The least-squares fitting will typically require a few thousand iterations, so the 'forward' calculation must be very efficient. We implement it using scalar diffraction theory with some simplifying approximations which make the calculation much faster than full electro-magnetic simulation, e.g. by a software package such as GRASP.

[1] I. Physics Institute, University of Cologne, Cologne, Germany.
[2] Electrical Engineering, Universidad de Chile, Santiago, Chile.
[3] Cavendish Laboratory, University of Cambridge, Cambridge, UK.
[4] Department of Astronomy, Cornell University, Ithaca, NY, USA.

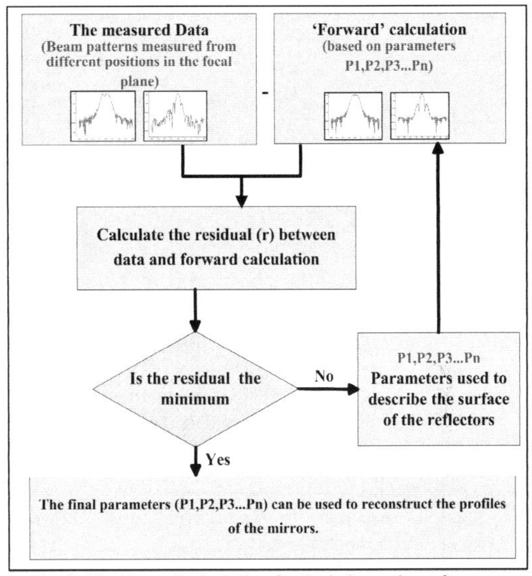


Fig. 1. The flow of calculation for the holography software.

The full solution, involving at least 3 and probably 5 beam maps, each with more than 1000 phase and amplitude data points, and fitting for 5 adjusters on each of ~70 panels on each mirror, as well as ~10 measurement parameters per beam map, is however a significant computational challenge.

The presentation will report on progress in achieving this by exploiting recent developments in software techniques, along with the results of simulations showing the accuracy that is expected.

### References

[1] S. C. Parshley et al., "CCAT-prime: a novel telescope for sub-millimeter astronomy", *SPIE Conference Series,* vol. 10700E..5XP, 2018.

[2] J. Baars, et al., "Near-Field Radio Holography of Large Reflector Antennas", *IEEE Trans. Antennas Propagat.* vol. 14, no. 4, pp. 24–41, Oct. 2007.

[3] S. C. Parshley et al., "The optical design of the six-meter CCAT-prime and Simons Observatory telescopes", *SPIE Conference Series,* vol. 10700E..41P, 2018.

NOTES:

# Characteristics Investigation on Thermal Deformation of Large Size Terahertz Reflector Antenna in Space

Y.Qian, X.F.Hao and Y.H.Shi,

***Abstract*—Microwave satellite X (MSX) is a geostationary satellite of meteorology for mainly surveying moisture in the atmosphere. In order to guaranteeing the observation efficiency, the surface accuracy of primary reflector with the size of 4.0m×3.0m is required to be 90 microns rms, and which is difficult to meet in the operating environment with the ambient temperature difference of over 200°C. In this paper, thermal deformation error of the CFRP reflectors with multiple structural parameters of thickness of sheet, normal direction of core and so on are simulated by finite element model (FEM) in the conditions of temperature difference. Displacements of model nodes in results data are used to calculate the surface rms, a parabola with fixed focal length is determined by using minimization method. Finally, the initial surface accuracy of prototype panel is measured by photogrammetry.**

***Index Terms*—CFRP, Reflector Panel, FEM, Thermal Deformation, Photogrammetry**

## I. INTRODUCTION

MICROWAVE satellite X(MSX) is one of the meteorology satellites under approval to mainly survey moisture in the atmosphere. This satellite is designed to operate in terahertz band in geosynchronous orbit with the environment of extreme large temperature difference. The surface accuracy of primary reflector for antenna is required to be less than 90 microns for guaranteeing the observation efficiency. The size of primary reflector is 4.0m×3.0m, which is too large make it so difficult to keep the high surface accuracy in the condition with a large temperature difference in space.

Light-weight and low-expansion material is more appropriate for the reflector panel with large size. Carbon fiber reinforced plastic (CFRP) is potentially the best material for precise structures where great thermal stability is required against a large temperature variation[1]. The sandwiched structure using CFRP sheets and honeycomb core has been widely applied on space telescopes in the past[2-5], but some of those surface shapes of panels were deformed into as an undesired shape at low temperature. All previous studies have shown that thermal stability of CFRP panel is not easily guaranteed although coefficient of thermal expansion (CTE) of carbon fiber is very low.

Characteristics with low thermal expansion coefficient (CTE) in all directions of reflector panel are difficult to design, because it is not guaranteed to maintain a high volume fraction of fibers in all directions in the plate-shell structure. Both the sheets and the core of reflector panel are anisotropic structures, and the effect of each part on the thermal deformation behavior of reflector panel has not been predicted yet.

Molding technologies of CFRP panel are unmatured in the domestic technology companies, the development is mainly limited by the low replication efficiency of surface accuracy between mold and panel in process. So a technology of rich resin coating is additionally applied to improve the surface accuracy of panel. A layer of room-temperature curable resin with low curing shrinkage is slightly pressed in front of sheet after the first molding. The surface of the rich resin coating is well adequate to eliminated most of the surface error[6].

Because of the poor surface accuracy of panel after first molding process, a thick and non-uniform resin layer is necessary to improve initial surface accuracy of the panel[7]. The defect of asymmetric performance of the upper and down sheet appears in the panel which is enlarged by the large CTE of domestic resin system.

In this paper, the thermal deformation behavior of large terahertz CFRP reflector panel is investigated by the method of numerical simulation, and the initial surface accuracy of prototype panel is measured by photogrammetry.

## II. NUMERICAL SIMULATION

Laminate element is used for sheet modeling in finite element software "ABAQUS", and solid element is applied to core structure, as seen in Fig 1. The sheets are made of CFRP M55J, and aluminum honeycomb and CFRP tube array made of T300 are two design schemes of the sandwiched structure. The performances of basic panel materials above were tested by the method of GB (Fig.2), which are listed in Table I.

This work has been supported by National Key Basic Research and Development Program (No.2018YFA0404702) and the Foundation of Graduate Innovation Center in NUAA (Grant No. kfjj20180107).

Y.Qian is with the Purple Mountain Observatory/Key Laboratory for Radio Astronomy, Chinese Academy of Sciences, Nanjing 210034, China (e-mail: yuanqian@pmo.ac.cn;).

X.F.Hao and Y.H.Shi are with Shanghai Composites Science &Technology Co.,Ltd, CASC.( e-mail: 121058763@qq.com; 1059779166@qq.com)

Fig. 1. Finite element model of primary reflector for MSX.

TABLE I

PERFORMANCE PARAMETERS OF SHEETS AND CORE

| Mechanical properties | Sheets (CFRP_M55J) | Honeycomb (aluminum) | Tube array (CFRP_T300) |
|---|---|---|---|
| $E_1$/MPa | 113000 | 0.06 | 1.2 |
| $E_2$/MPa | 113000 | 0.06 | 1.2 |
| $E_3$/MPa | / | 614.9 | 400.3 |
| $G_{12}$/MPa | 16700 | 0.13 | 0.33 |
| $G_{23}$/MPa | / | 53.4 | 390 |
| $G_{13}$/MPa | / | 83.6 | 390 |
| $\mu_{12}$ | 0.11 | 0.3 | 0.8 |
| $\mu_{23}$ | / | 0.005 | 0.004 |
| $\mu_{13}$ | / | 0.005 | 0.004 |
| $CTE/10^{-6}$ | 0.3 | 22.6 | 4.98 |
| *Density*/kg/m$^3$ | 1870 | 31.5 | 45 |

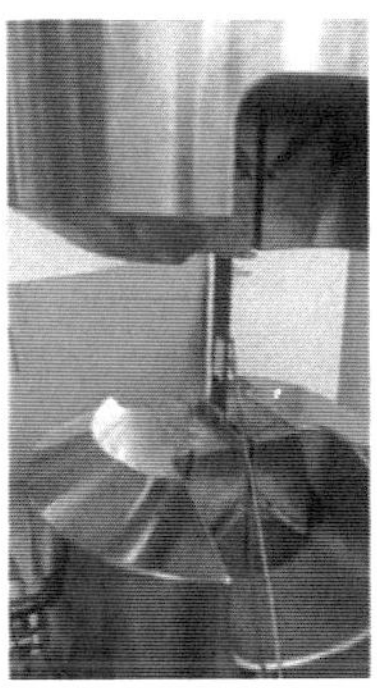

(a) Tensile test for M55J

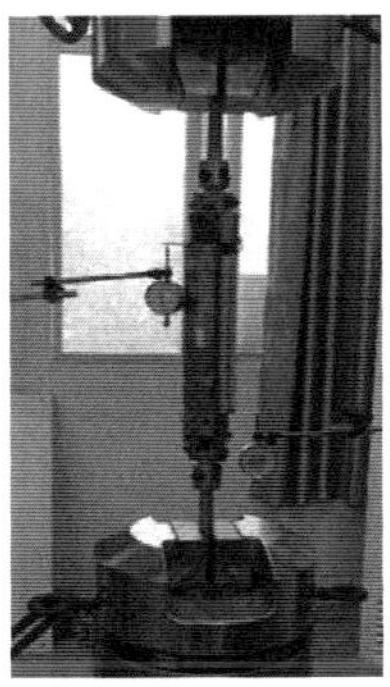

(b) Shear test for Al honeycomb

(c) Shear test for CFRP tube array

Fig. 2. Performances test of basic materials of reflector panel

The model is simulated in the conditions of temperature difference of 100°C, The effects of sheet thickness and core structure on thermal deformation behavior of reflector panel are discussed.

### *A. Condition I*

The normal direction of the core structure is perpendicular to the paraboloid of sheet in the FEM of condition I. The stress distribution of panel reflectors with aluminum honeycomb and CFRP tube array are shown in Fig.3.

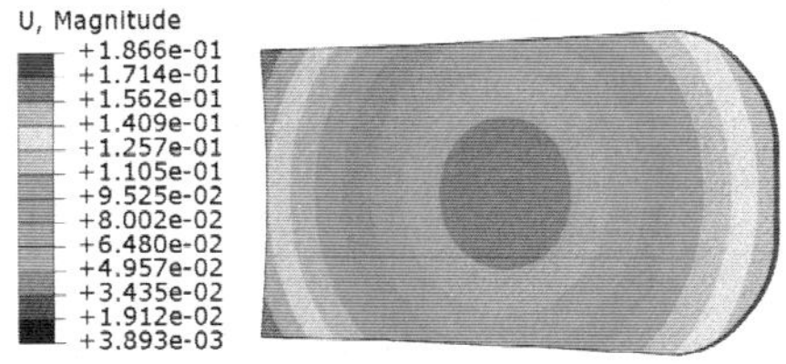


(a) Panel with CFRP tube array core

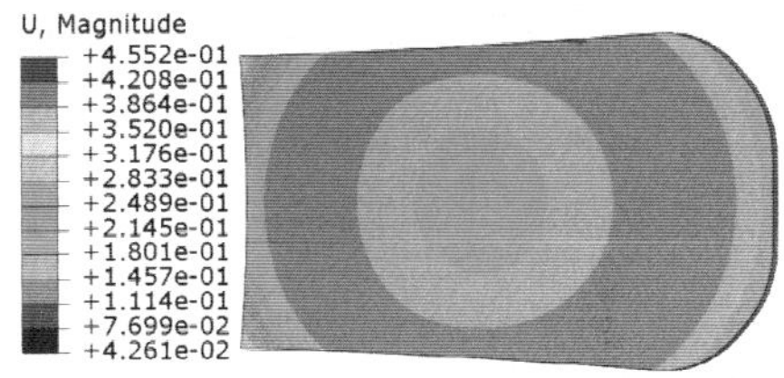


(b) Panel with Al honeycomb core

Fig. 3. Stress distribution of reflector panels in condition I

To calculate the thermal deformation error of panels, a best fitted parabola is determined using minimization method. The differences between the deformed nodes and the nodes on the best fitted surface are therefore derived, as listed in Table II.

TABLE II

THERMAL DEFORMATION ERROR OF PANELS WITH DIFFERENT CORES

| Thickness of sheet /mm | Thermal deformation error/μm rms | |
|---|---|---|
| | Honeycomb (aluminum) | Tube array (CFRP_T300) |
| 1 | 7.4 | 30.6 |
| 1.5 | 7.4 | 29.8 |
| 2 | 7.3 | 29.0 |
| 2.5 | 7.3 | 28.2 |
| 3 | 7.3 | 27.5 |

The deformation of in the directions of normal and in-plane of reflector surface are calculated by the results decomposed of absolute deformation data of nodes in FEM. The computational method refer to Eq.(1) and Eq.(2), and the results are listed in Table III.

$$D_n = \sqrt{\frac{\sum_1^N (\Delta)^2}{N}} \quad (1)$$

$$D_p = \sqrt{\frac{\sum_1^N (\Delta z)^2}{N}} \quad (2)$$

where $D_n$ and $D_p$ are the deformation of in the directions of normal and in-plane of reflector surface, $\Delta$ and $\Delta z$ the deformation of in the direction of normal and in-plane of nodes

TABLE III
THERMAL DEFORMATION OF PANELS WITH DIFFERENT CORES IN TWO DIRECTIONS

| Thickness of sheet /mm | Thermal deformation of reflector surface/mm | | | |
|---|---|---|---|---|
| | Aluminum honeycomb | | CFRP T300 tube array | |
| | $D_n$ | $D_p$ | $D_n$ | $D_p$ |
| 1 | 0.098161 | 0.041749 | 0.154588 | 0.108623 |
| 1.5 | 0.09789 | 0.041658 | 0.15471 | 0.108723 |
| 2 | 0.09776 | 0.041613 | 0.154828 | 0.108815 |
| 2.5 | 0.09767 | 0.041585 | 0.154942 | 0.108904 |
| 3 | 0.09762 | 0.041567 | 0.155053 | 0.108991 |

As shown in Table II, the results show that the surface accuracy of panel is little affected by the thickness variation of sheets when the ambient temperature changing. The thermal deformation of panel with aluminum honeycomb core is larger than the value of CFRP tube array. The thermal deformation error of panel with aluminum honeycomb is about 4 times the value of CFRP tube array, the thermal stability of the panel has been greatly improved after CFRP core application.

### B. Condition II

The normal direction of the core structure is parallel to the direction of optical axis in the FEM of condition II. The stress distribution of panel reflectors with aluminum honeycomb and CFRP tube array are shown in Fig.4.

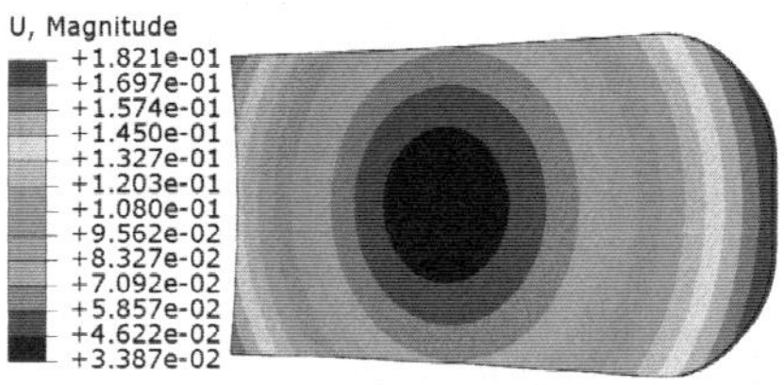


(a) Panel with CFRP tube array core

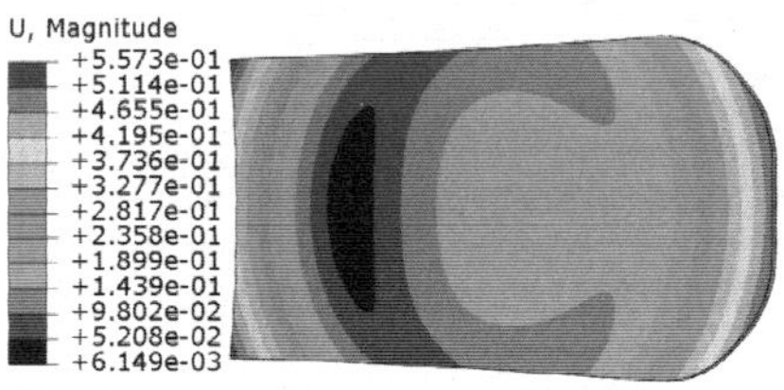


(b) Panel with Al honeycomb core
Fig. 4. Stress distribution of reflector panels in condition II

The thermal deformation errors of panels fitted are listed in Table IV, and the deformation of in the directions of normal and in-plane of reflector surface are listed in Table V. The thermal deformation error of panel with aluminum honeycomb is about 5 times the value of CFRP tube array in condition II. In two conditions, the trends of thermal deformation error caused by the thickness changing are consistent to each other.

TABLE IV
THERMAL DEFORMATION ERROR OF PANELS WITH DIFFERENT CORES

| Thickness of sheet /mm | Thermal deformation error/μm rms | |
|---|---|---|
| | Honeycomb (aluminum) | Tube array (CFRP T300) |
| 1 | 5.0 | 26.4 |
| 1.5 | 4.9 | 25.6 |
| 2 | 4.9 | 24.9 |
| 2.5 | 4.8 | 24.2 |
| 3 | 4.8 | 23.6 |

TABLE V
THERMAL DEFORMATION OF PANELS WITH DIFFERENT CORES IN TWO DIRECTIONS

| Thickness of sheet /mm | Thermal deformation of reflector surface/mm | | | |
|---|---|---|---|---|
| | Aluminum honeycomb | | CFRP T300 tube array | |
| | $D_n$ | $D_p$ | $D_n$ | $D_p$ |
| 1 | 0.101716 | 0.059144 | 0.228937 | 0.202718 |
| 1.5 | 0.101315 | 0.058982 | 0.22866 | 0.202471 |
| 2 | 0.101106 | 0.05887 | 0.228382 | 0.202218 |
| 2.5 | 0.100976 | 0.058821 | 0.228104 | 0.201962 |
| 3 | 0.100884 | 0.058768 | 0.227827 | 0.201704 |

Compared with the simulation results in condition I, it is shown that the thermal deformation behavior of panel is significantly affected when the normal direction of core structure changing. In condition II, the thermal deformation in the normal direction is smaller than the value simulated in condition I which strongly affect the surface accuracy of panel.

It is suggested that the molding method of CFRP tube should be surface processing after bonding instead of arranging directly on the sheet one by one.

## III. SURFACE MEASUREMENT OF PROTOTYPE PANEL BY PHOTOGRAMMETRY

A prototype panel has been manufactured recently, and the surface shape of the prototype panel are measured at room temperature. Since the coordinate measure machine (CMM) of company cannot measure the panel of such large size, photogrammetry is used instead with a high-resolution industrial camera. Photos of the prototype panel can be taken from a diversity of directions with a hand-holding camera (Fig. 5)

Fig.5. Surface measurement of prototype panel by photogrammetry

The surface accuracy of prototype panel is measured to be 57 microns rms, and the distribution of residual error is shown in Fig.6.

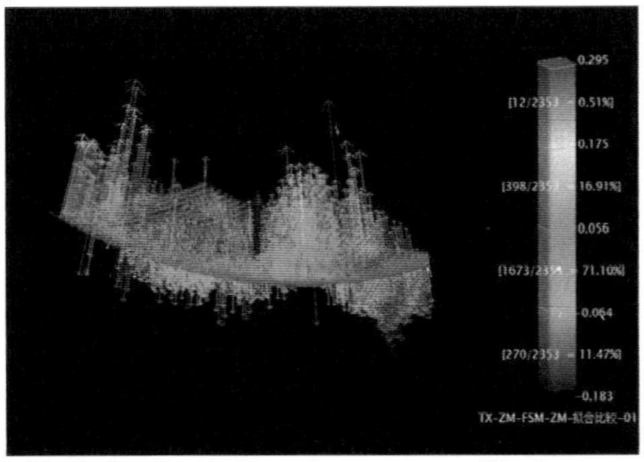


Fig.6. Distribution of residual error for prototype panel by photogrammetry

## IV. CONCLUSION

In this paper, characteristics on thermal deformation of large size terahertz reflector are investigated. Thermal deformation behaviors of reflector panels with different structural parameters are simulated by the method of finite element. A prototype panel has been manufactured and whose surface accuracy is measured by photogrammetry. The conclusions are as follow.

1) The CFRP tube array core is applied to the panel structure in place of aluminum honeycomb core for improving thermal stability.

2) Molding method of CFRP tube should be surface processing after bonding instead of arranging directly on the sheet one by one.

3) The surface accuracy of panel with CFRP tube array core is able to meet the requirement of MSX by considering the manufacture error and thermal deformation error of panel.

## REFERENCES

[1] R. N. Martin, R. C. Romeo, J.S. Kingsley. "CFRP panel concept design study for the CCAT," *SPIE International Society for Optical Engineering*. 2006. vol. 6273.p.21:SPIE.

[2] A. A. Abusafieh, D. R. Federico, S. J. Connell, et al. "Dimensional stability of CFRP composites for space based reflectors" *International Symposium on Optical Science and Technology*.2001.vol. 4444:9-16:SPIE.

[3] Y. Arao, J. Koyanagi, S. Utsunomiya, et al, "Analysis of thermal deformation on a honeycomb sandwich CFRP mirror, " *Mechanics of Advanced Materials and Structures*, vol.17, no.5, pp.328-334, 2010.

[4] S. Utsunomiya, T. Kamiya. "Development of CFRP mirrors for low-temperature application of satellite telescopes," 2012, *International Society for Optics and Photonics*,.

[5] S. Utsunomiya, T. Kamiya, R. Shimizu. "Development of CFRP mirrors for space telescopes," 2013, *Optical Engineering + Applications*, vol. 88370P.

[6] Massarello, J. Jack. "Fiber print-through mitigation technique for composite mirror replication", *Optical Engineering*, vol.45, no.12, pp.123401, 2006.

[7] Y. Qian, Z. Lou, X.F.Hao, et al. "Initial development of high-accuracy CFRP panel for DATE5 antenna", 2016, *Advances in Optical and Mechanical Technologies for Telescopes and Instrumentation II*, vol.99121D.

**Y. Qian** received the Ph.D degree in engineering mechanics from Nanjing University of Aeronautics and Astronautics, Nanjing, China, in 2013.

Since then, he was been an Assistant research fellow with the Astronomical Telescope Technology Lab, Purple Mountain Observatory, CAS. His research interests include the design of antenna structure for astronomical telescope and the application of CFRP on high precision reflector antenna.

**H.X. Hao** was born in Yuncheng, Shanxi Province, China in 1981. He received the B.E. degree in applied chemistry from North University of China in 2004, and the M.S. degrees in applied chemistry from Shanghai University in 2008. Since 2008, he has been working at Shanghai Composites Science &Technology Co.,Ltd. His major research interests and activities are in the fields of the fabrication of composites antenna reflector.

**Y.H. Shi** was born in Handan, Hebei Province, China in 1987. He received the B.E. degree in engineering mechanics from Hebei University of engineering in 2012, and the M.S. degree in engineering mechanics from Dalian University of Technology in 2015. Since 2015, he has been working as a structural engineer at Shanghai Composites Science & Technology Co., Ltd. His major research interests and activities are in the fields of structure design of composite materials and structural mechanics analysis, etc.

# Tuesday, April 16, 2019

# Session V. SIS Devices and Receivers

# Noise Analysis of SIS Receivers Using Chain Noise Correlation Matrices

C. Edward Tong[1], Paul Grimes[1], and Lingzhen Zeng[1]

Modern SIS receivers are built from a series of components including waveguide elements, multiple SIS devices, IF matching network, isolators and low noise amplifiers. A simple noise analysis of the receiver is generally performed with the well-known Friis formula for noise:

$$T_{eq} = T_1 + \frac{T_2}{G_1} + \frac{T_3}{G_1 G_2} + \cdots \quad (1)$$

Although this method provides the correct result if each of the components is well matched, this is generally not the case for SIS mixers. A more rigorous approach is based on the cascade of chain noise correlation matrices:

$$\boldsymbol{C}_{A_{eq}} = \boldsymbol{C}_{A_1} + \boldsymbol{A}_1 \boldsymbol{C}_{A_2} \boldsymbol{A}_1^+ + \boldsymbol{A}_1 \boldsymbol{A}_2 \boldsymbol{C}_{A_3} \boldsymbol{A}_2^+ \boldsymbol{A}_1^+ + \cdots \quad (2)$$

In the above equation, $\boldsymbol{A}_m$ and $\boldsymbol{A}_m^+$ are the chain circuit transmission matrix (also known as ABCD matrix) of the $m$-th component and its Hermitian conjugate respectively, while $\boldsymbol{C}_{A_m}$ is the corresponding chain noise correlation matrix. For an SIS mixer, there are at least 3 frequencies involved. A tri-frequency chain transmission matrix formalism has been introduced [1] to extend the single frequency circuit theory to cover the operation of series connected SIS mixer arrays. A similar treatment can be applied to extend the single frequency chain noise correlation matrix to general multi-junction mixer circuits [2]. In this case, the chain noise correlation matrix becomes a block matrix, with its entries appearing as 3x3 matrices,

$$\boldsymbol{C}_A = \begin{pmatrix} [\underline{V_n}\,\underline{V_n}^+] & [\underline{V_n}\,\underline{I_n}^+] \\ [\underline{V_n}^+\underline{I_n}] & [\underline{I_n}\,\underline{I_n}^+] \end{pmatrix} \quad (3)$$

where the noise voltages, $\underline{V_n}$, and noise currents, $\underline{I_n}$, are column vectors:

$$\underline{V_n} = [V_n^{\mathrm{USB}} \quad V_n^{\mathrm{IF}} \quad V_n^{\mathrm{LSB}}]^T \quad (4)$$

$$\underline{I_n} = [I_n^{\mathrm{USB}} \quad I_n^{\mathrm{IF}} \quad I_n^{\mathrm{LSB}}]^T \quad (5)$$

Note that the matrix $[\underline{I_n}\,\underline{I_n}^+]$ corresponds to the noise correlation matrix $\boldsymbol{H}$ for the SIS mixer in Tucker's Theory and a single SIS junction with one terminal connected to ground would have a chain noise correlation matrix of $\begin{pmatrix} 0 & 0 \\ 0 & \boldsymbol{H} \end{pmatrix}$. Therefore, this formalism allows a simple build-up of a noise analysis of a complete multi-junction SIS receiver using equation (2).

In this paper, we will develop the equations needed to develop a full noise model for an SIS receiver. This model incorporates the interaction between the SIS mixer, which has generally a high output impedance, and the IF stage, the noise temperature of which is a function of its input impedance. As a result, the model can easily be applied to the optimization of the SIS receiver with a given IF configuration. In addition, since the SIS mixer has generally a poor input match, it would interact with the waveguide components in front of it, causing frequency ripples in receiver noise temperature. The model can also be used to account for such effects. Finally, it can be shown that for an SIS mixer based on a distributed array, the chain noise correlation matrix for the array would have non-null matrix entries for each of its 4 constituting matrices. Thus, this model is well suited for the noise analysis of SIS receivers based on distributed series SIS junction array

[1] Harvard-Smithsonian Center for Astrophysics, Cambridge, MA 02138.

## References

[1] C.-Y. E. Tong and R. Blundell, "Theory of series-connected distributed SIS mixers with ultra-wide instantaneous bandwidth," in *Proc. 17th Int. Symp. Space THz Tech.*, pp. 35-38, Paris, France, May 2006.

[2] W. Wu, Z. Wang, and W. Shan, "An efficient noise analysis method of a mixing circuit based on linear network noise correlation matrices," presented at Applied Superconductivity Conference, Seattle, WA, Nov. 2018.

NOTES:

# A 1MM SIS Receiver Utilizing Different Intermediate Frequency (IF) Configurations

D.Meledin[1], V. Desmaris[1], E.Sundin[1], A. Pavolotsky[1], M. Fredrixon[1], I.Lapkin[1,3], M.Strandberg[1], A. Ermakov[1,4], J-D, Gallego[2], I. Lopez-Fernandez[2], C. Diaz[2], and V.Belitsky[1]

***Abstract* – We present experimental studies of the noise performance of a prototype heterodyne SIS receiver operating at wavelengths of about 1mm. The receiver employs different 4-12GHz intermediate frequency amplification chain configurations: a standalone low noise amplifier (LNA), the LNA cascading with a cryogenic isolator, and a low noise balanced amplifier.**

**From our experiments and measurements, we could conclude that the latter configuration demonstrates the best broadband noise performance. In fact, the receiver equipped with the balanced LNA does not have noticeable noise degradation caused by the IF hybrids of the balanced LNA scheme. Moreover, our results indicate that even broader IF bandwidth of the receivers could be prospectively reached using balanced LNAs in the IF amplification chain.**

## I. INTRODUCTION

Mm-wave heterodyne receivers for radio astronomy progress towards a wider (8 GHz and larger) intermediate frequency (IF) band. Since the IF circuitry introduces additional loss and noise to the receiver system, the high performance IF chains (wideband, low loss, compact) are required when designing a receiver system. Moreover, in modern receiver employing the side-band separating architecture for more efficient and accurate observations [1], the IF chain becomes more complicated as it includes the IF hybrid combining the signals from the two SIS mixers and providing sideband separation.

The most widely employed IF circuit configuration comprises cryogenic isolators followed by a low noise amplifiers (LNA). The isolator reduces the standing waves in the IF band. However, it introduces insertion loss affecting potentially the overall receiver noise as well as the wideband isolator being bulky to cover the 4-12 GHz frequency band. In addition, cost per unit could play a role when designing a dual polarization or a multipixel receiver.

Therefore, employing balanced amplifiers for the receiver IF chain has the potential of improving noise performance of the receiver, accounting for the lower loss of quadrature hybrid, especially when employing superconducting materials, as compared to the isolator cascaded with the LNA. In addition, the receivers could be more compact, which is important for array receivers where tight pixel packaging is required, or for designs with severe geometrical constrains as in ALMA cartridge receivers.

[1] Group for Advanced Receiver Development, Chalmers University of Technology, SE412 96, Gothenburg, Sweden.

[2] Observatorio de Yebes CDT (IGN), Guadalajara, Spain.

[3] Institute of Applied Physics of the Russian Academy of Sciences (IAP RAS), N. Novgorod, Russia.

[4] Kotelnikov Institute of Radioengineering and Electronics (IRE), Moscow, Russia.

In this work, we investigate the noise performance of a prototype SIS receiver operating at about 300 GHz, using different IF amplification chain configurations including a standalone commercial LNA, the LNA with the cryogenic isolator, and a low noise balanced amplifier.

## II. RECEIVER DESCRIPTION

The receiver prototype used in the experiments, employs the side-band separating scheme (2SB) and is intended to be a part of SEPIA heterodyne receiver installed at APEX telescope [2]. The receiver is dual polarization with an operating frequency range of 272-376 GHz and wide (4-12 GHz) IF bandwidth per sideband, i.e. 32 GHz instantaneous IF bandwidth. As mixer elements, Nb-AlxO-Nb SIS mixers, fabricated in-house [3], are employed. The receiver topology features, inside the mixer block, an input 3dB waveguide 90 degrees RF hybrid, SIS mixers along with output 90 degrees 3dB IF hybrid with integrated broadband impedance transformers and bias-T. The local oscillator (LO) power is injected to the mixers by integrated -18 dB directional couplers.

The same mixer block is used with three different IF multiplication chains: a standalone 4-16 GHz LNF-LNC4_16B amplifier from Low Noise Factory (referred in this paper as the configuration A); a 4-12 GHz Quinstar isolator cascaded with the LNF-LNC4_16B amplifier (referred as the configuration B, Fig.1) and a 4-12GHz balanced amplifier (configuration C). The balanced amplifier is realized in a modular configuration with two wideband miniature 3dB 90 degrees hybrids and two identical cryogenic LNAs, Fig. 2.

The hybrids are fabricated on alumina and employ the three-sections planar design [4]. The central section of the hybrid with a Lange coupler is placed between two sections with coupled line couplers. The Lange coupler provides the highest coupling coefficient while the other two sections are loosely coupled. The hybrids are produced using thin-film microfabrication technology that combines photolithography and dry etching processes for the formation of galvanically plated gold transmission lines and air bridges to connect the coupler fingers in the middle section. The overall size of the miniature hybrid chip is limited only by the pitch required between the through and coupled ports, in order to conveniently interface with coaxial contacts. The hybrids demonstrate excellent performance in terms of insertion loss, return loss, as well as amplitude and phase imbalance [4]. In the balanced amplifier, at the input, we used a 4-12 GHz hybrid and at the output, a hybrid with slightly extended bandwidth of 4-16 GHz [5].

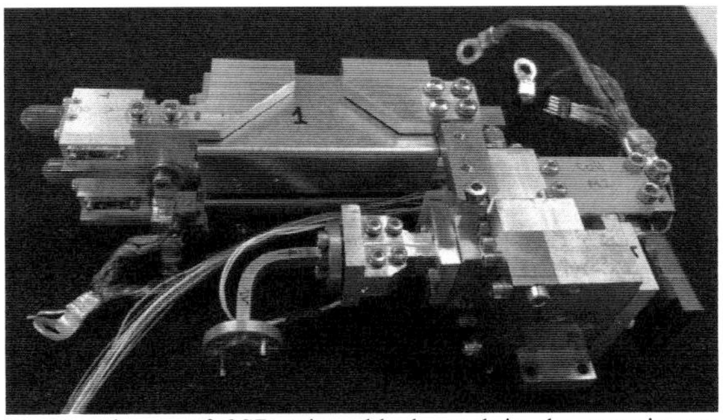

Fig. 1. Picture of 2SB mixer block used in the experiments including 4-12 GHz Quinstar isolator cascaded with the LNF-LNC4_16B amplifier (configuration B)

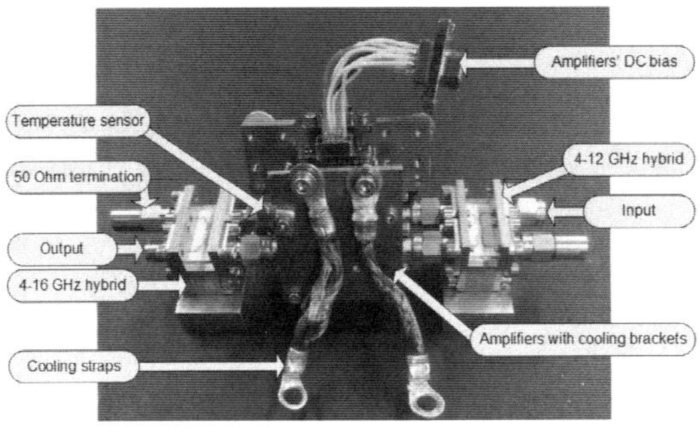


Fig. 2. Image of balanced modular amplifier.

The LNAs employed in the balanced amplifier were developed and fabricated at the Yebes Observatory. The amplifiers feature three stages with single 150 µm x 0.1 µm InP transistors in the first two stages and a GaAs commercial HEMT in the third. The design has been described in detail in [6].

The noise performance of the standalone Yebes LNAs and balanced amplifier were measured in a cryostat at about 4 K using a cold attenuator of about 20 dB. Later, the cold attenuator was removed and S-parameters measurements were conducted at the LNAs' optimum bias conditions for the best noise performance, thus requiring an additional cooling cycle. The amplifiers' characterization results are presented in Fig. 3a,b.

As could be seen from Fig.3, the noise and gain performance of the modular balanced LNA is similar to the single end amplifier. However, the input matching of the balanced amplifier is considerably improved as compared to the single end amplifier, and provides a return loss better than -13 dB over the 4-12 GHz frequency band.

In an earlier stage of the receiver development, we have designed and fabricated a mixer block where we could characterize the double side band (DSB) noise performance of the mixers. In this version, we have a single-end mixer arranged in the same manner as in a 2SB mixer block: with the same SIS mixer chip channel geometry, LO coupler, and IF transformer for the mixer to have a 50 Ohm output impedance.

## III. MEASUREMENT RESULTS

In order to deduce the noise temperature of the IF circuitry embedded in the receiver, we used the standard Y-factor technique, where the SIS mixers were biased in the normal part of their IV-curve and thus behave as a calibrated noise source providing loads with known equivalent temperatures. This method has been widely used, for instance in [7, 8].Those bias voltages (for example, $V_{cold}$=4 mV and $V_{hot}$=7mV) correspond to hot and cold loads at the input of IF network while the equivalent temperatures of these loads could be determined accounting for a rate of ≈5.8 K/mV.

a).

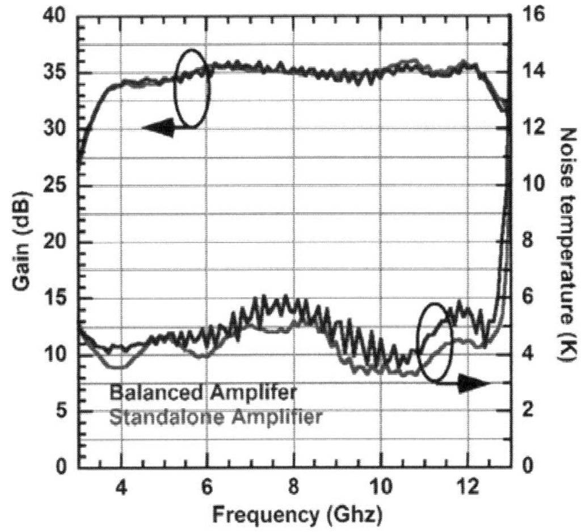


b).

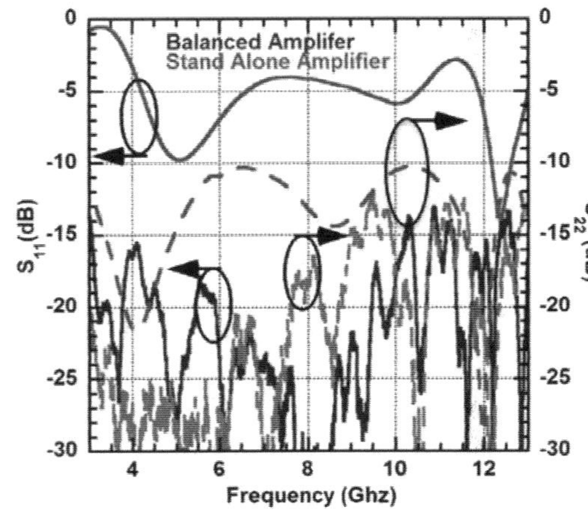


Fig. 3. Measured performance of YEBES standalone and balanced modular amplifiers as a function of frequency: a). noise temperature and gain values; b). S-parameters

The resulting noise temperatures of the whole IF-chain using DSB mixer block are plotted on Fig. 4. In Fig.5, the output IF power of the DSB receiver measured when SIS junctions are biased at $V_{cold}$ is demonstrated. These results show that the chain consisting of the standalone LNF-LNC4_16B amplifier (configuration A) is in average less noisy over the 4-12 GHz frequency band. However, it exhibits large standing waves which manifest as IF power ripples, as shown in Fig. 5. Therefore, configuration A of the IF chain was withdrawn from further tests with the 2SB receiver. The chains consisting of the isolator with the LNF-LNC4_16B amplifier (configuration B) and the balanced amplifier (configuration C) are very similar from a noise contribution perspective. However, the isolator in the configuration B seems to limit the performance at the lower end of the IF band (below 5 GHz). An improved flatness of the output power is also observed in Fig. 5 for the IF chain with balanced amplifier (configuration C).

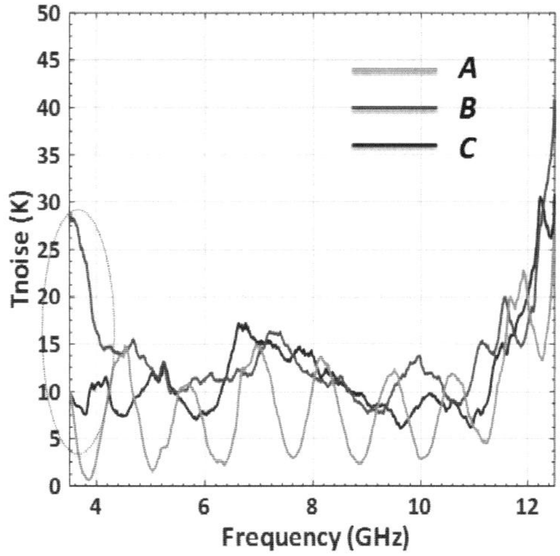


Fig. 4. Noise temperature of the IF chain of the receiver with DSB mixer block using different IF amplification chain: A, B and C (as described earlier).

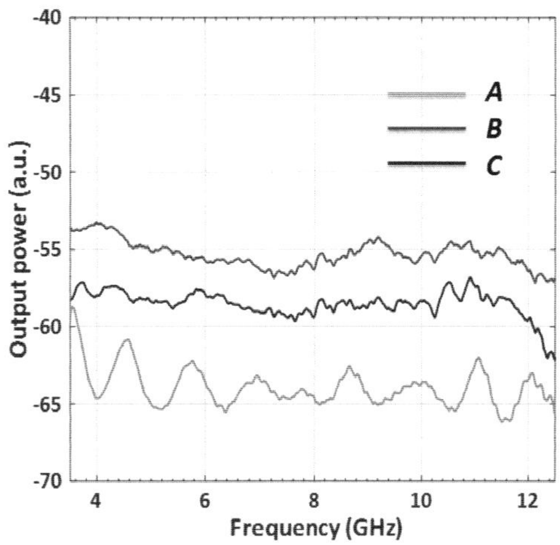


Fig. 5. Output power at $V_{cold}$ bias of the SIS junction with DSB mixer block using different IF amplification chain: A, B and C (described earlier).

The obtained results clearly indicated the advantage of a balanced scheme for the IF amplifier chain due to its superior wideband input matching.

The results shown in Fig.5 were completely consistent with the measurements of the IF chain noise temperature as a function of frequency using the 2SB receiver prototype with IF configuration B and C. Both configurations demonstrated the same performance except at frequencies below 5GHz where the configuration B showed an increase of the noise temperature that is still to be ascribed to the isolator.

The receiver noise temperature at the lower side band (LSB) averaged over 4-12 GHz as a function of the LO frequency is presented in Fig.6 with the IF chain configurations B and C. Both chains show very similar noise performances over the presented LO frequency range with the exception of the LO frequencies around 300 GHz, where receiver equipped with balanced amplifier has slightly lower noise. Since we have only one available balanced amplifier, switching of the side bands was conducted by bias reversing of one of the SIS junctions.

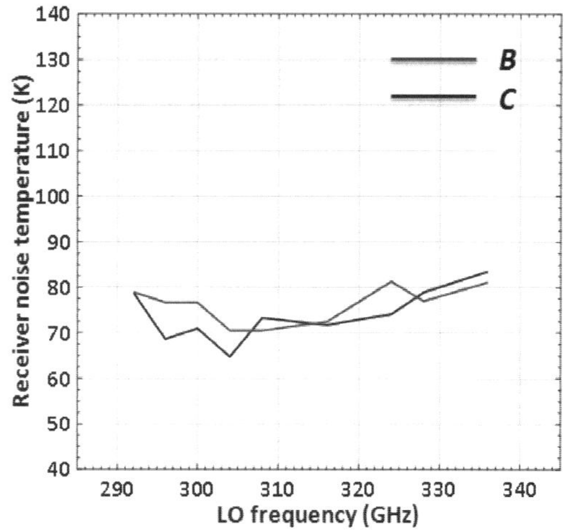


Fig. 6. Noise temperature of 2SB receiver at LSB with two IF configuration B and C (described earlier).

## IV. CONCLUSION

In this work, we have demonstrated the noise performance of the 1mm SIS receiver employing different configurations of the IF chains. We have shown that the cryogenic balanced IF amplifiers can be a very promising alternative to the conventional IF amplification chain comprising a cryogenic isolator cascaded with the LNA.

Our measurements indicate that both configurations (B and C) are very similar in terms of the receiver noise performance. However, currently available isolators with a passband wider than 4-12 GHz exhibit higher insertion losses above 10.5 GHz, whereas balanced amplifiers are likely to demonstrate even more broadband performance as the hybrids with the bandwidth 4-16 GHz have already done, and even wider bands are possible.

A more compact solution for a balanced amplifier instead of a modular architecture has been demonstrated [9]. Moreover, further reducing the number of interfaces and optimizing the amplifier design would allow us to further improve the input and output matching of the balanced amplifier. A broader band solution for compact hybrids would open the possibility to reach larger IF bands by employing balanced LNAs.

## ACKNOWLEGEMENTS

Part of the research leading to these results has received funding from the European Commission Seventh Framework Programme (FP/2007-2013) under grant agreement No 283393 (RadioNet3).

## REFERENCES

[1] V. Belitsky et al., "ALMA Band 5 receiver cartridge. Design, performance, and commissioning", *Astron. Astrophys.*, vol. 611, A98, 2018.

[2] V. Belitsky et al., "SEPIA - a new single pixel receiver at the APEX Telescope", *Astron. Astrophys.*, vol. 612, A23, 2018.

[3] A. B. Pavolotsky, D. Dochev, and V. Belitsky, "Aging- and annealing-induced variations in Nb/Al- AlOx/Nb tunnel junction properties," *J. Appl. Phys.*, vol. 109, no. 2, p. 024502, 2011.

[4] H. Rashid et al., "Wideband Passive Circuits for Sideband Separating Receivers", *Proceedings of the 29th*

*International Symposium on Space Terahertz Technology*, Pasadena, USA, 2018.

[5] H. Rashid, V. Desmaris, and V. Belitsky, "Wideband Passive Circuits for Sideband Separating Receivers", *Proceedings of the 29th International Symposium on Space Terahertz Technology*, Pasadena, USA, 2018.

[6] I. Lopez-Fernandez, J. D. Gallego Puyol, C. D. Gonzalez and A. B. Cancio, "Development of Cryogenic IF Low-Noise 4-12 GHz Amplifiers for ALMA Radio Astronomy Receivers" *IEEE MTT-S International Microwave Symposium Digest*, San Francisco, CA, pp. 1907-1910, 2006.

[7] V. Belitsky,, V. P. Koshelets, I.L Serpuchenko, M.A. Tarasov, L.V. Filippenko. , S.V. Shitov "Superconducting Tunnel Junction Noise Generator and SIS Mixer Noise Measurements", *20th European Microwave Conference 90, Budapest, Sept. 10-13, 1990, Conf. Proc.*- Tunbridge Wells, 1990, pp. 816 -820.

[8] D. P. Woody "Measurement of the noise contributions to SIS heterodyne receivers", *IEEE transactions on Applied Superconductivity,* vol. 5, No. 2, 1995.

[9] R. I. Amils, I. López-Fernández, J. D. Gallego et. al., "Compact Cryogenic Wide-Band Balanced Amplifiers with Superconducting 90° Hybrids for the IF of Submillimeter-Wave SIS Mixers", *Proceedings of the 30th International Symposium on Space Terahertz Technology*, Gothenburg, Sweden, 2019.

# Noise Characterisation of a Flux-Pumped Lumped-Element Josephson Parametric Amplifier using an SIS Mixer

Martina Esposito[1], John Garrett[2], Boon-Kok Tan[2], Peter Leek[1] and Ghassan Yassin[2].

Josephson parametric amplifiers (JPAs) are widely used in many ultra-sensitive experiments, due to their potential to achieve high gain and quantum-limited noise performance. They are essential for superconducting quantum information research [1] and recently have been considered for the dark matter axion searches [2]. We have recently developed a flux-pumped lumped-element JPA with a tunable operating frequency in the range of 8–10 GHz, as the first stage amplification for the readout of a superconducting qubit [3]. The layout of our JPA chip is shown in Fig. 1.

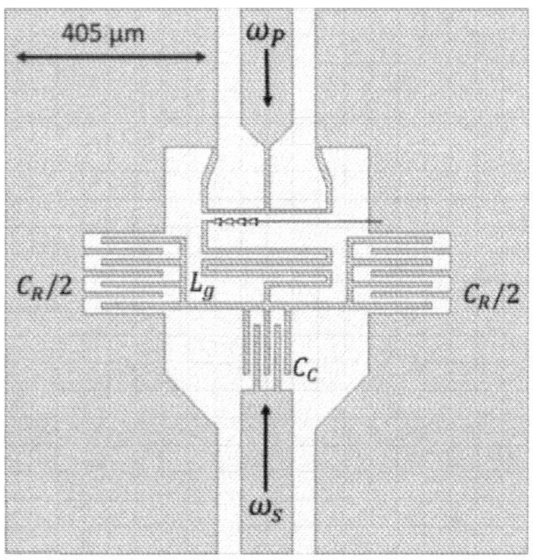


Fig. 1. Design layout of the JPA circuit. The signal is coupled to the *LC* resonator via a coupling capacitor $C_c$, and is read through a 50Ω CPW line. An array of four SQUIDs is cascaded in series to provide the nonlinear inductance required for parametric amplification. The pump signal is injected into the JPA with a flux line that is 15 μm away from the SQUIDs.

We have previously characterised our JPA at a base temperature of 10mK, and we managed to achieve a maximum gain of 25 dB with a 3dB bandwidth of 15 MHz and a 1dB compression point of −115dBm. Currently, we are in the process of characterising the noise performance of the JPA, using a superconductor-insulator-superconductor (SIS) tunnel junction as a calibrated noise source.

It is well known that the shot noise of an SIS tunnel junction above the gap is given by [4]:

$$T_{shot}(V_0) = \frac{e}{2k_B} I^0_{dc} R_{dyn} \coth\left(\frac{eV_0}{2k_B T_b}\right)$$

where $I_{dc}^0$ is the DC tunnelling current, $R_{dyn} = (dI_{dc}^0/dV_0)$ is the dynamic resistance, $V_0$ is the biased voltage and $T_b$ is the base temperature of the tunnel junction. Therefore, we can use the tunnel junction as a tunable noise source, by simply changing the bias voltage to provide noise power at different levels. In this experiment, we make use of one of the existing 220 GHz SIS mixer chip as the noise source. The SIS mixer block is connected to the input port of the JPA via a coaxial cable. The noise performance of the JPA is then measured by using the standard Y-factor method, with the SIS mixer biased at two separate distinct voltage points. In this conference, we will present a detailed design of the JPA, preliminary measured performance of gain and noise temperature, and analysis of the results.

## References

[1] B. Abdo, F. Schackert, M. Hatridge, C. Rigetti, and M. Devoret, "Josephson amplifier for qubit readout," *Applied Physics Letters*, vol. 99, no. 16, p. 162506, 2011.

[2] M. Malnou, D. Palken, W. Kindel, L. Vale, G. Hilton, and K. Lehnert, "Operating a josephson parametric amplifier for optimal squeezing in a dark matter axion search," in *APS Meeting Abstracts*, 2017.

[3] J. Rahamim, T. Behrle, M. Peterer, A. Patterson, P. Spring, T. Tsunoda, R. Manenti, G. Tancredi, and P. Leek, "Double-sided coaxial circuit QED with out-of-plane wiring," *Applied Physics Letters*, vol. 110, no. 22, p. 222602, 2017.

[4] D. Rogovin and D. Scalapino, "Fluctuation phenomena in tunnel junctions," *Annals of Physics*, vol. 86, no. 1, pp. 1–90, 1974.

[1] Clarendon Laboratory, Department of Physics, University of Oxford, OX1 3PU Oxford, UK.

[2] Department of Physics (Astrophysics), University of Oxford, Denys Wilkinson Building, Keble Road, OX1 3RH Oxford, UK.

NOTES:

# Multi-Tone Spectral Domain Analysis of a 230 GHz SIS Mixer

John D. Garrett, Boon Kok Tan, Faouzi Boussaha, Christine Chaumont, and Ghassan Yassin

*Abstract*—**We present a new software package for simulating the performance of Superconductor / Insulator / Superconductor (SIS) mixers. The package is called QMix ("Quasiparticle Mixing") and it uses multi-tone spectral domain analysis (MTSDA) to calculate the quasiparticle tunneling current through the SIS junction. This technique is very powerful and it allows QMix to simulate multiple strong tones and multiple higher-order harmonics. We have compared this software to the experimental data from a 230 GHz SIS mixer, both to validate the software and to explore the measured results. Overall, we found very good agreement, demonstrating that QMix can accurately simulate the performance of SIS mixers. We believe that QMix will be a useful tool for analyzing experimental data, designing new SIS mixers, and simulating new applications for SIS junctions, such as frequency multiplication.**

*Index Terms*—**Superconductor / Insulator / Superconductor (SIS) mixer, simulation software, millimeter-wave receivers, superconducting detectors**

We have developed a new software package for simulating the behavior of Superconductor / Insulator / Superconductor (SIS) mixers. The package is called QMix [1], short for "Quasiparticle Mixing", and it is now freely available online. The software package is currently hosted on the Python Package Index [2] and GitHub [3] under an open-source license (GNU General Public License v3), meaning that anyone is free to download, modify, distribute and utilize the software. Other researchers are also welcome to add new features to the QMix package by contributing to the GitHub repository.

The QMix software is based on multi-tone spectral domain analysis (MTSDA) [4-7]. To summarize, when a signal is applied to an SIS junction, it modulates the quasiparticle energy eigenstates on the ungrounded side by a phase factor. Using MTSDA, we convolve the phase factors from each signal applied to the mixer in the spectral-domain in order to calculate the total phase factor of the quasiparticles. This is then used to calculate the time-averaged tunneling current through the SIS junction, from which we can calculate the output power and the conversion gain of the mixer. (This technique is described in detail by [6, 8].)

Unlike other simulation methods, which normally use perturbation techniques, MTSDA can simulate multiple strong non-harmonically related frequencies and an arbitrary number of higher-order harmonics. Therefore, we can use QMix to simulate a wide variety of SIS mixer behavior, such as the effect of higher-order harmonics in the LO signal and gain saturation as a function of RF signal power. Beyond heterodyne mixing for radio astronomy, QMix can also simulate other applications for SIS junctions, including frequency multiplication [9] and potential high-frequency communications systems.

In order to validate the software, we compared QMix simulations to experimental data from a 230 GHz SIS mixer. This was also done as a means to analyze the experimental results. The SIS device that we used for this comparison has already been presented in [8]. It is a single-ended device that uses a finline to couple the RF+LO signals to the planar circuit, and it has a 1.5 μm$^2$ Nb/Al/AlO$_x$/Nb junction with tuning structures on either side to tune the capacitance of the junction across a wide RF frequency range (140—270 GHz). We have tested this device extensively [8] and it has provided noise temperatures down to 35 K at frequencies around 230 GHz.

We setup a QMix simulation specifically to recreate the experimental results from the 230 GHz SIS mixer. This simulation included 4 different tones: the local-osciallator (LO) frequency, the upper sideband (USB) frequency, the lower sideband (LSB) frequency, and the down-converted intermediate frequency (IF). Since this is a double-sideband (DSB) mixer, we applied equal input power to both the upper and lower sidebands. We also included 2 higher-order harmonics for each tone.

To run the QMix simulation, the software required: (a) the embedding circuit for each unique frequency that was applied to the mixer, and (b) the response function of the SIS junction. For (a), we simulated the planar circuit using electromagnetic simulation software. Since the planar circuit is entirely linear it can be reduced to a Thevenin equivalent circuit with one circuit for each individual signal. For (b), we used the measured DC current—voltage relationship of the junction (i.e., the DC I-V curve) to generate the imaginary component of the response function, and then we used the Kramers-Kronig transform of the DC I-V curve to calculate the real component.

We then simulated the conversion gain of the SIS mixer at different LO frequencies using the QMix software. In Fig. 1, we compare the simulated results to the experimental data from the SIS mixer at 225.0 GHz. There is very good agreement between the simulated and experimental results, suggesting that QMix accurately modeled the behavior of the SIS mixer.

When we tested this device, the experimental results displayed "broken photon steps" at frequencies between 236 GHz and 255 GHz. This effect is characterized by a notch in the pumped I-V curve and a sharp decrease in conversion

John D. Garrett, Boon Kok Tan and Ghassan Yassin are with the Department of Physics (Astrophysics), University of Oxford, Oxford, UK (email: john.garrett@physics.ox.ac.uk).

Faouzi Boussaha and Christine Chaumont are with GEPI, Observatoire de Paris—PSL, Paris, France.

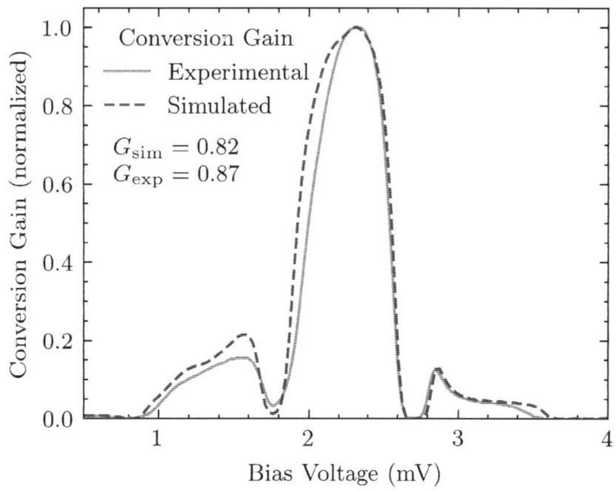


Fig. 1. The conversion gain of the SIS mixer at 225.0 GHz. The simulated results from QMix are shown in red, and the experimental results are shown in black. Both curves are normalized to their maximum values, which are listed below the legend.

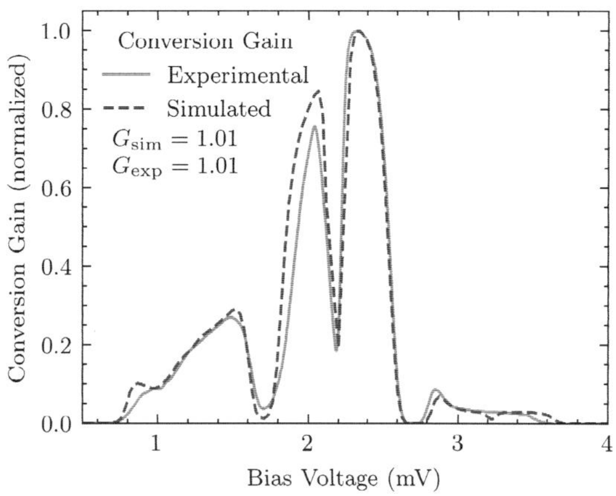


Fig. 2. Simulating the broken photon step effect at 240.5 GHz with the QMix software. This effect is characterized by a sharp decrease in conversion gain in the middle of the first photon step (at ~2.2 mV). We were able to recreate this effect with QMix by adding a spurious harmonic to the LO signal at $\frac{3}{2}f_{LO}$.

gain in the middle of the first photon step. Ermakov *et al.* [10] investigated this phenomenon and suggested that it could be caused by a spurious harmonic from the local-oscillator at either $\frac{1}{2}f_{LO}$ or $\frac{3}{2}f_{LO}$, where $f_{LO}$ is the LO frequency. Since $\frac{1}{2}f_{LO}$ is below the waveguide cutoff for this device we added a spurious harmonic at $\frac{3}{2}f_{LO}$ to the QMix simulation. The results are shown in Fig. 2 and again there is very good agreement between the simulated and experimental results, supporting the work of Ermakov *et al.* [10]. Based on these findings, we should filter the harmonics from the LO source to reduce the severity of the broken photon step effect. Overall, this example demonstrates how QMix can be used to investigate the experimental results from SIS mixers.

In conclusion, we have presented a new software package called QMix for simulating SIS devices. The software is based on MTSDA, which allows QMix to simulate multiple harmonics and multiple strong frequencies. We have compared the simulated results from this software to the experimental data from a 230 GHz SIS device. We found very good agreement, demonstrating that the software can accurately model the SIS mixer's behavior. We are now applying the software to design new SIS mixers and simulate new SIS applications, such as frequency multiplication. All of the QMix software is open-source and we invite other researchers to use the software and contribute to the project.

## References


[1] J. Garrett and G. Yassin, "QMix: A Python package for simulating the quasiparticle tunneling currents in SIS junctions," *J. Open Source Softw.*, vol. 4, no. 35, p. 1231, Mar. 2019.

[2] Online: https://pypi.org/project/QMix

[3] Online: https://github.com/garrettj403/QMix

[4] S. Withington and E. L. Kollberg, "Spectral-domain analysis of harmonic effects in superconducting quasiparticle mixers," *IEEE Trans. Microw. Theory Tech.*, vol. 37, no. 1, pp. 231–238, Jan. 1989.

[5] P. Kittara, S. Withington, and G. Yassin, "Theoretical and numerical analysis of very high harmonic superconducting tunnel junction mixers," *J. Appl. Phys.*, vol. 101, no. 2, p. 024508, Jan. 2007.

[6] P. Kittara, "The Development of a 700 GHz SIS Mixer with Nb Finline Devices: Nonlinear Mixer Theory, Design Techniques and Experimental Investigation," University of Cambridge, 2002.

[7] S. Withington, P. Kittara, and G. Yassin, "Multitone quantum simulations of saturating tunnel junction mixers," *J. Appl. Phys.*, vol. 93, no. 12, pp. 9812–9822, Jun. 2003.

[8] J. D. Garrett, "A 230 GHz Focal Plane Array Using a Wide IF Bandwidth SIS Receiver," University of Oxford, 2018.

[9] J. D. Garrett, H. Rashid, G. Yassin, V. Desmaris, A. Pavolotsky, and V. Belitsky, submitted to *AIP J. Appl. Phys.*

[10] Ermakov, V. Belitsky, P. Y. Aghdam, V. Desmaris, S. Ferm, M. Fredrixon, S. Krause, I. Lapkin, and D. Meledin, "Broken-step Phenomenon in SIS Mixers," in *Proc. 27th Int. Symp. Space THz Tech.*, Nanjing, China, Apr. 2016.

# Tuesday, April 16, 2019

# Session VI. Future Missions and Projects - I

# A Space Mission to Probe the Trail of Water

Paul Goldsmith[1], Youngmin Seo[1], Jon Kawamura[1], Jose Siles[1], William Langer[1], Dariusz Lis[2]

We present a concept for a submillimeter spectroscopic mission to probe the trail of water from the interstellar medium to habitable planets. Beyond its obvious biogenic importance, water is of great interest in many astronomical environments. It is an important coolant of warm interstellar clouds and is a significant reservoir of oxygen in the interstellar medium. It is a valuable tracer of the dynamics of interstellar clouds associated with star formation, including infall, outflows, and shocks. It is also a tracer of the thermo-chemical history of cloud material via the H2O ortho-para ratio. Finally, the question whether the Earth's water came from icy objects in the early history of the solar system is key for understanding planetary habitability. Observing gas-phase water in the local universe requires high spectral resolution because line widths may be less than 1 km/s in comets and dense cloud cores and a resolution of 0.1 km/s is compulsory. Thus, to study the water trail we need a heterodyne system, which is cryogenically cooled to minimize noise, but unlike the case for broadband photometry and low-resolution spectroscopy, the temperature of the telescope and associated optics is not critical. Thus, to follow the water trail, we can consider a relatively large, ambient temperature telescope to maximize sensitivity and angular resolution, coupled to a multiband receiver covering key spectral lines of water isotopologues. Large surveys of the submillimeter transitions of water vapor in conjunction with studies of water ice using JWST will revolutionize our understanding of the role of water, its distribution, and key ISM processes. The required submm observations will not be possible with either SPICA or OST as currently proposed. Two attributes for the receiver are: (1) to observe multiple bands simultaneously, and (2) to have modest-sized arrays in each band to increase the speed of imaging extended sources. Together these attributes dramatically increase the ability to determine conditions in sources, since multiple transitions are required to determine excitation conditions and sources are extended. The optical system first separates two linear polarizations, then separates the bands with frequency selective surfaces, and finally images beams coupling to the individual array feedhorns. We report on studies of a number of designs for a telescope of diameter between 2m and 7m. To observe lines between 500 GHz and 1200 GHz requires a surface accuracy of 13 microns rms or better. The thermal environment is a prime driver of telescope design and cost and many tradeoffs are possible. Other recent advances have improved the possibilities of observing water since Herschel, including: 1) SIS receiver performance and design have improved significantly relative to those used in HIFI, 2) frequency-multiplied local oscillator chains can now readily supply the LO power for modest arrays, and 3) CMOS ASIC digital spectrometers offer multi-GHz bandwidth per pixel with very low power consumption. These technical developments make a scientifically compelling Water Mission feasible and affordable.

[1] Jet Propulsion Laboratory, California Institute of Technology
[2] California Institute of Technology

NOTES:

# First Generation Heterodyne Instrumentation Concepts for the Atacama Large Aperture Submillimeter Telescope

Christopher Groppi, Andrey Baryshev, Urs Graf, Martina Wiedner, Pamela Klaassen, Tony Mroczkowski

***Abstract*— The Atacama Large Aperture Submillimeter Telescope (AtLAST) project aims to build a 50-meter-class submm telescope with >1-degree field of view, high in the Atacama Desert, providing fast and detailed mapping of the mm/submm sky. It will thus serve as a strong complement to existing facilities such as the Atacama Large Millimeter/Submillimeter Array (ALMA).**

**ALMA is currently the most sensitive observatory covering the atmospheric windows from centimeter through submillimeter wavelengths. It is a very powerful instrument for observing sub-arcminute-scale structures at high, sub-arcsecond spatial resolution. Yet its small field of view (< 15" at 350 GHz) limits its mapping speed for large surveys. In general, a single dish with a large field of view can host large multi-element instruments that can more efficiently map large portions of the sky than an interferometer, where correlator resources and the smaller fields of view of the antennas tend to limit the instantaneous number of beams any instrument can have on the sky. Small aperture survey instruments (typically much smaller than < 3 × the size of an interferometric array element) can mitigate this somewhat but lack the resolution for accurate recovery of source location and have small collecting areas. Furthermore, small aperture survey instruments do not provide sufficient overlap in the spatial scales they sample to provide a complete reconstruction of extended sources (i.e. the zero-spacing information is incomplete in u,v-space.)**

**Heterodyne instrumentation for the AtLAST telescope will take advantage of extensive developments in the past decade improving the performance and pixel count of heterodyne focal plane arrays. The current state of the art in heterodyne arrays are the 64-pixel Supercam instrument, the 16-pixel HARP instrument, the dual band SMART receiver with 8 pixels in two bands, and the GREAT instrument on SOFIA with 21 pixels (14 at 1.9 THz and 7 at 4.7 THz). Future receivers with larger pixel counts have been under development: CHAI for CCAT (64-pixels) and SHASTA for SOFIA (64 pixels) or under study, e.g. HERO for the Origins Space Telescope (2x9 to 2x64 pixels). Instruments with higher pixel counts have begun to take advantage of integration in the focal planes to increase packaging efficiency over simply stacking modular mixer blocks in the focal plane.**

**The authors believe that heterodyne instruments with pixel counts of approximately 1000 pixels per band could be considered for AtLAST on a decade timescale. The primary limiting factor in instrument capability (pixel count, instantaneous bandwidth, number of frequency bands, polarization capability, side-band separation etc.) is likely to be cost, rather than any fundamental technological limitation. As pixel counts increase, the cost and complexity of the IF system and spectrometer also rapidly increases, particularly if wide IF bandwidth, dual polarization and sideband separation is desired. Currently the IF and backend are limited by the cost and power consumption per unit bandwidth of the total processed science signal. While that cost is likely to decrease modestly in the next decade, no technology is likely to disrupt the scaling argument. Many of the front-end costs will also scale with pixel count, for example the size and cooling capacity of the cryostat, the complexity of the LO subsystem, and the I&T cost associated with developing, assembling and testing the focal plane units.**

**In this presentation, we review the state of the art for millimeter/sub-millimeter heterodyne instrumentation technology that could be suitable for AtLAST and attempt to forecast how the technologies will advance over the next decade. We then present a design concept for a potential first-generation AtLAST heterodyne instrument. These considerations meet the scientific demands and atmospheric considerations for a ground-based facility in the Atacama Desert.**

***Index Terms*—Astronomy, array receiver, radio telescope, terahertz, submillimeter**

## I. Introduction

THE Atacama Large Millimeter/Submillimeter Array (ALMA) is currently the most sensitive observatory that can cover the atmospheric windows from millimeter through submillimeter wavelengths (35-950 GHz). Yet its small field of view (<18" at 850 microns) limits its mapping speed for large surveys. In general, a large single dish with a large field of view can host large multi-element instruments that can more

Manuscript submitted June 25, 2019. The project leading to this publication has received funding from the European Union's Horizon 2020 research and innovation program under grant agreement No 730562 [RadioNet].

C.E. Groppi is with the School of Earth and Space Exploration, Arizona State University, Tempe, AZ 85287 USA (e-mail: cgroppi@asu.edu).

A.M. Baryshev is with the Kapteyn Astronomical Institute, University of Groningen, Groningen, 9700 AB NL (e-mail: andrey@astro.rug.nl).

U.U. Graf is with the Institute of Physics, University of Cologne, Cologne, 50937 Germany (e-mail: graf@ph1.uni-koeln.de).

M.C. Wiedner is with Sorbonne Université, Observatoire de Paris, Université PSL, CNRS, 75014 Paris, France (e-mail: martina.wiedner@obspm.fr)

P. Klaassen is with the UKATC, Royal Observatory Edinburgh, Edinburgh, Scotland, UK EH9 3HJ (email: pamela.klaassen@stfc.ac.uk).

T. Mroczkowski is with the European Southern Observatory, Garching, 85748 Germany (e-mail: tony.mroczkowski@eso.org).

efficiently map large portions of the sky than an interferometer, where correlator resources and the smaller fields of view of the array elements tend to limit the instantaneous number of beams the instrument has on sky.

Small aperture (6-meter) survey instruments like CCAT-prime [1] can mitigate this somewhat but lack the resolution for accurate recovery of source location. Furthermore, small aperture survey instruments do not provide sufficient overlap in the spatial scales they sample to provide a complete reconstruction of extended sources (i.e. the zero-spacing information is incomplete in u,v-space).

The Atacama Large Aperture Submillimeter Telescope (AtLAST) project aims to fill this technological capability gap (see [1]). AtLAST is a (domeless) 50-meter class dish with surface accuracy sufficient to provide good aperture efficiencies up to at least 950 GHz, covering the 350-micron window crucial for studies of both local and high-redshift star formation. It will feature a large field of view (> 1 degree, i.e. a factor of 225 times that of the 50-meter Large Millimeter Telescope) and a receiver cabin sufficiently large to host a broad suite of heterodyne and direct-detection instrumentation. AtLAST will strongly complement the high-resolution follow-up capabilities of ALMA, while delivering unique survey and targeted capabilities.

Basic considerations for the site were presented in [2] and include sites ranging from the planes to the peak of Chajnantor; i.e. this includes Llano de Chajnantor at 5100 meters above sea level, up to 5600 meters at Cerro Chajnantor. Upon first light, AtLAST will be fully outfitted with a number of instruments providing complementary capabilities such as broad instantaneous bandwidth, widefield survey capabilities, and the ability to explore new wavelength regimes. In this work, we discuss the state of the art for several heterodyne millimeter/sub-millimeter instrumentation technologies that could populate AtLAST's >1 square degree focal plane, and attempt to forecast how the technologies will advance over the next decade. Some factors considered are the bandwidth, spectral resolution, and multi-element capabilities, and how these couple to the AtLAST design concept advanced in the telescope design report. We then present a few on-paper design concepts for potential first-generation heterodyne instruments for AtLAST. These considerations are informed both the scientific demands and atmospheric considerations for a ground-based facility high in the Atacama Desert.

The past two decades have seen extensive surveys of the far-infrared to submillimeter continuum emission in the plane of our Galaxy. We line out prospects for the coming decade for corresponding molecular and atomic line surveys which are needed to fully understand the formation of the dense structures that give birth to clusters and stars out of the diffuse interstellar medium. We propose to work towards Galaxy wide surveys in mid-J CO lines to trace shocks from colliding clouds, Galaxy-wide surveys for atomic Carbon lines in order to get a detailed understanding of the relation of atomic and molecular gas in clouds, and to perform extensive surveys of the structure of the dense parts of molecular clouds to understand the importance of filaments/fibers over the full range of Galactic environments and to study how dense cloud cores are formed from the filaments. This work will require a large (50 m) Single Dish submillimeter telescope equipped with massively multipixel spectrometer arrays, such as envisaged by the AtLAST project.

## II. Potential First-generation Heterodyne Instruments

Heterodyne instrumentation for the AtLAST telescope will take advantage of extensive developments in the past decade improving the performance and pixel count of heterodyne focal plane arrays [4]. The current state of the art in heterodyne arrays are the 64-pixel Supercam instrument [5], the 16-pixel HARP instrument [6], the dual band SMART receiver with 8 pixels in two bands [7], and the GREAT instrument on SOFIA with 21 pixels (14 at 1.9 THz and 7 at 4.7 THz) [8]. Future receivers with larger pixel counts have been under development: CHAI for CCAT (64-pixels) and SHASTA for SOFIA (64 pixels). Instruments with higher pixel counts have begun to take advantage of integration in the focal planes to increase packaging efficiency over simply stacking modular mixer blocks in the focal plane. The AtLAST heterodyne instrument group believes that instruments with pixel counts as high as 1000 pixels could be considered for AtLAST on a decade timescale. The primary limiting factor in instrument capability (pixel count, instantaneous bandwidth, number of frequency bands, polarization capability, sideband separation etc.) is likely to be cost rather than any fundamental physical or engineering limit. As pixel counts increase, the cost and complexity of the IF system and spectrometer also rapidly increases, particularly if wide IF bandwidth, dual polarization and sideband separation is desired. Currently the IF and backend are limited by the cost per unit bandwidth of the total processed science signal. While that cost is likely to decrease modestly in the next decade, no technology is likely to disrupt the scaling argument. Many of the front-end costs will also scale with pixel count, for example the size and cooling capacity of the cryostat, the complexity of the LO subsystem, and the I&T cost associated with developing, assembling and testing the focal plane units.

Two areas where fundamental work is still required are mixer sensitivity and array packaging of complex mixer topologies. SIS receivers are reaching the few h/k noise level with the development from projects like ALMA and Herschel. But, packaging mixers into arrays results in unavoidable compromises due to size constraints in array packaging, widefield optics and relatively large cryostat windows and IR filters required. Further work is required to reduce the noise of pixel elements in large format arrays to the current state of the art or better for single pixel receivers (e.g. ALMA & Herschel) to fully take advantage of the expected AtLAST site.

In addition, more work is needed to efficiently package complex mixer topologies (e.g. 2SB mixers) in large format focal plane arrays. All the arrays mentioned above reach their pixel counts using relatively simple dual sideband mixers combined with quasi-optical LO injection, polarization diplexing and band separation. Heterodyne focal plane arrays with OMT based polarization diplexing and balanced, single sideband or sideband separating capabilities have yet to be demonstrated.

The AtLAST heterodyne instrument group also recommends close collaboration at this point in the project as requirements are generated for science instruments. Due to the cost limitation we expect to come into play, careful optimization of requirements is necessary to extract the maximum science from a future AtLAST heterodyne array. Participation of instrumentation specialists would be highly beneficial in optimizing the science vs. cost trade space for such an instrument.

### A. *Field of View*

The number of heterodyne pixels that can be accommodated in the telescope focal plane grows with the square of the telescope diameter and the square of the field of view diameter:

$$N \sim 2 \times 10^5 \left({FOV}/{1^\circ} \times {D}/{25m} \times {350\mu m}/{\lambda}\right)^2 \quad (1)$$

Several hundred thousand pixels would fit in the focal plane of AtLAST, but due to the complexity and cost of each pixel, it does not seem realistic to fully populate the focal plane on the timescale for AtLAST first light instrumentation. With current technology the cost of the IF chain for each channel from the low noise amplifier to the spectrometer is on the order of 25,000 US-$ or more. Quite likely this cost can be reduced if large quantities are produced, but we do not expect to see an order of magnitude price change. Together with the cost for common components like local oscillators, cryostats and refrigerators, a 1000-pixel array will likely require several tens of millions US-$. In the foreseeable future financial constraints will limit heterodyne arrays to approximately 1000 pixels, and thereby only filling ~1% of the focal plane of a large telescope. With this limitation, the science trades required are to determine how to allocate these pixels. The cost of an instrument with four 256 pixel frequency bands is of the same order of magnitude as a single band with 1000 pixels. The added optics and front-end complexity of a four-band instrument will be offset by savings realized in sharing IF and backend hardware, at least at the rough order of magnitude level presented here.

### B. *2SB vs. DSB for ground-based arrays*

Sideband separating (2SB) heterodyne receiver pixels are more complex than DSB pixels but have several key advantages. First, they allow for very accurate sideband ratio calibration because both the LSB and USB are available at different outputs, second 2SB receiver rejects atmospheric noise from the image sideband improving system noise temperature and the scanning speed.

Typically, a 2SB array pixel will have one input horn, two mixers, two low noise amplifiers and IF and RF hybrid circuits. In practice this is as complex and as expensive as two DSB array pixels. Therefore, 2SB pixels are only efficient to implement in an FPA if it they offer at least double the scanning speed relative to DSB pixels.

The scanning speed $V_s$ of heterodyne system can be expressed as:

$$V_s = \frac{\Delta f\, \Delta T^2}{{T_{sys}}^2}, \quad (2)$$

where $V_s$ is the frequency independent scanning speed in receiver beams per second, $\Delta f$ is the channel bandwidth, $\Delta T$ is

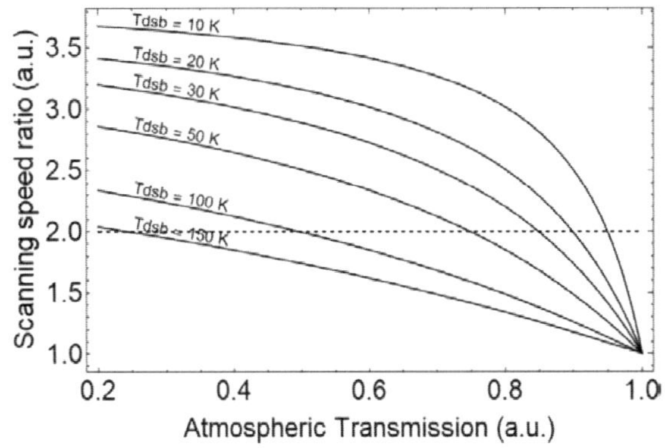


Fig. 1. Scanning speed ratio $R_{2SB/DSB}$ for typical atmospheric temperature $T_{atm} = 280\,K$ and different atmospheric transmission $\tau$ values and DSB receiver noise temperatures $T_{dsb}$. Threshold value where 2SB pixel is twice faster is indicated by the dashed line.

the required brightness noise level in K, and $T_{sys}$ is the system noise temperature including the contribution of both the receiver and atmosphere. One must note that scanning speed depends on ${T_{sys}}^2$ which is a strong dependence. Special attention should be paid not to degrade the $T_{sys}$ of a single pixel compared to its performance in the array to realize the benefit in scanning speed.

System noise temperature of a ground-based receiver can be calculated using following equation:

$$T_{sys} = 2\,\frac{T_{dsb}}{\tau} + \frac{T_{atm}(1-\tau)(1+SBR)}{\tau}, \quad (3)$$

where $T_{dsb}$ is the receiver equivalent DSB noise temperature, $T_{atm} =$ is the physical temperature of the lower layer of atmosphere, $T_{dsb}$ is the receiver DSB noise temperature, $\tau$ is atmospheric transmission and $SBR$ is the sideband ratio which is the ratio of receiver gain in the image sideband to the receiver gain in the signal sideband. For an ideal DSB receiver, the $SBR$ is equal 1 and for a 2SB receiver is in the range of 0.1 to 0.01 [refs].

Following equations (2) and (3) the scanning speed ratio $R_{2SB/DSB}$ of 2SB over DSB mixers can be expressed as:

$$R_{2SB/DSB} = \left(\frac{2\,T_{dsb}+2\,T_{atm}(1-\tau)}{2\,T_{dsb}+T_{atm}(1-\tau)(1+SBR)}\right)^2. \quad (4)$$

For an ideal 2SB receiver, ($SBR$=0) this depends only on the atmospheric brightness: ratio, $T_{br}/T_{dsb}$:

$$R_{2SB/SSB} = \left(\frac{2+2\,T_{br}/T_{dsb}}{2+T_{br}/T_{dsb}}\right)^2, \quad (5)$$

where $T_{br} = T_{atm}(1-\tau)$ is atmospheric noise seen by a ground-based receiver which represents the background limit for given atmospheric conditions. Without the presence of the atmosphere the $R_{2SB/SSB}$ equals unity and DSB and 2SB pixels are equivalent to each other. It is the influence of atmospheric noise makes a 2SB array pixel more efficient. With a fully background limited receiver $T_{br} \gg T_{DSB}$ the scanning speed of 2SB pixel can be significantly larger than a DSB pixel. Since a 2SB array pixel involves two DSB mixers, amplifiers and RF and IF hybrids, we estimate that it is approximately equivalent to two DSB pixels in terms of complexity and cost. Therefore we consider $R_{2SB/SSB} = 2$ as threshold value where 2SB pixel is equivalent to 2 DSB pixels in scanning speed, which corresponds to $T_{br}/T_{dsb} = \sqrt{2}$.

In reality, the $R_{2SB/SSB}$ depends on atmospheric conditions and DSB noise temperature of mixers as shown in figure 2. If $T_{dsb}$ is small, a 2SB pixel will significantly outperform 2xDSB pixels under large range of atmospheric transmission.

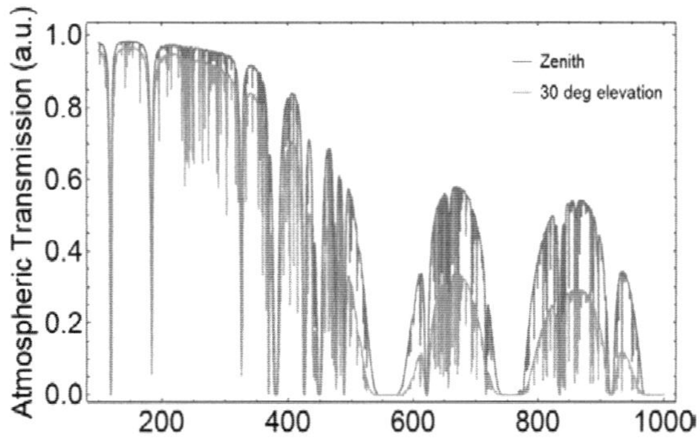


Fig. 2. Zenith and 30 Deg elevation atmospheric transmission $\tau$ for ALMA site at Chajnantor Plato, Atacama Desert in Chile. Values represents 25th percentile.

Let us consider atmospheric conditions on ALMA site shown in figure 2 for the 25th percentile. Atmospheric transmission is presented for zenith and 30 deg elevation which contains all typical observing elevations. The typical atmospheric brightness at ALMA site derived from fig. 2 is shown in fig. 3 in comparison with the average value of ALMA receiver band noise temperatures. While the most ALMA receivers are of 2SB type we present the equivalent DSB noise temperature which is half of the SSB noise temperature and is directly applicable in equations (1-4). Noise temperatures in units of the quantum limit are also shown for comparison. From this data, we can conclude that significant part of observing time even at the best available submm site, the system noise is dominated by atmosphere and thus the 2SB pixels are clearly beneficial over DSB for frequencies above 200 GHz.

## III. Potential Instrument Configurations

Section II.B makes a strong argument for the use of 2SB mixer pixels as long as their complexity in an array configuration can be handled (e.g. with on chip SSB technology). It is then feasible to consider ~500 SSB pixels given the cost constraints of a ~$25M USD instrument budget. How these pixels should be configured is then the main trade. Pixels configured as a dual polarization system adds some complexity to the design but is more well suited to making deeper, small area maps (e.g. for imaging of resolved extragalactic sources). For widefield mapping of the Galaxy, the mapping speed is identical for pixels all in one polarization or split between two polarizations. We will consider two possible configurations: a 512-pixel 2SB array covering ALMA band 6 implemented as a pair of 256 pixel sub-arrays in a dual polarization configuration, and four 128-pixel arrays covering four ALMA bands from bands 6 to 10 implemented as a pair of 64 pixel sub-arrays also in a dual polarization configuration.

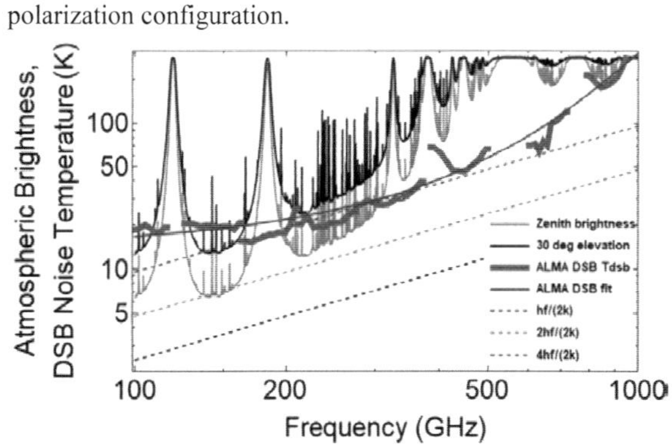


Fig. 3. Atmospheric brightness compared to equivalent DSB receiver noise temperature of ALMA receivers and quantum limit.

### A. *Large single band array*

For the purposes of a large Galactic plane survey with the goal of mapping the maximum area as fast as possible, a single band instrument operating at ALMA band 6 would be the most ideal choice. Mapping speed scales linearly with the number of pixels and is inversely proportional to the square of the system temperature. This favors a large number of pixels at a lower frequency band where both the receivers and the sky are the best. In addition, the field of view covered by the array scales inversely with the square of the frequency of operation simply because of the change in the diffraction limited beam size. This also strongly favors lower frequencies for maximum mapping speed. Assuming $T_{rec}=3h\nu/k$, atmospheric transmission of 0.95 (figure 2), and a 2SB scanning speed ratio of 3, such an instrument could map to a depth of 1 mJy in 100 kHz channels at a speed of approximately 30 hours per square degree not including overheads. Meeting this receiver temperature goal combined with 2SB mixer pixels in an array with 512 pixels still requires significant future technology development but is not unreasonable to consider on a decade timescale.

### B. *Smaller arrays in multiple bands*

While maximum mapping speed will be realized by implementing all pixels in one band, the scientific flexibility of smaller arrays covering multiple bands would be more useful for a larger number of projects. We estimate that the cost of implementing four 128-pixel arrays in four bands to be approximately the same (within a factor of two) as the single larger array covered in section III.A. We expect the cost of the multiple cryostats, additional integration and test and band selection optics to be offset by the savings realized in a smaller backend shared between the four bands. Such an instrument could cover four ALMA like bands between 230 GHz and 950 GHz, selected to maximize science output. Such a configuration would also allow relatively cost-effective incremental upgrades to add additional frequency bands as funds become available.

The 230 GHz channel would map at a speed four times slower than the instrument in section III.A, 120 hours per square degree to 1 mJy in 100 kHz channels. Mapping speeds of higher frequency bands would be significantly slower, scaling like $(\nu * Tsys)^{-2}$. Assuming receiver pixels still can be developed with $T_{rec}=3h\nu/k$ and the sky noise contribution scales roughly linearly with frequency, the mapping speed of higher frequency bands to a given depth unavoidably drops like $\nu^{4}$. This drop can only be mitigated by increasing the number of pixels at higher frequencies.

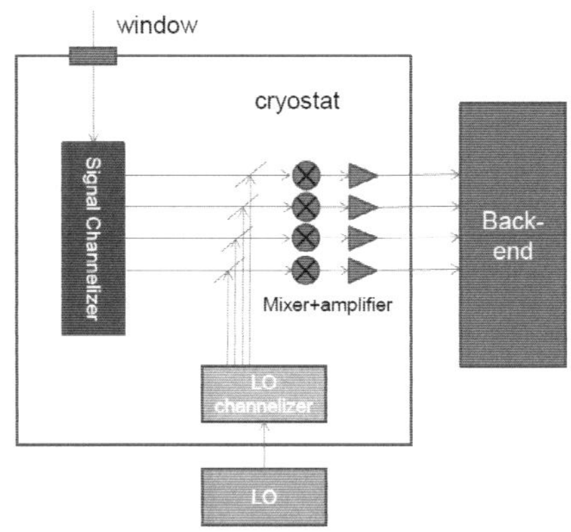


Fig. 4. Principle diagram of a quasi-optically combined frequency array receiver.

## IV. Frequency Array Receiver Concept

The concept of combining several spatial pixels into a focal plane array can be naturally extended in arranging pixels not in focal plane of a telescope, but in frequency space as presented in figures 4 and 5 resulting in a frequency array. The telescope signal from one spatial pixel on the sky is split into narrow frequency sub-bands using either quasioptical or planar filter. Each sub-band then feeds a SIS mixer. A comb of equally spaced LO signals are coupled to each of the mixer in such a way that continuous and simultaneous RF frequency coverage is achieved. The IFs of each frequency pixel are amplified and analyzed with the same type of back-end as for a conventional FPA. The input channelizing filter bandwidth of such a receiver should be matched with each pixel IF bandwidth. For current state of art mixers, this can be as wide as 16 GHz or more

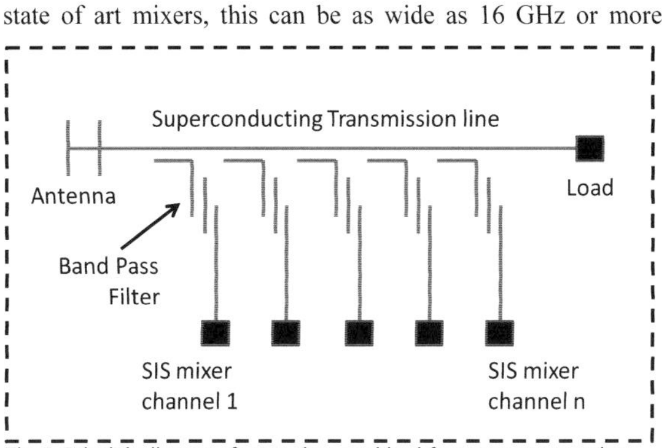


Fig. 5. Principle diagram of an on-chop combined frequency array receiver.

[9],[10]. In addition to quasi-optical free space filters, the channelization can be achieved in a very compact way on-chip as shown in figure 5. This concept is similar to filter bank on chip [11], [12] with the notable difference that the filter channel bandwidth is much larger, with high spectral resolution achieved through the use of a heterodyne receiver at the output of each channelizing filter. The frequency coverage of the frequency array is shown schematically in figure 6. In comparison with a direct detection receiver, the frequency information of incoming photons is retained in heterodyne reception. This design avoids efficiency loss due to channelization filter overlap, as the signal from neighboring bands can be recombined if the individual instantaneous coverage of each heterodyne mixer is slightly larger than 3dB bandwidth of each channel, i.e. the mixer IF bands have small frequency overlaps. The same argument allows significant relaxation of the requirement for filter edge sharpness, while maintaining high, close to unity overall coupling from input beam to each frequency pixel.

### A. *Sensitivity and utilization*

The frequency array receiver can be designed to have an extremely large instantaneous bandwidth. A 600 GHz band can be covered by a rather modest 40 frequency pixels of 16 GHz IF bandwidth. This bandwidth is more than 50 times larger than typical ALMA IF coverage, which means that for blind frequency searches of red shifted transmission lines this system can have the same speed as whole ALMA array if implemented on a 12m diameter dish. On AtLAST, the point source sensitivity would be increased by another factor of 4-17.

For total power observations, the total bandwidth of frequency array receiver is much larger than any direct detecting system, which covers typically only one atmospheric window while avoiding prominent atmospheric absorption lines. As illustrated by figure 2, the density of telluric absorption lines is significant, and their additional background contribution cannot be avoided by quasioptical filters. The frequency array receiver noise is not influenced by these lines and the spectral resolution is sufficient to resolve the lines and exclude the high background channels while still leaving large equivalent bandwidth for sensitive observation.

In comparison with direct detector systems, the quantum limit of heterodyne receivers presents significant limitation for low background systems. Direct detectors have no sensitivity limit and can be designed to have superior noise performance. However, as illustrated in figure 3, state of the art heterodyne ground-based receivers are sky background limited for all submm bands. Given the other advantages like bandwidth and frequency resolution, the frequency array receiver will be superior to both on-chip filter bank spectrometers, grating spectrometer and total power spatial pixels for ground-based applications.

Finally, when covering the full THz receiver window (100-1000GHz) the frequency array receiver will be the ultimate receiver for an radio interferometer, like ALMA or SMA as it will provide all complex signal information over the full ground based sub-mm band instantaneously, providing ultimate flexibility and sensitivity if such bandwidths are supported by a future correlator.

## V. Conclusion

Technology developments in the past decades make large format (~1000 pixel) heterodyne arrays with near quantum and background limited noise performance possible for telescopes like AtLAST. The primary limitation that will drive engineering vs. science output decisions is cost rather than technical readiness. We believe reasonable cost limitations on the order of tens of millions of USD to be dedicated to heterodyne instrumentation on AtLAST will limit the number of pixels to ~1000. These pixels can be deployed in many ways, from a single band large format array to yield the maximum possible mapping speed, or in smaller arrays covering multiple bands. 2SB mixer topologies have the potential to significantly improve performance for such receivers but do come at a cost increased complexity.

A novel frequency array receiver concept has been presented that could allow coverage of the entire sub-mm band instantaneously with heterodyne resolution. Such a receiver would be synergistic with a large focal plane array, allowing very deep and wideband point source observations. This instrument could also share much of the backend system required for a focal plane array, thereby reducing the cost of implementation.

## References


[1] Stephen C. Parshley, Jörg Kronshage, James Blair, Terry Herter, Mike Nolta, Gordon J. Stacey, Andrew Bazarko, Frank Bertoldi, Ricardo Bustos, Donald B. Campbell, Scott Chapman, Nicolas Cothard, Mark Devlin, Jens Erler, Michel Fich, Patricio A. Gallardo, Riccardo Giovanelli, Urs Graf, Scott Gramke, Martha P. Haynes, Richard Hills, Michele Limon, Jeffrey G. Mangum, Jeff McMahon, Michael D. Niemack, Thomas Nikola, Markus Omlor, Dominik A. Riechers, Karl Steeger, Jürgen Stutzki, Eve M. Vavagiakis, "CCAT-prime: a novel telescope for sub-millimeter astronomy," Proc. SPIE 10700, Ground-based and Airborne Telescopes VII, 107005X (6 July 2018)

[2] F. Bertoldi. The Atacama Large Aperture Submm/mm Telescope (AtLAST) Project. In Atacama Large-Aperture Submm/mm Telescope (AtLAST), page 3, Jan 2018. doi:10. 5281/zenodo.1158842.

[3] A. Otarola, C. De Breuck, T. Travouillon, S. Matsushita, L. Å. Nyman, A.Wootten, S .J. E. Radford, M. Sarazin, F. Kerber, and J. P. Pérez-Beaupuits. Precipitable Water Vapor, Temperature, and Wind Statistics At Sites Suitable for mm and Submm Wavelength Astronomy in Northern Chile. PASP, 131(998):045001, Apr 2019. arXiv:1902.04013, doi:10.1088/1538-3873/aafb78. vol. 134, pp. A635–A646, Dec. 1965.

[4] C. E. Groppi and J. H. Kawamura, "Coherent Detector Arrays for Terahertz Astrophysics Applications," in IEEE Transactions on Terahertz Science and Technology, vol. 1, no. 1, pp. 85-96, Sept. 2011. doi: 10.1109/TTHZ.2011.2159555

[5] Jenna Kloosterman, Tiara Cottam, Brandon Swift, David Lesser, Paul Schickling, Christopher Groppi, Michael Borden, Allison Towner, Per Schmidt, Craig Kulesa, Christian d'Aubigny, Christopher Walker, Dathon Golish, Sander Weinreb, Glenn Jones, Hamdi Mani, Jacob Kooi, Art Lichtenberger, Patrick Puetz, Gopal Narayanan, "First observations with SuperCam and future plans," Proc. SPIE 8452, Millimeter, Submillimeter, and Far-Infrared Detectors and Instrumentation for Astronomy VI, 845204 (27 September 2012)

[6] H. Smith, J. Buckle, R. Hills, G. Bell, J. Richer, E. Curtis, S. Withington, J. Leech, R. Williamson, W. Dent, P. Hastings, R. Redman, B. Wooff, K. Yeung, P. Friberg, C. Walther, R. Kackley, T. Jenness, R. Tilanus, J. Dempsey, M. Kroug, T. Zijlstra, T. M. Klapwijk, "HARP: a submillimetre heterodyne array receiver operating on the James Clerk Maxwell Telescope," Proc. SPIE 7020, Millimeter and Submillimeter Detectors and Instrumentation for Astronomy IV, 70200Z (18 July 2008)

[7] Urs U. Graf, Stefan Heyminck, Ernest A. Michael, S. Stanko, C. E. Honingh, Karl Jacobs, Rudolf T. Schieder, Juergen Stutzki, B. Vowinkel, "SMART: the KOSMA submillimeter array receiver for two frequencies," Proc. SPIE 4855, Millimeter and Submillimeter Detectors for Astronomy, (17 February 2003)

[8] GREAT: the SOFIA high-frequency heterodyne instrument S. Heyminck, U. U. Graf, R. Güsten, J. Stutzki, H. W. Hübers and P. Hartogh A&A, 542 (2012) L1 DOI: https://doi.org/10.1051/0004-6361/201218811

[9] T. Kojima, M. Kroug, K. Uemizu, Y. Niizeki, H. Takahashi, Y. Uzawa, Performance of a Wide IF SIS-Mixer-Amplifier Module for ALMA Band 8 (385-500 GHz), in 28th International Symposium on Space Terahertz Technology, Cologne, Germany, March 12-15 2017.

[10] P. K. Grimes, R. Blundell, S. Paine, J. Test, C-Y. E. Tong, R. W. Wilson, and L. Zeng, The wSMA Receivers - a New Wideband Receiver System for the Submillimeter Array, in 28th International Symposium on Space Terahertz Technology, Cologne, Germany, March 12-15 2017.

[11] Akira Endo, Kenichi Karatsu, Yoichi Tamura, Shun Ishii, Tatsuya Takekoshi, Tai Oshima, Akio Taniguchi, Tsuyoshi Ishida, Koyo Suzuki, David Thoen, Robert Huiting, Toshihiko Kobiki, Tom Bakx, Kaui Chin, Kazuyuki Fujita, Vignesh Murugesan, Sjoerd Bosma, Ozan Yurduseven, Tetsutaro Ueda, Masato Naruse, Shunichi Nakatsubo, Akira Kouchi, Jun Maekawa, Stephen Yates, Nuria Llombart, Teun Klapwijk, Ryohei Kawabe, Shin'ichiro Asayama, Paul van der Werf, Kotaro Kohno, Jochem Baselmans, "First light of DESHIMA on ASTE: on-chip filterbank spectrometer for submillimeter wave astronomy (Conference Presentation)," Proc. SPIE 10708, Millimeter, Submillimeter, and Far-Infrared Detectors and Instrumentation for Astronomy IX, 107080N (10 July 2018).

[12] Kirit S. Karkare, Peter S. Barry, C. Matt Bradford, Scott Chapman, George Che, Jason Glenn, Sam Gordon, Steven Hailey-Dunsheath, Matthew Hollister, Attila Kovacs, Henry G. Leduc, Philip Mauskopf, Ryan McGeehan, Chris McKenney, Theodore Reck, Joseph Redford, Erik Shirokoff, Carole Tucker, Jordan Turner, Samantha Walker, Jordan Wheeler, Jonas Zmuidzinas, "On-sky demonstration of the SuperSpec millimeter-wave spectrometer (Conference Presentation)," Proc. SPIE 10708, Millimeter, Submillimeter, and Far-Infrared Detectors and Instrumentation for Astronomy IX, 107081R (10 July 2018)

**Christopher E. Groppi** received the B.A. degree in astronomy (with honors) from Cornell University in Ithaca, NY, in 1997 and the Ph.D. degree in astronomy with a minor in electrical and computer engineering from the University of Arizona in Tucson, AZ, in 2003.

In 2003, he joined the National Radio Astronomy Observatory as a Director's Postdoctoral Fellow. He then moved to the University of Arizona as an assistant staff astronomer in 2005. In 2006, he received an Astronomy and Astrophysics Postdoctoral Fellowship from the National Science Foundation. In 2009, he joined the Arizona State University School of Earth and Space Exploration in Tempe, AZ as an assistant professor. He became an associate professor in 2015. He is an experimental astrophysicist interested in the process of star and planet formation and the evolution and structure of the interstellar medium. His current research focuses on the design and construction of state of the art terahertz receiver systems optimized to detect the light emitted by molecules and atoms in molecular clouds, the birthplace of stars. He also applies terahertz technology developed for astrophysics to a wide range of other applications including Earth and planetary science remote sensing, hazardous materials detection and applied physics.

**Andrey M. Baryshev** received the M.S. degree (summa cum laude) in physical quantum electronics from the Moscow Physical Technical institute, Moscow, Russia, in 1993, and the Ph.D. degree in superconducting integrated receiver combining SIS mixer and flux flow oscillatoron into one chip from the Technical University of Delft, Delft, The Netherlands, in 2005.

He is currently an associate professor at the Kapteyn Astronomical Institute, University of Groningen, Groningen, and was previously a Senior Instrument Scientist with the SRON Low Energy Astrophysics Division, Groningen, The Netherlands. In 1993, he was an Instrument Scientist with the Institute of Radio Engineering and Electronics, Moscow, involved in the field of sensitive superconducting heterodyne detectors. In 2000, he joined an effort to develop an SIS receiver (600–720 GHz) for the atacama large millimiter array, where he designed the SIS mixer, quasi-optical system, and contributed to a system design.

His current main research interests include application heterodyne and direct detectors for large focal plane arrays in THz frequencies and quasi-optical systems design and experimental verification.

Dr. Baryshev was the recipient of the NWO-VENI Grant for his research on heterodyne focal plane arrays technology in 2008 and, in 2009, he was the recipient of the EU commission Starting Researcher Grant for his research on focal plane arrays of direct detectors.

**Martina C. Wiedner** received the Ph.D. degree in physics from the University of Cambridge, Cambridge, U.K., in 1999.

She was a Postdoctoral Fellow with the Harvard Smithsonian Center for Astrophysics, Boston, MA, USA, and she also led a junior research group with the University of Cologne, from 2003 to 2009. Since 2009, she has been a Permanent Researcher with the Laboratoire d'Etude du Rayonnement et de la Matiere en Astrophysique, Observatory of Paris, ´ Paris, France. Her main research is focused on the development of heterodyne instrumentation for astronomy and earth sciences.

**Pamela D. Klaassen** received a B.Sc and M.Sc in astronomy from the University of Calgary in 2002 and 2004 respectively. She then moved to McMaster University in Hamilton, Ontario, Canada for her Ph.D. which she obtained in 2008. During that time, she also spent a year at the Harvard-Smithsonian Center for Astrophysics as an SMA pre-doctoral fellow.

She did her post-doctoral fellowships in the European ALMA regional Centres at ESO in Garching, Germany (2008-2011) and Leiden Observatory, in Leiden, the Netherlands (2011-2014). Since then, she has been an Instrument Scientist at the UK Astronomy Technology Centre, where she is the science lead for the JWST/MIRI simulator, and for the team developing the Observatory Science Operations software for the SKA. She is the lead of the science working groups for the AtlAST project, and her scientific interests focus on the formation of the most massive stars in the Milky Way.

**Tony Mroczkowski** received his B.S. in engineering from the Cooper Union for the Advancement of Science and Art in 2001 and his Ph.D. in astronomy from Columbia University in New York City, NY in 2009 (defended in 2008). From 2008-2011, Mroczkowski was a postdoctoral researcher at the University of Pennsylvania in Philadelphia, PA, the last year of which was funded by the NASA Einstein Fellowship program. Mroczkowski continued this fellowship at NASA Jet Propulsion Laboratory in Pasadena, CA, in 2011-2013. Mroczkowski was then awarded the National Research Council Fellowship to work at the Naval Research Laboratory in Washington, D.C. from 2013-2016. In 2016, he became assistant level faculty at the European Southern Observatory in Garching, Germany, working as an astronomer / (sub)millimeter instrument scientist on ALMA development. He is an experimental cosmologist primarily interested in the Sunyaev-Zel'dovich effect from galaxy clusters, groups, and large scale structure.

# Millimetron Space Observatory: progress in the development of payload module

A. Smirnov[1], E. Golubev[1], M. Arkhipov[1], E. Filina[1], V. Pyshnov[1], N. Myshonkova[1], S. Fedorchuk[1], T. Kosmovich[1], I. Vinogradov[1], A. Baryshev[1,2], Th. de Graauw[1], S. Likhachev[1] and N. Kardashev[1]

Millimetron Space Observatory (MSO) is mission addressed to creation a space cryogenic telescope with aperture about 10-m [1]. Such telescope will allow scientific community to have an astronomical instrument with enormous sensitivity and angular resolution in the submillimeter and far-infrared wavelength ranges. We plan to install at the telescope several FIR and sub-millimeter scientific instruments, which will enable high-resolution imaging and spectroscopy observations with unprecedented sensitivity. At the same time, MSO will enable observations with an extremely high angular resolution (up to $0.1\times10^{-6}$arcsec) as an element of a ground-space very long baseline interferometry system (S-VLBI). Thereby the observatory will contribute breakthrough capability into solution a number of cosmology and fundamental astrophysics questions about the origin and evolution of our Universe, galaxies, stars and other objects [2].

The MSO is divided into two parts: the payload module and the bus module. Due to the complexity of the payload module, most of the recent years of work are focused on it. This module includes an antenna of the telescope, scientific receivers, functional and service systems and a high-gain radio system for transmitting scientific data to Earth.

The primary mirror of the telescope will be deployable and consist from of a 3-m aperture central part surrounded by 24 deployable petals. The concept of petals deployment is based on the successfully launched and currently working Radioastron project [3]. The surface accuracy of the deployable 10-m primary mirror of Radioastron achieves about 1 mm in space conditions. The telescope of MSO would have much better surface accuracy - less than 10 μm (rms). In order to achieve this we plan to use an active surface control system based on a wave front sensing. This system will be periodically employed to correct inaccuracies in the positions of the panels caused by different factors.

A combination of a high modulus carbon fiber reinforced plastic (CFRP) and a cyanate ester resin as a binder provides a lightweight structure with low moisture absorption, high thermal stability and high stiffness. This combination has been chosen for the material of the primary mirror of telescope and many parts of it. The panels are mounted on the back support structure (Fig. 1) made from CFRP via precision cryogenic actuators.

Fig. 1. Photo of the full-scale deployable mock-up of back up structure of the primary mirror with six installed panels.

To achieve the required sensitivity of the telescope in the submm/FIR we need to cool antenna down to the temperature less than 10K (goal). It may be possible to do this on-orbit only by a combination of effective radiation cooling and additional active mechanical cooling. A cold space antenna requires minimization and stability of external thermal radiation. This is one of the reasons why MSO will be placed into orbit around the second Earth-Sun Lagrange point (L2). The MSO antenna into L2 will be cooled passively to a temperature about 30 – 60K by a suite of the deployable multi-layer V-groove shields. The following steps to reduce the temperature of the antenna are based on active reducing the thermal loads applied to it. Active mechanical cooling is based on existing close cycling space mechanical coolers.

In this work, we will focus on the progress in the development of payload module.

[1] Astro Space Center of P.N. Lebedev Physical Institute, 84/32 Profsoyuznaya str., Moscow, Russia.

[2] Kapteyn Astronomical Institute, University of Groningen, Landleven 12, 9747 AD, Groningen, The Netherlands.

References

[1] Smirnov A.V. et al. "Space mission Millimetron for terahertz astronomy," *Proc. SPIE 8442, Space Telescopes and Instrumentation: Optical, Infrared, and Millimeter Wave,* 84424C, 2012.

[2] Kardashev N.S., et al, "Review of scientific topics for Millimetron space observatory" *Phys. Usp.*, vol.57, Is. 12, pp. 1199-1228, 2014.

NOTES:

[3] Kardashev N.S., et al. ""RadioAstron"– A Telescope with a Size of 300 000 km: Main Parameters and First Observational Results," *Astronomy Reports*, 57, 153-194, 2013.

NOTES:

# Tuesday, April 16, 2019

## Session VII. THz Optics and Antennas

# Wide-Field Designs for Off-Axis Telescopes: Application to the Optics of CCAT-prime.

Richard E. Hills[1]

The advantages of using off-axis optical designs for radio telescopes have been recognized for many years. These include the absence of aperture blocking, which lowers the side-lobe level, and a large reduction in multi-path reflections, which greatly improves spectral baselines. In addition, two-mirror designs meeting the Mizuguchi-Dragone criterion have good polarization properties. For applications requiring a large field of view and a reasonably flat focal plane, the crossed-Dragone configuration (upper part of Fig. 1) is particularly advantageous [1].

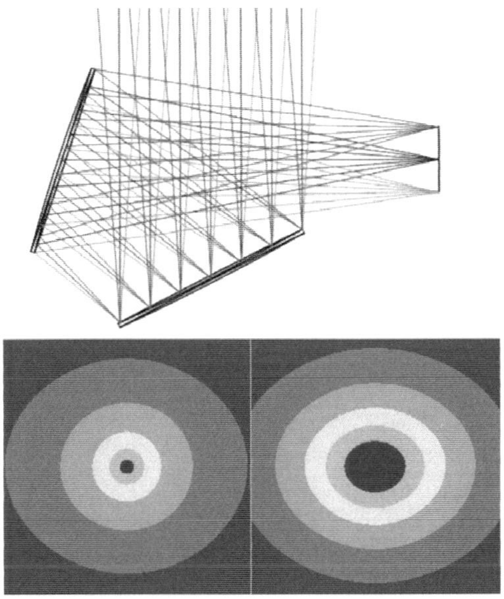

Fig. 1. Upper: Crossed-Dragone f/2.5 configuration with a 6m aperture and an 8-degree diameter field of view. Lower: Contours of 80% Strehl ratio in the focal plane (8deg x 8deg) at (blue to red) 75, 150, 300, 600 and 1500 GHz. Left: classical design. Right: with correction for coma aberrations.

These designs are however still based on the classical Cassegrain telescope, e.g. the primary and secondary reflectors are, respectively, segments of a paraboloid and of a hyperboloid. We describe here how a small modification to the shapes of the mirrors, analogous to the Ritchey-Chrétien design long used for optical telescopes, can greatly increase the field of view, especially at high frequencies.

The dominant aberration in such optical systems is usually coma and this can be corrected by adding higher-order terms to one mirror and compensating for them in the other. Dragone [2] did in fact set out the principles for making such a correction as long ago as 1983, but this option does not seem to have been widely adopted, perhaps because it was expressed in a purely analytical form. With modern optical design packages, however, the relevant optimization is relatively straight-forward to achieve. Practical ways of doing this, along with some of the pit-falls to be avoided, will be explained in the presentation.

This approach has been applied to the optics of CCAT-prime [3]. The lower part of Fig. 1 illustrates the very large increase in the useable field of view that is achieved in the sub-mm wavebands. This improvement is gained without significant sacrifice of the other aspects of the optical performance, although there are some modest penalties in terms of manufacturing the reflectors.

The presentation will include a general description of the CCAT-prime telescope design (which has also been adopted for the Simons Observatory CMB Large-Aperture Telescope) as well as a brief discussion of other designs and applications of the coma-correction approach.


### References

[1] M. D. Niemack. "Designs for a large-aperture telescope to map the CMB 10X faster," *Applied Optics*, vol. 55, no. 7, pp. 1688–1696, 2016.

[2] Dragone, C., "First-Order Correction of Aberrations in Cassegrainian and Gregorian Antennas", *IEEE, Antennas Propagat.*, Vol. 31, 764-775, (1983)

[3] S. C. Parshley et al., "The optical design of the six-meter CCAT-prime and Simons Observatory telescopes", *SPIE Conference Series*, vol. 10700E..41P, 2018.


[1] Cavendish Laboratory, University of Cambridge, Cambridge, UK, CB3 0HE.

NOTES:

# In Flight Measurements System of Millimetron Primary Mirror Surface

A. M. Baryshev[1,2], E. Golubev[2], M. Arkhipov[2], E. Filina[2], V. Pyshnov[2], N. Myshonkova[2], A. Smirnov[2], S. Fedorchuk[2], A. Andrianov[2], Th. de Graauw[2], S. Likhachev[2] and N. Kardashev[2]

Millimetron Space Observatory (MSO) is a mission dedicated to far infrared astronomical observation from space. The key element of MSO is its 10 m primary mirror which is cooled to cryogenic temperatures: 10 K is goal and below 20 K is specification. The primary mirror accuracy specification is 10 μm and goal is 5 μm RMS which would allow for effective observations at wavelength above 150 μm. Due to its mirror temperature, aperture size and state of art instrumentation on board, that MSO will be several orders of magnitude more sensitive than previous space missions in far infrared like Herschel Space Observatory (HSO) and it will have 3 times more spatial resolution in single dish mode than HSO.

Main mirror (see figure 1) of MSO will have to be deployed in space because solid 10 m aperture does not fit under fairing of available launching systems. Primary mirror consists of central fixed part of 3 m diameter and 24 deployable petals located around central part. Each petal contains support structure and three panels attached to it by means of mechanical actuators, in such a way that both position and curvature of each panel can be adjusted while in space. Central part of the mirror also consist of 24 identical panels supported by mechanical actuators.

Panels of primary mirror are made from carbon reinforced plastic composite material designed to have zero coefficient of thermal expansion. The panels are replicated during curing off the AstroSetal glass negatives which are polished to high accuracy, of 1..2 μm RMS. The CRFP panel design allow to achieve very low specific mirror mass of <10 kg/m$^2$.

The main dish is designed to deploy with accuracy of 1 mm PV. Final surface accuracy will be achieved by placing the panels around optimal paraboloid surface with common focus by means of mechanical actuators. Both piezo and step motors solutions for actuators are being considered.

One of the key technological challenges in order to achieve final surface accuracy is ability to measure surface deviations in flight conditions. In addition the surface accuracy of the panels will not allow to use visual light optical methods to measure mirror accuracy. In this contribution we will discuss three main methods considered to achieve required measurement accuracy:

- Photogrammetry, using on-board high resolution cameras and photogrammetry targets on panels. This method achieves 100 μm typical accuracy
- Laser ranging. This method will involve fiber – optics based absolute distance laser measure with accuracy of 1 μm which allows to measure positions of special targets, located on each panel in 3D space. This information will be used to place the panels at desired positions and change their curvature if needed.
- Coherent signal adjustment using signal form bright astronomical source and successive approximation approaches, staring with lower frequencies and then propagating to the higher frequencies. Astronomical receiver would be used for this purpose.

We will provide numerical estimate for efficiency of each method and its limiting accuracy at realistic MSO main dish design and available astronomical sources.

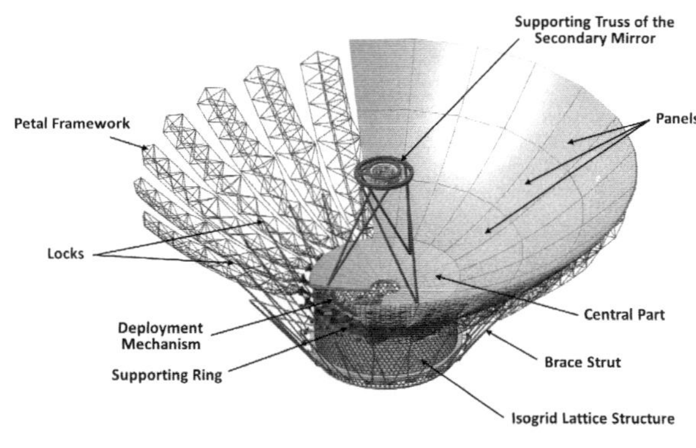


Figure 1, Construction of MSO's primary mirror

[1] Kapteyn Astronomical Institute, University of Groningen, Landleven 12, 9747 AD, Groningen, The Netherlands.

[2] Astro Space Center of P.N. Lebedev Physical Institute, 84/32 Profsoyuznaya str., Moscow, Russia.

NOTES:

# Far-field beam pattern technique for high pointing accuracy characterization of GUSTO HEB mixer arrays

J. R. Silva[1,2], Stephen Yates[1], W. Laauwen[1], M. Finkel[1], J. R. Gao[1,3]

Until now HEB mixer arrays employed in supra-THz heterodyne detection consist of multiple individual mixers (e.g. UpGreat) that could be aligned with the respective individual optics. In GUSTO [1], a NASA ultra-long duration balloon borne Terahertz observatory, that will employ three HEB mixer arrays to simultaneously measure the THz relevant molecular lines of [NII], [CII] and [OI] lines at 1.4, 1.9 and 4.7 THz, it requires the use of compact monolithic arrays consisting of 8 pixels, in a 4x2 configuration, that may not be independently moved or aligned. This new geometry imposes the need for a requirement on the pointing accuracy of each individual pixel. In the case of GUSTO all pointing vectors should be aligned within 0.1 degrees in relationship to the array mixer block normal. The pointing requirement implies two different challenges. The first is to set proper component tolerances (e.g. lens fabrication) and its assembly. The second is the characterization of the final assembly. The latter is the focus of this work.

In order to properly characterize the pointing of any individual pixels, an obvious approach is to characterize its beam pattern at the desired frequency. A near field phase and amplitude or a far-field amplitude measurement can simultaneously characterize not only the pointing but the beam pattern of the individual mixer, a very important information for the validation of the arrays. Due to the complexity and not yet developed phase and amplitude measurements for higher frequencies, e.g. 4.7 THz, we have been focusing on the amplitude only measurement.

In the past a widely used technique was the rotation of the mixer at the beam waist position whilst facing a THz beam, which is typically from the LO. In our opinion there are a few drawbacks of this technique to characterize with sufficient resolution the GUSTO mixer such as: a) the need for a highly reliable system to align the HEB mixer to the rotation axis of the stage, affecting the side lobe level and balance; b) the mechanical (in)stability of the rotation stage moving such a big mass, which could be solved by longer waiting times but would lead to unrealistic long scans. c) the impossibility for such technique to co-register the pointing of all the pixels in the array assembly; Therefore, we explore another technique that makes use of the scanning of an Hot/Cold load in the far field of the mixer, which is operated in heterodyne mode similarly as reported in [2].

We report on a beam pattern technique setup with sub 0.1 degrees resolution for single pixel characterization. For this the HEB is operated in the heterodyne mode, the LO provided by a FIR Gas Laser or QCL, and the measurable signal from a chopped Hot Source aperture that is scanned 1 meter away from the HEB. The hot source setup used is similar to the one used in [3]. For improved pointing resolution multiple planes are measured. The same technique is employed while using the HEB as a direct detector to simultaneously characterize the pointing of the entire array vs the normal reference. The characterization measurement is in progress and we will update the outcome at the conference.

[1] SRON Netherlands Institute for Space Research, Groningen/Utrecht, the Netherlands

[2] Kapteyn Astronomical Institute, University of Groningen, 9747 AD, Groningen, The Netherlands

[3] Kavli Institute of NanoScience, Delft University of Technology, Delft, the Netherlands

## References

[1] J. R. Silva et al. " 4×2 HEB receiver at 4.7 THz for GUSTO", *Proceedings of SPIE*, Vol. 10708, 107080Z, 2018.

[2] A. D. Semenov et al, "Terahertz Performance of Integrated Lens Antennas With a Hot-Electron Bolometer" *IEEE Trans. Micro. Theory and Techniques*, vol. 55, no. 2, pp. 239–247, Feb. 2007.

[3] L. Ferrari et al, "Antenna Coupled MKID Performance Verification at 850 GHz for Large Format Astrophysics Arrays", *IEEE Trans. Terahertz Sci. and Tech.*, vol. 8, no. 1, pp.127-139, Jan. 2018.

NOTES:

# Tuesday, April 16, 2019

# Session VIII. HEBs and KIDs

Mid-infrared heterodyne receiver based on a superconducting hot electron bolometer and a quantum cascade laser

Yuan Ren[1,2], Daixi Zhang[1,2], Kangmin Zhou[1,2], Wei Miao[1,2], Wen Zhang[1,2], ShengCai Shi[1,2]
*1.Purple Mountain Observatory, Chinese Academy of Sciences, 8 Yuanhua Road, Nanjing, China, 210034*
*2.Key Lab of Radio Astronomy, Chinese Academy of Sciences, 8 Yuanhua Road, Nanjing, China, 210034*

The mid-infrared frequency region plays a vital role in the modern astronomic research, which includes early cosmic evolution, star and galaxy formation, and the planet's atmosphere research. However, the high-resolution spectrometer in this frequency region is still under developed. This paper focused on the development of a heterodyne receiver based on a superconducting hot electron bolometer as a mixer and a quantum cascade laser at 10.6 μm as a local oscillator.
A superconducting NbN hot electron bolometer working at 4 K was utilized as the mixer, with a combination of a hyper-hemispherical lens and a spiral antenna employed as the coupling element. A distributed feedback quantum cascade laser providing more than 30 mW power at 10.6 μm served as the local oscillator. The double sideband receiver noise temperature ($T_{DSB,Rec}$) was characterized with a Hg lamp as hot load and room temperature blackbody as cold load, and the $T_{DSB,Rec}$ was measured to be about 5000 K with an intermediate frequency bandwidth of 2.8 GHz. This development paves a new way to the high resolution spectroscopy at mid-infrared frequencies.

# 2 THz Hot Electron Bolometer Mixer using a Magnetic Thin Film

Akira Kawakami*, Yoshihisa Irimajiri
National Institute of Information and Communications Technology, Kobe, Hyogo, 651-2492, Japan
*Contact: kawakami@nict.go.jp

***Abstract*— To expand the intermediate frequency (IF) band and improve the sensitivity of a hot electron bolometer mixer (HEBM), we have proposed and examined a new HEBM structure using a nickel (Ni) magnetic thin film (Ni-HEBM). We found that it was possible to suppress the superconductivity under the electrodes of the HEBM caused by the niobium nitride (NbN) thin film for construction of the superconducting strip by the addition of a Ni thin film. By using Ni-HEBM structure, superconductivity exists only in the region between both electrodes and we think that it is possible to further miniaturize the HEBM. The miniaturization acts to expand the IF band, improve the sensitivity and is expected to reduce the required LO power. By using the Au (100nm) / Ni (0.6 nm) bilayer for the electrodes, we fabricated Ni-HEBM with a NbN strip of 0.1 μm-length. The IF bandwidth of the fabricated Ni-HEBMs was evaluated at 1.9 THz. We confirmed that the IF bandwidth expands, and the evaluated bandwidths was about 6.9 GHz at 4 K.**

***Index Terms*—IF bandwidth, HEBM, NbN, Ni, THz.**

## I. Introduction

In the terahertz band, there are plenty of emission lines from atmospheric constituents, that are applicable to observational studies of atmospheric dynamics and chemistry. For such applications, a heterodyne receiver with a high frequency resolution is necessary. Up to 1 THz, superconductor–insulator–superconductor (SIS) mixers show excellent performance and have already been used [1]-[3]. However, in the frequency region above 1.5 THz, it is difficult to realize, and superconducting hot-electron bolometer mixers (HEBMs) are expected as low-noise mixer elements. Several reports have already been made on the low noise operation of HEBMs with less than ten times the quantum noise limit in the terahertz frequency range [4]-[6]. However, the IF bandwidth of an HEBM is not sufficient when compared to that of an SIS mixer. Consequently, the usable IF band of receivers with an HEBM still remains limited to typically 3–5 GHz [7], [8]. Therefore, in recent years, several studies have focused on broadening the IF band. To broaden the IF band, efficient cooling of the hot electrons is required. In general, there are two cooling mechanisms of the HEBM, the heat dissipation process of excited electrons: lattice cooling and diffusion cooling. Lattice cooling releases the excitation energy to the substrate via the lattice, and diffusion cooling diffuses the excited electrons to the metal electrode directly [9]-[11]. In recent years, much research has focused on lattice cooling. However, we have proposed a new HEBM structure using a magnetic thin film that actively uses diffusion cooling to expand the IF band. Here, Ni thin films were used as the magnetic thin film; therefore, the new HEBM structure is denoted as Ni-HEBM. In this report, first, we influence of the Ni thin film on the superconductivity of the NbN thin film was investigated, and Ni-HEBMs with a strip length of 0.1 μm were fabricated and characterized.

## II. Fabrication of Ni-HEBM

In Ni-HEBM proposed by this research, suppression of superconductivity by Ni is important. To suppress the superconductivity of NbN strip under the electrode, a Ni magnetic thin film places between the superconducting strip and the gold (Au) metal electrode. The details of the element structure and the manufacturing method are described in [12]. As a result, superconductivity exists only between both electrodes, and we think that it is desirable for further miniaturization of HEBM. To expand the IF band and improve the sensitivity, we tried to fabricate Ni-HEBMs with a strip length of 0.1 μm.

### *A. Optimization of the Ni thickness for Suppression of NbN Superconductivity*

For miniaturization of Ni-HEBM, reducing the influence of the Ni addition is preferred. We have reported that superconductivity of the NbN strip near the electrodes was also suppressed by the Ni thin film and the region was to be series resistance of several Ω [12]. The resistance caused by Ni addition is expected to act as loss and reducing dR/dT of HEBM. Therefore, we tried to optimize the thickness of Ni for miniaturization of Ni-HEBM.

To investigate the influence of Ni thickness dependency on the superconductivity, three types of samples were prepared and

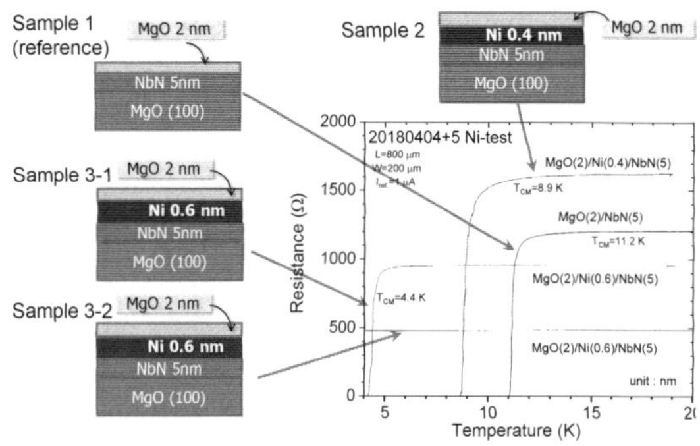


Fig. 1. Ni thickness dependency of NbN superconductivity in Ni/NbN bilayers. To investigate the influence of Ni thickness dependency on the superconductivity, three types of samples were prepared and tested.

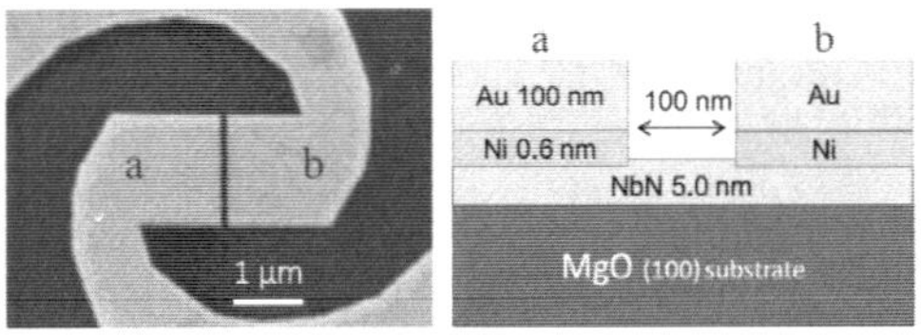


(a) (b)

Fig. 2. A SEM image and a schematic of the cross section of the Ni-HEBM. The superconducting strip length and width were 0.1 and 0.5 μm, respectively. (a) A SEM image. (b) Cross section.

tested. Sample-1 had a bilayer of MgO (2 nm) and NbN (5 nm) which was fabricated as a reference of NbN superconductivity. Samples-2 and -3 are both three-layer films of MgO (2 nm)/Ni /NbN (5 nm), and only the Ni film thickness was changed. The Ni film thickness of the Sample 2 was 0.4 nm, and the Sample 3 was 0.6 nm. Here, the MgO layer which was deposited by an ion-beam sputtering was used as a passivation layer on the Ni or NbN surface [13], and all films were continuously deposited in high vacuum. Fig. 1 shows the schematics and the temperature dependences of resistance of each sample. Sample-1 exhibited a $T_{CM}$ of 11.2 K. Here, $T_{CM}$ is the temperature at which the resistance is halved. Sample-2 showed $T_{CM}$ of 8.9 K, and it showed that 0.4 nm thick Ni thin film was insufficient. Sample-3 was fabricated two samples, and one showed that one was $T_{CM}$ of 4.4 K and the other was no superconductivity. We think that 0.6 nm thick of Ni thin film is critical thickness to suppress the superconductivity of 5 nm thick NbN thin film. In actual device fabrication, it is considered that the superconductivity of the NbN thin film under the electrode is degraded because ion beam etching of the NbN surface of about 1 nm is performed before forming the Ni film. Therefore, we thought that 0.6 nm thick Ni thin film is enough, and this film thickness was adopted to fabricate Ni-HEBMs with the 0.1 μm strip length.

### *B. Fabrication and evaluation of the Ni-HEBM with a strip length of 0.1 μm*

HEBMs comprise a structure in which two metal electrodes are connected by an extremely thin superconducting strip. In general, to ensure a good electrical contact, the superconducting strip and both electrodes are usually connected via an overlapped region on the strip. In our proposal, a magnetic thin film is placed between the metal electrode and the superconducting strip in the overlapping region, and the superconductivity in that region is suppressed by the spin electron diffusion from the magnetic thin film. Thus, a HEBM structure with superconductivity only between the two metal electrodes can be realized stable. Details of the Ni-HEBM fabrication process have been published elsewhere [12].

A SEM image (a) and a schematic of its cross section (b) of the Ni-HEBM are shown in Fig. 2. The electrode interval was set at 0.1 μm and the strip width was set to 0.4 – 0.7 μm. A log spiral antenna was adopted as a plane antenna. For both electrodes connected to the superconducting strip, an Au (100 nm)/Ni (0.6 nm) bilayer film was used. The electrode pattern at about 7 μm from the center was drawn by an electron beam lithography system, and the pattern for a lift-off process was formed.

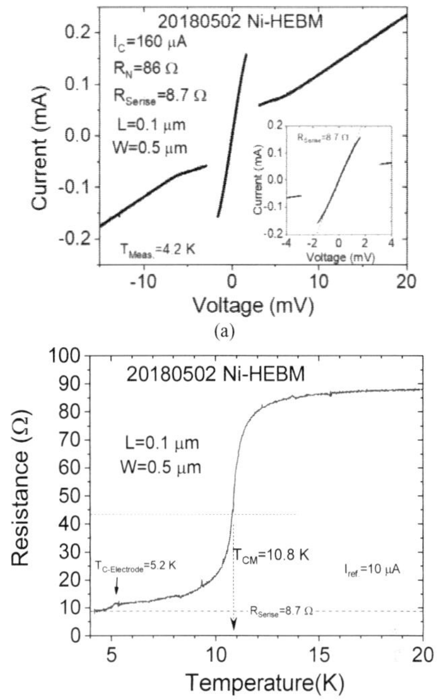


(a)

(b)

Fig. 3. I–V and Resistance–temperature characteristics of a Ni-HEBM with a strip length of 0.1 μm. (a) I–V characteristics of a typical Ni-HEBM at 4.2 K. (b) A typical resistance–temperature characteristics.

Fig. 3 shows the *I–V* characteristics (a) and resistance–temperature characteristics (b) of the typical Ni-HEBM measured by the four-terminal method. The NbN strip length and width were 0.1 and 0.5 μm, respectively. In Fig. 3(b), the transition temperature $T_{CM}$ was about 10.8 K. The antenna pattern made of the Au(150nm) /Nb(5nm)/NbN(5nm) trilayer film showed superconductivity at about 5.2 K. However, the metal electrodes in the central region of about 7 μm did not show superconductivity due to the presence of Ni. The resistance of the metal electrodes ($R_{Electrode}$) was evaluated to be about 0.8 Ω [12]. Meanwhile, the Ni-HEBM showed a series resistances ($R_{Series}$) of 8.7 Ω at 4.2 K. It is considered that the superconductivity of the NbN strip near both electrodes was also suppressed by the effect of the Ni. As a result, both regions of NbN strip near both electrodes are considered to be series resistances ($2R_{Strip}$) and $2R_{Strip}$ was estimated to be about 7.9 Ω. We think that the optimization of Ni influence was insufficient and it is necessary to reduce $R_{Strip}$.

## III. Evaluation of the Ni-HEBM

The IF bandwidth of the fabricated Ni-HEBMs was evaluated at 1.9 THz. As a signal, a stable source that was generated by a unitraveling carrier photodiode using the difference frequency component of two optical comb signals with an appropriate frequency interval of 106 was used. As a local oscillator, 144 times VDI multiplier was used. The HEBM

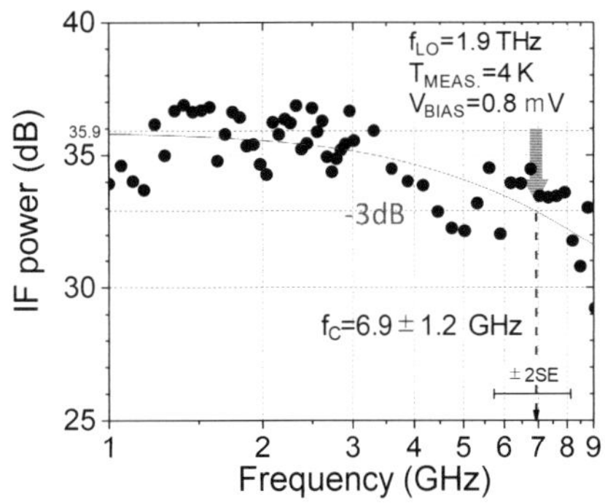


Fig. 4. The IF gain bandwidth of the Ni-HEBM. The superconducting strip length and width were 0.1 μm and 0.5 μm, respectively. The LO frequency and the measurement temperature were 1.9 THz and 4 K, respectively.

was biased with a voltage source, and the irradiation power of LO was controlled to keep a constant current value with the attenuator. Details of the evaluation setup for the IF gain bandwidth have been published elsewhere [12].

Fig.4 shows the evaluation of the IF gain bandwidth of the Ni-HEBM with strip length of 0.1 μm. Here, ±2 times the standard error (SE) is written as $f_C$ error bars in the figure. The IF bandwidth was evaluated about 6.9 GHz at 4 K. The uncorrected receiver noise temperature of same Ni-HEBM was also evaluated at 4 K, and it was about 1220 K(DSB) at 2 THz.

## IV. Conclusion

To expand the IF band of a HEBM, we have proposed and examined a new structure of Ni-HEBM. To realize the miniaturization of Ni-HEBM, the thickness dependency of Ni to NbN superconductivity was evaluated and we found that 0.6 nm thick of Ni thin film was critical thickness to suppress the superconductivity of 5 nm thick NbN thin film. By using the Au (100 nm) / Ni (0.6 nm) bilayer for the electrodes, Ni-HEBMs with a NbN strip of 0.1 μm-length were fabricated. However, the Ni-HEBM fabricated showed the series resistance of 7.9 Ω which was caused by Ni influence and further optimization is needed. The IF bandwidth of the fabricated Ni-HEBMs was evaluated at 1.9 THz and it was about 6.9 GHz at 4 K.

## References

[1] W. Shan, T. Noguchi, S. Shi, and Y. Sekimoto, "Design and development of SIS mixers for ALMA band 8," *IEEE Trans. Appl. Supercond.*, vol. 15, no. 2, pp. 503–506, Jun. 2005.

[2] C. E. Tong, P. K. Grimes, A. Barbier, K. Schuster, and M.Wang, "Design and Performance of a 3-Junction series distributed SIS mixer for wide IF applications," *IEEE Trans. Appl. Supercond.*, vol. 23, no. 3, Jun. 2013, Art. no. 1400404.

[3] A. Karpov et al., "Low Noise 1 THz–1.4 THz mixers using Nb/Al-AlN/NbTiN SIS junctions," *IEEE Trans. Appl. Supercond.*, vol. 17, no. 2, pp. 343–346, Jun. 2007.

[4] J. L. Kloosterman et al., "Hot electron bolometer heterodyne receiver with a 4.7-THz quantum cascade laser as a local oscillator," *Appl. Phys. Lett.*, vol. 102, 2013, Art. no. 011123.

[5] J. R.Gao et al., "Terahertz superconducting hot electron bolometer heterodyne receivers," *IEEE Trans. Appl. Supercond.*, vol. 17, no. 2, pp. 252–258, Jun. 2007.

[6] F. M. Boussaha, J. H. Kawamura, J. A. Stern, A. Skalare, and V. White, "A Low Noise 2.7 THz waveguide-Based superconducting mixer," *IEEE Trans. THz Sci. Technol.*, vol. 2, no. 3, pp. 284–289, May 2012.

[7] K. M. Zhou, W. Miao, S. C. Shi, R. Lefevre, Y. Delorme, "Noise temperature and IF bandwidth of a 1.4 THz superconducting HEB mixer," URSI AP-RASC., pp. 2010-2012, 2016.

[8] D. Meledin, C. E. Tong, R. Blundell, N. Kaurova, K. Smirnov, B. Voronov, G. Goltsman, "Study of the IF bandwidth of NbN HEB mixers based on crystalline quartz substrate with an MgO buffer layer," IEEE Trans. on Appl. Supercond. vol. 13, pp. 164-167, 2003.

[9] D. E. Prober, "Superconducting terahertz mixer using a transition-edge microbolometer," Appl., Phys., Lett., 62, 2119, 1993.

[10] P. J. Burke, R. J. Schoelkopf, D. E. Prober, "Length scaling of bandwidth and noise in hot-electron superconducting mixers," Appl., Phys., Lett., 68, 3344, 1996.

[11] I. Siddiqi, D. E. Prober, A. Skalare, B. S. Karasik, W. R. McGrath, P. Echternach, H. G. LeDuc, "Noise and conversion efficiency of aluminum superconducting hot-electron bolometer mixer," IEEE Trans. on Appl. Supercond. vol. 11, pp. 958-961, 2001.

[12] A. Kawakami, Y. Irimajiri, T. Yamashita, S. Ochiai, Y. Uzawa, "Broadening the IF band of a THz hot-electron bolometer mixer by using a magnetic thin film," *IEEE Trans. THz Sci. Technol.*, vol. 8, no. 6, pp. 647–653, Nov. 2018.

[13] A. Kawakami, S. Miki, Z. Wang, "Fabrication of epitaxial NbN/MgO/NbN tunnel junctions using an ion-beam sputtered tunnel barrier," Physica C 392-396, pp. 1387-1391, 2003.

# Demonstration of a TACIT Heterodyne Detector at 2.5 THz

Changyun Yoo[1], Mengchen Huang[1], Jonathan Kawamura[2], Ken W. West[3], Boris. S. Karasik[2], Loren N. Pfeiffer[3], and Mark S. Sherwin[1]

The field of THz mixers for astrophysics is dominated by superconducting hot-electron bolometers (HEBs), whereas Schottky-diode mixers have been the only devices suitable for planetary instruments. The Schottky mixers operate at ambient temperature, which is a great advantage for planetary applications, but are much less sensitive than the state-of-the art HEBs and require a $10^3$ higher local oscillator (LO) power. Here, we have demonstrated a novel THz mixer which offers the best of both worlds: it operates at ~ 60 K (accessible by passive cooling on space), requires ~ μW LO power, and has a potential to be as sensitive as the HEB mixer [1].

Our THz device – the Tunable Antenna-Coupled Intersubband Transition (TACIT) mixer – is essentially an HEB based on a 2-dimentional electron gas (2DEG) in an GaAs/AlGaAs quantum well (QW). HEB mixers based on a high-mobility 2DEG have been investigated since 1990 [2]. In two-terminal versions of these 2DEG HEB mixers, the RF and LO electric fields are oriented along the 2DEG plane and the radiation couples to the 2DEG directly, by means of Drude conductivity. Since the kinetic inductance of high-mobility electrons is large, this results in large conversion losses, placing a practical upper frequency limit on such mixers to be 500 GHz [3]. The TACIT mixer functions as a four-terminal HEB (Fig. 1a). Electrons in a 2DEG are efficiently heated by the THz RF and LO electric fields oriented perpendicularly to the plane of the 2DEG to couple resonantly to an intersubband transition, overcoming the conversion losses associated with the two-terminal 2DEG HEB mixers. The gates used to apply the THz field can also be used to electrically tune the intersubband transition frequency and impedance. The intermediate frequency (IF) response is coupled out through source and drain (Fig. 1a).

We have fabricated TACIT mixers using a flip-chip process in which lithography is performed on both sides of a sub-micron thick membrane (Fig. 1b). The wafers from which these mixers were made contain a 40 nm QW, in which the intersubband absorption frequency, in the absence of voltages applied to both gates, is near 2.5 THz. The 2DEG in this wafer has a mobility that exceeds $10^7$ cm$^2$/V-s at 2 K, and is strongly temperature-dependent above that temperature ($1/R*dR/dT \approx 0.02$ K$^{-1}$ at 50-77 K).

Direct detection measurements in response to monochromatic radiation show that the responsivity is sensitive to applied gate voltages at temperatures between 20 and 90 K. At 60 K, the peak responsivity is tunable between at least 2.52 THz and 3.11 THz with gate voltages, consistent with expectations for a 40 nm GaAs QW. Mixing of two monochromatic signals near 2.5 THz was also observed at 60 K (Fig 1c). By tuning the difference between those signals, the IF bandwidth was measured to be 6 GHz. The demonstration of THz mixing in a TACIT device lays a strong foundation for future development of this technology.

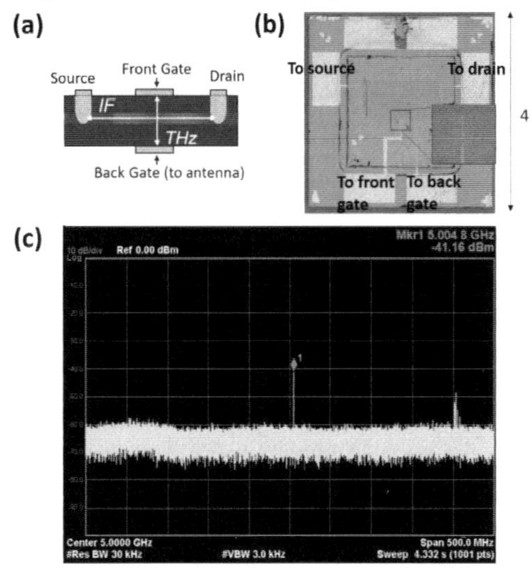


Fig. 1. 2.5 THz TACIT mixer. (a) schematic of four-terminal TACIT mixer, (b) fabricated TACIT mixer (c) IF response at differential frequency of 5 GHz between two sources emitting near 2.5 THz, observed at 60 K.

REFERENCES

[1] M. S. Sherwin et al, "Tunable Antenna Coupled Intersubband Terahertz (TACIT) Mixers: The Quantum Limit Without the Quantum Liquid," *Proc. Far-IR, Sub-mm and mm Detector Technology Workshop*, 2002.

[2] J. -X. Yang et al, "High mobility hot electron mixers for millimeter waves and submillimeter waves," *Proc. SPIE* vol. 1514, p. 15141P, Dec. 1990.

[3] M. Lee et al, "Ballistic cooling in a wideband two-dimensional electron gas bolometric mixer," *Appl. Phys. Letts.*, vol. 81, pp. 1243-1245, Aug. 2002.

[1] Physics Dept. & Inst. for Terahertz Science and Technology, University of California at Santa Barbara, Santa Barbara, CA 93106, USA.

[2] Jet Propulsion Laboratory, California Institute of Technology, Pasadena, CA 91109, USA.

[3] Dept. of Elec. Eng., Princeton University, Princeton, NJ 08544, USA.

NOTES:

# Understanding dissipative behaviour in superconducting microresonators over a wide range of readout power

T. Skyrme[1], S.Withington[1], C.N. Thomas[1], D.J. Goldie[1], A.L. Hornsby[2], S. Doyle[2], P. Barry[3]

High-quality-factor superconducting transmission lines and microresonators can be used as ultra-low-noise detectors for a broad range of astronomical measurements: CMB, galaxy evolution, red shift and line emission spectroscopy. Superconducting devices are ideal for astronomy, as well as quantum computing and Earth observation, for being highly sensitive, low loss and, in the case of Kinetic Inductance Detectors, easily multiplexible [1] [2].

Understanding the mechanisms responsible for loss and decoherence in superconducting microresonators, their physical origins, and their relationship with applied readout power is essential. We describe a unique combination of theory, experimental methodology and data analysis techniques employable for understanding superconducting microresonators when several non-linear dissipative loss mechanisms are present simultaneously. We explore behaviour over a wide range of readout powers, spanning 6 orders of magnitude, in the case where two-level systems and sub-gap quasiparticle heating are present and interact dynamically. Our method attributes quality factors to different loss mechanisms, and considers the steady state values of dielectric energy and quasiparticle population. Many phenomena are seen, which are predicted and verified experimentally: for example, the distortion of resonance curves in the I-Q plane, bistability, and under certain circumstances, the rapid switching on of resonance curves at low readout powers. The measurement of quality factor as a function of readout power, even when the resonance curve is highly distorted, turns out to be a particularly valuable way of uncovering information about the dissipative processes present. We show that the relationship between quality factor and readout power ultimately determines the best operating point of many devices, and warn against the consequences of ignoring non-linear dissipative loss in superconducting resonators used for low-noise and high-quality-factor applications.

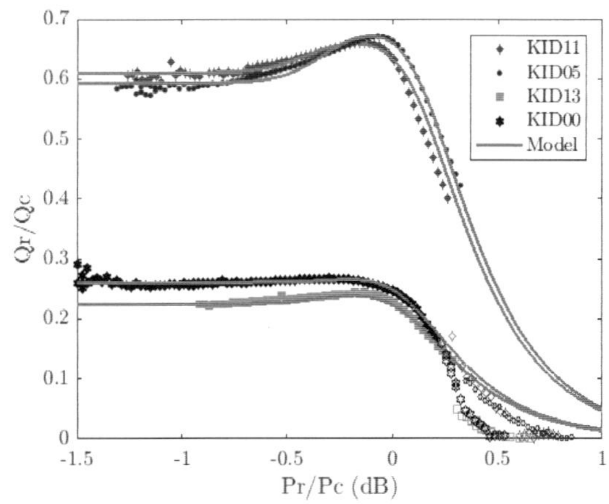


Fig. 1. Our comprehensive model (solid line) simulates the change in quality factor ($Q_r/Q_c$) with readout power ($P_r/P_c$) of a superconducting microresonator subject to several non-linear dissipative loss mechanisms. The model is shown to successfully describe the large signal behaviour of lumped element kinetic inductance detectors of varying designs [1].


REFERENCES

[1] Simon Doyle, Jack Naylon, Philip Mauskopf, Adrian Porch, Stafford Withington, David Goldie, Dorota M. Glowacka, Jochem J. A. Baselmans, Stephen J. C. Yates, Henk Hoevers, "Lumped element kinetic inductance detectors for far-infrared astronomy", Proc. SPIE 7020, Millimeter and Submillimeter Detectors and Instrumentation for Astronomy IV, 70200T (19 July 2008); doi: 10.1117/12.788941;

[2] Peter Hargrave, Stafford Withington, Stefan A. Buehler, Lukas Kluft, Ben Flatman, Prateek Kumar Dongre, "THz spectroscopy of the atmosphere for climatology and meteorology applications", Proc. SPIE 10210, Next-Generation Spectroscopic Technologies X, 1021010 (5 May 2017); doi: 10.1117/12.2263626;



[1] University of Cambridge, Cambridge, England, CB3 0HE

[2] Cardiff University, Cardiff, Wales, CF10 3AT

[3] University of Chicago, 5801 S Ellis Ave, Chicago, IL 60637, USA


NOTES:

# Increased multiplexing of kinetic-inductance detector arrays by post-characterization adaptation of the individual detectors

Shibo Shu[1], M. Calvo[2], J. Goupy[2], S. Leclercq[1], A. Catalano[2,3], A. Bideaud[2], A. Monfardini[2], and E.F.C. Driessen[1]

Kinetic Inductance Detectors (KIDs) have been proven to be an interesting technology for continuum detection from the mm-wave to infrared frequencies. Their intrinsic multiplexibility makes the fabrication of large arrays relatively simple, and a number of instruments have shown high quality performance on telescope, while many more instruments employing this technology are being developed.

A major challenge in fabricating large KID arrays is the frequency scatter of individual detectors, due to fabrication imperfections. This frequency scatter inevitably causes cross talk when two pixels get too close in resonance frequency. This problem can be mitigated at the expense of increasing the available frequency bandwidth per pixel, but this approach significantly limits the possible number of pixels, and is therefore not preferred especially when readout bandwidth is a scarce resource.

In this work, we follow a different approach, inspired by the work of Liu et al. [1]. We demonstrate that it is possible to improve the frequency scatter and readout bandwidth of an existing KID array, by individually adapting the on-chip capacitors of the individual pixels. We show the viability of this approach on a small (112 pixel) prototype array, optimized for detection in the 230 GHz atmospheric window.

After fabrication of the array, we characterize the optical response of all pixels using an optical cryostat and a sky simulator. This allows us to identify each individual pixel with its position on the array and its resonance frequency. As shown in Fig. 1 (top), the resonance frequencies show an irregular frequency comb, with a scatter of a few percent around the design frequency. [2]

We use these characterization results to define a unique adaptation mask. This mask allows to trim the capacitor fingers of each individual pixel, such that after this trimming, the resonance frequencies form a regular frequency comb. The resulting feedline transmission of this array, after adaptation, is shown in Fig. 1 (bottom). It can be clearly seen that not only the necessary readout bandwidth is reduced by ~ 15%, but more importantly, the frequency scatter is reduced by approximately 2 orders of magnitude.

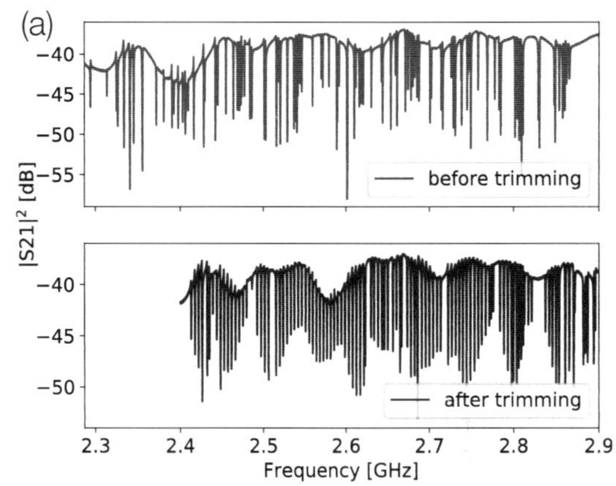


Fig. 1. Feedline transmission of the KID array before (up) and after adaptation of the individual pixels. The necessary readout bandwidth is reduced from 562 MHz to 490 MHz, whereas the frequency scatter with respect to the design value is reduced from ~ 2% to ~ 0.02 %.

In this contribution, we will discuss in detail the causes of the observed frequency scatter, the methodology to improve, and the limitations of our current procedures. Besides that we will focus on the feasibility of this trimming method for larger arrays, such as the NIKA2 1.3 mm arrays, that currently host 1140 pixels, and can be estimated to go up to 2500 pixels per array, using the same readout electronics. We will also discuss possible improvements on the characterization method and the trimming procedure.

### References

[1] X. Liu et al., "Superconducting micro-resonator arrays with ideal frequency spacing and extremely low frequency collision rate", *Appl. Phys. Lett.* **111**, 252601 (2017).

[2] S. Shu et al., "Increased multiplexing of superconducting microresonator arrays by post-characterization adaptation of the on-chip capacitors", *Appl. Phys. Lett.* **113**, 082603 (2018).

[1] Institut de Radioastronomie Millimétrique, Grenoble, France
[2] Institut Néel, CNRS and Université Grenoble-Alpes, Grenoble, France
[3] LPSC, CNRS and Univerité Grenoble-Alpes, Grenoble, France

NOTES:

# Tuesday, April 16, 2019

# Session IX. Future Missions and Projects - II

# Prospects of High Angular Resolution Terahertz Astronomy from Antarctica

Hiroshi Matsuo[1], Sheng-Cai Shi[2], Scott Paine[3], Qi-Jun Yao[2], Zhen-Hui Lin[2]

Based on atmospheric transmission spectra measured using Fourier transform spectrometer from Dome A, Antarctica [1], prospects of terahertz astronomy through atmospheric windows are discussed. Special focus is on high angular resolution observations which are not easy to realize using space telescopes. High altitude sites in Antarctica provide rare opportunities to observe terahertz sources with large THz telescopes and interferometers.

Figure 1 shows an example of the measured atmospheric transmission spectrum from Dome A, which is one of the best transmittances measured at 12-18h UTC on August 9th, 2010 [2]. When compared with the Atacama site at 5000 m altitude, supra-THz windows from 1 to 1.5 THz are transparent more than twice.

Terahertz astronomy from Antarctica with high angular resolution are of interests, especially for atomic emission lines at 1.46 THz from [NII], 3.4 and 5.8 THz from [OIII] and a water ice feature at 7 THz from protoplanetary disks.

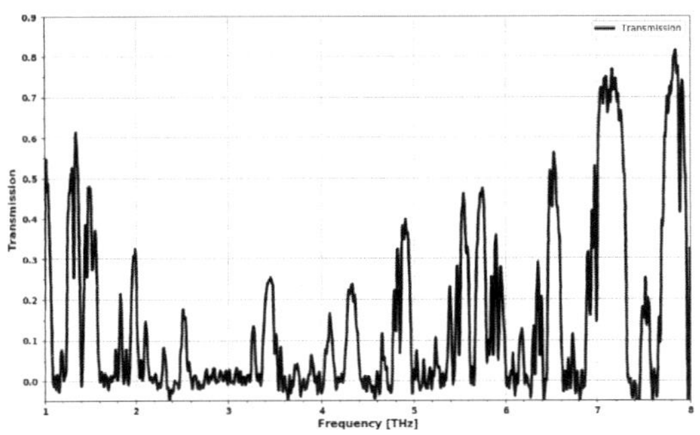


Fig. 1. Atmospheric transmission spectrum measured with Fourier transform spectrometer in Dome A, Antarctica [2].

To achieve high angular resolution, either large single dish telescopes such as DATE5 and THz interferometers will be used. Angular resolution better than 1 arcsecond will resolve many catalogued far-infrared sources; active galaxies, massive star-forming regions, late-type stars and protoplanetary disks. With angular resolution better than 10 milli-arcsecond exceeding ALMA resolution, you can image broad-line region in AGN, late-type stars and inner region of protoplanetary disks.

There are several options for interferometer technologies, such as heterodyne interferometry, Michelson-type beam combining interferometry, Fizeau-type image plane interferometry and Hanbury-Brown and Twiss intensity interferometry. For ease of installation on independent telescopes, heterodyne and intensity interferometry is discussed. The heterodyne interferometry such as ALMA uses SIS mixers in submillimeter frequencies and cross correlation analysis is made on electric fields. The intensity interferometry uses fast detectors, including SIS mixers, and cross correlation analysis is made on intensity (electric field squared). The intensity interferometry can use fast direct detectors such is discussed in Ezawa et al. [3]. The direct detectors do not suffer quantum limited receiver noise, but atmospheric noise dominates for ground-based observations. The difference of heterodyne and intensity interferometry is their dependence on phase fluctuation. The correlation on electric field requires higher phase stability compared with intensity interferometry. The correlation on intensity is stable but requires longer integration for a delay time calibration. Imaging array with direct detector is relatively easy to install. We will be discussing early installation of heterodyne interferometry and later with fast direct detectors for THz interferometry in Antarctica.

### References

[1] S. C. Shi, S. Paine, Q. J. Yao et al., "Terahertz and far-infrared windows opened at Dome A in Antarctica," *Nature Astronomy*, 1, 0001, 2016.

[2] H. Matsuo, S. C. Shi, S. Paine et al., "Terahertz Atmospheric Windows for High Angular Resolution Terahertz Astronomy from Dome A," *Advances in Polar Science*, in press.

[3] H. Ezawa, H. Matsuo, K. Shibasaki et al., "Towards the Intensity Interferometry at Terahertz Wavelengths," *Proc ISSTT-2015*, W2-2, 2015.

[1] National Astronomical Observatory of Japan, Mitaka, Tokyo 181-8588, Japan.

[2] Purple Mountain Observatory, Key Laboratory of Radio Astronomy, Chinese Academy of Sciences, Nanjing 210034, China.

[3] Smithsonian Astrophysical Observatory, Cambridge 02138, Massachusetts, USA.

NOTES:

# New stage of the Suffa Submm Observatory in Uzbekistan Project

Yury N. Artemenko, Yury Yu. Balega, Andrey M. Baryshev, Grigoriy M. Bubnov, *Member,* IEEE, Andrey Etnaer, Nikolay S. Kardashev, Vadim G. Gimmelman, Thijs M. de Graauw, Alexandra Gunbina, *Member*, IEEE, Shuhrat A. Eghamberdiev, Boris L.Kogan, Nikolay N. Kolachevsky, Valery P. Koshelets, Gerbert Lagerweij, Ilya V. Lesnov, Larissa N. Likhacheva, Sergey F. Likhachev, Maria Mansfeld, Gianpietro Marchiori, Francesco Rampini, Gennady I. Shanin1, Yury A. Schekinov, Igor I.Zinchenko and Vyacheslav F. Vdovin, *Member,* IEEE

**Abstract— The International Radio Astronomy Observatory (IRAO) is the project carried out by an international collaboration led by the Astro Space Center of the Lebedev Physical Institute of the Russian Academy of Sciences [1] to provide fundamental and applied astrophysical, geophysical and space research in the centimeter, millimeter and submillimeter wavelengths. The main instrument of the observatory is the radio telescope RT-70 of the centimeter wave band with a mirror of 70 meters in diameter - the only large radio telescope in the Eastern hemisphere located in the center of the Eurasian continent in Uzbekistan already built to at least of 50% readiness. It is able of highly effective long-duration operations due to its location at the altitude of 2,500 m and the unique astroclimate of the Suffa Plateau. Since 2018 the new stage of the project started following the three visits of the RAS President A. Sergeev to Uzbekistan and the Road Map of the Suffa project signed.**

***Index Terms* - radio astronomy, millimeter waves, sub millimeter waves, subTHz observatory, radiotelescope**

"This work have been supported by Academy of Sciences of the Republic Uzbekistan and RAS, projects of the state programs of PI RAS and by Institute of Applied Physics RAS state program 0035-2019-0005", grant RSF-19-19-00499.

Yury N. Artemenko artemenko.akc@yandex.ru , Nikolay S. Kardashev, Nikolay N. Kolachevsky, Sergey F. Likhachev, and Yury A. Schekinov are with Lebedev Physical Institute RAS, Moscow, Russia.

Yury Yu. Balega balega.yuyu@gmail.com is with Russian Academy of Sciences, Moscow Russia.

Andrey M. Baryshev a.m.baryshev@gmail.com is with University of Groningen, the Netherlands.

Grigoriy M. Bubnov, bubnov@ieee.org, Aleksandra Gunbina aleksandragunbina@mail.ru, Ilya V. Lesnov ss18f@mail.ru, Maria Mansfeld murmyu@mail.ru are with Institute of Applied physics RAS and with Nizhny Novgorod State Technical University n.a. R.E. Alekseev, Nizhniy Novgorod, Russia.

Andrey Etnaer aetnaer@vertexant.de and Gerbert Lagerweij are with Vertex Antennentechnik GmbH, Duisburg, Germany.

Thijs M. de Graauw thijs.de.graauw@alma.cl is with Lebedev Physical Institute RAS, Moscow, Russia and with Kapteyn Astronomical Institute, the Netherlands.

Shuhrat A. Eghamberdiev is with Astronomical Institute of Academy of Sci., Repub. Uzbekistan, shuhrat@astrin.uz

Gennady I. Shanin, is with RT-70 observatory, Repub.Uzbekistan, suffa@mail.ru

Boris L.Kogan koganbl@gmail.com is with Bauman Moscow State Technical University, Russia.

Valery P. Koshelets is with Kotelnikov IRE RAS, Moscow, Russia, valery@hitech.cplire.ru

Francesco Rampini and Gianpietro Marchiori gmarchiori@eie.it are with EIE group, Italy,

Igor I. Zinchenko is with Institute of Applied physics RAS, Nizhniy Novgorod, Russia, zin@appl.sci-nnov.ru

Vyacheslav F. Vdovin is with Lebedev Physical Institute RAS, Moscow, Russia, IAP RAS and Nizhny Novgorod State Technical University n.a. R.E. Alekseev, Nizhniy Novgorod, Russia.vdovin_iap@mail.ru

## I. Introduction

A new stage of a rather old project of the SubMM observatory has been presented (Fig. 1). It will be an extremely important tool in ground-based and terrestrial VLBI observations with the prospect of adding mm wavelengths [2]. In these studies, the RT-70 will provide the highest sensitivity and angular resolution as concerns continuous spectrum observations, spectral lines observations, polarization measurements and the study of rapidly varying processes occurring in the Universe. Instrumentation of the Suffa Observatory will be extended by relatively small (10-15 m of diameter) telescope working in Submm waves and installed near the main mirror no later then 2024. Radioastronomical observations in atmospheric windows around 1.3 and 0.8 mm will become possible. First results of atmospheric opacity measurements provided some data presented here which give some grounds to moderate optimism. Both the telescopes will be equipped by extremely high sensitive cryogenically cooled receivers [3].

Fig.1. Suffa Project 1980-th

## II. Historical And Geographical Review

The Suffa project (Fig.1) started in the early 80's as the formal opening ceremony of construction took place on the Suffa plateau with the leader of the Uzbek Republic Sh. Rashidov and Vice-President of the Academy V. Kotelnikov present (Fig.2). In 10 years when more than a half of the project had been fulfilled following the disintegration of the USSR the project was frozen.

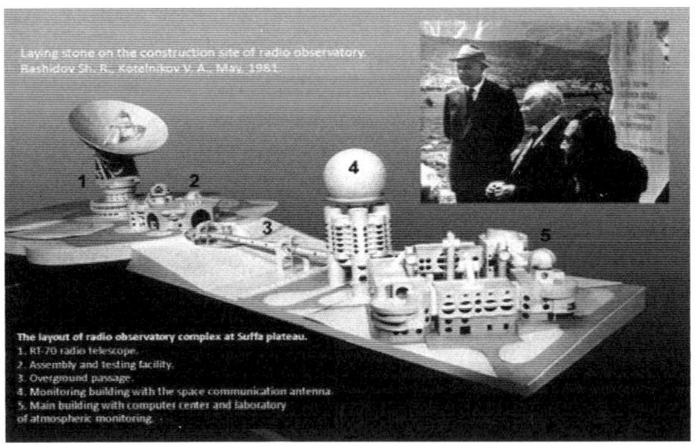


Fig.2. Suffa Project in 1981: formal opening.

In 2005 Uzbek and Russian authorities decided that the project should continue and be finalized [2]. Since the appearance of the first publications concerning the Suffa observatory, 10 years have passed without any visible progress. Only in 2018 thanks to the initiative of the Russian Academy of Sciences and the National Academy of the Republic of Uzbekistan significant progress was made in promoting the project, including the promise of some substantial financing starting from 2019. The current status of the project and results of the project development in 2018 as well as some plans for 2019/20 are being presented in this paper.

### *A. Suffa plateau, pic.1:*

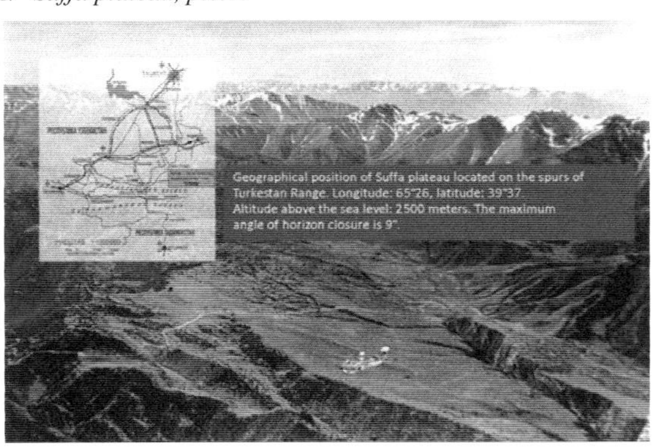


*Fig.3 Location of Suffa*

Geographic coordinates: 39°37'28.46“N 68°26'54.16“E;
Measurements period: permanently since 2012;
Altitude above sea level: 2400m
Climate zone: Dry sharply continental

### *B. Nearest mountains for small mirror to be installed*

Altitude above sea level: 3200m

### *C. Maidanak is the place of the Uzbek national optical observatory as the alternative place for the small SubTHz mirror Main Performances*

Main specifications of the projected 70 m telescope are listed below:

Main mirror diameter, m 70
Surface geometrical shape Paraboloid of rotation
Focal length, m. 21
Opening angle, deg. 160
Optical scheme Gregory two mirror system with periscopic mirror
Wavelength range, λ S - millimeter (λ= 0.87 - 10 mm)
M- centimeter (λ = 1 - 6cm)
Secondary mirror diameter, m 3 (5)

TABLE I
WORKING BANDS OF SUFFA RECEIVERS

| Operating wavelength ranges (In priority: II, III, IV) | I | II | III | IV | V | VI | VII |
|---|---|---|---|---|---|---|---|
| Average wavelength, (mm) | 0.87 | 1.2 | 1.9 | 3.3 | 7.5 | 13 | 61 |
| Frequency range, (GHz) | 275-373 | 211-275 | 125-211 | 67-116 | 26.5-50 | 18-26.6 | 4-8 |
| Beam, 1.02(λ/D) (")at level of 0.5 | 2.6 | 3.5 | 5.6 | 9.7 | 22 | 38 | 180 |
| Antenna effective area (m^2) | 1350 | 1350 | 2000 | 2000 | 2700 | 2700 | 2700 |
| Antenna aperture efficiency | 0.3 | 0.4 | 0.5 | 0.6 | 0.7 | 0.7 | 0.7 |
| Sensitivity (RMS) in μJy (integration time 1 minute /8 hours) | 480/20 | 220/10 | 140/7 | 100/5 | 170/8 | 560/30 | 800/40 |
| System noise temperature (K°) (with 3 mm of precipitated water) | 380 | 230 | 190 | 160 | 110 | 80 | 60 |
| Receiver noise temperature (K°) | 100 | 100 | 100 | 100 | 50 | 30 | 10 |
| Maximum receiver bandwidth (GHz) | 30 | 30 | 30 | 30 | 10 | 2 | 0.1 |

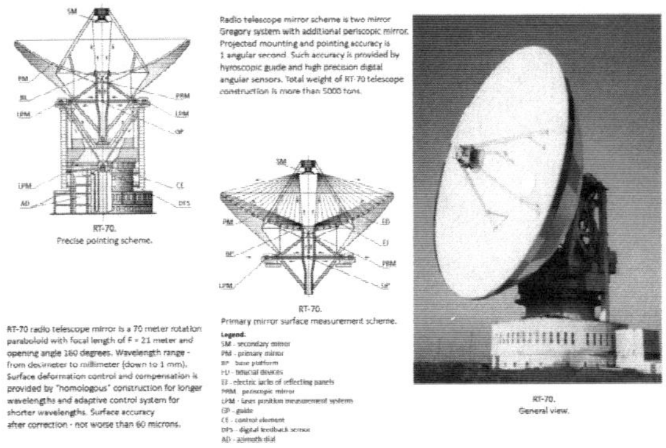


Fig. 4. Suffa RT-70 telescope

Geometric shape of the mirror surface: Ellipsoid of rotation

Inter focal distance of ellipsoid, m 242

Size of diagonal periscopic mirror, mm 600 mm

Equivalent focus of Gregory system, m 571 (345)

The method of the compensation of the main mirror weight deformation is a combination the homologous and active forms of adjustment (so-called "adaptive mirror").

The standard deviation of the main mirror paraboloid from approximated shape, mm 0.062

Operating panel profile errors (RMS), mkm 50

Antenna installation type Full-circle

Base type Tower

Telescope mounting Azimuth, with mutually perpendicular non-intersecting axes,

Provided mounting and pointing accuracy at wavelength ranges: M, S, angular sec. <2– by detector «angle-code», 22 rank, <1 - by additional system using guide.

Elevation, m: 2,324.

Antenna installation type Full-circle.

The reflective surface of the radio telescope is formed by panels of trapezoidal shape with the maximum size of 2.5 by 2 meters; the total number of panels of 14 different sizes is 1,188. The panels have a special design that allows using pre-alignment of the reflective surface with on average 50 points, RMS better than 50 microns. To provide the operation of the radio telescope at short millimeter wavelength ranges, the shape of the reflecting surface (parabola) must be preserved during the observation under the action of gravity, wind and heat, with an accuracy of $\lambda$ /D = 20, i.e., 50-70 microns. To ensure this, each panel is installed on special electrical jacks in its corners, which are mounted on the frame truss of the telescope. The total number of electrical jacks is 1,440. During the observation, the control system tracks the position of each panel and, if necessary, adjusts them by their relative position to create the optimal shape of the radio telescope reflecting surface (adaptive method). It is supposed that in the wavelength range of 6 cm to 8 mm the shape of the reflecting surface will be maintained through the use of the homologous method, and at shorter wavelengths by the adaptive method.

The secondary mirror with the diameter of 3 or 5 meters has the shape of an ellipsoid of rotation with five degrees of freedom of movement in space. Depending on the diameter it is either a one-piece mirror (3 meters) or collected of individual panels on the frame (5 meters). The mirror position is controlled by the computation and control complex.

The periscope (diagonal) mirror has a flat elliptical shape. The major axis is 600 mm. The mirror has four degrees of freedom. Subsequently, it can be made adaptive. The telescope control system consists of the electric drive of the traditional for radio astronomy antennas scheme, and the precision-pointing contour, providing operation at the millimeter wavelength range. The electric drive provides pointing at the source and its tracking with required accuracy and speed. It uses a computation and control complex and digital 22-bit (0."3) feedback sensors, installed on the elevation and azimuth axes, and provides pointing at the source within the RMS of 1 arc seconds with allowance for the errors in the drive mechanisms etc. At 1 mm wavelength, the calculated radiation pattern of the radio telescope is 3 arc seconds. There is a special high-precision pointing system of the electrical axis of the antenna aimed at the source under observation which has a range of angles of ± 10" from the current direction determined by the feedback sensors of the electric drive, and provides pointing accuracy of at least 0.3 arc seconds.

Table 2 shows the operating frequency ranges of the radio telescope. Radiometers will be placed at the primary and secondary focus. In the primary focus the change of radiometers will be done using a fixed service tower, in the secondary 7 radiometers with the fixed mounting and beam switching is done by periscopic mirror. The set of radiometers and its radiophysical characteristics are determined by specific scientific tasks of observations cycle. The preferences are the short-wave part of the millimeter range, the search of weak sources and deep surveys in the continuous spectrum, polarimetry of cosmological background (CMB), molecular radiospectroscopy, and rapidly changing processes. Within this range bolometers are mainly used (or bolometer arrays) which are cooled down to 4 - 0.3 K. The super heterodyne receivers are also cooled down to 4 - 20 K.

The Gregory telescope optical system has a field of view in the secondary focus of 15'–40', depending on the diameter of the secondary mirror (5 or 3 m). With the radiation pattern of the 3", 1,000-element (bolometers or mixer receivers) or more cooled matrix can be placed in the focus.

## III. ASTROCLIMAT

Since 1981, we have carried out astroclimate monitoring as presented in pic.66. We can see that observations are possible even in atmospheric windows 1.3 and 0.8. However, for 0.8 mm there are only some dozens of days from December till March when the zenith absorption is less than 50%. See table 2.

At present, some results of direct SubTHz measurements ($\lambda$ = 2 & 3 mm) made over past 6 years added to our optimism. These are presented here by G.Bubnov at al [4].

Definitely, direct 1.3 and 0.8 mm (similar to presented in [4] 2 and 3 mm measurements) measurements should be fulfilled before the adjustment of 70-m mirror is started for operation in these wavebands.

TABLE II
ZENITH ATMOSPHERE TRANSPARENCY (DAYS)

| % absorbtion in zenit | Wavelength, (mm) | | | | |
|---|---|---|---|---|---|
| | 0.88 | 1.36 | 2.2 | 3.15 | 8 |
| <10 | - | - | 9/18 Dec.-Feb. | 46/68 all year | 46/68 all year |
| <25 | - | 9/18 Dec.-Feb. | 46/68 all year | | |
| <50 | 2/4 Dec.-Feb. | 34/57 Sept.-May | | | |

## IV. Fundamental Questions To Be Answered

Despite the optimistic new status of the Project, revival of the Observatory construction in the previous year and the results of astroclimate research presented above, there are still many problems and fundamental question (FQ) to be solved as soon as possible.

*A. FQ1: Will the RT-70 operating at wavelengths down to 0.8 mm (RT-70-mm) be interesting from the scientific point of view?*

The answer that could be heard from astronomers at the 1st International workshop "Present Status and Future Prospects of the Radioobservatory at Plateau Suffa in Uzbekistan" in Tashkent, Uzbekistan, August 27-29, 2018 was definitely YES [5].

*B. FQ2 Does the Suffa plateau enjoy the conditions for transmission good enough for 0.8 mm observations, "good enough" being the key word?*

The current answer is: This is not quite clear at the moment. The answer will be extremely important and will have huge consequences. There are rumors that the Maidanak site (also in Uzbekistan, 250 km S-W direction) is much better. It should be (and will be) proved by direct measurements soon.

Action required: Carrying out of a careful analysis of available meteo- and direct transmission of atmosphere data. Making comparison between the Suffa and Maidanak sites. Do we have enough data on water vapor for the final decision? If not, let's get more (through measuring, meteo sats; etc..)

*C. FQ3. Are we sure that a completed RT-70-mm structure design can be made to operate well at 0.8 mm?*

Answer: Probably yes. Sooner or later this instrument will work at 0.8 mm, but definitely the Suffa observatory should be extended by addition of a 12-15 m SubMM to the main 70-meter telescope. Following the ALMA path (Pathfinder telescope) making a small mirror commercially available a smaller telescope will be built much sooner than the 70-m one and will provide short mm radiostronomical observations rather soon. Now EIE group (Italy) and Vertex (Germany) are actively involved into preparation of this part of the project and there are no doubts that such kind of pathfinder will be in an operational condition before the 2024.

*D. FQ4. Is the present RT-70 Suffa structure at the Suffa plateau still usable and is completion of the antenna is worthwhile to be carried out?*

Answer: yes. Start-up telescope: a 12-15-meter 0.8 mm antenna with receivers and equipment for bands ALMA 2-3, 6, 7 equipment for VLBI observations with Vertex (Germany) or another winner of a further tender.

Main requirements for the tender are: Establishing an error budget for a 0.8 mm telescope wavefront performance: Overall 60 microns (1/13 lambda),

panels 30-micron rms, (secondary 15microns),
structure (including secondary support) 40 microns,
wind loads 30 microns.

The current RT-70 status of the antenna design for a 70-meter diameter antenna, "existing" design of two similar telescopes already built (Evpatoria, Ussuriysk), and lots of critical details mainly connected with the adaptive surface should be considered to provide transition from RT-70 as the CM telescope to RT-70 as the MM (with SubMM) telescope.

The new concept of instrumentation should be developed on the ALMA++ ideology based on wide international collaboration. It is part of the answer to FQ5.

*E. FQ5: Are the Suffa Observatory team ready to undertake a mm astronomy project of world-class size and scale? Do they have the expertise?*

The current answer: Not yet, but there are special items in the Road Map aimed at this specific problem. The new extended team should be collected around the project, and new staff trained.

## V. Conclusion

We must say that the Suffa project has been restarted now and the new stage of it is in progress. There are no doubts that it will be fulfilled thanks to worldwide cooperation. There are still lots of problems and fundamental questions to be solved and results will depend on solutions and answers to them.

## Acknowledgment

Authors are very obliged to the team of the RT-70 observatory for their support of the project over 25 years, and to presidents of two national Academies, Uzbek and Russian, B. Yuldashev and A. Sergeev, for their initiative in revival of the project.

## References

[1] http://asc-lebedev.ru/index2.php?engdep=16&engsuffa=1

[2] A.Hojaev , G.I., Shanin and Y.Artyomenko Suffa Radio Observatory in Uzbekistan: progress and radio-seeing research plans. Proceedings of the International Astronomical Union, vol.2, Aug 2006, pp. 177-182 , https://doi.org/10.1017/S1743921307006965

[3] V. Vdovin, I. Zinchenko, Yu. Artemenko, V. Dubrovich, A. Baryshev New Challenge for 0.1 - 0.3 THz Technology: Development of Apparatus for Radio Telescope RT-70. 19th International Symposium on Space Terahertz Technology, Groningen, 28-30 April 2008

[4] G.Bubnov, V. Vdovin, V. Grigorev, P. Zemlyanukha, I. Zinchenko, "Astroclimatic studies of the sites for forthcoming radio astronomical observatories" To be published in this issue.

[5] 1 st International workshop "Present Status and Future Prospects of the Radioobservatory at Plateau Suffa in Uzbekistan" Tashkent, Uzbekistan, August 27-29, 2018 https://sites.google.com/view/1-st-international-workshop-su

**Yury Yu. Balega** was born on January 8, 1953, in the Kolchyno village of the Mukachevo district, Transcarpathian region. Since 1975, he has held a research position in the Special Astrophysical Observatory of the Russian Academy of Sciences. He received the Ph.D. degree in 1985 and the Doctor of Science degree in 1996. In 1993, he was elected Director of the Observatory. In 1997 he was elected as the Corresponding Member of the Russian Academy of Sciences in the Department of General Physics and Astronomy. He holds a chair of Professor in several universities. His field of scientific interests includes registration and processing of astronomical images in the visible and infrared, interferometry methods in astronomy, study of the fundamental properties, origin and evolution of multiple stellar systems, and study of the gas/dust shells around stars at late evolution stages.

In 1991, Dr. Balega received the USSR State Award in Science and Technology, and in 2003, the State Award of Ukraine.

**Andrey M. Baryshev** received the M.S. degree (summa cum laude) in physical quantum electronics from the Moscow Physical Technical institute, Moscow, Russia, in 1993, and the Ph.D. degree in superconducting integrated receiver combining SIS mixer and flux flow oscillator into one chip from the Technical University of Delft, Delft, The Netherlands, in 2005.,He is currently an Associate Professor with the Kapteyn Astronomical Institute, University of Groningen, Groningen, The Netherlands. He was previously a Senior Instrument Scientist with SRON—Low Energy Astrophysics Division, Groningen, The Netherlands, from 1998 to 2017. In 1993, he was an Instrument Scientist with the Institute of Radio Engineering and Electronics, Moscow, Russia, involved in the field of sensitive superconducting heterodyne detectors. In 2000, he joined an effort to develop an SIS receiver (600–720 GHz) for the Atacama Large Millimeter Array, where he designed the SIS mixer, quasi-optical system, and contributed to a system design. His current main research interests include application heterodyne and direct detectors for large focal plane arrays in terahertz frequencies and quasi-optical systems design and experimental verification.,Dr. Baryshev was the recipient of the NOW-VENI Grant for his research on heterodyne focal plan arrays technology in 2008 and, in 2009, he was the recipient of the EU commission Starting Researcher Grant for his research on focal plane arrays of direct detectors. *(Based on document published on 19 September 2018).*

**Grigoriy M. Bubnov** (IEEE Member' 2016) was born 29.10.1989 in Gorky (now Nizhny Novgorod), Russia. Graduated from Nizhny Novgorod State Technical University (NNSTU) in 2013 with master-engineer's degree of Radiotechnics. The theme of master's thesis is "MM-wave radiometer for atmospheric absorption measurements". Finished the postgraduate school of NNSTU in 2017. The field of scientific interests includes methods and techniques of millimeter-submillimeter wave spectroscopy and atmospheric radiometry, laboratory and field investigations of atmospheric absorption. Took part and organized a several expeditions of astroclimate research. Project leader of two grants on astroclimate research. Since 2011 to the present time works at the Institute of Applied Physics RAS. Current position is Junior Researcher.

**Nikolay S. Kardashev** graduated from Moscow State University in 1955, following up at Sternberg Astronomical Institute. He studied under Shklovskii and finished his PhD in 1962. In 1963 Kardashev examined quasar CTA-102, the first Soviet effort in the search for extraterrestrial intelligence (SETI). In this work he came up with the idea that some galactic civilizations would be perhaps millions or billions of years ahead of us, and created the Kardashev classification scheme to rank such civilizations. Kardashev defined three levels of civilizations, based on energy consumption: Type I with "technological level close to the level presently attained on earth, with energy consumption at ≈4×1019 erg/sec (4 × 1012 watts)". Type II, "a civilization capable of harnessing the energy radiated by its own star", and Type III, "a civilization in possession of energy on the scale of its own galaxy".[2] Serious Russian efforts in SETI predate similar programs in the US by some years. Other notable experts in the USSR were Vsevolod Troitskii and Iosif Samuilovich Shklovskii (Kardashev's former professor). Kardashev became a corresponding (associate) member of the USSR Academy of Sciences, Division of General Physics and Astronomy on December 12, 1976. He became a full member of the Russian Academy of Sciences on March 21, 1994 and was awarded the Demidov Prize in 2014.

**Thijs M. de Graauw** studied astronomy at Utrecht University and received there his Ph.D. in 1975 under H. van Buren with a dissertation on heterodyne instrumentation applied to infrared observations. From 1975 to 1983 he worked as a scientist for the Space Science Department of ESA (European Space Agency). At ESA's largest facility, ESTEC in Noordwijk, he worked on the development of microwave receivers. In 1983 he became the director of the Groningen branch of SRON (Stichting Ruimte Onderzoek Nederland).[3] From 2008 to 2013 he was the director of the Atacama Large Millimeter/submillimeter Array (ALMA). In 2012 he won the Joseph Weber Award for his work on the short wavelength spectrometer on the Infrared Space Observatory[3] and also for his work on the HIFI camera, which was launched on board Herschel, ESA's infrared space observatory.

**Shuhrat A. Eghamberdiev** Shukhrat Abdumannapovich Egamberdiev was born in Tashkent, on December 18, 1953, and graduated from the Astronomy Department of the Physics Faculty of Moscow State University (MSU) in 1977. Five years later, in 1982, he successfully defended his thesis "The Study of Solar X-ray Bright Points" for the degree of Candidate of Physical and Mathematical Sciences (Ph.D. equivalent) at Moscow State University. In 1994, Shukhrat Egamberdiev at Moscow State University successfully defended his thesis for the degree of Doctor of Physical and Mathematical Sciences. In 2000, Professor Egamberdiev was elected Chief Scientific Secretary of the Academy of Sciences of Uzbekistan. In 2003-2005, he was appointed Deputy Adviser to the President of Uzbekistan for Science, Education and Health.

**Valery P. Koshelets** received the M.S. degree in physics from Lomonosov Moscow State University, Moscow, Russia, in 1973, and the Ph.D. degree in radio physics and D.Sc. (Habilitation) degree in physical electronics from the Kotel'nikov Institute of Radio Engineering and Electronics, Russian Academy of Sciences, Moscow, in 1978 and 1990, respectively.,Since 1973, he has been with the Kotel'nikov Institute of Radio Engineering and Electronics, Russian Academy of Sciences, Moscow, Russia, where he is currently the head of the Laboratory of superconducting devices for signal detection and processing. (Based on document published on 24 December 2015).
Gerbert Lagerweij

**Ilya V. Lesnov** received the M.S. and Ph.D. degrees from Nizhny Novgorod State Technical University n.a. R.E. Alekseev in 2009 and 2017 respectively. Since 2012, he has been with the Institute of Applied Physics of the Russian Academy of Sciences in positions from Junior Research Fellow up to Researcher. His research interests include cryogenics, THz communication systems and experiment automation.

**Yury A. Schekinov** received a higher education at the Department of Space Physics, Rostov State University. The topic of his thesis is "Towards a theory of the gravitational stability of the isotropic world". Yuri Shchekinov graduated in 1972. Among the main areas of scientific activity are cosmology, interstellar physics, dusty plasma physics, and physics of protoplanetary disks. Author of textbooks "Numerical Methods in Astrophysics" and "Physics of the Interstellar Medium." Head of the Department of Space Physics of the Southern Federal University. He defended his doctoral dissertation on September 16, 1992. Consists in the Council on Astronomy of the Russian Academy of Sciences.

**Igor I.Zinchenko** was born in Gorky (now Nizhny Novgorod), Russia, in 1950. He received an equivalent of the M.S. degree in radio physics from the Lobachevsky University of Nizhny Novgorod in 1972, the Candidate of Science (Ph.D.) degree in astrophysics and radio astronomy from the Radiophysical Research Institute (Nizhny Novgorod) in 1982, and the Doctor of Science degree in the same field in 1998.,From 1972 to 1977, he was a junior researcher at the Radiophysical Research Institute. Since 1987, he has been Head of Division of Millimeter Wave Astronomy and Radio Engineering at the Institute of Applied Physics of the Russian Academy of Sciences. Since 2002, he has also been a Professor at the Lobachevsky University of Nizhny Novgorod. He is an author of more than 100 articles. His research interests include millimeter and submillimeter wave radio astronomy, physics of the interstellar medium, star formation, and radio waves propagation in atmosphere.,Prof. Zinchenko is a member of the International Astronomical Union (IAU), European Astronomical Society (EAS), Euro-Asian Astronomical Society (EAAS), and International Union of Radio Science (URSI). (Based on document published on 14 January 2015).

**Vyacheslav F. Vdovin** Professor, Doctor of Sciences. Graduated Radio engineering faculty of Gorki Polytechnical Institute (nowadays - the Nizhniy Novgorod State Technical University - NNSTU) in 1978, radio engineering. 1993 – Ph.D. in Radio physics, 2006 – Doctor of Sci. on Devices and methods of experimental physics. Scope of professional interests: Radio physics, a radio engineering of low temperatures, development of high-sensitivity cryogenically cooled receivers for millimeter and submillimeter frequency ranges for spectral astronomical researches, and purposes of atmospheric spectroscopy, medical and biologic researches and other applications. Since 1978 on the present with Institute of applied physics: Since 2008 on the present – collaboration with Astrospace Centre of FIAN. Since 2010 professor of Nizhniy Novgorod State Technical University, head of the Center of Cryogenic Nanoelectronics of NNSTU. Vice-president of the World federation of science workers (WFSW), Member of the Cryogenic Society of America (CSA).

# Wednesday, April 17, 2019

# Session X. Future Missions and Projects - III

# COMETS – Comets Observation & Mapping Enhanced THz Spectrometer at 210-580 GHz: Objectives and Development Status

Jose V. Siles[1], Jonathan H. Kawamura[1], Maria Alonso-del-Pino[1], Goutam Chattopadhyay[1], Choonsup Lee[1], Robert H. Lin[1] and Mathieu Choukroun[1].

Comets are one of the most intriguing and fascinating objects of the Solar System not only because of their potential for providing information on conditions in the pre-Solar and Solar Nebula, but also for their possible connections with the origin of life on Earth. They are indeed among the most primitive least processed bodies, and hence they may provide reliable information on the composition and thermo-dynamical conditions in the Solar Nebula. Understanding the nature of the cometary nucleus, outgassing from the nucleus, and development of the comet coma is essential to understand the evolution of comets over time, and how primordial the materials they expose are. The Microwave Instrument on the Rosetta Orbiter (MIRO), through its high spatial and high spectral resolution observations has been able to identify morphological features as small as 5 m on the comet nucleus and correlate them with regions of outgassing. Outgassing rates from the nucleus for a number of molecules including water could be measured directly with MIRO [1].

However MIRO had a very important limitation: A single-pixel receiver topology which made it impossible to map the comet before it significantly rotates. In fact, retrieving thermal properties of the nucleus requires maximal coverage of the nucleus as function of location and time of day. In addition, the cometary coma are extremely asymmetric, especially close to the nucleus, making the retrieval of coma properties strongly model-dependent and computationally intensive, particularly when each point is acquired at different times (and thus rotational phase). A "straight forward" solution to these limitations is to use multi-pixel systems for instantaneous and efficient mapping of the cometary comas. But for planetary science missions, where the power budget and compactness of the instrument is crucial to be considered a part of the baseline payload, this mapping capability was not feasible until now.

As a response to this problem, we are developing COMETS (Comets Observation & Mapping Enhanced THz Spectrometer), the first multi-pixel broadband all-solid-state submillimeter-wave heterodyne spectrometer & radiometer to enable instantaneous mapping of cometary comas and surfaces with very high-spectral resolution ($\lambda/\Delta\lambda \sim 10^7$). COMETS will feature a 16-pixel dual-band 210-245 GHz/500-560 GHz array receiver integrated on a single front-end channel for continuum & spectroscopic measurements, including instantaneous mapping of water, H/D ratios, and other key molecular species (see Fig. 1).

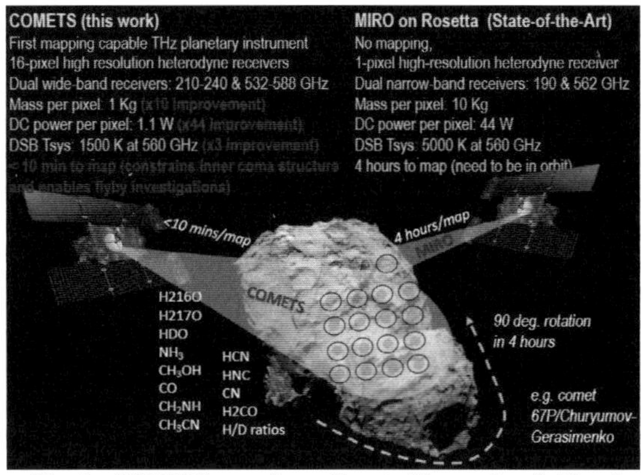


Fig. 1. COMETS overview - Comets Observation & Mapping Enhanced THz Spectrometer at 210-580 GHz (16-pixel array).

The receiver front-end consist of a combination of fundamental/subharmonic mixers and Schottky diode based frequency multiplied local oscillators [2], introducing a new design concept that increases the bandwidth of traditional frequency multipliers by a factor of two, allowing to combine the two-bands on a single channel. This simplifies considerably the instrument optics. A novel dual-band leaky-wave micro-lens array antenna [3] has been designed to provide the same beam width for both bands so that the high and low frequency pixels are collocated in the field of view, allowing identical foot-prints in the comet. Based on tests performed on a 2-pixel proof-of-concept version, the current-base estimate (CBE) of the dc power consumption of COMETS is ~54 W for the 16-pixel array. This is already lower than MIRO's numbers. An overview of COMETS science, overall instrument architecture design and preliminary receiver results will be discussed at the conference.

## References

[1] Gulkis, S., Frerking, M., Crovisier, *J. et al. Space Sci Rev (2007) 128: 561. https://doi.org/10.1007/s11214-006-9032-y.*

[2] J. V. Siles, et al., "A New Generation of Room Temperature Frequency Multiplied Sources With up to 10× Higher Output Power in the 160 GHz – 1.6 THz Range," *IEEE Trans. on THz Science and Technology*, Vol. 8, No. 6, pp. 596-604, November 2018.

[3] S. Bosma, M. Alonso-del-Pino, et al., "Dual-Band Leaky-Wave Lens Antenna for Submillimeter-Wave Heterodyne Instruments", *to appear on 2019 EuCAP Proc.* 2019.

[1] Jet Propulsion Laboratory, California Institute of Technology, Pasadena, CA 91109.

*The research was carried out at the Jet Propulsion Laboratory, California Institute of Technology, under a contract with the National Aeronautics and Space Administration. NASA's PICASSO program acknowledged.* 

NOTES:

# The Origins Space Telescope and the HEterodyne Receiver for Origins (HERO)

M. C. Wiedner, A. Baryshev, V. Belitsky, V. Desmaris, A. DiGiorgio, J,-D. Gallego, M. Gerin, P. Goldsmith, F. Helmich, W. Jellema, A. Laurens, I. Mehdi, C. Risacher, HERO technical team, HERO science team, the heterodyne receiver roadmap team, A. Cooray, M. Meixner and the Origins Study Team

***Abstract*—The Origins Space Telescope is one of four large mission concept studies carried out by NASA for the 2020 Decadal survey. Origins is a far-infrared telescope designed to understand the evolution of galaxies and black holes, to follow the trail of water from protostars to habitable planets and to search for biosignatures in the atmospheres of exoplanets. The Heterodyne Receiver for Origins (HERO) is the high spectral resolution receiver. It is the first heterodyne array receiver designed to fly on a satellite and an example for possible future focal plane arrays for space. HERO has focal plane arrays with nine pixels in two polarization. HERO covers a large frequency range between 486 and 2700 GHz in only 4 frequency bands, requiring local oscillators with fractional bandwidth of 45%. HERO uses the best superconducting mixers with noise temperatures between 1 and 3 hf/k and an intermediate bandwidth of 6 to 8 GHz. HERO can carry out dual polarization and dual-frequency observations. The major challenges for the HERO design are the low cooling power and the low electrical power available on a spacecraft, which impact the choice of the cryogenic amplifiers and backends. SiGe cryogenic amplifiers with a consumption of less than 0.5 mW, as well as CMOS spectrometers with a power consumption below 2W are the baseline for HERO. The development plan includes broadband (45%) multiplier-amplifier chains, low noise mixers (1-3 hf/k), low-power consuming (< 05.mW) cryogenic amplifiers and low-power consuming spectrometer backends (< 2W).**

***Index Terms* — Astronomy, array receiver, terahertz, submillimeter, space technology**

Manuscript submitted June 25, 2019. The project leading to this

M. C. Wiedner and M. Gerin are with Sorbonne Université, Observatoire de Paris, Université PSL, CNRS, 75014 Paris, France (e-mail: martina.wiedner@obspm.fr and maryvonne.gerin@obspm.fr).

A. Baryshev is with Kapteyn Astronomical Institute, University of Groningen, Postbus 800, 9700 AV, Groningen (e-mail: andrey@astro.rug.nl).
A. Baryshev is supported by the European Commission Seventh Framework Programme (FP/2007-2013) under grant agreement No 283393 (RadioNet3).

V. Belitsky and V. Desmaris are with the Group for Advanced Receiver Development (GARD), Department of Space, Earth and Environment, Chalmers University of Technology, 41296, Gothenburg, Sweden. (e-mail: victor.belitsky@chalmers.se and vincent.desmaris@chalmers.se).

A. Giorgio is with the Istituto Nazionale di Astrofisica-Istituto di Astrofisica e Planetologia Spaziali, Via Fosso del Cavaliere 100, I-00133 Roma, Italy (e-mail: anna.digiorgio@iaps.inaf.it).

J. D. Gallego is with the Centro Astronómico de Yebes, Observatorio Astronómico Nacional, Apdo. 148, 19080 Guadalajara, Spain (e-mail: jd.gallego@oan.es).

P. Goldsmith and I. Mehdi are with the Jet Propulsion Laboratory, which is operated for NASA by the California Institute of Technology (e-mail: paul.f.goldsmith@jpl.nasa.gov and imran.mehdi@jpl.nasa.gov).

F. Helmich and W. Jellema are with the SRON Netherlands Institute for Space Research, Groningen, The Netherlands and Kapteyn Astronomical Institute, University of Groningen, The Netherlands (e-mail: f.p.helmich@sron.nl and W.Jellema@sron.nl).

A. Laurens is with the Centre National d'Études Spatiales, 18 Avenue Edouard Belin, 31400 Toulouse, France (e-mail: Andre.Laurens@cnes.fr).

C. Risacher was with the Max-Planck-Institut für Radioastronomie, Auf dem Hügel 69, 53121, Bonn, Germany and is now at IRAM, 300 rue de la Piscine, 38406, St. Martin d'Hères, France (e-mail: risacher@iram.fr).

A. Coooray is at the Department of Physics & Astronomy, University of California, Irvine, 92697, USA (e-mail: acooray@uci.edu).

M. Meixner is Space Telescope Science Institute, 3700 San Martin Drive, Baltimore, MD 21218, USA, Department of Physics and Astronomy, the Johns Hopkins University, 366 Bloomberg Center, 3400 North Charles Street, Baltimore, MD 21218, USA and currently on sabbatical at NASA Goddard Space Flight Ctr., United States (e-mail: meixner@stsci.edu).

## I. ORIGINS SPACE TELESCOPE

THE Origins Space Telescope [1][2] is one of four large mission studies NASA has carried out for submission to the 2020 Decadal Survey. Origins addresses three large questions: How does the universe work? How did we get here? and Are we alone? To answer these questions Origins observes the evolution of galaxies, the formation of dust and the feedback mechanisms of galaxies over cosmic time. Origins follows the trail of water from protostars, via planetary disks to debris disks, weighs disk masses and measures the D/H ratio of comets in order to understand how water, a prerequisite for life, arrives on planets. Last but not least Origins searches for markers of life by looking for biosignatures in temperate exoplanets with transparent atmospheres. All these observations require a very sensitive mid to far-infrared telescope in space. Origns has a 5.9 m antenna cooled to 4.5 K and has 3 principal instruments as well as 2 upscope instrument. The 3 principal instruments are 1) the Origins Spectral Surveyor (OSS) [3] that covers a wavelength range of 25 to 588 μm at resolving power of 300, 43000 or 325000; 2) the large field Far-infrared Imaging Polarimeter (FIP) [4] with a 50 and a 250 μm channel allowing polarimetry and 3) the Mid-Infrared Spectrometer Camera Transit (MISC-T) [5], an ultra-stable transit spectrometer for 2.8 to 20μm. The upscope options include the Heterodyne Receiver for Origins (HERO) described below and the Mid-Infrared Spectrometer Camera Imager (MISC-I)

[5] allowing spectral imaging in the mid-infrared between 5 and 28 μm.

## II. Heterodyne Receiver for Origins (HERO)

### A. Motivation

HERO complements the Origins instrument suit by providing extremely high spectral resolving power up to $10^7$. The high resolving power enables line tomography where the observed spectra together with simple models allow us to deduce the distribution of gas at special scales much smaller than any telescope would allow.

HERO was designed for the trail of water science case. It has a moderate field of view with a footprint of 3x3 pixels and covers many water lines emitting between 586 and 2700 GHz. HERO can carry out dual-polarization and dual-frequency observations. The characteristics of HERO are given in Table 1.

TABLE 1
Little-HERO Design Parameters (for OST Concept 2)

| Band | $F_{min}$ | $F_{max}$ | Pixel | Trx | Beam | $T_{rms}$ [a] | Line Flux [b] |
|---|---|---|---|---|---|---|---|
| | GHz | GHz | | K | " | mK | $Wm^{-2}$ |
| 1 | 486 | 756 | 2x9 | 50 | 20.3 | 2.6 | 6.4 E-21 |
| 2 | 756 | 1188 | 2x9 | 100 | 12.9 | 4.2 | 1.6 E-20 |
| 3 | 1188 | 1782 | 2x9 | 200 | 8.5 | 6.8 | 4.0 E-20 |
| 4 | 1782 | 2700 | 2x9 | 300 | 5.6 | 8.4 | 7.3 E-20 |

[a]Receiver noise for 1h integration at $10^6$ resolution (0.3 km/s) using one polarization.

[b]Detectable line flux at 5 sigma, for 1h pointed integration (on+off source) in two polarizations, with a 5.9 m primary mirror as designed for OST Concept 2.

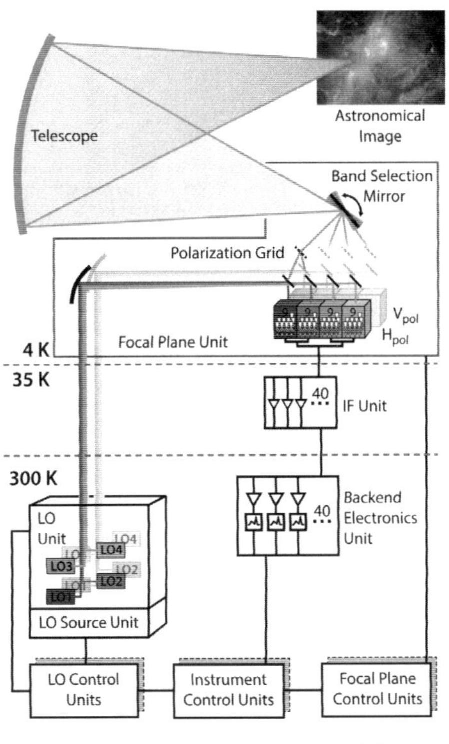


Figure 1: HERO instrument architecture closely follows architectures of successful heterodyne instruments like HIFI. Novel coupling optics and advances in component technologies allows for mapping speeds that are orders of magnitude faster than HIFI. (Image Credit: Britt Griswold, NASA.)

### B. Instrument Design

The HERO design is shown in Figure 1.

The design follows the standard heterodyne layout, but is specifically adapted for space. In order to reduce the weight each frequency band is ~45% wide and HERO covers the entire frequency range of 586 to 27000 GHz in only 4 bands. HERO also uses low-power components, in particular the cryogenic low noise amplifiers only consume 0.5 mW, a tenth of those of Herschel, and the backends consume less then 2 W for 6 GHz to 8 THz bandwidth about 1/35th of the backends commonly in use. In spite of these savings the sensitivity is close to quantum limit and the intermediate frequency (IF) bandwidth is at least 6 GHz (goal 8 GHz). These substantial reductions of weight and power enable the design of the first heterodyne focal plane array for space application.

### C. Components

**Optics:** A low loss and compact design with very high fractional bandwidth has been achieved for HERO. Figure 2 shows the cold optics.

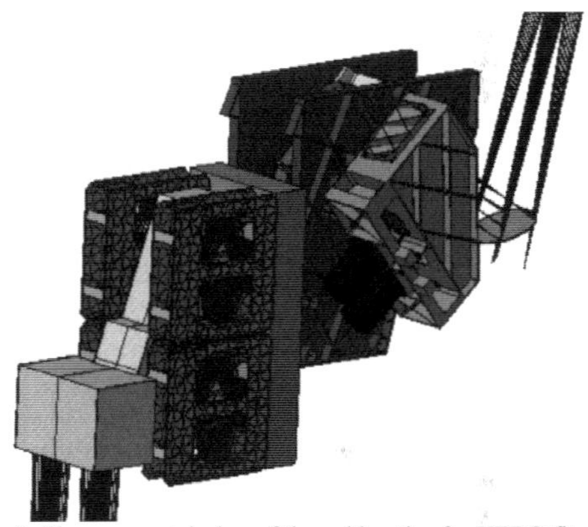

Figure 2: The compact design of the cold optics for HERO fits easily in the Origins space craft.

Radiation from the sky arrives from the top right. A pick-off mirror directs it to the HERO instrument. The Offner Relay (between the green plates) will direct either the light from the sky or from the internal calibration loads (blue cylinders on right) to one of the four bands. Within the bands the light from the sky / calibration loads will be split in polarization and superimposed with the local oscillator (LO) reference signal coming from the warm space craft (bottom left). After superposition of LO and sky with a wire grid, ellipsoidal mirrors refocus the beam and a lenslet array matches it to the mixer array.

To minimize infrared radiation coming from the space craft bus via the LO beam 1) the LO beams are superimposed and only two beams pass through the sun shields and 2) infrared bandpass filters are inserted in the beam blocking all radiation except at that of the LOs.

**Local Oscillators:** Local oscillators (LO) are a critical item, as they need to be tunable over a very wide frequency range, reach high frequencies (for HERO up to 2.7 THz), pump many pixels, and have low power consumption. Schottky diode-based frequency multiplier chains have made considerable progress recently [6][7] and are the baseline for the HERO design. By utilizing high-power GaN amplifiers at W band and power-combining multiplication technology in the submillimeter-wave range, more than 1 mW of power has been demonstrated at 1.6 THz. The LO signal is split in waveguide to 3x3 beams to match the focal plane mixer array. HERO has two LO chains for each frequency band, one for each polarization.

An alternative to the multiplier-amplifier chains are quantum cascade lasers. They have the advantage of having high output power, but require cooling and are more difficult to tune over a wide bandwidth. However, considerable progress has been made [8][11].

**Mixers:** HERO uses the most sensitive mixers: Superconducting Insolating Superconducting (SIS) mixers for the two lower frequency channels and Hot Electron Bolometer (HEB) mixers for the two upper frequency channels. SIS mixers have already reached a noise levels around 2 hf/k and intermediate frequency bandwidth of 8 GHz required for HERO. HEB mixers still have slightly higher noise and lower IF bandwidth, but rapid progress is made and bandwidth of 7.5 GHz have been reported [14], as well as noise temperatures of 3.3 hf/k [9].

The mixers employ horns that are followed by orthomode transducers to separate the LO from the sky signal, as suggested by Belitsky [12]. All mixers have two junctions and are balanced to reduce the LO power requirements and to enhance stability by suppressing LO AM noise [13]. One mixer of each array is sideband separating (2SB); the others are double sideband (DSB) mixers. The 2SB mixer is used to help calibrate the sideband ratio of the DSB mixers. We did not select 2SB mixers everywhere in the array, because for most of the science drivers the lines are sparse (either in the upper or the lower sideband), and because we want to limit the required IF power.

**Intermediate Frequency Chain**: In order to be able to observe lines that are up to 500km/s wide, the HERO requires a bandwidth of at least 6 GHz with a goal of 8 GHz.

The weak IF signal from the mixers is first amplified directly behind the mixers, a second time at 35K and then before the spectrometers at around 300K. The cryogenic amplifiers are allocated only 0.5 mW while they need to be low noise (< 5K) and wideband (> 6 GHz). Currently, SiGe [15][16][17] amplifiers are the most promising candidates. They consume only 0.3mW albeit with only 4 GHz of bandwidth and further development is required.

An alternative are the well established InP amplifiers [18][19]. They are usually operated with 5mW power, but still show good performance at reduced power [20][21].

**Backends:** HERO requires 36 backends to allow dual polarization and dual frequency operation with 9 pixel focal plane arrays. The excellent Digital Fourier Transform Spectrometers (DFTS) commonly used in ground based telescopes consume around 70W, unfortunately too much for a space mission.

As a baseline, HERO will use CMOS-based spectrometers, which are advancing quickly with the telecommunication industry and are predicted to reach the required bandwidth and power within a few years. Current versions have six GHz bandwidth, are extremely lightweight (<120 g), and require little power (< 1W) per backend [22][23].

An autocorrelation spectrometer (ACS) is another viable option, as it has been used already in space missions (ODIN) [24], balloon mission TELIS [25], and low power ASIC versions are becoming available. For HERO, it is essential that backend power consumption is reduced from about 40W to less than 2 W per 8 GHz IF.

**Control Electronics:**

HERO has three control units: the LO control unit commands the frequency synthesizer and the LO chains, the focal plane control unit commands and powers all components mounted on the 4K stage and the instrument control unit (ICU) is the overall control unit of HERO with spacewire connections to the other units. In addition it is responsible for the IF chain and the backends and collects and compresses the data. The ICU connects to the spacecraft computer via MIL1553STDB. All control units will use next generation space qualified processors. HERO flies two units of each control unit for redundancy.

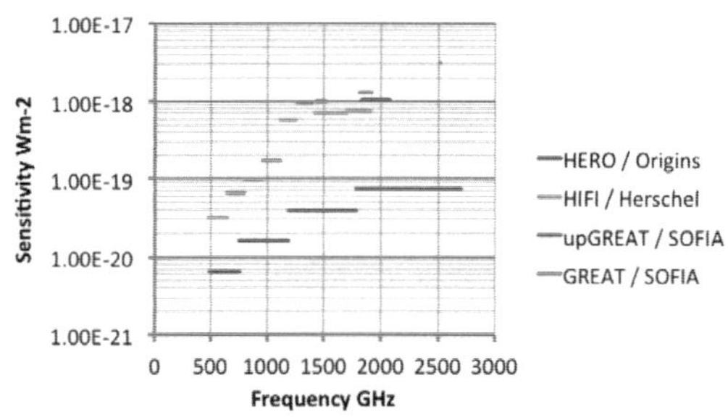


Figure 3: HERO is more than ten times as sensitive as the best current and past airborne heterodyne receivers. With its nine pixels, dual polarization and dual frequency modes HERO is also an efficient mapping instrument..

## III. Enabling Technologies and Required Developments

The enabling technologies are the LOs, the mixers, the amplifiers and the spectrometers. The LOs need to be very broadband with a fractional bandwidth up to 45%, to allow covering a large frequency range with few bands, i.e. save weight. HERO requires near quantum limited mixers with less than 1 to 3 hf/k. SIS mixers already have noise temperatures around 2hf/k, but some development is needed to reduce the noise and increase the IF bandwidth of HEB misers. In order to be able to cool heterodyne arrays in space, we have only attributed 0.5 mW of power for each amplifier with about 25dB of gain. It is also critical to reduce the power consumption of the spectrometric backend to about 2W per 8 GHz. For HERO the components need to be Technology Readiness Level (TRL) 5 by 2025 and TRL 6 by 2027. Table 2 shows the technology development plan.

| | | 2019 | 2020 | 2021 | 2022 | 2023 | 2024 | 2025 | 2026 | 2027 |
|---|---|---|---|---|---|---|---|---|---|---|
| | | | | | | | | TRL 5 | EM | TRL 6 |
| Optics | Lenslet arrays, filters, calib source, etc | | | | Develop braodband design | | Envirn/Qual testing | Characterization in relevant environment | Subsystem level integration | System level testing |
| Mixers | SIS and Hot Electron Bolometer | Single pixel with required sensitivity and frequency coverage, search for materials for high freq. SIS | | Single pixel with sensitivity and IF BW | Array proof of concept, balanced mixers | Envir. test | Qual. test | | | |
| Local Oscillator | Schottky Multipliers | Single pixel with fractional BW | | | Array proof of concept | | | | | |
| | QCL | | Powerful and efficient source, large and continuous coverage, high operating temp | | Development of mode selection optics, PLL, beam divider | Efficient and long distance (few m) LO and mixer coupling scheme | Down select. Envirn/Qual testing | | | |
| | Parametric multipliers | | Single pixel with output power | Single pixel with BW | Array proof of concept | | | | | |
| IF | LNAs | | | | Low noise, wideband and Low DC power | | Down select. Envirn/Qual testing | | | |
| Backends | SoC ASIC | | Bandwidth, DC power, calibration | | | Envir/Qual testing and Down selection | | | | |
| | Autocorrelators | | Bandwidth, DC power | | | | | | | |

Table 2 Technology Development.

## IV. Conclusion

Origins is a very powerful far-IR satellite concept that will revolutionize our understanding of the universe. With its 5.9 m cooled dish it is much more sensitive than any prior mission and will help us to understand the evolution of galaxies and the trail of water to planets, as well as search for biosignatures of exoplanets.

HERO is the first focal plane array designed for space. It is expected to have a performance that is more than ten times more sensitive than any current heterodyne receiver (Figure 3). HERO allows efficient mapping at high spectral resolution with its 9 pixel arrays, the dual polarization and the dual frequency modes.

## References


[1] M. Meixner et al., "Overview of the Origins Space telescope: science drivers to observatory requirements," in Proceedings of the SPIE, Volume 10698, id. 106980N 11 pp. (2018).

[2] D. Leisawitz et al., "The Origins Space Telescope: mission concept overview," in Proceedings of the SPIE, Volume 10698, id. 1069815 13 pp. (2018).

[3] M. Bradford et al.. "The Origins Survey Spectrometer (OSS): a far-IR discovery machine for the Origins Space Telescope," in Proceedings of the SPIE, Volume 10698, id. 1069818 17 pp. (2018).

[4] J. Staguhn et al., "Origins Space Telescope: the far infrared imager and polarimeter FIP," in Proceedings of the SPIE, Volume 10698, id. 106981A 6 pp. (2018).

[5] I. Sakon, "The mid-infrared imager, spectrometer, coronagraph (MISC) for the Origins Space telescope (OST)," in Proceedings of the SPIE, Volume 10698, id. 1069817 9 pp. (2018).

[6] I. Mehdi, J. V. Siles, C. Lee and E. Schlecht, "THz Diode Technology: Status, Prospects, and Applications," in Proceedings of the IEEE, vol. 105, no. 6, pp. 990-1007, June 2017. DOI: 10.1109/JPROC.2017.2650235

[7] T. W. Crowe, J. L. Hesler, S. A. Retzloff and D. S. Kurtz, "Higher power terahertz sources based on diode multipliers," 2017 42nd International Conference on Infrared, Millimeter, and Terahertz Waves (IRMMW-THz), Cancun, 2017, pp. 1-1. DOI: 10.1109/IRMMW-THz.2017.8067091

[8] Krause, S., Meledin, D., Desmaris, V., Pavolotsky, A., Rashid, H., Belitksy, V., "Noise and IF Gain Bandwidth of a Balanced Waveguide NbN/GaN Hot Electron Bolometer Mixer Operating at 1.3 THz," IEEE Transactions on Terahertz Science and Technology, vol 8, Issue 3, (2018). DOI: 10.1109/TTHZ.2018.2824027

[9] C. Risacher et al., "The upGREAT 1.9 THz multi-pixel high resolution spectrometer for the SOFIA Observatory," in Astronomy & Astrophysics, Volume 595, id.A34, 7 pp.

[10] H.-W. Hübers, T. Hagelschuer, H. Richter, M. Wienold, L. Schrottke, X. Lü, B. Röben, K. Biermann, and H. T. Grahn "Compact and efficient 4.7-THz local oscillator with a GaAs/AlAs quantum-cascade laser," 29TH IEEE International Sympopsium on Space Terahertz Technology, 26-28 March 2018

[11] N. van Marrewijk, B. Mirzaei, D. Hayton, R. J. Gao, T. Y. Kao, Q. Hu, J. L. Reno "Frequency Locking and Monitoring Based on Bi-directional Terahertz Radiation of a 3rd-Order Distributed Feedback Quantum Cascade Laser," Journal of Infrared, Millimeter, and Terahertz Waves, Volume 36, Issue 12, pp.1210-1220 (2015) DOI: 10.1007/s10762-015-0210-4.

[12] V. Belitsky, V. Desmaris, D. Dochev, D. Meledin, A. Pavolotsky, "Towards Multi-Pixel Heterodyne Terahertz Receivers", Proceedings of the 22nd International Symposium on Space Terahertz Technology, Tucson, AZ, USA, April 26-28, p. 15-18 (2011).

[13] Meledin, D. et al, "A 1.3-THz Balanced Waveguide HEB Mixer for the APEX Telescope," IEEE Transactions on Microwave Theory and Techniques, 57, No 1 (January 2009). DOI: 10.1109/TMTT.2008.2008946.

[14] Krause, S., Meledin, D., Desmaris, V., Pavolotsky, A., Rashid, H., Belitksy, V., "Noise and IF Gain Bandwidth of a Balanced Waveguide NbN/GaN Hot Electron Bolometer Mixer Operating at 1.3 THz," IEEE Transactions on Terahertz Science and Technology, vol 8, Issue 3, (2018). DOI: 10.1109/TTHZ.2018.2824027

[15] S. Montazeri, W. T. Wong, A. H. Coskun and J. C. Bardin, "Ultra-Low-Power Cryogenic SiGe Low-Noise Amplifiers: Theory and Demonstration," in *IEEE Transactions on Microwave Theory and Techniques*, vol. 64, no. 1, pp. 178-187, Jan. 2016. doi: 10.1109/TMTT.2015.2497685

[16] C. Bardin, S. Montazeri, and Su-Wei Chang, "Silicon Germanium Cryogenic Low Noise Amplifiers," 12th International Workshop on Low Temperature Electronics, J. Phys.: Conf. Ser Conf. Series 834 (2017) 012007. DOI :10.1088/1742-6596/834/1/012007

[17] Montazeri, S., Grimes, P. K., Tong, C-Y. E., Bardin, J. C., "A Wide-Band High-Gain Compact SIS Receiver Utilizing a 300μW SiGe IF LNA," IEEE Transactions on Applied Superconductivity, Vol. 27, NO. 4 , June 2017. DOI: 10.1109/TASC.2016.2631441.

[18] I. López-Fernández, J. D. Gallego, C. Diez, A. Barcia, J. M. Pintado, "Wide Band, Ultra Low Noise Cryogenic InP IF Amplifiers for the Herschel Mission Radiometers," Millimeter and Submillimeter Detectors for Astronomy, Proc. SPIE, vol. 4855, pp. 489-500, 2003

[19] Low Noise Factory, 412 63 Göteborg, SWEDEN, https://www.lownoisefactory.com/

[20] Wadefalk, Niklas, et al. "Cryogenic wide-band ultra-low-noise IF amplifiers operating at ultra-low DC power." IEEE Transactions on Microwave Theory and Techniques 51.6 (2003): 1705-1711

[21] Joel Schleeh, private communication (2018)

[22] Y. Zhang, Y. Kim, A. Tang, J. Kawamura, T. Reck and M. C. F. Chang, "A 2.6GS/s Spectrometer System in 65nm CMOS for Spaceborne Telescopic Sensing," *2018 IEEE International Symposium on Circuits and Systems (ISCAS)*, Florence, Italy, 2018, pp. 1-4. DOI: 10.1109/ISCAS.2018.8351690

[23] Kim, Y., Zhang, Y., Tang, A., Reck, T., and Chang, M-C. F, "A 1.5W 3 GHz Back-end Processor in 65 m CMOS for Sub-millimeter-wave Heterodyne Receiver Arrays," International Symposium for Space Terahertz Technology (2018).

[24] Emrich, A., " Autocorrelation Spectrometers for Space Borne (sub) Millimetre Astronomy", The Far Infrared and Submillimetre Universe. Edited by A. Wilson. Noordwijk, The Netherlands : ESA, p.361, (1997).

[25] Vogt Peter, Birk Manfred, Wagner Georg, Geiger Felix, Lange Gert, Golstein Hans, Kiselev Oleg, Emrich Anders. "Characterisation of the TELIS autocorrelator spectrometer", 21st International Symposium on Space Terahertz Technology (2010).

[26]

# The Terahertz Intensity Mapper (TIM): an Imaging Spectrometer for Galaxy Evolution Studies at High-Redshift

Joaquin Vieira, James Aguirre, C. Matt Bradford, Jeffrey Filippini, Christopher Groppi, Dan Marrone, Matthieu Bethermin, Tzu-Ching Chang, Mark Devlin, Oliver Dore, Jianyang Fu, Steven Hailey Dunsheath, Gilbert Holder, Garrett Keating, Ryan Keenan, Ely Kovetz, Guilaine Lagache, Philip Mauskopf, Desika Narayanan, Gergo Popping, Erik Shirokoff, Rachel Somerville, Isaac Trumper, Bade Uzgil, and Jonas Zmuidzinas

***Abstract*— Understanding the formation and evolution of galaxies over cosmic time is one of the foremost goals of astrophysics and cosmology today. The cosmic star formation rate has undergone a dramatic evolution over the course of the last 14 billion years, and dust obscured star forming galaxies (DSFGs) are a crucial component of this evolution. A variety of important, bright, and unextincted diagnostic lines are present in the far-infrared (FIR) which can provide crucial insight into the physical conditions of galaxy evolution, including the instantaneous star formation rate, the effect of AGN feedback on star formation, the mass function of the stars, metallicities, and the spectrum of their ionizing radiation. FIR spectroscopy is technically difficult but scientifically crucial. The FIR waveband is impossible to observe from the ground, and spans a crucial gap in the spectroscopic coverage between the Atacama Large Millimeter/submillimeter Array (ALMA) in the sub/mm, and the James Webb Space Telescope (JWST) in the mid-IR. Stratospheric balloons offer a platform which can outperform current instrument sensitivities and are the only way to provide large-area, wide bandwidth spatial/spectral mapping at FIR wavelengths.**

**NASA recently selected TIM, the Terahertz Intensity Mapper, with the goal of demonstrating the key technical milestones necessary for FIR spectroscopy. TIM will provide a technological stepping stone to the future space-borne instrumentation such as the Origins Space Telescope (OST, formerly the Far-IR Surveyor) or a Probe mission. TIM will address the two key technical issues necessary to achieve this:**

**1. Low-emissivity, high-throughput telescope and spectrometer optics for the FIR;**

**2. Background-limited detectors in large format arrays, scalable to >10,000 pixels.**

**We will do this by constructing a integral-field spectrometer from 240 - 420 microns with 3600 kinetic-inductance detectors (KIDs) coupled to a 2-meter low-emissivity carbon fiber telescope.**

**In addition to the development and demonstration of crucial technologies for the FIR, TIM will perform groundbreaking science. We will survey two fields centered on GOODS-S and the South Pole Telescope Deep Field, both of which have rich ancillary data. Scientifically, we will:**

**1. Obtain spectroscopic line detections of ~100 galaxies in the atomic fine structure lines [CII] (158 microns) (at 0.5<z<1.5), [NII]**

Manuscript submitted June 25, 2019.

J. Vieira is with the Department of Astronomy, University of Illinois Urbana-Champagne, Urbana IL 61801 USA (email: jvieira@illinois.edu).

J. Aguirre is with the Department of Physics and Astronomy, University of Pennsylvania, Philadelphia, PA 19104 USA (email: jaguirre@sas.upenn.edu).

C.M. Bradford is with the NASA Jet Propulsion Laboratory, Pasadena CA 91109 USA (email: bradford@submm.caltech.edu)

J. Filippini is with the Department of Astronomy, University of Illinois Urbana-Champagne, Urbana IL 61801 USA (email: jpf@illinois.edu).

C. Groppi is with the School of Earth and Space Exploration, Arizona State University, Tempe, AZ 85287 USA (e-mail: cgroppi@asu.edu).

D. Marrone is with Steward Observatory, University of Arizona, Tucson, AZ 85721 USA (email: dmarrone@email.arizona.edu).

M. Bethermin is with the Laboratoire d'astrophysique de Marseille, 13388 Marseille FRANCE (email: matthieu.bethermin@lam.fr).

T-C. Chang is with the NASA Jet Propulsion Laboratory, Pasadena CA 91109 USA (email: tzu-ching.chang@jpl.nasa.gov)

M. Devlin is with the Department of Physics and Astronomy, University of Pennsylvania, Philadelphia, PA 19104 USA (email: devlin@physics.upenn.edu).

O. Dore is with the NASA Jet Propulsion Laboratory, Pasadena CA 91109 USA (email: olivier.p.dore@jpl.nasa.gov).

J. Fu is with the Department of Astronomy, University of Illinois Urbana-Champagne, Urbana IL 61801 USA (email:fu32@illinois.edu)

S. Hailey Dunsheath is with the Department of Astronomy, California Institute of Technology, Pasadena, CA 91125 USA (email: haileyds@caltech.edu).

G. Holder is with Department of Astronomy, University of Illinois Urbana-Champagne, Urbana IL 61801 USA (email: gholder@illinois.edu).

G. Keating is with the Harvard Smithsonian Center for Astrophysics, Cambridge, MA 02138 USA (email: garrett.keating@cfa.harvard.edu).

R. Keenan is with the Steward Observatory, University of Arizona, Tucson, AZ 85721 USA (email: rpkeenan@email.arizona.edu).

E. Kovetz is with the Department of Physics and Astronomy, Johns Hopkins University, Baltimore, MD 21218 USA (email: elykovetz@jhu.edu).

G. Lagache is with the Laboratoire d'astrophysique de Marseille, 13388 Marseille FRANCE (email: guilaine.lagache@lam.fr).

P. Mauskopf is with the School of Earth and Space Exploration, Arizona State University, Tempe, AZ 85287 USA (e-mail: Philip.Mauskopf@asu.edu).

D. Narayanan is with the Department of Astronomy, University of Florida, Gainsville, Gainsville, FL 32611 USA (email: desika.narayanan@ufl.edu).

G. Popping is with the Max Planck Institute for Astronomy, 69117 Heidelberg, Germany (email: popping@mpia.de).

E. Shirokoff is with the Department of Astronomy and Astrophysics, University of Chicago, Chicago, IL 60637 USA (email: shiro@chicago.edu).

R. Somerville is with the Department of Physics and Astronomy, Rutgers University, Piscataway, NJ 08854 USA (email: somerville@physics.rutgers.edu).

I. Trumper is with Steward Observatory, University of Arizona, Tucson, AZ 85721 USA (email: itrumper@optics.arizona.edu).

B. Uzgil is with the Max Planck Institute for Astronomy, 69117 Heidelberg, Germany (email: uzgil@mpia.de).

J. Zmuidzinas is with the Department of Astronomy, California Institute of Technology, Pasadena, CA 91125 USA (email: jonas@caltech.edu).

**(205 microns) (at 0.2<z<1 ), [OI] (63 microns) (at 2.8<z<5.7 ) and [OIII] (88 microns) (at 1.7<z<3.8);**

**2. Establish the mean star formation rate (proportional to [CII] luminosity), metallicities (proportional to the [CII]/[NII] ratio), and AGN content (proportional to the [OIII] luminosity) of galaxies using a stacking analysis of known sources in the field;**

**3. Produce deep maps of the 3D structure of the Universe by redshift tomography ("intensity mapping") with [CI], and [CII] × [NII] cross-spectra, to constrain the cosmic star formation history at cosmic noon and lay the important groundwork for extending this technique to even higher redshifts to eventually explore the epoch of reionization.**

**In this paper, we will summarize plans for the TIM experiment's development, test and deployment for a planned flight from Antarctica in Austral summer of 2022-2023.**

***Index Terms*—Astronomy, suborbital, balloon, intensity mapping, kinetic inductance detector.**

## I. Introduction

Explaining the history of cosmic star formation through the evolution of galaxies is one of the most important challenges in modern astrophysics. A wealth of data has been assembled in the last decade showing clearly that the total star formation rate density has fallen dramatically since its peak 7−10 Gyr ago (z∼1-3; [1]). The nature of the galaxies responsible for the bulk of star formation has also changed over cosmic time, with star formation previously dominated by luminous, dust-obscured, star-forming galaxies (DSFGs) that are almost absent in the local universe [2].

Half of the total energy output from the cosmic star formation has been absorbed by interstellar dust and re-emitted in the FIR [3],[4]. Moreover, there are a variety of un-extincted FIR diagnostic lines that can reveal the physics of galaxy evolution by tracing the star formation rate (SFR), black hole accretion rate, mass function of stars, spectrum of ionizing radiation, and metallicity of the interstellar medium (ISM). The path to understanding galaxy evolution will necessarily run through observations in the far infrared (FIR).

Spectroscopy in the FIR is technically difficult but essential to the study of galaxy evolution. We are building the Terahertz Intensity Mapper (TIM), with the goal of demonstrating balloon-borne FIR spectroscopy limited by the photon noise from the atmosphere. TIM will be a vital technological, data analysis, and scientific stepping stone to future orbital missions, and will also advance our understanding of galaxy evolution through observations that cannot be replicated with current FIR instruments. TIM combines a long-slit spectrometer operating from 240−420 μm with a 2-m low-emissivity carbon-fiber telescope to provide a substantial increase in sensitivity over existing instruments. We will survey one 0.1 $deg^2$ field centered on GOODS-S and one wider field (∼1 $deg^2$) within the South Pole Telescope (SPT) Deep Field, both of which have rich multi-wavelength ancillary data.

The science goals of TIM are:

1. Produce deep tomographic maps of the 3D structure of the Universe to measure the power spectrum of [CII] and [CII]×[NII]. This will be a pioneering demonstration of the technique of "intensity mapping," which provides a new method to constrain the cosmic star formation history and measure its relation to the underlying dark matter distribution;
2. Perform a blind spectroscopic survey for [CII] line emitters within an enormous cosmic volume, $10^7$ $Mpc^3$, at 0.52<z<1.67. We expect to detect ∼100 galaxies, which will be a powerful observational constraint on models of galaxy evolution.
3. Capture the star formation contribution of galaxies too faint to be detected individually, by measuring the [CII] luminosity function across the peak of cosmic star formation;
4. Use stellar mass-selected galaxies with spectroscopic redshifts from the GOODS-S field to stack on [CII] and [NII], and develop the theory to relate this to the total star formation rate ([CII]), star formation mode ([CII]/$L_{FIR}$), metallicity ([NII]/[CII]), and specific star formation rate ([CII]/$M_{star}$);
5. Cross-correlate the [CII] data cube (which provides redshift information) with Herschel/SPIRE maps (SFR) to calibrate the [CII]/SFR relation, and Spitzer/IRAC maps (stellar mass) to measure the specific star formation rate versus redshift.

TIM is a wholly unprecedented experiment to study the cosmic star formation history. It will map a volume spanning 4.5 billion years of cosmic history (0.52<z<1.67), on scales from 1−50 Mpc (30" to ∼1°) with complete spectroscopic information. In the coming decades this will be a powerful cosmological tool for charting the 3D structure of the universe. There is significant discovery potential with TIM, since it will be probing an under-explored wavelength range with unprecedented sensitivity using a new astrophysical technique. TIM will be the first generation of experiments using intensity mapping in the FIR regime and fills in a crucial wavelength regime only accessible from either space or a balloon platform. TIM fills a unique and vital scientific niche not filled by Herschel, SOFIA, ALMA, JWST, or even SPICA as currently conceived.

## II. Instrument

### *A. Design Considerations and Sensitivity*

To achieve the scientific and technical goals of TIM, we must be able to demonstrate atmosphere limited performance of the telescope and detectors. Photon loading this low is only possible from a (sub)orbital platform and motivates the necessity of a balloon program. To estimate the atmosphere background, the (proprietary) ATM model of Juan Pardo was used. We have assumed a flight at mid-latitudes, an altitude of 37 km, and observations at 45" elevation. The ATM model calculates the opacity due to all relevant atmospheric species. We calculate the noise equivalent power (NEP) due to photon noise from a greybody in the usual manner assuming an instrument transmission of 25% and two photon modes (both polarizations, horn coupled) and require that the intrinsic noise in the detectors be sub-dominant to this photon noise. The line sensitivity on the sky then incorporates the point-source efficiency, ~64% with the horn architecture we choose.

We have considered the performance of various potential telescope architectures for TIM, incorporating the atmospheric transmission and loading, a range of telescope sizes, and the possibility of actively cooling the telescope to minimize its thermal emission. While a cooled aperture performs better, it is very costly for a given aperture size, and the performance improvement is modest because even at balloon altitudes there is ~1% emission from the 250 K atmosphere. In light of these calculations, we believe we can demonstrate near atmosphere-limited performance using the on-axis telescope described below with carefully controlled primary illumination, cold stops at pupils in the system, and baffling to prevent spillover to warm surfaces.

In principle, the required spectral resolution and large area mapping could be achieved with either a Fabry-Perot (FP) or FTS spectrometer design. However, both of these incur sensitivity penalties. The FP does not cover the entire frequency range instantaneously and must be scanned, and the FTS places the full optical bandwidth of the entire band on each detector, increasing the noise. The best approach is a reflective, blazed diffraction grating, which orders large instantaneous bandwidth and good sensitivity. We target a spectral resolving power R = 250 (830 kms$^{-1}$), a value that is well-suited to intensity mapping because it enables instantaneous coverage of a wide band (32%) with a single spectrometer with a modest number of resolution elements (<100). While R = 250 under-resolves galaxies' intrinsic line widths, the sensitivity degradation is only modest ($\propto \sqrt{\delta v}$) for a photon-noise limited instrument, and the wider bandwidth recovers this factor through the speed increase for blind survey experiments, as demonstrated with Z-Spec (e.g. [5]). The instrument parameters and resulting sensitivity for our design are given in Table 1.

*B. Telescope*

The TIM telescope design must be lightweight and compact, with low overall emissivity (i.e., any coupling to warm, non-astrophysical radiation loads) and high efficiency coupling to the spectrometer. In principle, an off-axis telescope can achieve a lower emissivity than an on-axis one, due to the lack of scattering from the secondary mirror and its support legs. However, off-axis mirrors are more costly and present fabrication challenges. We have designed a segmented, 2.0-meter diameter, carbon fiber primary mirror, as well as a fully carbon fiber secondary and support structure, with gold metallization of the reflecting surfaces to minimize the emissivity.

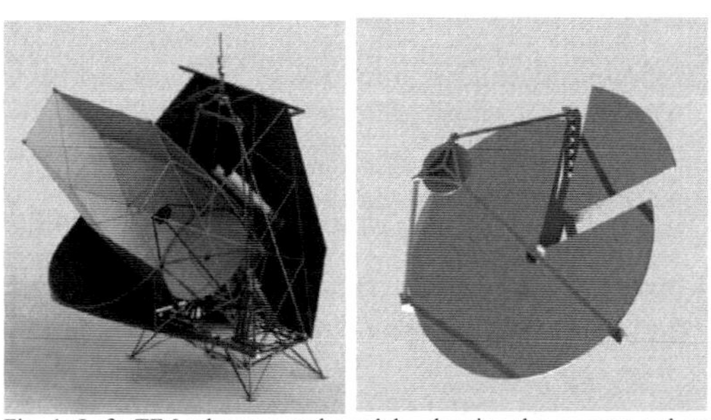

Fig. 1: Left: TIM telescope and gondola, showing the cryostat and star cameras. We will re-use the design for the gondola, cryostat, readout electronics, and star camera design from BLAST-TNG, which have all been constructed and are being prepared for a flight in late 2019. Right: The design for the segmented CFRP mirror from UA and CMA.

There are three key specifications that lead to our current design. First, we are driven to the largest aperture we can accommodate by sensitivity and resolution requirements, particularly for source stacking and single-object detections in our survey fields. The mirror diameter defines the resolution and therefore the maximum wavenumber (k) accessible to the experiment. A 2.0-meter mirror is the minimum size at which we can study both the clustered and Poisson portions of the power spectrum and should also retain appreciable signal to noise for stacking. Second, our telescope must be both cost and weight efficient. The BLAST aluminum 1.8-meter primary mirror had a surface mass density of 45 kg/m$^2$; this would be 110 kg for the mirror alone for a 2.0-meter aperture. Our carbon fiber reinforced polymer (CFRP) mirror will be 7-10 kg/m$^2$, making the telescope a marginal contributor to the mass and moment of inertia of the payload. Fabrication of a 2.0-meter mirror poses significant challenges, however. FIR-quality mirrors are usually machined from aluminum on diamond lathes, but there are very few machines that can produce a mirror this large. Even CFRP, which is assembled rather than machined, becomes costlier and riskier at this large diameter. For this reason, we have elected to divide the mirror into 6 segments, reducing the fabrication scale to 1.0 m where we have access to key components (e.g., oven, metallization chamber, machines for CF pressing) of appropriate size. The third constraint, low emissivity, drives the requirement of a gold reflective surface. The emissivity of gold is half of that of aluminum, which will provide a useful sensitivity boost. The total system emissivity will also be lowered by using three feedlegs to support the secondary (as in the current BLAST-TNG design, but unlike the four in BLAST), and by covering the feedlegs with reflective baffles to direct light incident on the legs to the cold sky. The final expected emissivity is 2-2.5%, compared to 4% in measured by BLAST. We assume 2.5% emissivity for sensitivity calculations, as given in Table 1.

The telescope concept has been prepared in collaboration with Composite Mirror Applications (CMA), a Tucson company that specializes in CFRP reflectors for optical wavelengths. CMA has demonstrated highly repeatable replication of spherical and paraboloidal optical mirrors, with surface figure accuracies of ~100 nm (50-100× better than needed) for mirrors as large as 0.8 m (Figure 1). For TIM wavelengths, accuracies of a few microns will be more than adequate, and there are many published examples of meter-scale CFRP mirrors by CMA and other vendors that exceed this requirement (e.g., [6], [7], [8]). The CFRP mirror segments will be made by replicating against a glass mandrel, whose precision should determine the accuracy of the mirror. The mandrel for this project will be cast, ground and polished to the appropriate figure using the tools developed for making large optical mirrors. The mirror segments will be built on top of the mandrel by laying unidirectional CFRP sheets in many orientations and curing under pressure. Each individual mirror segment will be subjected to precision metrology at the University of Arizona, using laser trackers and the Software Configurable Optical Test System [9] to confirm their figure. The six segments, after gilding, will be aligned and pinned at Arizona using similar validation tools.

The entire telescope, including the structural support for the secondary mirror feedlegs, which will anchor to the rear hub at the center of the primary mirror, will be made from CFRP. The operating temperature of this structure is not very low (250 K), and thermal contraction for CFRP is 20× smaller than aluminum, but matching materials in this way should lead to conformal contraction of the telescope (to first order). Residual correction can be made by focusing the telescope upon reaching flight altitude, where BLAST has observed just ±1.5 K diurnal variations. We will use the BLAST focusing system which has three precision actuators behind the secondary to provide 3-micron positioning. The entire system worked flawlessly during the BLAST 2006, 2010, and most recently 2012 [10], [11]. The secondary focus is located in the same location as for BLAST-TNG to allow re-use of the gondola and inner frame design, also shown in Figure 1. The maximum field of view of the telescope (defined as when the beam at 300 μm drops to a Strehl ratio of 0.95) is 0.5" in diameter. This is well-matched to TIM, and the design could be reused for other imaging submillimeter balloon missions.

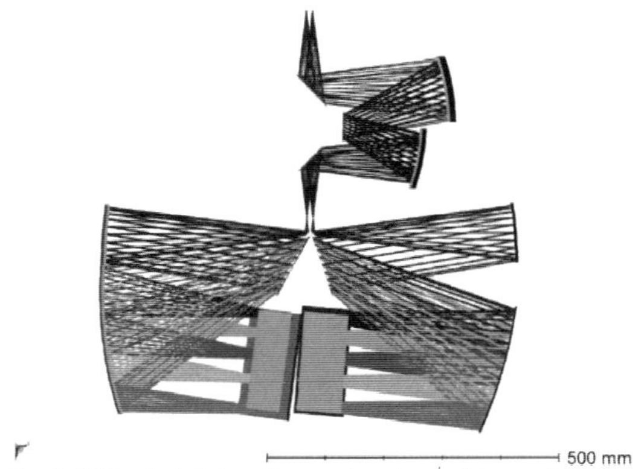


Fig. 2: TIM optical design. Two spectrometer modules are fed with a common Offner-style relay that provides a cold pupil mirror. Pickoff mirrors placed directly behind the slits redirect the light the individual modules. The full package is sized for the BLAST-TNG cryostat.

## C. *Spectrometer Architecture*

To efficiently perform spectroscopy over the full 240 – 420 μm band, we will deploy two independent spectrometer modules: a short wavelength (SW) module covering 240–317 μm, and a long wavelength (LW) module covering 317 – 420 μm. The two modules follow the same basic design; each consists of a plane diffraction grating mounted between concave collimating and camera mirrors in a Czerny- Turner configuration. The optics are designed to accept an f/4 cone at the entrance slit and produce a telecentric f/4 image at the output. The grating is operated in first order and is sized to provide a resolving power of R ≈ 250. The diffraction gratings will be cut from M1 mold plate and machined with diamond tooling, similar to ZEUS-2 [12], [13].

The full optical layout is shown in Figure 2. A set of cold relay optics (M3 – M6) reimages the f/5 telescope focus to an f/4 image at the entrance slits of the spectrometer modules. An image of the secondary mirror is formed on M4, and this optic serves as a Lyot stop to block stray radiation. The central circular area of M4 conjugate to the hole in the primary mirror will also be painted black to further reduce the optical loading. The LW and SW slits are 350 and 260 in length, respectively, are aligned in azimuth, and separated by 2.6'

Pickoff mirrors placed directly behind the entrance slits direct the radiation away from the common optical axis, allowing the spectrometer modules to be well separated. Another set of pickoffs (SW3 and LW3) placed after the camera mirrors position the focal planes adjacently, allowing the two detector modules to be mounted in close proximity on a sub-Kelvin stage. The full set of optics occupies a 770 mm × 610 mm × 350 mm volume and is sized to approximately match the instrument volume available in the BLAST-TNG cryostat.

The spectrometer modules are designed for modularity to allow separate testing and optimization in a test cryostat smaller than the full TIM cryostat.

## D. *Array format, Horn Coupling*

Each focal plane will be sampled by an array of 1800 hexagonally close-packed, straight-walled conical feed horns (see Figure 3). We utilize a 25 (spatial) × 72 (spectral) element array of 1.5Fλ horns (though the optical design above accommodates ~42 spatial beams). The Strehl ratio over much of this focal plane is (0.90, with some degradation at the array edges. The optics provide an instantaneous spectral coverage of $\lambda_{max}/\lambda_{min} \approx 32\%$ at R ~250.

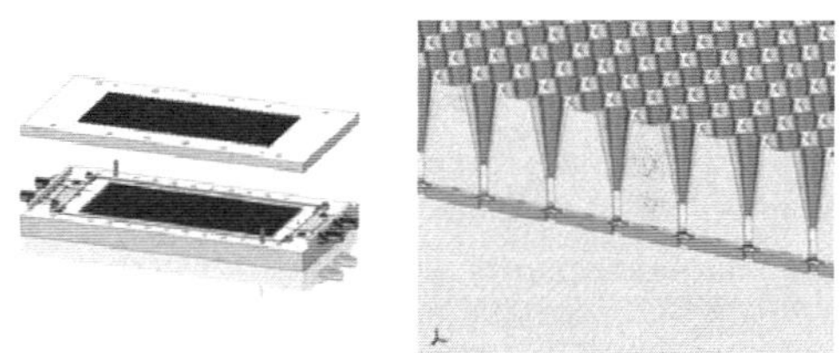

Fig. 3: TIM horn coupled focal-plane array architecture. Conical multi-flare-angle horns are drilled into the metal substrate with a custom tool. Alignment between the metal and the silicon die is via a pin and a pin and-slot, accommodating the differential CTE.

While optimized for intensity mapping, the spectrometer is sensitive to individual galaxies. The width of the entrance slit impacts the monochromatic point source sensitivity by controlling the amount of background power seen by a detector, as well as the fraction of the point source radiation entering the spectrometer. For a 1.5Fλ horn this sensitivity is maximized for a slit width of ≈Fλ, which yields a point source coupling efficiency of ≈0.64. Most of the rest of the horn power terminates on the entrance slit or the Lyot stop, with only a small loss in point source coupling due to a finite beam taper at the edge of the pupil. The spectrometer optics produce a small amount of anamorphic magnification, but this has negligible effect on the point source coupling.

The spectrometer modules are housed in separate 1K enclosures, with blackened surfaces and optical baffling control stray light and minimize loading on the detectors. A set of IR blocking filters at 77K, 40K, and 4K is used to reduce the loading on the cold stages, and a final set of capacitive low-pass metal mesh filters mounted to the horn arrays are used to define the bandpass of each module. We estimate the optical efficiency of the cold instrument, including the transmission and illumination efficiencies of all filters and mirrors, as well as the finite point source coupling provided by the horns, at 25%.

### *E. Kinetic-Inductance Detector (KID) Arrays*

KIDs have emerged in the last decade as a straightforward approach to very large detector arrays for astrophysics. These devices use thin-film, high-Q micro-resonators that absorb incident radiation and respond by changing resonance frequency and linewidth. Due to the high resonance quality factors Q~105 = $f/\delta f$ can be obtained (corresponding to narrow line widths), large numbers of KIDs may be read out on a single RF/microwave circuit, and the only cryogenic electronics necessary is a single cold (4–20 K) RF/microwave amplifier per readout circuit. Each circuit is simply a single RF line down to the focal plane and another line returning via the amplifier, and it carries the signals of $\sim 10^3$ detectors.

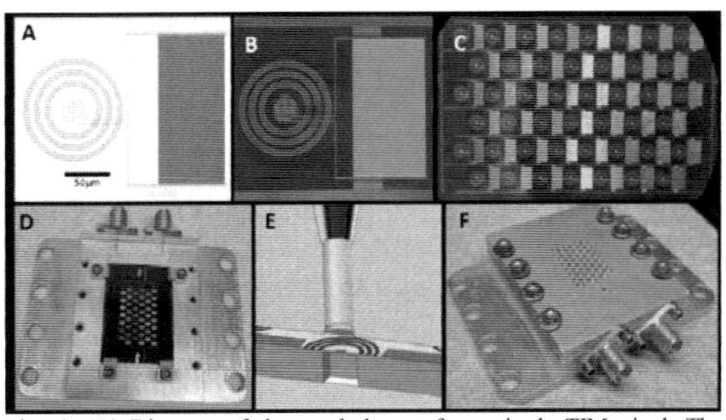


Fig. 4: (A) Diagram of the mask layout for a single TIM pixel. The meandered inductor (green) is surrounded by an optical choke structure (blue). An interdigitated capacitor (red) sets the resonance frequency of the pixel, and two coupling capacitors (yellow) couple microwave signal onto microstrip feedlines. (B) A microscope image of a single pixel as fabricated. All pixel elements of the prototype array are patterned out of 40 nm Al film. (C) A microscope image of the 45-pixel prototype array, as fabricated. (D) The fabricated array in its enclosure. The back side of the die is bare silicon and lies flat on the gold-plated package surface. The full size of the die is 30mm×22mm. (E) A CAD model of the detector package. The optical power is coupled into a feedhorn, and travels through a circular waveguide that is terminated by the inductor of the LEKID. A backshort is formed by deep trench etching from the backside to a 27-micron buried oxide layer, then metallizing. (F) The prototype feedhorn block installed above the 45-pixel array.

KID technology is now rivaling the performance levels of the SQUID-multiplexed bolometer systems in ground-based instruments. One example is the dual-band 150/240 GHz, 2896-pixel NIKA-2 camera fielded at the IRAM 30-m telescope by European groups at SRON Utrecht (Baselmans), Institut NEEL (Benoit), and Cardiff (Doyle), which has demonstrated sensitivities approaching the photon background limit. KIDs are being delivered now for flight with BLAST-[14], [15] and are under development for the TolTEC mm-wave camera to be deployed to the Large Millimeter Telescope (LMT). At Caltech / JPL, we have fielded the 350-micron camera MAKO [16], [17] at the CSO before its shutdown. More recently, we have demonstrated high yield and detector sensitivity below $10^{-18}$ W $Hz^{-1/2}$ in our mm-wave on-chip spectrometer SuperSpec ([18], [19], [20], [21]). A SuperSpec demonstration instrument is being prepared now for deployment to the LMT in the winter of 2018–2019.

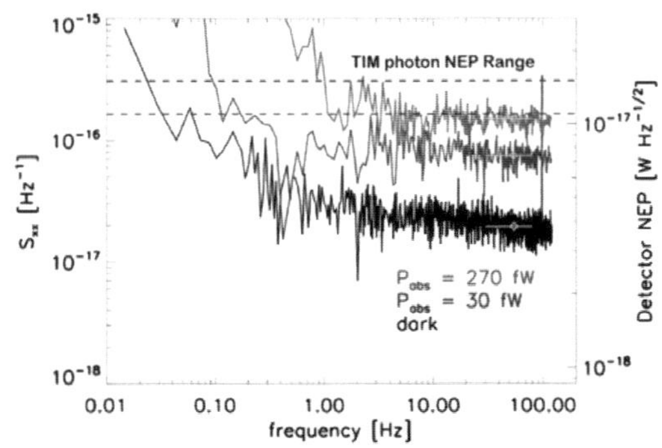


Fig. 5: Noise equivalent power (NEP) of a typical TIM KID on the prototype array shown in Figure 4. The left axis is fractional fractional frequency noise variance $S_{xx}$, the right is the NEP = $\sqrt{Sxx}/R$, where R is the power to fractional frequency responsivity ($1.2\times10^9 W^{-1}$ for these devices). The detector noise is dominated by photons for loading at and below that expected on float with TIM. The dark noise indicates a detector 1/f which produces knee at <0.1 Hz when compared with the expected photon noise.

Achieving photon-noise-limited performance on the TIM balloon-borne spectrometer requires more sensitive KIDs than have yet been fielded scientifically. We have designed the TIM pixel, prototyped small arrays in the JPL microdevices laboratory, and demonstrated detector sensitivities which outperform the TIM photon noise specification by a factor of ~2.5. The approach is shown in Figure 4, and top-level parameters are provided in Table 1. We employ the same basic single-layer lumped-element architecture as shown in Figure 4. We use a similar interdigitated capacitor design; this sets the two-level-system (TLS) frequency noise, ensuring that it is sub-dominant to the other noise sources. We use the same 100–250 MHz resonator frequencies, enabling us to use digitize directly in baseband without mixing. To meet the lower NEP required for TIM (as well as the other low-background applications), we need only reduce the volume of the inductor to increase the power-to-fractional-frequency responsivity, $R = (df/f)/dP_{inc}$, which scales as 1/V. The low volume, together with the requirement to create an impedance-matched absorber from the meandered inductor drives us to aluminum rather than titanium nitride.

The inductor volume is too small to permit a full MAKO-style 2-D meandered absorber, so each TIM detector will use a circular waveguide fed by a feedhorn (Figure 3), an approach broadly similar to that for super-BLAST-Pol [22], but with design enhancements to support our low-NEP device. The electromagnetic structure is designed to couple to both $TE_{11}$ modes in the circular guide, and the design includes a flare at the bottom of the waveguide and a lithographically patterned choke structure (3 concentric annular rings on the wafer surface) to help eliminate conversion into substrate modes. The backshort is integral to the device – it is created by simply etching from the backside to a buried oxide layer at the appropriate depth (27 μm for the long-wave device), then depositing aluminum.

At the heart of the device is the inductor / absorber—a single meander of 0.4-micron-wide aluminum patterned into the 40-nm-thick film with a total volume of 76 $\mu m^3$. The single meander couples as a 'mesh' to both polarizations by allowing the various segments of meander line to come close enough to one another at the corners to create capacitive shorts at the optical frequencies (715–1250 GHz). This requires a 0.5 μm gap and a 1.5 μm overlap length for each of the intersections. Our measurements, as well as those with the MAKO devices, indicate that this indeed couples well to both polarizations, and does not impact yield or readout frequency.

We have fabricated 45-pixel TIM prototype sub-arrays and matching feedhorn blocks shown in Figure 4. To demonstrate performance, we use our cryogenic test facility in which a cryogenic blackbody illuminates the array through with a 350-micron bandpass filter. The results are shown in Figure 5. We find that the devices meet the TIM target already, showing photon noise 2.5× better. Our characterization is based on the run of noise and responsivity with stage temperature and optical loading, as well as dark noise measurements. The observed optical efficiency of the prototype system is 20%, exactly as modeled in HFSS for this backshort-less design. The same HFSS models indicate 80% once the backshort is incorporated. Our characterization also shows that the system is well-explained by the superconductivity theory and the simple model for the quasiparticle recombination rates in the aluminum (see [23], [24]) The low-loading (dark) noise floor is due to generation-recombination (GR) of residual quasiparticles at the 220 mK operating temperature. We refer the reader to [25] for more details. The device yield in the current prototype is 89% (40/45), which exceeds our 80% requirement, but we expect to improve yield prior to TIM array delivery as we fine tune the fabrication recipe.

*F. Readouts*

In order to achieve frequency domain multiplexing of KID arrays, two tasks must be accomplished: 1) A waveform consisting of a sum of frequency tones (each at an individual pixel frequency) must be generated and transmitted to the array and 2) after interacting with the pixels, the complex transmission of the individual tones must be extracted from the waveform. The first task is easily accomplished using a cyclic memory buffer and a DAC to continuously play back a pre-calculated periodic waveform. The second task can be accomplished using advanced digital-signal processing hardware. A fast, high dynamic-range ADC is followed by a Field Programmable Gate Array (FPGA) to detect the phase and amplitude modulation of the signal tones by changes in the KID inductance.

TIM will use the system developed for BLAST-TNG. This system leverages the Reconfigurable Open Architecture Computing Hardware 2 (ROACH2) platform developed by the Berkeley CASPER group which features a Xilnix Virtex-6 FPGA. An additional daughter card provides two 1 GSPS DACs and two 500 MSPS ADCs. DRAM on the ROACH2 enables waveform playback by both DACs simultaneously. The TIM system is simpler in that we do not require mixing to reach the KID resonant frequency. The readout uses custom FPGA firmware developed specifically for KID readout. The firmware implements a polyphase filter bank (PFB) followed by a fast Fourier transform (FFT) to achieve an initial stage of coarse frequency separation into 1024 channels. This is followed by multiplication of the FFT channels by pre-specified sinusoids and then low-pass filtering. This readout is working in the laboratory and is being deployed with BLAST-TNG for its flight in late 2016 or early 2017.

The total noise from the cold amplifier and readout electronics is a comfortable factor of 10 lower (in $S_{xx}$ units) than the noise from the devices themselves. A single readout can perform simultaneous complex transmission measurements of 1024 tones at a rate of 488 Hz, plenty of margin on the 900 detectors per readout chain we intend to use with TIM. We require 4 ROACH2 readout chains reading out 1024 detectors each, 2 per spectrometer. Additionally, a readout computer is used to perform post-processing of the data and storage of the time stream.

*G. Cryogenics*

The TIM cryostat and cryogenic design will copy BLAST-TNG, which builds from the flight heritage of BLAST and BLASTpol. The design is intended to hold for 28 days, based on the prototype, long enough for an LDB flight, with an exceptionally large cryogenic volume. Unlike previous designs, the BLAST-TNG cryostat uses only LHe in a single 250 L tank which forces the boiled-off He gas through two heat exchangers that cool two vapor-cooled shields, which operate at 66 K and 190 K. A separate pumped helium pot maintains a 1 K stage with 20 mW of cooling power, which contains the entire spectrometer. The detectors are cooled using pumped $^3He/^4He$ sorption fridges which provide a 300 mK sink during flight with 30 μW of cooling power for 3 days. It is backed and cycled by a $^4He$ stage. It can be recycled within 2 hrs.

*H. Gondola and Pointing System*

The TIM gondola and pointing system is designed around the successful BLAST heritage. The gondola is shown in Figure 1. It consists of a precision-pointed inner frame (composed of the primary, secondary, near-field baffle, and cryostat) supported by an external gondola. The outer frame is pointed in azimuth by a flywheel and an active pivot. The inner frame has an elevation mount with direct drive servo motors driving it relative to the outer frame. Balance of the inner frame is maintained by pumping liquid from the bottom of the frame to the top to compensate for cryogen boil-off.

The attitude determination system uses an array of pointing sensors including two sophisticated star cameras, two sets of fast, low-drift gyroscopes, a quad-GPS system, a digital Sun sensor, encoders, tilt sensors, and a magnetometer. The software is written to take full advantage of the abilities of each sensor in a hierarchical scheme where the fast, high-drift sensors (gyros) are continuously updated by the slower, absolute sensors (star cameras) and is robust against sensor failure. Optical encoders report the relative position of the inner frame to the outer frame. Motion sensing for the inner frame is provided by two sets of three orthogonally mounted, high bandwidth gyroscopes.

The absolute pointing sensors are two integrating star cameras [26] that are mounted above the receiver on the inner frame. Each star camera has an internal computer that calculates a real-time pointing solution at 1 Hz by comparing measured star separations with an on-board catalog of stars. The cameras are capable of dead reckoning. The two cameras run independently, providing failsafe redundancy. The camera system has been tested extensively on the ground and has flown five times on balloon payloads (four times on BLAST and once on the x-ray telescope InFOCμS). A comparison of simultaneous pointing solutions from both cameras gives a rms uncertainty of <2". To meet the absolute pointing requirements for TIM, we will reduce the field of view of the cameras by a factor of two and use new CCDs with enhanced quantum efficiency that roughly doubles the sensitivity in the far red (where TIM uses them). This will allow us to obtain high-accuracy, continuously updated pointing solutions for our observations.

## III. Flight Operations

TIM will perform its science from an Antarctica long duration balloon (LDB) flight. The richest field available from an Antarctic LDB flight is GOODS-S (03hr32m30s, -27d48m17s), the same field covered by BLAST in 2006 (Devlin et al. 2009). This field has deep coverage at optical through mm wavelengths. It is the target for the deep fields with HST, Chandra, Spitzer, Herschel, ALMA, and planned surveys with JWST and Euclid. There exist thousands of spectroscopic redshifts in the field, making it optimal for multi-wavelength counterpart identification and stacking analyses. For Sun avoidance, we will need to choose two fields, roughly 12 hours in RA apart. For optimal survey strategy for a power spectrum analysis for the clustering term, this field should be wide and narrow to access the low k modes in Figure 4A. For pointing reconstruction, counterpart identification, line extraction, stacking analyses, and cross correlations we will want as much ancillary data as possible. We will place our second field in the SPT-Deep field (23hr30m00s -55d00m00s) which has Herschel/SPIRE, Spitzer/IRAC, XMM, radio, and optical data. Maps, catalogs, and raw data will be made publicly available to the community to increase the legacy value of the TIM data set.

We plan for a mission length of two weeks, for a total time aloft of 336 hours. Significantly longer flight times have been achieved; up to 28 days could be used given the cryostat hold time. We assume 2 hours in every 40 are taken up with recycling of the $^3$He fridges and other overhead operations, and ~10% of the time is given to calibration and pointing observations. This leaves 280 hours to be divided between two fields, for an approximate integration time per field of ~140 hours. Our current baseline plan is to map a small deep field (0.1 deg$^2$) in GOODS-S and a wider, shallower field (~1 deg$^2$) within the SPT Deep field.

The TIM scan strategy will be very similar to that of BLAST, with scans at fixed elevation while the telescope and spectrometer slits are moved back and forth in azimuth. The inner frame is then stepped to cover the entire field, by a larger amount for GOODS-S and smaller for SPT. The nominal offset between the star camera and spectrometer slits is determined on the ground. In flight, the offset is determined on bright targets. In both cases the approach is to map the source via small raster scans. The accuracy of the mechanical alignment and the blind pointing solution from the star cameras should be better than 1', easily allowing the target to appear in the small field map. The spectrometer data are summed spectrally to produce a continuum flux and reduced in real-time to locate the source. This offset will need to be periodically re-checked throughout the flight. These bright sources will also provide calibration of the PSF and focus checks for the secondary mirror.

The flight software for TIM will be designed so that it can operate autonomously after launch. The survey fields will be decided before the flight. The autonomous scheduling system (developed for BLAST) will use schedule files that consist of a sequential list of observations or actions as a function of the Local Sidereal Time. This system is robust against temporary system failures because the telescope only needs to know the current time and location to resume operation upon recovery. Using a local sidereal clock rather than a clock fixed in some time zone, it is possible to account for purely astronomical visibility constraints (such as the RA of the Sun and of the astronomical targets) using a static description. For every launch opportunity, six schedule files are generated, which account for 3 different cases of flight latitude and longitude, and two cases of measured instrument sensitivity. The gondola uses the GPS to decide which schedule file to use, appropriate for the declination of the target field. At the beginning of the flight, the sensitivities and beam size are estimated from scans across calibrators. Based on this information the ground station team can decide which of the two sets of schedule files the instrument should use, and switch between the two using a single command.

We will use several methods for primary calibration, most based on the successful approaches used by the direct-detection millimeter-wave spectrometer Z-Spec (e.g., [27], [28]). The TIM bands are sufficiently wide that a continuum calibrator can be used to calibrate both the absolute and relative response of the channels, and channels may also be co-added. For frequency calibration we will begin with a Fourier transform spectrometer (FTS) as a laboratory calibrator, as was done for Z-Spec. Observations of sources with rich, known spectra, such as evolved stars like NGC7027 [29], [30] may be used as spectral templates. A similar technique was used successfully for Z-Spec using IRC+10216. For TIM, we have the additional frequency calibration scale of line emission from the atmosphere itself (since these lines will be narrow and do not suffer the severe pressure broadening present in ground-based observations). The telescope modulation as well as telluric-line-dominated spectral channels may also be used to reduce any effects of time-varying atmospheric emission.

The data rate from the TIM detectors will be substantial, but not prohibitive. Even at the full sample rate of the readout, 488 Hz, the 3600 detectors sampled at 32 bits produces a data rate of 7 MB/s, or 25 GB per hour. Over the course of a 14-day flight, this results in a total data volume less than 10 TB, even including overheads for housekeeping data.

## References

[1] Madau, P. & Dickinson, M. 2014, ARAA, 52, 415
[2] Casey, C. M., Narayanan, D., & Cooray, A. 2014, Phys. Rep., 541, 45
[3] Hauser, M. G. & Dwek, E. 2001, ARAA, 39, 249

[4] Lagache, G., Puget, J.-L., & Dole, H. 2005, ARAA, 43, 727

[5] Lupu, R. E. et al. 2012, ApJ, 757, 135

[6] Barber, G. J. et al. 2008, Nuclear Instruments and Methods in Physics Research A, 593, 624

[7] Martin, R. N., Romeo, R. C., & Kingsley, J. S. 2006, in Proc. SPIE, Vol. 6273, Society of Photo-Optical Instrumentation Engineers (SPIE) Conference Series, 62730P

[8] Woody, D. et al. 2008, in Proc. SPIE, Vol. 7018, Advanced Optical and Mechanical Technologies in Telescopes and Instrumentation, 70180T

[9] Su, P., Parks, R. E., Wang, L., Angel, R. P., & Burge, J. H. 2010, in International Optical Design Conference and Optical Fabrication and Testing (Optical Society of America), OTuB3

[10] Pascale, E. et al. 2012, in Society of Photo-Optical Instrumentation Engineers (SPIE) Conference Series, Vol. 8444, Society of Photo-Optical Instrumentation Engineers (SPIE) Conference Series

[11] Pascale, E. et al. 2008, ApJ, 681, 400

[12] Ferkinhoff, C. et al. 2014, ApJ, 780, 142

[13] Parshley, S. C., Ferkinhoff, C., Nikola, T., Stacey, G. J., Ade, P. A., & Tucker, C. E. 2012, in Society of Photo-Optical Instrumentation Engineers (SPIE) Conference Series, Vol. 8452, Society of Photo- Optical Instrumentation Engineers (SPIE) Conference Series

[14] Dober, B. et al. 2016, Journal of Low Temperature Physics, 184, 173

[15] Galitzki, N. et al. 2016, in Proc. SPIE, Vol. 9914, Millimeter, Submillimeter, and Far-Infrared Detectors and Instrumentation for Astronomy VIII, 99140J

[16] McKenney, C. M., Leduc, H. G., Swenson, L. J., Day, P. K., Eom, B. H., & Zmuidzinas, J. 2012, in Society of Photo-Optical Instrumentation Engineers (SPIE) Conference Series, Vol. 8452, Society of Photo-Optical Instrumentation Engineers (SPIE) Conference Series

[17] Swenson, L. J. et al. 2012, in Society of Photo-Optical Instrumentation Engineers (SPIE) Conference Series, Vol. 8452, Society of Photo-Optical Instrumentation Engineers (SPIE) Conference Series

[18] Hailey-Dunsheath, S. et al. 2014, Journal of Low Temperature Physics, 1

[19] Hailey-Dunsheath, S. et al. 2016, Journal of Low Temperature Physics, 184, 180

[20] Shirokoff, E. et al. 2014, Journal of Low Temperature Physics, 1

[21] Wheeler, J. et al. 2016, in Proc. SPIE, Vol. 9914, Millimeter, Submillimeter, and Far-Infrared Detectors and Instrumentation for Astronomy VIII, 99143K

[22] Hubmayr, J. et al. 2013, Applied Superconductivity, IEEE Transactions on, 23, 2400304

[23] Gao, J. et al. 2008, Applied Physics Letters, 92, 212504

[24] Zmuidzinas, J. 2012, Annual Review of Condensed Matter Physics, 3, 169

[25] Hailey-Dunsheath, S. et al. 2018, Journal of Low Temperature Physics, submitted, arXiv/1803.02470

[26] Rex, M., Chapin, E., Devlin, M. J., Gundersen, J., Klein, J., Pascale, E., &Wiebe, D. 2006, in Presented at the Society of Photo-Optical Instrumentation Engineers (SPIE) Conference, Vol. 6269, Groundbased and Airborne Instrumentation for Astronomy. Edited by McLean, Ian S.; Iye, Masanori. Proceedings of the SPIE, Volume 6269, pp. 62693H (2006).

[27] Kamenetzky, J. et al. 2011, ApJ, 731, 83

[28] Naylor, B. J. et al. 2010, ApJ, 722, 668

[29] Groenewegen, M. A. T. et al. 2011, A&A, 526, A162

[30] Wesson, R. et al. 2011, in Astronomical Society of the Pacific Conference Series, Vol. 445, Why Galaxies Care about AGB Stars II: Shining Examples and Common Inhabitants, ed. F. Kerschbaum, T. Lebzelter, & R. F. Wing, 607

# Study for proposal of SMILES-2 to JAXA M-class mission

Satoshi Ochiai[1,*], Philippe Baron[1], Yoshihisa Irimajiri[1], Toshiyuki Nishibori[2], Yutaka Hasegawa[2], Yoshinori Uzawa[3], Hiroyuki Maezawa[4], Makoto Suzuki[2], Akinori Saito[5], and Masato Shiotani[5]

[1]*National Institute of Information and Communications Technology, Koganei, Tokyo 184-8795, Japan*

[2]*Japan Aerospace Exploration Agency, Tsukuba, Ibaraki 305-8505, Japan*

[3]*National Astronomical Observatory of Japan, Mitaka, Tokyo 181-8588, Japan*

[4]*Osaka Prefecture University, Sakai, Osaka 599-8531, Japan*

[5]*Kyoto University, Kyoto 611-0011, Japan*

*Contact: ochiai@nict.go.jp

***Abstract*—The Superconducting Submillimeter-Wave Limb-Emission Sounder 2 (SMILES-2) is a spaceborne mission concept for the Earth's atmospheric observation using limb sounding technique. The uniqueness of SMILES-2 among spaceborne sensors for the observation of the middle and upper atmosphere is its wide altitude range of observed atmosphere and the capability to retrieve diurnal cycle of atmospheric change. To realize those distinctive features it is required to be equipped with low noise receivers at submillimeter waves including 2 THz atomic oxygen band aboard a satellite on a low Earth non-sunsynchronous orbit. SMILES-2 will use SIS receivers at 638 GHz and 763 GHz and HEB receiver at 2 THz. Those are cooled below 4.8 K by a JT cryocooler. The power consumption of the cryocooling system is expected 150 – 181 W or more. The power consumption must be reduced to balance with available power of the satellite bus. The SMILES-2 mission will be proposed to the next competition of the M-class mission.**

## Introduction

Submillimeter-wave observation provides valuable data to the Earth's atmospheric science. Wind measurement in the middle and upper atmospheres is one of the strongest motivation for using the submillimeter-wave technique. No other remote sensing technique can observe wind in such wide range of altitude. Comparing with the submillimeter wind measurement, optical technique using an atomic oxygen lines has limitations on the observation altitude range in night time [1]. The night-time wind particularly in an altitude between 105 km and 200 km is difficult to observe with optical sensors. The altitude region of 100 – 200 km is a transition layer between the upper and middle atmospheres. The upper and middle atmospheres conspicuously differ in the time scale of their dynamics. A diurnal variation is dominant in zonal wind variation of the upper atmosphere above 200 km, while the thermal and dynamical structure in the middle atmosphere varies with a time scale longer than a day or a seasonal time scale. It is shown by a numerical simulation that the zonal wind in the transition layer of 100 – 200 km has a semidiurnal variation, which propagates upward with a shorter vertical wavelength [2]. The region is important for knowing the energy transfer from the lower to upper atmosphere as atmospheric waves. Despite its importance the global measurement of the region is insufficient for revealing the vertical connection of the atmosphere. Submillimeter-wave limb sounding measuring atomic oxygen band at 2.06 THz as well as $O_2$, $H_2O$, and $O_3$ bands can largely contribute to studying the vertical connection between the middle and upper atmosphere by providing wind observation of those altitude range. Until today submillimeter-wave limb sounders, e.g. Odin/SMR, Aura/MLS, and JEM/SMILES, are dedicated to the stratospheric and mesospheric observation without atomic oxygen band. None of them was designed to measure wind speed. Consequently the night-time wind variation in an altitude of 105 – 200 km has not been measured from space.

The Superconducting Submillimeter-Wave Limb-Emission Sounder 2 (SMILES-2) is a proposed satellite mission to observe wind, temperature, and chemical compositions in the stratosphere, mesosphere, and lower thermosphere. SMILES-2 will have a hot-electron bolometer (HEB) mixer at 2 THz and SIS mixers at 638 and 763 GHz to achieve enough sensitivity to limb emissions from atomic oxygen and other stratospheric and mesospheric molecules to retrieve their horizontal movement.

## SMILES heritage

The Superconducting Submillimeter-Wave Limb-Emission Sounder (SMILES) is a payload attached to the Japanese Experiment Module (JEM) on the International Space Station (ISS). SMILES was successfully operated there from October 2009 to April 2010. SMILES has two SIS mixers at 625 GHz and 650 GHz cooled by a mechanical cooler consisting of Joule-Thomson cycle (JT) cooler and two-stage Stirling (2ST) cooler as a precooler. The local oscillator at 637.32 GHz, having no redundancy and shared by two mixers, was failed after 6 month operation on orbit. The SMILES JT cooler was healthy even after the LO failure and was working more than 8 month since the start in October 2009. After the suspension of the cryogenic system due to a breakdown of the JEM coolant system, with which the cooled stage temperature raised to room temperature level, the JT cooler could not resume its function because of clogging of the JT cycle with solidified $CO_2$ [3].

SMILES has made excellent observation on diurnal variations by taking advantage of the ISS non-sunsynchronous orbit

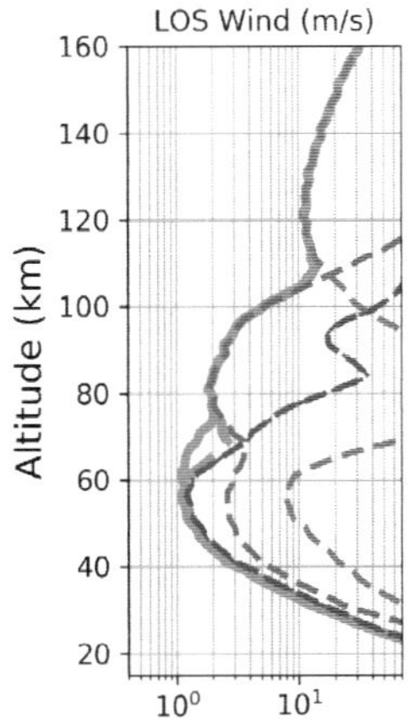


Fig. 1. Retrieval error for wind measurement estimated from a simulation study. The dark orange thick line shows error of composition of three band measurement. Dashed lines show the daytime errors for measurement of 638 GHz (red), 763 GHz (green), and 2 THz (purple). Vertical resolution of the measurement is 2.5 km. Tropical atmosphere is assumed [8].

in photochemistry related species as well as tropospheric ice clouds [4]. The SMILES measurement also revealed significant diurnal variation of the stratospheric ozone [5]. Moreover, SMILES demonstrated the successful measurement of wind between 8 and 0.01 hPa (~35–80 km) [6].

## Science Objectives of SMILES-2

We propose the SMILES-2 mission to observe wind, temperature and distribution of atmospheric minor constituents in the middle and upper atmosphere. The SMILES-2 mission has the following four science objectives [7].

(MO.1) To investigate the 4-D space-time structure of the diurnal variations in view of dynamics, chemistry, and electromagnetic processes.

(MO.2) To unveil the vertical propagation of synoptic-to-planetary scale disturbances from the middle atmosphere to the upper atmosphere.

(MO.3) To understand atmospheric variations due to energy inputs from the magnetosphere.

(MO.4) To provide benchmarks for whole atmosphere models and climate models with detailed description of the background thermal structure and distribution of minor species.

## Observation Requirements for SMILES-2

To investigate the aforementioned science objectives, wide coverage of the observation altitude range and the capability of diurnal variation of the atmosphere are the essential requisite for the SMILES-2 mission. The frequency bands of the SMILES-2 receiver are selected to cover the whole atmospheric layers from the tropopause to an altitude of about 150 km in the lower thermosphere. The orbit of the SMILES-2 satellite is a circular orbit at an altitude of 550 km with an orbit inclination of 66°, with which the local time shifts 24 hours in 3 months.

The frequency bands of SMILES-2 will be 638 GHz (LO is fixed at 638.075 GHz. IF is 10.975 – 18.975 GHz), 763 GHz (LO is fixed at 763.5 GHz. IF is 7.5 – 13.5 GHz), and 2 THz [9]. LO of the 2 THz band can be tunable to observe atomic oxygen line at 2.06 THz, OH lines around 1.83 THz, and other lines. All submillimeter mixers are used in double sideband (DSB) receiver. DSB is necessary because we need to observe both sidebands simultaneously, for example, $H_2O$ line at 752.03 GHz in LSB and $O_2$ line at 773.84 GHz in USB of 763 GHz band are to be observed. Assuming those frequency bands, the retrieval errors are estimated [8]. Figure 1 shows the retrieval error of horizontal wind speed from limb measurement with three receivers scanned over a tangent-height range between 0 km and 185 km in a scan time of 43 s. In the estimation of Fig. 1, SIS and HEB mixers are assumed for the receivers. Because the amplitude of the diurnal or semidiurnal cycle of zonal wind in the transition layer of the upper and middle atmospheres is roughly 30 m/s, the precision of the wind measurement is required to be less than 5 or 10 m/s. Although the required precision is smaller than the estimation in Fig. 1, daily zonal average or vertically lower resolution data will well satisfy the requirement. Lower noise receiver may be preferable for the lower thermospheric wind measurement. The observation requirements for SMILES-2 are summarized in [9].

Each three band contributes to wind measurement in different altitude range. Wind measurements above 110 km, an altitude range between 70 km and 110 km, and below 70 km are mainly made with 2 THz, 763 GHz, and 638 GHz receivers, respectively, as shown in Fig. 1. The $O_2$ line at 773.84 GHz largely contributes to the wind and temperature measurements in 70 – 100 km. The uncertainty of the magnetic field knowledge may affect the retrieval errors through the Zeeman effect of the $O_2$ line. The error due to the Zeeman effect is extensively studied in [10]. The SMILES-2 763 GHz receiver will observe a polarization (linear vertical) that minimizes the wind measurement error due to the Zeeman effect. The wind and temperature measurement error due to the atomic oxygen Zeeman effect in 2 THz receiver is considered to be lower than that in 763 GHz receiver. The magnetic field is potentially measured using the 773.84 GHz $O_2$ line. The retrieval errors of the vertical and horizontal magnetic field are estimated 30 – 100 nT and 100 – 300 nT in high latitude, respectively [10].

## SMILES-2 Instrument

The SMILES-2 mission consists of two 75-cm aperture antennas, calibration system, local oscillators (LO), superconducting heterodyne receivers at frequencies of 638 GHz, 763 GHz, and 2 THz, IF chains, microwave spectrometers, and other subsystems. The superconducting mixers (Nb-SIS, NbTiN-SIS, and NbN-HEB for 638 GHz, 763 GHz, and 2 THz mixers, respectively) are cooled down below 4.8 K by a mechanical cooler. LOs are injected through a beam splitter on the outside of the cryostat, and not introduced directly into the cold stage via waveguide in order to avoid excess heat flux into the cold stage. More detailed description of the SMILES-2 instrument with a block diagram is found in [11].

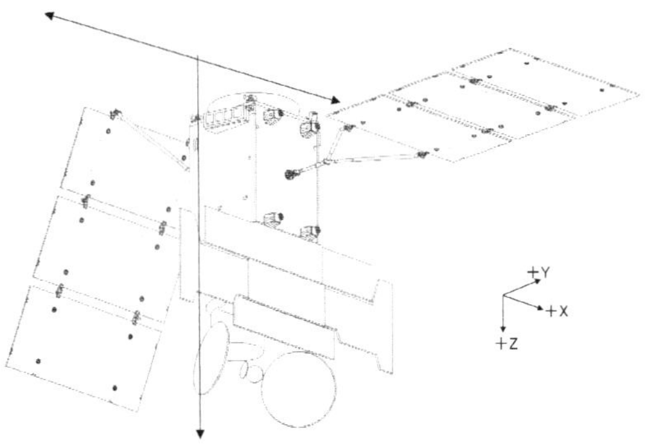


Fig. 2. Conceptual image of the SMILES-2 satellite

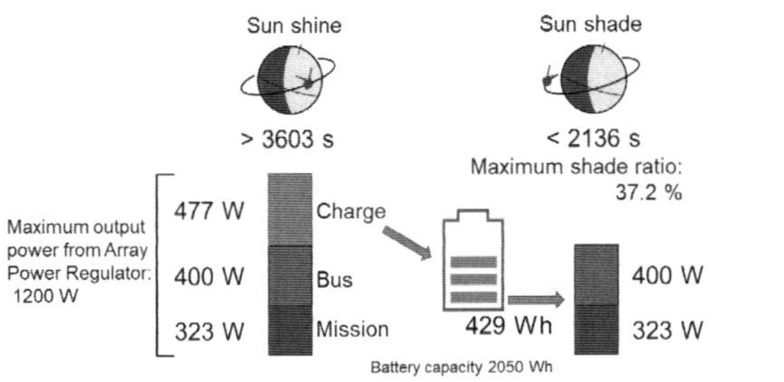


Fig. 3. Schematic description of the available power for the SMILES-2 mission. The efficiency of the battery is assumed 0.9.

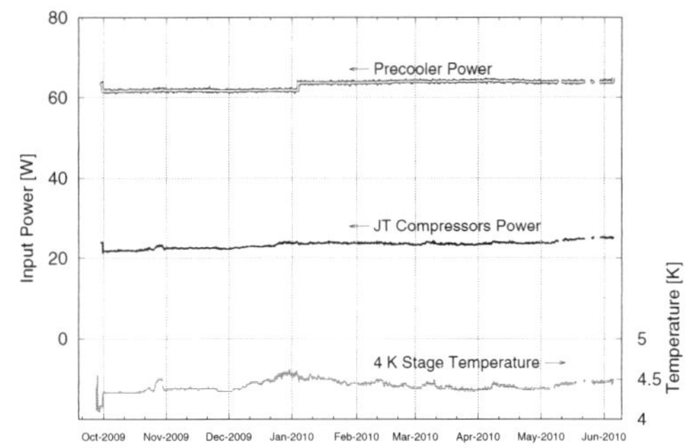


Fig. 4. Input power to JT compressors and 2ST precooler and the temperature of the 4 K stage of the JEM/SMILES cryogenic system in 2009 – 2010.

SMILES-2 is being designed to board Japan Aerospace Exploration Agency (JAXA) M-class satellite, which should be launchable with JAXA's Epsilon rocket, and has a mass of about 500 kg or less. An M-class satellite is usually supposed to use a standard M-class satellite bus unit that has a simplified interface with mission unit (Fig.2). There is a limitation of available power at the Array Power Regulator (APR) on the satellite bus unit. The total output power of APR is 1,000 – 1,200 W depending on temperature and other conditions. For the SMILES-2 orbit the ratio of the time where the satellite in the Earth's shade to the orbit period can be 37.2 % at worst. Figure 3 describes the available power for the mission at the worst case. The available power for the SMILES-2 mission will be 323 W as Fig. 3, or 203 W when the output power of APR is 1,000 W.

### Cryocooling system

The 4 K cryocooling system consists of a Joule–Thomson (JT) cycle cooler and two-stage Stirling (2ST) cooler as a precooler. The precooler is also used to cool the low noise amplifiers on 20 K stage and thermal shields at 20 K and 100 K in the receiver cryostat. The cooler configuration and the structure of the cryostat basically follow the design of those in JEM/SMILES. The design lifetime of the JEM/SMILES cryocooling system was 1 year. The cryocooling system for SMILES-2 is expected to have a lifetime of 3 years, or hopefully 5 years. The ground demonstration model of JT cryocooling system works more than 3 years without remarkable degradation of performance [3].

The cryocooling system of SMILES-2 is required to work with power as low as possible, even though JT cooler is regarded as the most efficient 4 K cooling system. Figure 4 shows input power to JT compressors and 2ST precooler of JEM/SMILES with the temperature of 4 K stage. JT compressors and 2ST cooler consumed 22 W and 62 W, respectively, at the beginning of operation in space. They increased to 25 W and 64 W after 8 month. The total power consumption of the cooling system including driver electronics for the coolers is estimated 153 W at the beginning, and 160 W after 8 month. If the SMILES-2 cryostat and the receiver system in it are almost comparable with those of JEM/SMILES, the power consumption of the SMILES-2 cooling system can be expected to be the same level. The power consumption expected at the end-of-life (EOL), that is 3 years for SMILES-2, will be much larger than the power consumption after 8 month. In the ground demonstration model the input power to keep the cooling power increased by 50 % for JT compressors and 15 % for 2ST precooler. The total power consumption is estimated to be 181 W with that increment. Due to an increase in the number of receivers from 2 for JEM/SMILES, i.e. 625 GHz SIS and 650 GHz SIS, to 3, i.e. 638 GHz SIS, 763 GHz SIS, and 2 THz HEB, required input power to the coolers of SMILES-2 may increase further. Considering the total available power for the mission unit at EOL, the power consumption must be smaller than the JEM/SMILES cryocooling system. To reduce the increased power consumption lowering the ambient temperature of the cryocooling system will be necessary. Another possibility of power reduction will be to reduce the heat dissipation of low noise amplifiers on 20 K stage.

### Status of mission proposal

In the JAXA schedule, they plan to launch M-class mission every two years. A proposal of the SMILES-2 mission submitted to JAXA in 2018 was not successful. The SMILES-2 mission still stays under Pre-Phase A1a. The main criticisms to the proposal are that the required power for SMILES-2 is larger than the output power of a standard M-class satellite bus, and that the estimated cost of SMILES-2 overruns the limitation of M-class projects. Solving the power balance problem is ongoing. The cost reduction is discussed in the SMILES-2 working group. The schedule of the next opportunity for M-class mission is unknown.

## Conclusions

The proposed SMILES-2 mission has a capability of measurement of diurnal variation of horizontal wind in the stratosphere, mesosphere, and lower thermosphere. SMILES-2 can also measure temperature field, distribution of atmospheric minor constituents, and magnetic field. SMILES-2 has 638 GHz, 763 GHz, and 2 THz superconducting receivers that are cooled below 4.8 K by JT cooler with 2ST precooler. The power consumption of the cryocooling system needs to be reduced to fit the power availability of the JAXA M-class satellite. We are preparing a SMILES-2 mission proposal to the next opportunity of M-class mission.

## Acknowledgment

The authors would like to acknowledge the SMILES-2 working group. The Superconducting Submillimeter-Wave Limb-Emission Sounder (SMILES) on the Japanese Experiment Module (JEM) is a mission developed and operated jointly by the National Institute of Information and Communications Technology (NICT) and the Japan Aerospace Exploration Agency (JAXA).

## References

[1] C. R. Englert, J. M. Harlander, C. M. Brown *et al.*, "Michelson Interferometer for Global High-Resolution Thermospheric Imaging (MIGHTI): Instrument Design and Calibration," *Space Sci. Rev.*, vol. 212, no. 1–2, pp. 553–584, 2017, 10.1007/s11214-017-0358-4.

[2] Y. Miyoshi, H. Fujiwara, H. Jin, and H. Shinagawa, "A global view of gravity waves in the thermosphere simulated by a general circulation model," *J. Geophys. Res. Space Physics*, vol. 119, pp. 5807–5820, 2014, 10.1002/2014JA019848.

[3] K. Narasaki, S. Tsunematsu, K. Otsuka, K. Kanao, A. Okabayashi, S. Yoshida, H. Sugita, Y. Sato, K. Mitsuda, T. Nakagawa, and T. Nishibori, "Lifetime Test and Heritage On-Orbit of SHI Coolers for Space Use," *Cryocoolers*, vol. 19, pp. 613–622, 2016.

[4] J. H. Jiang, *et al.*, "Evaluating the Diurnal Cycle of Upper-Tropospheric Ice Clouds in Climate Models Using SMILES Observations," *J. Atmos. Sci.*, vol. 72, pp. 1022–1044, 2015, 10.1175/JAS-D-14-0124.1.

[5] T. Sakazaki, M. Fujiwara, C. Mitsuda, K. Imai, N. Manago, Y. Naito, T. Nakamura, H. Akiyoshi, D. Kinnison, T. Sano, M. Suzuki, and M. Shiotani, "Diurnal ozone variations in the stratosphere revealed in observations from the Superconducting Submillimeter-Wave Limb-Emission Sounder (SMILES) on board the International Space Station (ISS)," *J. Geophys. Res. Atmos.*, vol. 118, pp. 2991–3006, 2013, 10.1002/jgrd.50220.

[6] P. Baron, D. P. Murtagh, J. Urban, H. Sagawa, S. Ochiai, Y. Kasai, K. Kikuchi, F. Khosrawi, H. Körnich, S. Mizobuchi, K. Sagi, and M. Yasui, "Observation of horizontal winds in the middle-atmosphere between 30S and 55N during the northern winter 2009–2010," *Atmospheric Chemistry and Physics*, vol. 13, no. 13, pp. 6049–6064, 2013, 10.5194/acp-13-6049-2013.

[7] M. Shiotani, A. Saito, T. Sakazaki, S. Ochiai, P. Baron, T. Nishibori, M. Suzuki, T. Abe, H. Maezawa, and S. Oyama, "A proposal for satellite observation of the whole atmosphere – Superconducting Submillimeter-Wave Limb-Emission Sounder (SMILES-2)," *Proc. IGARSS 2019*, submitted.

[8] P. Baron, S. Ochiai, D. Murtagh, H. Sagawa, A. Saito, M. Shiotani, and M. Suzuki, "Performance assessment of Superconducting Submillimeter-Wave Limb-Emission Sounder-2 (SMILES-2)," *Proc. IGARSS 2019*, submitted.

[9] S. Ochiai, P. Baron, Y. Irimajiri, Y. Uzawa, T. Nishibori, M. Suzuki, A. Saito, and M. Shiotani, "Superconducting Submillimeter-Wave Limb-Emission Sounder, SMILES-2, for middle and upper atmospheric study," *Proc. IGARSS 2018*, July, 2018, 10.1109/IGARSS.2018.8518300.

[10] P. Baron, S. Ochiai, E. Dupuy, R. Larsson, H. Liu, N. Manago, D. Murtagh, S. Oyama, H. Sagawa, A. Saito, T. Sakazaki, M. Shiotani, and M. Suzuki, "Potential for the measurement of MLT wind, temperature, density and geomagnetic field with Superconducting Submillimeter-Wave Limb-Emission Sounder-2 (SMILES-2)," *Atmos. Meas. Tech.*, submitted.

[11] S. Ochiai, P. Baron, Y. Irimajiri, T. Nishibori, Y. Hasegawa, Y. Uzawa, H. Maezawa, T. Manabe, A. Mizuno, T. Nagahama, K. Kimura, M. Suzuki, A. Saito, M. Shiotani, "Conceptual study of Superconducting Submillimeter-Wave Limb-Emission Sounder-2 (SMILES-2) receiver," *Proc. IGARSS 2019*, submitted.

# Wednesday, April 17, 2019

# Session XI. QCL THz Sources

# A Double-Metal QCL with Backshort Tuner

M.R. Mertens[1], M. Justen[1], C. Bonzon[2], K. Ohtani[2,3], M. Beck[2], U.U. Graf[1] and J. Faist[2]

The development towards many-pixel arrays for astronomic terahertz heterodyne detection has led to an ever-increasing demand for high power local oscillator sources. Quantum Cascade Lasers provide high optical power in particular at frequencies above ~2 THz, where frequency multiplication techniques are severely limited. The main drawback with QCLs is the fixed operating frequency with only a few GHz of tuning bandwidth.

We have recently developed a new tuning technique based on mechanical variation of the cavity length. The system consists of a double-metal QCL with a design frequency of 1.9 THz which is terminated with a patch antenna array on one side [1,2] and is coupled by a taper section to a microstrip transmission line. On top of this microstrip we position a Bragg-reflector made from a series of perpendicular gold stripes on a thin silicon membrane. In this way the total cavity length can be controlled by changing the tuner membrane position using piezo actuators.

Our measurements confirm that a tuning range over a complete FSR of 30 GHz is attainable with the fabricated device (L~1mm). Although the q-factor without tuner membrane is sufficiently high that lasing occurs at high currents, both optical output power and threshold current improve significantly when the membrane is applied.

The QCL provides enough CW power to pump an heterodyne array while consuming a moderate electrical power of 1.6W. We operate the device with a low vibration Stirling cooler at 50K.

[1] 1. Physikalisches Institut, University of Cologne, Cologne, Germany.
[2] Institute for Quantum Electronics, ETH-Zürich, CH-8093 Zürich, Switzerland
[3] now with: Niels Bohr Institute, Condensed Matter Physics, Copenhagen, Denmark.

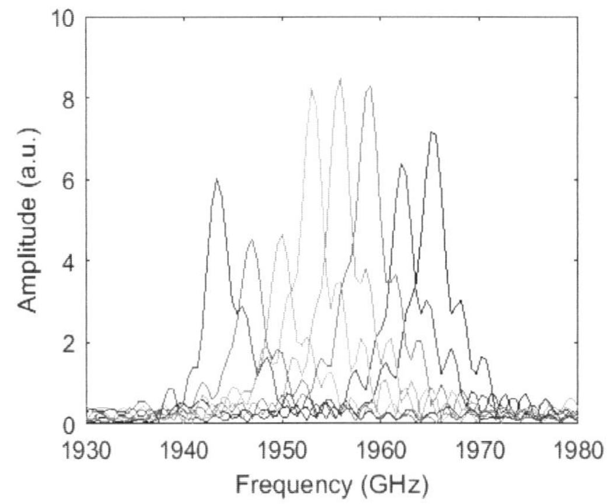


Fig. 1. Recorded spectra for different tuner membrane positions. The tuner membrane was moved between each measurement to change the cavity length, resulting in a shift of the laser emission frequency. Peak shape is influenced by zero-padding of the FFT.

### References

[1] M. Justen, "2D patch antenna array on a double metal quantum cascade laser with >90% coupling to a Gaussian beam and selectable facet transparency at 1.9 THz," *Optics Letters*, 4590 Vol. 41, No. 19, Oct. 2016.

[2] L. Bosco, "A patch-array antenna single-mode low electrical dissipation continuous wave terahertz quantum cascade laser" *Applied Physics Letters 109, 201103,* Nov. 2016.

NOTES:

# Frequency tuning of terahertz quantum-cascade lasers by optical excitation

M. Wienold[1, 2], T. Alam[2], X. Lü[3], K. Biermann[3], L. Schrottke[3], H. T. Grahn[3], and H. W. Hübers[1, 2]

The ability to tune the frequency is an important requirement for a local oscillator (LO) in a heterodyne spectrometer. When employed as LO, terahertz (THz) quantum-cascade lasers (QCLs) are usually tuned by changing their driving current or temperature [1]. This yields a frequency tunability of a few GHz. In addition to the small frequency coverage, the output power of the QCL does not remain constant using this method.

Light-induced frequency tuning is an alternative approach which is based on the optical excitation of carriers in the QCL cavity. We demonstrate the feasibility of this approach by molecular laser absorption spectroscopy. For a 3.1-THz QCL, we obtain a frequency tuning range of about 40 GHz for continuous-wave operation, which represents a ten-fold improvement over the usually employed tuning by current.

In our experiments, we illuminated the back facet of a THz QCL with light from a near-infrared diode laser improving a recently reported approach by Hempel *et al.* [2]. For a well-defined excitation, we used a confocal microscope setup with a 10× objective lens. The largest tuning of almost 40 GHz was obtained by exciting the substrate underneath the QCL active region with a high-power multimode diode laser at 808 nm with up to $P$=3.5 W output power as shown in Fig. 1. Using a single-mode diode laser, we obtained for the same device a tuning range of more than 12 GHz with as little as 200 mW.

We explain the tuning effect by the excitation of an electron-hole plasma in the vicinity of the back facet, which locally changes the dielectric constant inside the QCL cavity. The calculated frequency shift follows a square root power dependence as shown in the inset of Fig. 1, which is also observed experimentally [3].

In addition to wideband laser absorption spectroscopy, the method is of interest for frequency alignment and frequency stabilization of QCLs employed in heterodyne spectrometers [1]. By replacing the microscope optics by a fiber coupling scheme, very compact configurations become feasible, which are straightforward to implement in instrumentation hardware. Since no dedicated QCL development is required, light-induced frequency tuning can be readily applied to a large class of devices.

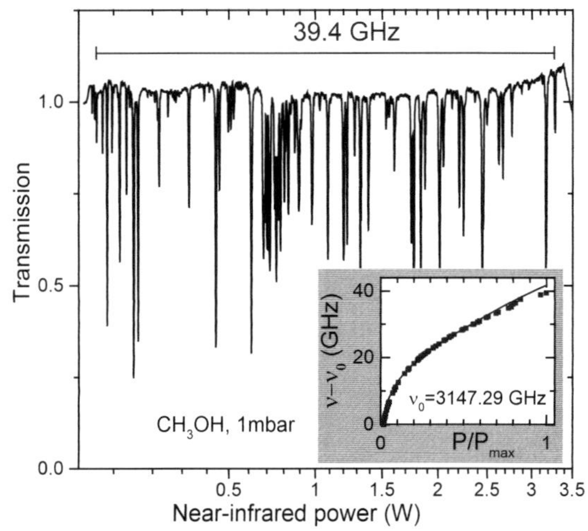


Fig. 1. Transmission through a methanol gas cell for near-infrared excitation with a high-power diode laser. Inset: Frequency shift according to the JPL molecular catalog. The solid line represents a square root fit to the data.

[1] German Aerospace Center (DLR), Institute of Optical Sensor Systems, Rutherfordstr. 2, 12489 Berlin, Germany

[2] Humboldt-Universität zu Berlin, Department of Physics, Newtonstr. 15, 12489 Berlin, Germany

[3] Paul-Drude-Institut für Festkörperelektronik, Leibnitz-Institut im Forschungsverbund Berlin e. V., Hausvogteiplatz 5 – 7, 10117 Berlin, Germany.

## References

[1] H. Richter, M. Wienold, L. Schrottke, K. Biermann, H. T. Grahn, and H.-W. Hübers, "4.7-THz local oscillator for the GREAT heterodyne spectrometer on SOFIA," *IEEE Trans. Terahertz Sci. Technol.* vol. 5, pp. 539–545, Jul. 2015.

[2] M. Hempel, B. Röben, L. Schrottke, H.-W. Hübers, and H. T. Grahn, "Fast continuous tuning of terahertz quantum-cascade lasers by rear-facet illumination," *Appl. Phys. Lett.* vol. 108, 191106, May 2016.

[3] T. Alam, M. Wienold, X. Lü, K. Biermann, L. Schrottke, H. T. Grahn, and H. W. Hübers, "Wideband high-resolution terahertz spectroscopy by light-induced frequency tuning of quantum cascade lasers," *Opt. Express* vol. 27 , in press (2019).

NOTES:

# A compact 4.7-THz source based on a high-power quantum-cascade laser with a back-facet mirror

T. Hagelschuer[1], H. Richter[1], M. Wienold[1,2], X. Lü[3], B. Röben[3], K. Biermann[3], L. Schrottke[3], H. T. Grahn[3], and H.-W. Hübers[1,2]

A key challenge for heterodyne receivers operating above approximately 3 THz is the local oscillator (LO), which has to be a compact source requiring high output power and low electrical input power. The neutral atomic oxygen (OI) fine structure line at 4.7448 THz is of particular interest for several spaceborne missions that are proposed for measuring this transition. Examples are OST (Origins Space Telescope), LOCUS (Low-Cost UpperAtmosphere Sounder) and FIRSPEX (Far-Infrared Spectroscopic Explorer). THz quantum-cascade lasers (QCLs) are compact semiconductor lasers, which are very attractive as LO. During the last five years, a 4.7-THz LO based on a QCL has been in operation on SOFIA (the Stratospheric Observatory for Infrared Astronomy). It is part of the heterodyne receivers GREAT (German REceiver for Astronomy at Terahertz frequencies) and upGREAT [1]. While the QCL is only about 1 mm long, the mechanical cryocooler, which is required for laser operation, determines the mass and power budget of the LO. The LO of GREAT and upGREAT has a mass of about 40 kg and a power consumption of about 400 W. For spaceborne applications, an LO with the same performance has to be significantly more compact and to consume less power than for airborne instruments.

We report on the development of a compact, easy-to-use source, which is based on a QCL with a mirror at its back facet. The QCL is operated in a compact, low-input-power linear Stirling cooler (AIM SL400). The QCL has been developed for optimum LO performance. The active region consists of a GaAs/AlAs heterostructure [2], which provides a three times higher wall plug efficiency than QCLs based on GaAs/(Al,Ga)As heterostructures such as currently used for GREAT and upGREAT [1]. This leads to reduced QCL pump powers and a minimum of dissipated heat, which opens the path for operation in a miniature cryocooler. For improvement of the LO power, the QCL has a mirror at its back facet.

[1] German Aerospace Center (DLR), Institute of Optical Sensor Systems, Rutherfordstr. 2, 12489 Berlin, Germany.

[2] Humboldt-Universität zu Berlin, Department of Physics, Newtonstr. 15, 12489 Berlin, Germany.

[3] Paul-Drude-Institut für Festkörperelektronik, Leibniz-Institut im Forschungsverbund Berlin e. V., Hausvogteiplatz 5–7, 10117 Berlin, Germany.

Figure 1 presents a photograph of the 4.7 THz source. The miniature cryocooler consists of two components, a cold finger mounted in a vacuum housing and a cylindrical compressor unit operating at a cycling frequency of 37 Hz. A copper submount with the QCL is mounted to the cold finger. The THz beam, indicated by the red arrow, passes the vacuum housing through an exit window made of poly-4-methylpentene-1 (TPX). The whole system covers a volume of approximately 300×120×140 mm$^3$ with a mass of less than 4 kg. The QCL provides up to 10 mW output power in an almost Gaussian-shaped beam profile and fulfills the requirements for observation of the OI transition as demonstrated by measuring the absorption spectrum of $CH_3OH$ gas in an absorption cell.

The results indicate that a compact LO based on a high-power QCL with a back-facet mirror and a miniature cryocooler is feasible for spaceborne applications.

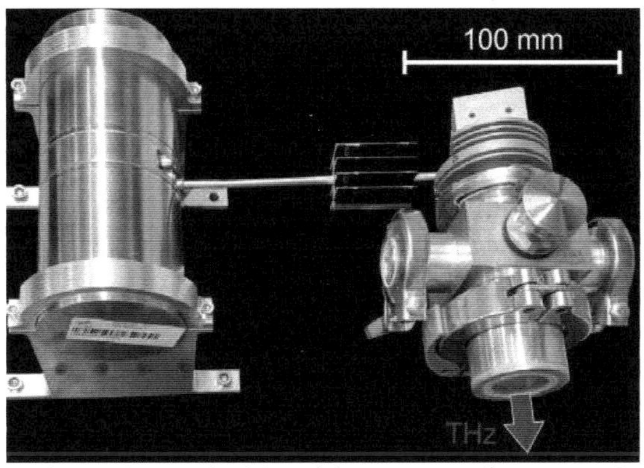


Fig. 1. Photograph of the miniature cryocooler system. The red arrow indicates the direction of the exiting THz beam.

REFERENCES

[1] H. Richter, M. Wienold, L. Schrottke, K. Biermann, H. T. Grahn, and H.-W. Hübers, "4.7-THz local oscillator for the GREAT heterodyne spectrometer on SOFIA", *IEEE Trans. Terahertz Sci. Technol.*, vol. 5, 539–545, July 2015.

[2] L. Schrottke, X. Lü, G. Rozas, K. Biermann, and H. T. Grahn, "Terahertz GaAs/AlAs quantum-cascade lasers," *Appl. Phys. Lett.*, vol. 108, 102102, Mar. 2016.

NOTES:

# 81-beam supra-THz local oscillator by a phase grating and a quantum cascade laser

Y. Gan[1,2], B. Mirzaei[1,3], J. R. G. Silva[1,2], A. Khalatpour[4], Qing Hu[4] , C. Groppi[5], J. V. Siles[6] and

J.R. Gao[1,3],

THz heterodyne detection, combining the light intensity measurement with an exceptionally high spectral resolution, has been widely used to study astronomic fine structure lines at THz frequencies (0.3-1 THz) as well as the lines at supra-THz frequencies (>1 THz). In order to effectively map the lines from a large area of the sky, future heterodyne receivers need large arrays. Generating multiple beam local oscillators (LOs) is thus one of the key technologies demanded to realize such a goal.

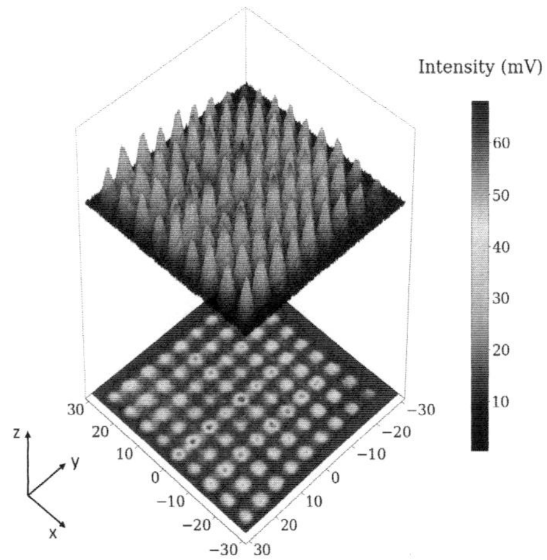


Fig. 1. Measured beam pattern from a 81-pixel Fourier grating using a 3.8 THz unidirectional photonic wire quantum cascade laser as the input source. The incident angle of the beam is 12° in y direction and 0° in x direction.

In supra-terahertz region, quantum cascade laser (QCL) is the most promising LO source. However, generating multiple QCLs with the exactly same frequency and phase is quite challenging. A phase grating[1], illuminated by a single THz QCL, has demonstrated multiple beam LOs at supra-THz[2]. However, so far it has been limited to less than 10 beams. Here we report for the first time a 81-beam LO using a newly simulated and fabricated Fourier grating and a novel quantum cascade laser that is based on an unidirectional photonic wire concept and that emits single mode line at 3.8 THz[3].

We succeeded in measuring 81 diffraction beams at 3.8 THz. The result is shown in figure 1, where the bottom panel plots the measured far field output beams in the 2D format, while the top panel illustrates the beams in a 3D format. Thanks to a high diffraction efficiency of the grating and a high output power of the QCL ( >10 mW), we have achieved a high signal-to-noise ratio in the measurement, allowing us to do all the analysis including the diffraction efficiency. The latter is a ratio of the total diffraction beam power to that of the incoming beam. We can compare an experimental efficiency with the calculated one reliably. We believe our result is a milestone for the development of multiple beam supra-THz LOs towards ultimately a practical 100-pixel large heterodyne array.

## References

[1] Urs U. Graf, Stefan Heyminck, "Fourier Grating as Submillimeter Beam splitter," IEEE transaction on antennas and propagation, Vol. 49, 542 (2001).

[2] B. Mizaei, J. R. G. Silva, D. J. Hayton, C. Groppi, T. Y. Kao, Q. Hu, J. L. Reno, and J. R. Gao, "8-beam local oscillator array at 4.7 THz generated by a phase grating and a quantum cascade laser," Opt. Express 25, 29587-29596 (2017).

[3] Khalatpour, Ali, John L. Reno, Nazir P. Kherani, and Qing Hu . "Unidirectional photonic wire laser." Nature Photonics 11.9 (2017): 555.

[1] SRON Netherlands Institute for Space Research, Groningen/Utrecht, the Netherlands

[2] Kapteyn Astronomical Institute, University of Groningen, 9747 AD, Groningen, The Netherlands

[3] Faculty of Applied Science, Delft University of Technology, Delft, the Netherlands

[4] Department of Electrical Engineering and Computer Science, Massachusetts Institute of Technology, Cambridge, MA 02139, USA

[5] School of Earth and Space Exploration, Arizona State University, AZ, USA

[6] NASA Jet Propulsion Laboratory, Pasadena, CA, USA

NOTES:

# Wednesday, April 17, 2019

# Session XII. Radars, Systems, Backend

# Digital high-resolution wide-band Fast Fourier Transform Spectrometer

Bernd Klein[1,2], Stefan Hochgürtel[1], Ingo Krämer[1], Andreas Bell[1], and Gerrit Grutzeck[1]

Since a few years digital Fast Fourier Transform Spectrometers (FFTS), based on high-speed analog-to-digital converter (ADC) and high-performance field-programmable gate array (FPGA) chips, have become a standard for heterodyne receivers, particularly in the millimeter and submillimeter wavelength range [1]. The high dynamic range of today's high-speed ADCs with 10 or 12-bit allows observing strong continuum sources and bright maser lines without signal loss. FFT spectrometers are calibration- and aging free and operate very stable with long Allan-variance times of several 1000 seconds. In addition, FFT spectrometers have proven to be extremely reliable and robust, even in very harsh environmental conditions such as at the APEX telescope at an altitude of 5100 meters or on board SOFIA [2].

At the Max-Planck-Institut für Radioastronomie, the FFTS technology has been advanced over the last 15 years from 50 MHz to 4 GHz instantaneous bandwidth today. Our latest dFFTS4G spectrometer enables spectral analysis of two synchronously sampled signal inputs with an instantaneous bandwidth of 2 x 4 GHz with 2 x 64k spectral channels on one single 160 x 100 mm euro-sized board. Up to eight dFFTS4G boards can be housed in one 19" crate together with power-supplies and one FFTS-controller, so that a total bandwidth of 64 GHz per crate can be processed. Figure 1 shows a photo of the dFFTS4G spectrometer board.

Because even the fastest ADCs available today are not yet sufficient for broadband applications in radio astronomy, two 4 GS/s ADCs are time-interleaved on the dFFTS4G board to synthesize the behavior of an 8 GS/s converter. The main challenges with time-interleaving are accurate phase alignment of sampling-clock edges between channels and compensating for manufacturing variations that inherently occur between ADC chips. Accurately matching the gain, offset, and clock phase between separate ADCs is very challenging, especially because these parameters are frequency-dependent. For the dFFTS4G we optimized our adaptive FPGA-based calibration routine that measures an injected fixed frequency line and calculates the best parameter for gain, offset and clock phase. Applying this calibration scheme, no interleaving artefacts are noticed, even in long integrations.

The design of a new generation of FFT spectrometer for the requirements of future heterodyne multi-pixel receiver arrays, such as CHAI for CCAT-prime, is the goal of our further FFTS development. Especially with large array receivers, the simplest possible system layout of all components is particularly important so that these projects can be realized at all. In particular, the analog mixing of the IF bandwidth (4 – 8 GHz) to baseband (0 – 4 GHz), which is still required today, is to be eliminated. The availability of new RF-ADCs, which can sample analog signals from DC to 8 GHz, enables direct bandpass sampling and thus significantly simplified analog IF processing for future heterodyne receivers.

Fig. 1. Photograph of the dFFTS4G spectrometer board. The dFFTS4G uses four 12-bit 4 GS/s ADCs (TI ADC12D4000) and a high-performance Xilinx Virtex-7 960 FPGA for digital signal processing. Two ADCs each represent an ADC pair and sample the same input signal with 180 degree phase shift. By this time-interleave method an 8 GS/s ADC can be built up, which can acquire 4 GHz signal bandwidth.

[1] Max-Planck-Institut für Radioastronomie, Bonn, Germany.
[2] University of Applied Sciences Bonn-Rhein-Sieg, Sankt Augustin, Germany.

## References

[1] B. Klein, et al., „High-resolution wide-band Fast Fourier Transform Spectrometer" A&A 542, L3, 2012

[2] C. Risacher, et al., „First Supra-THz Heterodyne Array Receivers for Astronomy with the SOFIA Observatory", IEEE Transactions on Terahertz Science and Technology 6, 199, 2016

NOTES:

# Validation Measurements of Humidity Profiling in Rain Using a 170 GHz Differential Absorption Radar

Ken B. Cooper[1], Richard J. Roy, Matthew Lebsock, Robert Dengler, Raquel Rodriguez-Monje, Jose Siles, Luis Millan, and Maria Alonso-Delpino

A tunable 167-174.8 GHz differential absorption radar has been developed as a remote sensing tool for measuring range-resolved absolute humidity inside of clouds and rain [1]. This new capability complements the established humidity profiling technique of passive radiometric remote sensing, since the latter suffers from biases induced by scattering by cloud and precipitation droplets [2].

The VIPR (Vapor In-Cloud Profiling Radar) system, shown in Fig. 1a, has been designed based on prior short-range, low-power radar systems that the Jet Propulsion Laboratory has developed at 95, 340, and 680 GHz. Its salient features are a Schottky diode frequency-doubler source with nearly 500 mW of continuous-wave transmit power [3]; ultra-high-isolation quasioptical transmit/receive duplexing; digital chirp generation and FFT-based range compression; and a 60 cm diameter primary aperture with nearly 58 dB antenna gain.

To validate VIPR's ability to measure absolute water vapor concentration, relative-humidity and temperature sensors were deployed on the ground during light rain at a distance of 820 m from the radar. VIPR's beam was pointed into the sky just above the sensors. Absolute humidity averaged over a 60 m wide swath centered over the sensors' range was retrieved from VIPR's differential absorption signal using the methods previously described in [1]. Fig. 1b shows a comparison of the radar measurements (blue) and the mean value of the in situ results (black) over approximately three hours of observation in moderate rain.

The data of Fig. 1b show that the radar measurements provide an accurate measure of the local water vapor content with a mean value within ~10-20% of the sensor value. However, the significant scatter in the radar measurements, which is approximately ten times higher than the theoretical expectation from [1], is not fully understood yet. One possible contribution to this uncertainty is poorly decorrelated speckle arising from a too-short (1 ms) pulse repetition interval with respect to the beam width and rainfall rate. Another is and frequency-dependent dispersion in the rain drop scattering cross-section that competes with the magnitude of change in water vapor absorption over the radar bandwidth. Additional measurements and analysis of these effects will be presented at the ISSTT meeting, including a discussion of how they will affect an upcoming airborne deployment of VIPR in late 2019.

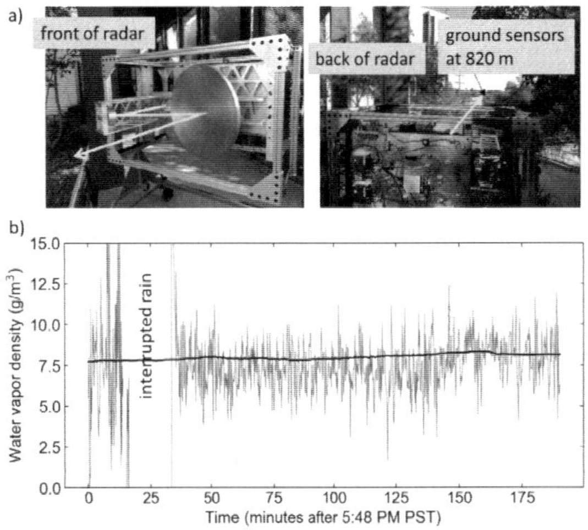


Fig. 1. a) VIPR system hardware showing how its 167-174.8 GHz beam is pointed above a distant hillside. b) Retrieved absolute water vapor content over ground sensors at 820 m (blue) are in good agreements with in situ measurements (black), but with higher than anticipated levels of measurement scatter.

[1] All authors are with Jet Propulsion Laboratory, California Institute of Technology, Pasadena CA 91109.  Government Sponsorship Acknowledged.

## References

[1] Richard J. Roy, et al., "Boundary-layer water vapor profiling using differential absorption radar," *Atmospheric Measurement Techniques*, vol. 11, no. 12, 2018.

[2] E.J. Fetzer, et al., "Biases in total precipitable water vapor climatologies from Atmospheric Infrared Sounder and Advanced Microwave Scanning Radiometer," *Journal of Geophysical Research: Atmospheres*, vol. 111 (D9), 2006.

[3] Jose V. Siles, et al., "A New Generation of Room-Temperature Frequency Multiplied Sources with up to 10x Higher Output Power in the 160 GHz-1.6 THz Range," *IEEE Transactions on Terahertz Science and Technology*, vol. 8 (6), 2018.

NOTES:

# Cold-Source Noise Temperature Measurements with a Vector Network Analyzer Frequency Extender at WR-6.5

Theodore J. Reck[1], Steve Durant[1] and Jeffrey Hesler[1]

The cold-source method is adapted to Vector Network Analyzer (VNA) frequency extenders to enable measurement of both noise temperature and S-parameters with the same test setup. With LNA MMICs steadily moving upward in frequency, a single testing solution for characterizing a device's sensitivity and s-parameters would accelerate on-wafer testing. Particularly for cryogenic measurements, such an approach could save many hours of cryo-cycling and improved measurement accuracy by capturing both characteristics within seconds of each other.

Noise temperature is typically measured at Terahertz frequencies with the Y-factor technique, which presents the DUT with noise power of two different, well known temperatures. While this technique is well-proven, it requires either free-space optics to a temperature controlled black-body or a noise source. Y-factor measurements only capture noise temperature and conversion loss, so if s-parameters are required, another test setup must be applied. To avoid this second setup, the cold-source technique relies on accurate knowledge of the s-parameters of the DUT to calculate the noise temperature from the noise power emitted by the device [1,2]. With proper characterization of the receiver's bandwidth and noise temperature beforehand, the noise temperature of the DUT can then be calculated as:

$$T_D = \frac{P_{out}^D}{kG_rB_rG_D} - \frac{T_r}{G_D} - T_0$$

Where $P^D_{out}$ is the noise power of the DUT measured by the receiver, $T_r$ is the noise temperature of the receiver, $G_r$ is the gain of the receiver, $B_r$ is the bandwidth of the receiver, $G_D$ is the gain of the DUT, $T_0$ is the temperature of the load presented to the DUT and $k$ is Boltzmann's constant.

With VNA frequency extenders now available, high accuracy s-parameters are relatively easy to measure up to 1 THz [3]. To accurately measure noise temperature with the cold-source technique, the receiver temperature must not dominate the total power measured. To achieve this, the frequency extender modules are modified to minimize the receiver's noise temperature at the expense of test-port power. In this study, the modification of VDI's WR-6.5 module achieves a noise temperature between 5000K and 8000K with a test-port power of 10uW.

Figure 1 compares the noise temperature measurements of a standard quasi-optical Y-factor to the cold-source technique for a WR-6.5 power amplifier. While the high noise of this device makes it easier to measure with the cold-source technique, an analysis will be presented that shows that noise temperatures down to 100K can be reliably measured is the DUT has gain over 10dB.

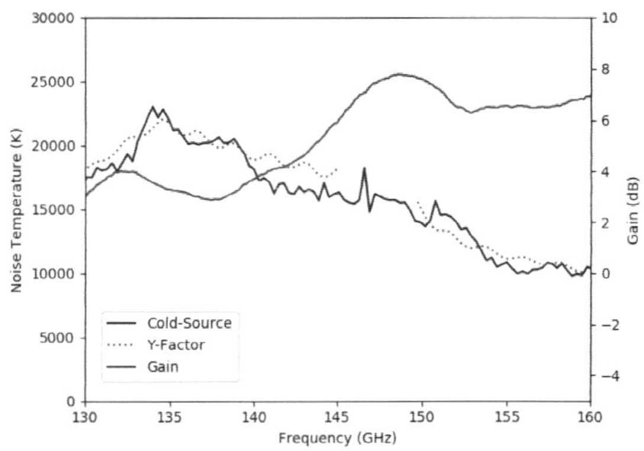


Fig. 1. A comparison between the noise temperatures measured with the cold-source and Y-factor techniques. The gap in the Y-factor measurement around 148GHz is due to the high gain creating an error in the software designed for mixer measurements.


## References

[1] Keysight Technologies, "High-Accuracy Noise Figure Measurements Using the PNA-X Series Network Analyzer", Application Note 5990-5800EN.

[2] D. Pepe, C. Barnett, G. D'Amore and D. Zito, "On-Chip Millimeter-Wave Cold-Source Noise Figure Measurements With PNA-X," in IEEE Transactions on Instrumentation and Measurement, vol. 66, no. 12, pp. 3399-3401, Dec. 2017.

[3] T. W. Crowe, B. Foley, S. Durant, K. Hui, Y. Duan and J. L. Hesler, "VNA frequency extenders to 1.1 THz," 2011 International Conference on Infrared, Millimeter, and Terahertz Waves, Houston, TX, 2011, pp. 1-1.


[1] Virginia Diodes Inc, Charlottesville, VA, 22902, USA.

NOTES:

# On the Comparison Between Low Noise Amplifiers and Photonic Upconverters for Millimeter and Terahertz Radiometry

Gabriel Santamaría Botello[1], Kerlos Atia Abdalmalak[1], Daniel Segovia-Vargas [1], Axel Murk[2], Luis Enrique García Muñoz[1]

***Abstract*—We analyze the feasibility of upconverting THz radiation to the optical domain for high sensitivity room temperature radiometry, as an alternative to radiometers based on low noise amplifiers (LNAs). Following a semiclassical approach, the noise performance of the upconverters is studied. A similar analysis is followed to model the thermal and quantum sources of noise in low noise amplifiers. A comparison between both schemes is done, showing the potential of the upconversion approach if high efficiencies are accomplished. This is due to the low thermal occupation achievable in whispering-gallery mode upconverters and the fact that the minimum introduced noise is not fundamentally bounded by a quantum limit when direct detection is performed.**

***Index Terms*—WGM, upconverters, radiometers.**

## I. Photonic upconversion for THz radiometry

Many applications of technological and scientific interest require the accurate detection and measurement of the electromagnetic power radiated by sources of thermal nature. It is in general desired to retrieve the temperature of the source from such power measurements that are performed with radiometers. Radiometers collect the thermal radiation with antennas and then a receiver stage measures the average power by integrating over an interval $\tau$. As will be shown in the following section, even in the ideal assumption that the radiometer does not introduce noise, the randomness of the source's instantaneous intensity and photon arrival unavoidably leads to an uncertainty in the power measurement that decreases with the observation time $\tau$. Non-ideal mechanisms present in a real radiometer such as impedance mismatchings, dissipation losses and internally generated thermal noise worsen the uncertainty of the measurements done by the instrument.

Conventional high sensitivity radiometers consist of a power detector whose input is pre-amplified by a low noise amplifier (LNA) which contributes significantly to the overall noise of the instrument. While cryogenic LNAs exhibit much lower noise temperatures than room temperature ones, their performance is severely degraded at high frequencies in the millimeter and sub-millimeter wave range. Indeed, it has been suggested that there is a limit in the minimum noise temperature of field effect transistors (FETs) in general. For indium phosphide (InP) high-electron-mobility-transistors (HEMTs) this limit is about 4.5 times the quantum limit [1], which is close to the state-of-the-art. The quantum limit is the minimum noise temperature that any amplifier can exhibit and is given (in Rayleigh-Jeans units) by $T_e = h\nu_0/k_B$ in the high gain limit, where $\nu_0$ is the operation frequency, and $h$ and $k_B$ the Planck and Boltzmann's constants respectively. The existence of such limit has a fundamental origin and comes from the Heisenberg's uncertainty principle: if amplification occurred without noise, the output of the amplifier could be measured with an uncertainty in energy and time lower than the minimum enforced by Heisenberg's uncertainty's principle. This is a consequence of the fact, that the number of photons at the output is higher than the number of photons at the input.

A different approach potentially useful for high sensitivity radiometry is the nonlinear upconversion of the thermal THz radiation to the optical domain, and its subsequent detection with non-cooled photodetectors. Under certain conditions the upconversion process is intrinsically noiseless as no photon multiplication occurs [2]. Indeed, the number of photons at the output and at the input of an ideal upconverter matches, corresponding to a unity photon conversion efficiency $\eta = 1$. In this case, even though there is power amplification since photons are more energetic at the output than at the input, Heisenberg's uncertainty principle holds at the output with no need of added noise. Therefore, the insertion of an ideal upconverter does not worsen the signal to noise ratio of the input, in contrast to the insertion of an ideal LNA. Even though not fundamentally limited, a real upconverter introduces noise since it exhibits a non-ideal efficiency $\eta < 1$ and is thermally occupied due to its physical temperature above 0 K. Nevertheless, efficient upconverters can be designed with resonant structures made of low absorption nonlinear crystals that reduce significantly the upconverted thermal noise at room temperature [3]–[6]. These facts justify the study of nonlinear upconverters for potential high sensitivity detection in the millimeter and sub-millimeter wave range with less stringent cooling requirements than LNAs.

## II. Radiometer equation in upconverters

We model an upconverter with $\eta < 1$ as an ideal upconverter whose input is passed through a beamsplitter with coupling coefficient $\eta$. Some thermal radiation generated inside the upconverter due to its physical temperature $T_p$ is converted to the optical domain along with the antenna signal. This

[1] Signal Theory and Communications Department, Charles III University of Madrid, Spain.
[2] Institute of Applied Physics, University of Bern, Switzerland.

is modeled by means of an artificial input thermal source at temperature $T_{\text{eff}}$ superimposed to the antenna signal. In upconverters based on whispering-gallery mode resonators (WGM) for most cases $T_{\text{eff}} < T_p$ provided that the resonator is sufficiently overcoupled to the antenna [3]. The parameters, $\eta$ and $T_{\text{eff}}$ along with the upconversion frequency $\nu_0$ and bandwidth $\Delta\nu$ characterize the upconverter. The bandwidth is defined as

$$\Delta\nu = \frac{\int_0^\infty H(\nu)\,\mathrm{d}\nu}{H(\nu_0)} \tag{1}$$

where $H(\nu)$ is the one-sided power transfer function of the upconverter, such that the total thermal power received by the upconverter can be written as $P = k_B T \Delta\nu$, with $T$ the temperature of the observed source in Rayleigh-Jeans units. This implicitly assumes $H(\nu)$ is sufficiently narrowband to consider the power spectral density of the thermal radiation constant over the transfer function. Similarly, the noise equivalent bandwidth $B$ is defined as [7]

$$B = \frac{\left[\int_0^\infty H(\nu)\,\mathrm{d}\nu\right]^2}{\int_0^\infty H^2(\nu)\,\mathrm{d}\nu} \tag{2}$$

For typical filter shapes, the relation $a = B/\Delta\nu$ is a constant on the order of unity. Similarly, $B$ and $\Delta\nu$ are commonly on the same order of the full width half power bandwidth.

### A. Direct detection: Semiclassical radiometer equation

We assume the filter function has a Lorentzian shape $H(\nu) = \frac{\gamma/2\pi}{(\nu-\nu_0)^2+(\gamma/2\pi)^2}$ since this is the case for ultra high-Q WGM upconverters. Thermal radiation passed through $H(\nu)$ is equivalent to Gaussian-Lorentzian chaotic light with coherence time $\tau_c = 1/\gamma = 1/(2\Delta\nu) = 1/B$ whose photodetection statistical properties can be obtained from Mandel's formula [8]. A noiseless photon counter with quantum efficiency $\eta_q \leq 1$ directly detecting the Lorentzian-filtered thermal radiation at temperature $T_A = \langle P_A \rangle / (k_B \Delta\nu)$ where $\langle P_A \rangle$ is the average received power, will count on average $\langle m \rangle = \frac{\tau}{h\nu_0}\eta_q \langle P_A \rangle$ photons during the interval $\tau$. For arbitrary observation time $\tau$, the variance in the photon counts is given by [8]:

$$\operatorname{var}(m) = \langle m \rangle + \langle m \rangle^2 \left(\frac{\tau_c^2}{2\tau^2}\right)\left[e^{-\frac{2\tau}{\tau_c}} - 1 + \frac{2\tau}{\tau_c}\right] \tag{3}$$

Rewriting the variance in photon counts as a variance in measured power $P_A$ during $\tau \gg \tau_c$, yields a semiclassical radiometer equation:

$$\operatorname{var}(P_A) = \frac{\langle P_A \rangle^2}{B\tau}\left(1 + \frac{h\nu_0 B}{\eta_q \langle P_A \rangle}\right) \tag{4}$$

When $\eta_q = 1$, Eq. (4) gives the minimum measurement uncertainty of thermal radiation achievable by any detection scheme [9], [10]. It the converges to the classical radiometer equation in the large photon number limit when $\eta_q k_B T_A \gg h\nu_0$ and photon shot noise is negligible compared to excess noise. One of the consequences of non-negligible photon shot noise, is that the signal to noise ratio $\langle P_A \rangle^2 / \operatorname{var}(P_A)$ is not independent from the input power.

### B. Detection with an upconverter

In a real upconverter, some thermal radiation due to the ambient temperature couples to it. This is accounted by a thermal source at temperature $T_{\text{eff}}$, added to the antenna temperature $T_A$. Due to the Gaussian and additive nature of thermal radiation and the fact that both sources are uncorrelated, it is expected this superposition of thermal radiation has the same statistics of a single thermal source whose temperature is $T_A + T_{\text{eff}}$. This photon stream is passed through a lossless beamsplitter with photon number transmission $\eta$ to account for the non-ideal efficiency, and then to an ideal upconverter. The output is then measured by an optical photon counter with quantum efficiency $\eta_q$. The ideal upconverter after the beamsplitter does not change the photon statistics but only boosts their energy. Detecting with a noiseless photodetector with quantum efficiency $\eta_q$ is equivalent to detecting with an ideal photodetector after passing the optical signal through a beamsplitter with photon number transmission $\eta_q$ which results in the replacement of $\eta$ by $\eta\eta_q$. The input mean photon number is given by $\langle m_i \rangle = \frac{\tau}{h\nu_0}\langle P_T \rangle$ where $\langle P_T \rangle = \langle P_A \rangle + \langle P_{\text{eff}} \rangle$ and $\langle P_{\text{eff}} \rangle$ is the average power of the artificial source accounting for coupled thermal noise inside the upconverter. On the other hand, the input variance results $\operatorname{var}(m_i) = \langle m_i \rangle + \langle m_i \rangle^2 / (B\tau)$ for $\tau \gg 1/B$ (see Eq. (3)). This results in a measured photon mean $\langle m \rangle = \eta\eta_q \langle m_i \rangle$ with variance

$$\operatorname{var}(m) = \eta\eta_q \langle m_i \rangle + (\eta\eta_q)^2 \frac{\langle m_i \rangle^2}{B\tau} \tag{5}$$

Rewriting Eq. (5) in terms of the measured incoming power knowing that $\operatorname{var}(m) = \left(\eta\eta_q \frac{\tau}{h\nu_0}\right)^2 \operatorname{var}(P_T)$, we have

$$\operatorname{var}(P_T) = \frac{\langle P_T \rangle^2}{B\tau}\left(1 + \frac{h\nu_0 B}{\eta\eta_q \langle P_T \rangle}\right) \tag{6}$$

Hence, the power estimation of the source of interest $P_A$ can be retrieved from the measurement of $m$ photons during $\tau$, as $P_A = \frac{h\nu_0}{\tau\eta\eta_q} m - \langle P_{\text{eff}} \rangle$, assuming $\langle P_{\text{eff}} \rangle$ is a known offset that can be removed. This way, the expected value of the measurement is exactly the mean power of the source $\langle P_A \rangle$ and its variance

$$\operatorname{var}(P_A) = \frac{(\langle P_A \rangle + \langle P_{\text{eff}} \rangle)^2}{B\tau}\left[1 + \frac{h\nu_0 B}{\eta\eta_q (\langle P_A \rangle + \langle P_{\text{eff}} \rangle)}\right] \tag{7}$$

The photon shot noise term in Eq. (7) might be significant for sufficiently high frequencies and low efficiencies $\eta\eta_q$.

## III. Comparison with Low Noise Amplifiers

A similar approach can be followed to estimate the variance in the power measured by an LNA based radiometer. Normally, LNAs are only characterized by an equivalent noise temperature referred to its input $T_e$ which is superimposed to the signal to account for all intrinsic LNA noise. This is analogous to $T_{\text{eff}}$ in the upconverter. $T_e$ is commonly measured by means of the Y factor method. Then, the radiometric variability is assumed to follow the classical radiometer equation $\operatorname{var}(P_A)\,B\tau =$

$(\langle P_A\rangle + \langle P_e\rangle)^2$ where $\langle P_e\rangle = k_B T_e \Delta\nu$. It is not clear whether this assumption still holds for LNAs in the millimeter and sub-millimeter wave range when observing cold sources such as the cosmic microwave background with cryogenic LNAs. The reason is that according to Eq. (4), photon shot noise might be significant and the classical radiometer equation is not a good approximation. Therefore, it can be important to quantify the losses introduced by the LNA, which like the upconverter's efficiency, would contribute to the photon shot noise term. These losses are not immediately available since they are masked by the gain of the amplifier in standard measurement setups.

We account for internal thermal noise in an LNA as originating from the ohmic dissipation losses $\alpha$ of the circuits prior to amplification. This is modeled through a beamsplitter whose inputs are the antenna signal and a thermal source at the physical temperature of the LNA $T_p$, with transmission coefficient $\alpha$ and $1-\alpha$ respectively. Other photon losses which do not reciprocally lead to coupled thermal noise from the ambient are accounted by a beamsplitter with coupling coefficient $\eta_i$. In an LNA such losses can be due to an impedance mismatching where part of the incoming power is reflected and radiated back through the antenna. The resulting power feeds an ideal LNA, which still has an intrinsic noise source at the quantum limit level, due to the amplification of the zero point fluctuations [11]. Fundamentally, the minimum noise power at the output of an ideal amplifier of gain $G$ is on average $h\nu_0\Delta\nu(G-1)$ [11], which referred to the input, corresponds to $\langle P_{\mathrm{ASE}}\rangle = h\nu_0\Delta\nu\frac{G-1}{G}$, that is, $1-1/G$ photons per second per Hertz of bandwidth and has thermal statistics [11], [12].

Finally, the output of the ideal LNA feeds a power detector (photon counter) with quantum efficiency $\eta_q$. The noise contribution of the photodetector will be negligible as long as the amplifier gain $G$ is large enough. However, $\eta_q$ can be taken into account.

The total mean power received by the ideal amplifier is

$$\langle P_i\rangle = \alpha\eta_i\langle P_A\rangle + (1-\alpha)\langle P_p\rangle + \langle P_{\mathrm{ASE}}\rangle \tag{8}$$

where $\langle P_p\rangle = k_B T_p \Delta\nu$ is the mean power of thermal noise due to the physical temperature of the amplifier. Then, the average output power detected is

$$\langle P_o\rangle = \eta_q G\langle P_i\rangle \tag{9}$$

and its variance for $\tau \gg \tau_c$

$$\mathrm{var}(P_o) = \frac{\eta_q^2 G^2\langle P_i\rangle^2}{B\tau}\left[1 + \frac{h\nu_0 B}{\eta_q G\langle P_i\rangle}\right] \tag{10}$$

Therefore, for a given power measurement $P_o$, the temperature of the antenna can be estimated as

$$P_A = \frac{P_o}{\eta_q G\alpha\eta_i} - \left(\frac{1-\alpha}{\alpha\eta_i}\right)\langle P_p\rangle - \frac{1}{\alpha\eta_i}\langle P_{\mathrm{ASE}}\rangle, \tag{11}$$

assuming the offset $\langle P_e\rangle = \left(\frac{1-\alpha}{\alpha\eta_i}\right)\langle P_p\rangle + \frac{1}{\alpha\eta_i}\langle P_{\mathrm{ASE}}\rangle$ is known and can be removed and the total gain $G_t = \eta_q G\alpha\eta_i$ is also known. The resulting variance in the measured power received by the antenna $P_A$ is given by

$$\mathrm{var}(P_A) = \frac{(\langle P_A\rangle + \langle P_e\rangle)^2}{B\tau}\left[1 + \frac{h\nu_0 B}{G_t(\langle P_A\rangle + \langle P_e\rangle)}\right] \tag{12}$$

Equation (12) is analogous to Eq. (7) for the upconverter. In that case the parameters $\eta\eta_q$ and $T_{\mathrm{eff}}$ can be known from theoretical models and verified experimentally [3]. For the LNA, the parameters $G_t$ and $\langle P_e\rangle$ can be determined experimentally via the Y factor method. Indeed, by using hot and cold calibrated loads at temperatures $T_h = \langle P_h\rangle/(k_B\Delta\nu)$ and $T_c = \langle P_c\rangle/(k_B\Delta\nu)$ respectively, the ratio $Y$ between measured output mean powers in each case is calculated:

$$Y = \frac{P_{o(h)}}{P_{o(c)}} \approx \frac{\eta_q G\alpha\eta_i(\langle P_h\rangle + \langle P_e\rangle)}{\eta_q G\alpha\eta_i(\langle P_c\rangle + \langle P_e\rangle)} \tag{13}$$

from which $\langle P_e\rangle$ can be obtained as

$$\langle P_e\rangle \approx \frac{\langle P_h\rangle - Y\langle P_c\rangle}{Y-1} \tag{14}$$

The approximations in (13) and (14) are better for longer observation times $\tau$. Similarly, the total gain of the amplifier can be obtained experimentally from

$$G_t \approx \frac{P_{o(h)} - P_{o(c)}}{\langle P_h\rangle - \langle P_c\rangle} \tag{15}$$

Hence, the variance in the antenna temperature estimation of an LNA based radiometer, follows the semiclassical radiometer equation of (4), where besides the antenna temperature $T_A$, the system temperature $T_e$ as measured with the conventional Y factor method must be included. The photon shot noise factor is signal dependent, but in principle can be made arbitrarily small for sufficiently high amplifier gain $G$ that surpasses the overall losses in the LNA. It is worth noting that any additional gain introduced after the photodetector does not affect the validity of Eq. (12) (as long as the additional introduced noise is negligible). However, such additional gain cannot be included in the definition of $G_t$, so the experimental determination of $G_t$ via Eq. (15) is only valid when measurements are done right after the power detector. Otherwise, Eq. (15) must be divided by the post-detection gain.

## IV. Discussion

Since the effective thermal noise temperature of a WGM-based upconverter is in principle lower than the physical temperature of the resonator $T_{\mathrm{eff}} \leq T_p$ [3], its noise is mainly determined by the low photon conversion efficiencies achieved so far. This is evidenced in the inverse proportionality to the efficiency of the shot noise term in Eq. (7). The simplified LNA model presented in this work considers the sources of noise as being of thermal and quantum origin. The conclusion is that from conventional Y factor measurements, the variance at the output of the LNA can be estimated by means of the semiclassical radiometer equation. It was not clear whether the internal losses and mismatches of the LNA which are not characterized might lead to a significant photon shot noise contribution for the output variance, invalidating the use of the classical radiometer equation. Our theoretical result of Eq. (12) shows that this is not the case provided that the amplifier gain is large enough. This might seem counter intuitive for the following: The penalty carried by amplification is the amplified spontaneous emission (ASE) noise which practically does not change for gains ranging from moderate values $G \approx 10$ to $G \to \infty$. Superficially this can lead to the erroneous conclusion that strong photon loss in an LNA with arbitrarily large gain does not increase the photon shot noise effect in contrast to an upconverter which does not exhibit gain. This is a fallacy, since even though ASE noise is constant, its contribution to $\langle P_e \rangle$ is inversely proportional to the overall loss coefficient $\alpha\eta_i$. In fact, since the term $\langle P_e \rangle$ is quadratic in Eq. (12), whereas the $1/\eta\eta_q$ dependence of photon shot noise term in Eq. (7) is linear, photon loss fundamentally has a stronger negative effect in LNAs than in upconverters.

In order to show the potential of the upconversion approach for submillimeter wave and THz radiometry, we can compare the normalized radiometer variability $\sqrt{\mathrm{var}\,(P_A)\,B\tau}/(k_B\Delta\nu)$ of the upconverter with that of state-of-the-art low noise amplifiers. In the latter case $\langle P_e \rangle = k_B T_e \Delta\nu$ where $T_e$ is the system temperature of the LNA referred to the input in Rayleigh-Jeans units, obtained from Y-factor measurements. We assume the best case for the LNAs when $G \to \infty$. Figure 1 shows these results for some millimeter and submillimeter wave LNAs reported in the literature, and compared with those achievable by an upconverter with $\eta = 10^{-2}$. We have assumed $T_{\text{eff}} = 290\,\text{K}$ for the upconverter as it has been shown that the WGM resonator can always be overcoupled such that $T_{\text{eff}}$ is below the physical temperature of the crystal, provided that sufficiently high intrinsic microwave $Q \gtrsim 20$ is realized with low azimuthal mode order $n \approx 4$ [3]. We also plot the upconverter results for $T_e = 100\,\text{K}$, which is achievable by overcoupling if $Q > 50$ [3]. In lithium niobate, intrinsic Q factors ranging from 400 to 40 are achievable from 100 GHz up to 2 THz [13]. The higher the intrinsic Q, the lower the $T_{\text{eff}}$ that can be realized via overcoupling [3].

It can be seen from Figure 1 how the theoretically expected sensitivity of the upconversion approach significantly surpasses state of the art HEMT LNAs in the submillimeter range. Indeed, room temperature overcoupled WGM resonators leading to $T_{\text{eff}} \leq 100\,\text{K}$ could lead comparable sensitivities to state of the art LNAs cooled down to 50 K [14].

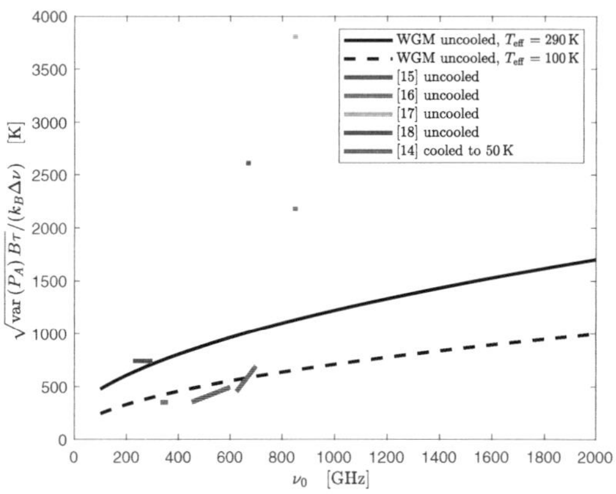


Fig. 1. Radiometer variability of the room temperature upconverter for two different $T_e$ achievable by overcoupling, depending on the intrinsic microwave Q factor of the resonator. A comparison is made with several state of the art receiver schemes reported in the literature.

## References


[1] M. W. Pospieszalski, "On the limits of noise performance of field effect transistors," in 2017 IEEE MTT-S International Microwave Symposium (IMS), Honolulu, HI, USA, 2017, pp. 1953–1956.

[2] W. H. Louisell, A. Yariv, and A. E. Siegman, "Quantum Fluctuations and Noise in Parametric Processes. I.," Physical Review, vol. 124, no. 6, pp. 1646–1654, Dec. 1961.

[3] G. Santamaría Botello et al., "Sensitivity limits of millimeter-wave photonic radiometers based on efficient electro-optic upconverters," Optica, vol. 5, no. 10, p. 1210, Oct. 2018.

[4] A. Rueda et al., "Efficient microwave to optical photon conversion: an electro-optical realization," Optica, vol. 3, no. 6, p. 597, Jun. 2016.

[5] D. V. Strekalov, A. A. Savchenkov, A. B. Matsko, and N. Yu, "Efficient upconversion of sub-THz radiation in a high-Q whispering gallery resonator," Optics Letters, vol. 34, no. 6, p. 713, Mar. 2009.

[6] A. B. Matsko, D. V. Strekalov, and N. Yu, "Sensitivity of terahertz photonic receivers," Physical Review A, vol. 77, no. 4, Apr. 2008.

[7] E. Tiuri, "Radio Astronomy Receivers," p. 9.

[8] R. Loudon, The quantum theory of light, 3rd ed. Oxford ; New York: Oxford University Press, 2000.

[9] R. Nair and M. Tsang, "Quantum Optimality of Photon Counting for Temperature Measurement of Thermal Astronomical Sources," The Astrophysical Journal, vol. 808, no. 2, p. 125, Jul. 2015.

[10] J. Zmuidzinas, "On the Use of Shot Noise for Photon Counting," The Astrophysical Journal, vol. 813, no. 1, p. 17, Oct. 2015.

[11] E. Desurvire and M. N. Zervas, "Erbium-Doped Fiber Amplifiers: Principles and Applications," Physics Today, vol. 48, no. 2, pp. 56–58, Feb. 1995.

[12] H. A. Haus, "The noise figure of optical amplifiers," IEEE Photonics Technology Letters, vol. 10, no. 11, pp. 1602–1604, Nov. 1998.

[13] C. Cochard, T. Spielmann, N. Bahlawane, A. Halpin, and T. Granzow, *Broadband characterization of congruent lithium niobate from mHz to optical frequencies,* Journal of Physics D: Applied Physics, vol. 50, no. 36, p. 36LT01, Sep. 2017.

[14] G. Chattopadhyay et al., *A 340 GHz cryogenic amplifier based spectrometer for space based atmospheric science applications,* in 2017 42nd International Conference on Infrared, Millimeter, and Terahertz Waves (IRMMW-THz), Cancun, Mexico, 2017, pp. 1–2.

[15] W. R. Deal et al., *A 670 GHz Low Noise Amplifier with ¡10 dB Packaged Noise Figure,* IEEE Microwave and Wireless Components Letters, vol. 26, no. 10, pp. 837–839, Oct. 2016.

[16] K. M. K. H. Leong et al., *A 0.85 THz Low Noise Amplifier Using InP HEMT Transistors,* IEEE Microwave and Wireless Components Letters, vol. 25, no. 6, pp. 397–399, Jun. 2015.

[17] K. M. K. H. Leong et al., *850 GHz Receiver and Transmitter Front-Ends Using InP HEMT,* IEEE Transactions on Terahertz Science and Technology, vol. 7, no. 4, pp. 466–475, Jul. 2017.

[18] A. Tessmann et al., *243 GHz low-noise amplifier MMICs and modules based on metamorphic HEMT technology,* International Journal of Microwave and Wireless Technologies, vol. 6, no. 3–4, pp. 215–223, Jun. 2014.

# A broadband down-conversion module for the extended W-Band

David Monasterio, Claudio Jarufe, Diego Gallardo, Nicolás Reyes, Patricio Mena, Leonardo Bronfman

***Abstract*—We have developed a broad band receiver module in the extended W-band (70 - 116 GHz). The module is a sideband separation receiver based on a high-frequency amplification stage followed by a broadband downconverter. A general description of the architecture will be presented describing its most important components. Some of its more important figures of merit will be discussed.**

***Index Terms*—Microwave integrated circuits, Multichip modules, Frequency conversion, Microwave mixers, Low-noise amplifiers, Focal plane arrays, Multi-beam systems.**

## I. INTRODUCTION

Interest in constructing focal plane arrays (FPAs) has been growing over the last years. The potential of synchronous/ simultaneous detection using a large number of receiver elements make them very interesting for applications such as radio-astronomy, millimeter wave imaging systems, satellite communications and Earth remote sensing [1].

We have focused our work in the design and construction of compact down-converters modules in the band from 67 to 116 GHz, equivalent to a fractional bandwidth of 53%. This frequency range is motivated by several astronomical projects, including the Atacama Large Millimeter Array (ALMA) [2] but multiple non-astronomical systems can benefit from such large bandwidth and feasibility of integration in large imaging systems. These modules are designed with a scalable and compact architecture and good input return loss that allows easy integration into large systems.

This work has been funded by the Chinese Academia off Science, South American Center for Astronomy for Astronomy. The team acknowledges support from CONICYT project Basal AFB-170002.

D. Monasterio is with Electrical Engineering Department, Universidad de Chile, Santiago, Chile (e-mail: david.monasterio@raig.uchile.cl).

C. Jarufe was with Electrical Engineering Department, Universidad de Chile, Santiago, Chile. He is now with the School of Electrical and Electronic Engineering, University of Manchester, UK. (e-mail: Claudio.jarufe@raig.uchile.cl).

D. Gallardo is with Electrical Engineering Department, Universidad de Chile, Santiago, Chile (diego.gallardovives@gmail.com).

N. Reyes is with Electrical Engineering Department, Universidad de Chile, Santiago, Chile and with the Max Planck Institute fur Radioastronomie, auf dem Hugel 69, Bonn, Germany (nireyes@uchile.cl).

F. P. Mena is with Electrical Engineering Department, Universidad de Chile, Santiago, Chile (fpmena@uchile.cl).

L. Bronfman is with Astronomy Department, Universidad de Chile, Santiago, Chile (leo@das.uchile.cl).

## II. MODULE DESIGN AND CONSTRUCTION

The general design of the module is shown in figure 1. The input stage of the module is a quadrature hybrid implemented in WR10 waveguide with the isolated port terminated in a waveguide load. Each signal is amplified by a commercial MMIC low noise amplifier (CGY2190 of OMMIC) with a gain of 20dB and a noise of 2.8 dB. The signal is then down-converted using sub-harmonic mixers which were designed by us and built in MMIC technology using the GaAs Schottky diodes process from UMS, this process has shown good performance in W-band [3]. Previous measurements of the mixers show conversion loss lower than 15 dB when driven by a 5 dBm local oscillator (LO) source. The output IF signal have a broadband IF range of 0-12GHz. The LO signal required by the down-converter is delivered by a Wilkinson divider followed by a lowpass filter that acts as controlled termination for RF signals leaking into the LO path.

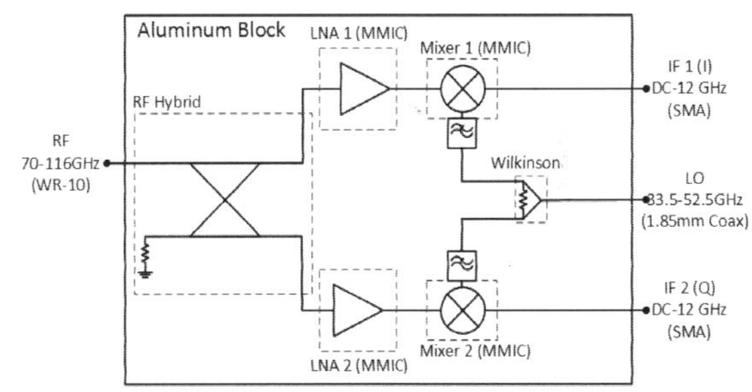


Fig. 1. Proposed broadband downconverter architecture. The input quadrature coupler allows the module to be used as sideband separation mixer. The first amplification stage is a 20 dB MMIC from OMMIC. The second stage is the downconverter itself which is implemented as a subharmonic mixer. The LO is distributed using a Wilkinson coupler followed by a LO filter.

The module was built and assembled at our workshop. A picture of the complete module is shown in figure 2. The housing, including the waveguide structures was fabricating by CNC milling on Aluminum 6061. The complete block size is 50mm x 25mm x 20mm. The input RF connector is a rectangular waveguide (WR10) with a standard flange UG387/U. The LO is fed by a 1.85mm connector and the IF outputs corresponds to SMA connectors. The module uses a

15 pin bias connector for the amplifiers.

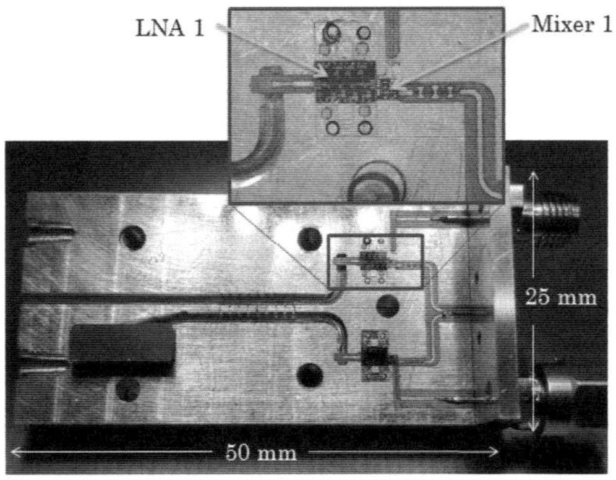


Fig. 9. Picture of the interior of the assembled module. An amplification of the picture shows the LNA and mixer MMICs.

## III. Module Measurements

The input return loss for the module was measured and shown in figure 3. The measurements show return loss above 10 dB in the complete band above 70 GHz. Several measurements with different Bias setting for the amplifiers where made showing similar performance. This good return loss allows the module to operate without an isolator.

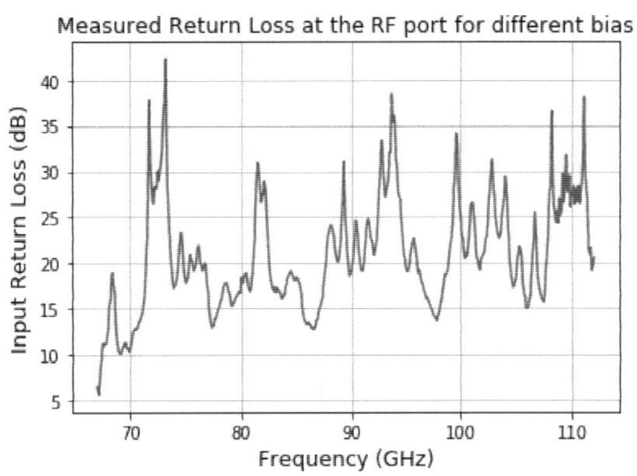


Fig. 3. Measured input return loss of the module for a LO of 40 GHz with and equal bias setting for the two amplifiers.

The block requires an external IF coupler to accomplish with sideband separation. During testing we use a 0.8-4.2GHz unit which limit our IF bandwidth. The operating points of the amplifiers were chosen in order to maximize Side-band Rejection Ratio (SRR), effectively cancelling small amplitude imbalances in the system. SRR > 10 dB has been archived in over 70% of the band. This value could be improved using digital techniques [3]. Figure 4 shows the measured conversion gain in a 2SB configuration. I/Q conversion gain were also measured with the removal of the IF Hybrid.

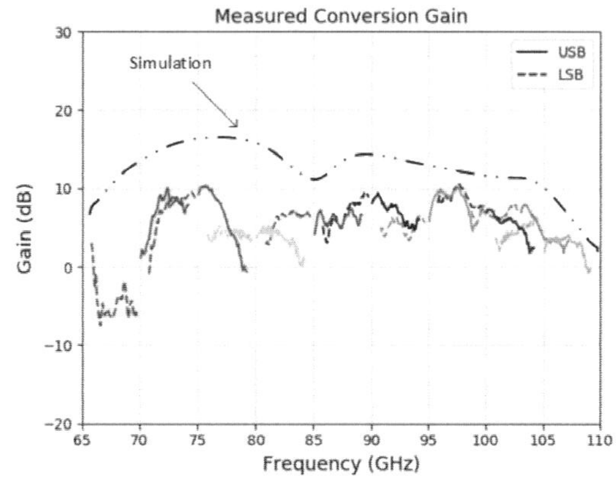


Fig. 4. Measured conversion gain with sideband separation, for a LO frequencies of 35, 37.5, 40, 42.5, 45, 47.5, 50 and 52.5 GHz with an average power of 12 dBm and comparison with the simulation in NI AWR Microwave office.

The results show good performance in the majority of the band. The decrease in conversion gain and return losses at the beginning of the band is due to mechanical constrains in the RF hybrid. Additional measurements of noise have also been made showing an average of ~1500K at room temperature. Additionally, all components of the module had been measured separately at cryogenic temperatures showing no downgrade in performance, so the module should work at cryogenics if necessary.

## References

[1] A. Kerr, et al, "In support of instrument technology development for THz astronomy," Astro2010: The Astronomy and Astrophysics Decadal Survey, 2009. [Online] Available: http://adsabs.harvard.edu/abs/2009astro2010T..29K

[2] P.Yageoubov et al., "67-116 GHz optics development for ALMA Band 2+3 receivers" Proceedings of the 41st international conference on Infrared Millimeter and Terahertz waves, 2016. DOI: 10.1109/IRMMW-THz.2016.7758971

[3] R. Gawande et al., "W-band IQ sub-harmonic mixers with low LO power for cryogenic operation in large arrays" Proceedings of the 9th Microwave Integrated Circuits Conference, 2014. DOI: 10.1109/EuMIC.2014.699785

[4] R. Finger, P. Mena, N. Reyes, R. Rodriguez, L. Bronfman, "A calibrated digital sideband separating spectrometer for radio astronomy applications", Publications of the Astronomical Society of the Pacific, 125, pp. 263-269, 2013. DOI: 10.1086/670175

# Wednesday, April 17, 2019

# Session XIII. Novel Devices and Technologies

# Quantum transport at Dirac point enables graphene for terahertz heterodyne astronomy

Sergey Cherednichenko[1], Andrey Danilov[1], Dmitriy Golubev[3], Hans He[1], K.H. Kim[1], Rositsa Yakimova[4], Floriana Lombardi[1], Thilo Bauch[1], Samuel Lara-Avila[1,2] and Sergey Kubatkin[1]

Further leaps in astronomy demand new detector materials and devices reaching the fundamental detection limit[1]. Superconducting hot-electron bolometer (S-HEB) mixers form the baseline for modern astronomical receivers above 1 THz. In these, the wave beating between the Local Oscillator (LO) and the THz signal causes temperature oscillations in a metal around the transition temperature, at the Intermediate Frequency (IF), enabling read-out through changes in electrical resistance R (resistive read-out) as long as the temperature can follow the signal modulation. Despite huge efforts, the instantaneous bandwidth in practical niobium nitride (NbN)-based S-HEB mixers does not exceed 4-5GHz, limited by the electron temperature relaxation rates. The search for new materials lead to MgB2 devices,[2] where an 11 GHz bandwidths and a 1000K noise temperature are possible but at the expense of high LO power requirements, which is particularly detrimental for array applications. Beyond superconducting materials, charge-neutral graphene has been discussed as an ideal platform for terahertz bolometric direct detectors due to its small heat capacity and weak electron-phonon coupling. However, absence of large-area graphene homogeneously doped to Dirac point hinders any prospects for practical detectors in astronomy and other sensing applications. Furthermore, negligible temperature dependent resistance has kept this approach as not acceptable for bolometric mixers where voltage read-out is required.

Here we investigate graphene that is doped to the Dirac point by assembly of molecular dopants on its surface with a high uniformity across the wafer and the long-though temperature-dependent resistance. With the resistance dominated by quantum localization, and thermal relaxation of carriers governed by electron diffusion, we demonstrate a graphene bolometric terahertz mixer with a gain bandwidth (presently) of 9 GHz (relaxation time 20ps) (see Fig.1) and a mixer noise temperature of 475 K. We conclude that with the present quality of graphene, optimization of the device layout will result in a mixer noise temperature as low as 36 K and a gain bandwidth exceeding 20 GHz, with a Local Oscillator power of < 100 pW for operation temperatures <1K. Given the scalability of the material and in conjunction with emerging quantum-limited amplifiers in the GHz domain, we envisage large arrays of quantum–limited sensors in the THz domain for radio astronomy, potentially surpassing superconductor-based heterodyne detectors.

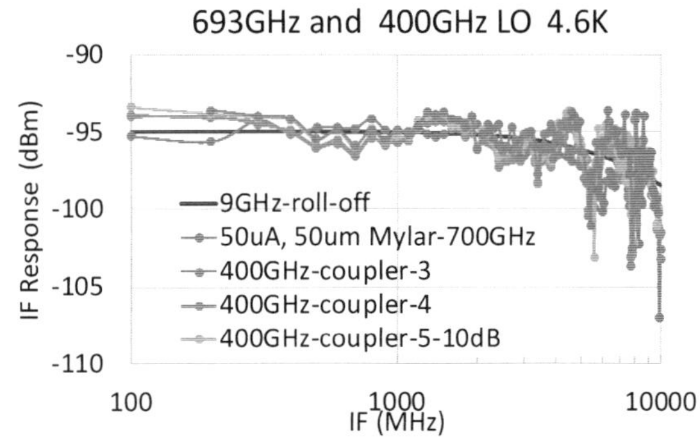


Fig. 1. Intermediate frequency response of graphene bolometric mixer. LOs at 693GHz and 400GHz were utilized with matching tunable signal sources. The gain of the IF chain has been removed from the measured data. The residual ripples are probably coming from the long bonding wires and the mixer unit itself.

[1] Department of Microtechnology and Nanoscience, Chalmers University of Technology, SE-412 96, Gothenburg, Sweden.

[2] National Physical Laboratory, Hampton Road, Teddington, TW11 0LW, UK

[3] Department of Applied Physics, Aalto University, FI-00076 Aalto, Finland.

[4] Department of Physics, Chemistry and Biology, Linkoping University, S-581 83, Linköping, Sweden.

## References

[1] M. Rowan-Robinson, "Astronomy. Probing the cold universe" *Science* 325, 546–7 (2009).

[2] H. He, et al. "Uniform doping of graphene close to the charge neutrality point by polymer-assisted spontaneous assembly of molecular dopants" *Nat. Commun.* 9, 3956 (2018).

[3] E. Novoselov and S. Cherednichenko, "Low noise terahertz MgB2 hot-electron bolometer mixers with an 11 GHz bandwidth" *Appl. Phys. Lett.* 110, 032601 (2017).

NOTES:

# Design and Evaluation of SIS Photon Detectors at Terahertz Frequencies

Hajime Ezawa[1,2], Hiroshi Matsuo[1,2], Masahiro Ukibe[3], Go Fujii[3], and Shigetomo Shiki[3]

Demands for better sensitivity and higher angular resolution are increasing in observational astronomy and astrophysics. ALMA has successfully revealed its high observation capability among millimeter and submillimeter wavelengths, which includes detection of high redsifted [OIII] 88 μm emission line from distant galaxies, or resolving the detailed structure of proto-planetary disks. Photon counting detectors would be one of the next generation technologies for high sensitivity terahertz observation. Fast detectors may resolve each photons to realize high signal to noise ratio. Moreover, they will enable us to introduce "photon statistics" to terahertz astronomy to realize high precision measurements [1].

We are considering SIS junctions with low leakage current to be integrated into photon counting detectors at terahertz frequencies. The requirements to the SIS junction to realize photon counting capability has been discussed recently [2]. When we consider to observe an astronomical source of 1 Jy through a 10 m telescope at 1 THz with 100 GHz bandwidth, the photon rate is expected to be in the order of 100 M photons/s. In order to resolve these photons, the detector should be as fast as 1 GHz with the noise equivalent power (NEP) of $< 3 \times 10^{-17}\ \mathrm{W}/\sqrt{\mathrm{Hz}}$. When the SIS junction operates in photon-assisted tunneling mode, this NEP requirement can be achieved with leakage current of 1 pA (assuming $\eta$ =0.5).

Recently we have successfully developed an SIS junction with Nb/Al/AlOx/Al/Nb which exhibits its leakage current as low as 1 pA at a cryogenic temperature of $T \leq$ 0.7 K. The junction was developed using the CRAVITY facility in AIST [3]. Following the success, we have integrated this junction into an antenna coupled SIS detector, in order to realize photon counting experiments in the lab. The detector consists of a twin slot antenna which feeds the RF signal into the SIS junction through a coplanar wave guide. The photo-current from the SIS junction will be fed to the first-stage readout circuit (FET) through a choke filter which suppresses the RF signal. The SIS is designed as a parallel connected twin junction (PCTJ) in order to match its impedance with that of the coplanar wave guide. The detector design is tuned for lab experiments: The antenna is optimized for 500 GHz, and the critical current density of the SIS junction is designed to be 300 A/cm$^2$. This design aims for relatively narrow bandwidth to limit the contribution of background photons in lab experiments.

The designed SIS detector was fabricated at CRAVITY, and its evaluation is undergoing. The temperature dependence of the leakage current was evaluated utilizing a $^3$He-$^4$He sorption cooler, to confirm the low leakage feature of 1-2 pA at $T < 0.7$ K. The photo-response will be evaluated with the Fourier Transform Spectrometer.

Considering the system design, we are planning to utilize a $^4$He sorption cooler, which exhibits a larger cooling capacity compared with $^3$He based sorption fridges. This allows us to locate the first-stage FET adjacent to the SIS detector on the cryogenic stage. For this purpose we are tuning the SIS junction for higher operation temperature to exhibit low leakage current even at $T \geq 0.8$ K.

The concept of the detection system, as well as the design and performance of the developed detector will be discussed in the presentation.

[1] National Astronomical Observatory of Japan, Mitaka, Tokyo 181-8588, Japan.

[2] Department of Astronomical Science, The Graduate University for Advanced Studies (SOKENDAI), Mitaka, Tokyo 181-8588, Japan

[3] National Institute of Advanced Industrial Science and Technology (AIST), Tsukuba, Ibaraki 305-8568, Japan

## References

[1] H. Matsuo, J. Low Temp. Phys. 167, 840-845 (2012), DOI: 10.1007/s10909-012-0579-6.

[2] H. Ezawa, H. Matsuo, M. Ukibe, G. Fujii, S. Shiki, J. Low Temp. Phys. 184, 244-249 (2016), DOI: 10.1007/s10909-015-1465-9.

[3] M. Ukibe, S. Shiki, Y. Kitajima, M. Ohkubo, JJAP 51, 010115 (2012), DOI: 10.1143/JJAP.51.010115.

NOTES:

# Near infrared photon detectors using titanium-based superconducting transition-edge sensors

W. Zhang[1,2], Y. Gen[1,2,3], P. Z. Li[1,2,3], J.Q. Zhong[1,2], Z. Wang[1,2], W. Miao[1,2], Y. Ren[1,2], Q.J. Yao[1,2], J.F. Wang[4], and S.C. Shi[1,2]

Transition edge sensor (TES) detectors are showing great promise as single photon detectors in quantum information [1] and as fast bolometers in astronomical applications [2]. In the near infrared/optical range TESs offer good broadband quantum efficiency, high time resolution, and modest energy resolution. Therefore TESs provide the potential capabilities for the study of rapidly varying compact object systems. We are developing Ti-based TES as optical/near-infrared photon detectors [3].

The TES sensors are photolithographically patterned ~50 nm thick Ti films deposited on quartz substrate. These films exhibit a sharp superconducting to normal transition at about 400 mK. Niobium films are used as superconducting leads for the devices.

The sensors were mounted on the mixture stage of a Triton 400 dilution refrigerator. The sensor temperature is maintained within its superconducting to normal transition via the Joule heating produced by the voltage bias due to strong negative electrothermal feedback. A series array of dc SQUIDs, working as current-sensitive amplifiers readout the signal. The photons were introduced through a single-mode fiber with a 9 μm core and a 125 μm cladding. The fiber is aligned to the active area of the TES with the help of IR microscope to provide high coupling efficiency.

We have measured the current-voltage curves at different bath temperatures, based on which the calculated thermal conductance is ~300 pW/K for the 10μm×20μm TES (2C1) fabricated on a 1550 nm dielectric mirror. The effective time constant is about 20 μs. The input power is measured by a power meter and then attenuated by 42-45 dB. The absorbed power is the integral of the corresponding drop in current multiplied by the bias voltage. Thus system efficiency is the ratio of absorbed power to the input power. Fig. 1 shows the measured system efficiency as a function of their normalized resistance. The highest system efficiency is 33 % at 1550 nm, and drops down to 14 % as the resistance increases. In comparison, the TES devices (3B1 and 3C1) without dielectric mirror shows lower system efficiency, in good agreement with prediction.

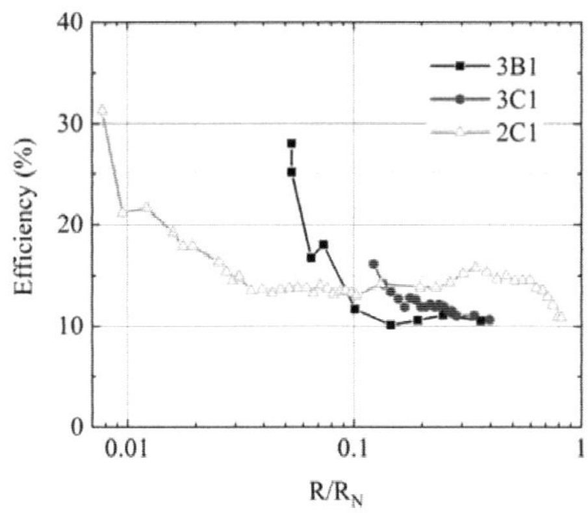


Fig. 1. Measured system efficiency of Ti-based TES detectors at 1550 nm.

In conclusion, we have designed and fabricated near infrared photon detectors using Ti-based superconducting transition edge sensor. The first generation devices have a system efficiency of ~30% and effective response time of 10 μs. The combination of system efficiency and high time resolution allows the construction of a wide-band high-speed spectrophotometers for a number of interesting applications, such as rapid time varying sources.

[1] Purple Mountain Observatory, Chinese Academy of Sciences, Nanjing, Jiangsu, 210034, China.

[2] Key Lab of Radio Astronomy, Chinese Academy of Sciences, Nanjing, Jiangsu, 210034, China.

[3] University of Science and Technology of China Hefei, Anhui, 230026, China.

[4] Nanjing Institute of Astronomical Optics & Technology, National Astronomical Observatory, Chinese Academy of Sciences, Nanjing, 210042, China.

## REFERENCES

[1] A.E. Lita, A.J. Miller, and S.W. Nam, "Counting near-infrared single photons with 95% efficiency," Opt. Express, vol. 16, no.2, 3032, 2008.

[2] R.W. Romani, A.J. Miller, B. Cabrera, S.W. Nam, and J.M. martinis, "Phase resolved crab studies with a cryogenic transition-edge sensor spectrophotometer," ApJ, 563, 221-228, 2001.

[3] W. Zhang, Y. Gen, Z. Wang, J.Q. Zhong, P.Z. Li, W. Miao, Y. Ren, Q.J. Yao, J.F. Wang, and S.C. Shi, "Development of titanium based transition-edge single photon detector," *IEEE Trans. Appl. Supercond.*, submitted.

NOTES:

# On-chip refrigerator integrated into a photon-noise-limited detector for high-performance Cosmology missions

A.L. Pankratov*[1,2], L.S. Kuzmin[1,3], A.V. Gordeeva[1,2], V.O. Zbrozhek[1], V.A. Shamporov[1,2], L.S. Revin[1,2], S. Masi[4], P. de Bernardis[4]
alp@ipmras.ru
[1]Nizhny Novgorod State Technical University, Nizhny Novgorod, Russia
[2]Institute for Physics of Microstructures of RAS, Nizhny Novgorod, Russia
[3]Chalmers University of Technology, Gothenburg, Sweden,
[4]University La Sapienza, Roma, Italy

The SIN tunnel junctions are known for their ability to remove the heat from the electron system of the normal metal electrode. Two SIN junctions, connected in SINIS structure, remove the heat twice more effectively than a single junction. This property is intensively used in Cold-Electron bolometers.

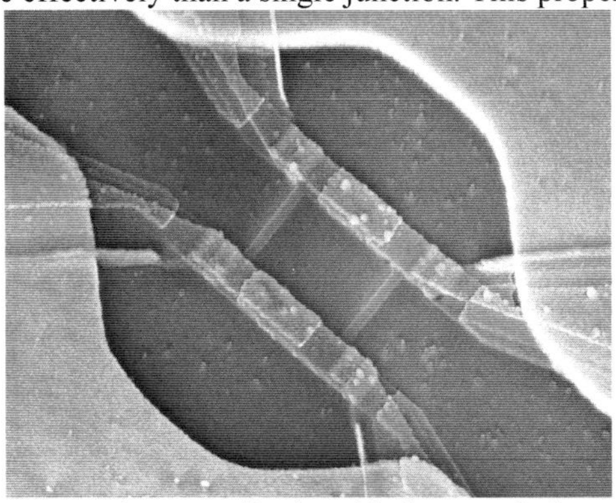

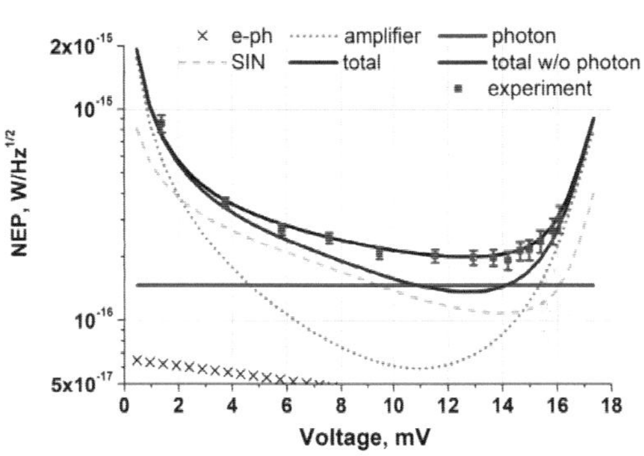


We have realized cold-electron bolometers (CEB) with direct electron self-cooling of the nanoabsorber by SIN (Superconductor-Insulator-Normal metal) tunnel junctions. We have made several improvements in the bolometer design, that decreased the return power of hot quasi-particles from the vicinity of the tunnel barrier from 30% to just 1-2%.

The effective electron self-cooling acts as a strong negative electrothermal feedback, improving noise and dynamic properties. Due to this cooling the photon-noise-limited operation of CEBs was realized in array of bolometers developed for the 345 GHz channel of the OLIMPO Balloon Telescope in the power range from 10 pW to 20 pW at phonon temperature $T_{ph}$=310 mK. The negative electrothermal feedback in CEB is analogous to TES but instead of artificial heating we use cooling of the absorber. The high efficiency of the electron self-cooling is achieved by:

- small volume of the nanoabsorber (0.02 um3) and a large area (up to 80%) of the SIN tunnel junctions,
- effective removal of hot quasiparticles by arranging double stock at both sides of the junctions and close position of the normal metal traps,
- self-protection of the 2-dimensional (2D) array of CEBs against interferences by dividing them between N series CEBs (for voltage interferences) and M parallel CEBs (for current interferences),
- suppression of Andreev reflection by a thin layer of Fe in the hybrid S/F AlFe absorber.

Due to these improvements, the internal bolometer noise (including room temperature amplifier noise) is 1.3 times smaller than the receiving signal noise (photon noise) at 20 pW power load and base temperature 310 mK.

Thus, we have developed, manufactured and tested a bolometric receiving system designed for a large received power, showing record sensitivity (the internal noise is less than the photon noise) due to the effect of electronic cooling, i.e., operating at electron temperature about 2 times less than the phonon temperature of the sample. It is shown that the NEP of a single bolometer of a given pixel at a temperature of 300 mK and the room temperature amplifiers used is 8×10^-18 W/Hz^1/2. If we replace the room temperature amplifiers with standard cooled ones, then the NEP decreases to a record 3×10^-18 W/Hz^1/2, unattainable for competitors - TES and KID bolometer for this temperature range of about 0.3 K, obtained in He3 cryostats without necessity of using dilution cryostats, which is especially important for space applications.

The work is supported by Russian Science Foundation (Project 16-19-10468).

# List of Registered Symposium Participants
## as of 15/03/2019

| Name | Affiliation | e-mail address |
|---|---|---|
| Acharya, Narendra | Chalmers University | narach@chalmers.se |
| Adema, Joost | NOVA | j.adema@astro.rug.nl |
| Alonso, Maria | JPL | Maria.Alonso@jpl.nasa.gov |
| Barkhof, Jan | NOVA | J.Barkhof@sron.nl |
| Baryshev, Andrey | NOVA/Kapteyn Astronomical Institute | andrey@astro.rug.nl |
| Bekema, Marielle | Netherlands Research School for Astronomy | M.E.Bekema@rug.nl |
| Belitsky, Victor | Chalmers University | victor.belitsky@chalmers.se |
| Blundell, Raymond | Smithsonain Astrophysical Observatory | rblundell@cfa.harvard.edu |
| Bubnov, Grigoriy | IAP RAS | bubnov@ieee.org |
| Böhm, Johanna | University of Cologne | boehm@ph1.uni-koeln.de |
| Caselli, Paola | MPE | caselli@mpe.mpg.de |
| Chattopadhyay, Goutam | NASA-JPL/Caltech | goutam@jpl.nasa.gov |
| Chen, Christine | JPL | christine.p.chen@jpl.nasa.gov |
| Cherednichenko, Sergey | Chalmers University | serguei@chalmers.se |
| Chin, Kah Wuy | NAOJ | kahwuy.chin@nao.ac.jp |
| Conway, John | OSO/Chalmers | John.Conway@chalmers.se |
| Crowe, Tom | Virginia Diodes Inc. | crowe@vadiodes.com |
| De Beck, Elvire | Chalmers University | elvire.debeck@chalmers.se |
| De Graauw, Thijs | ESO | tdegraau@alma.cl |
| Desmaris, Vincent | Chalmers University | vincent.desmaris@chalmers.se |
| Driessen, Eduard | IRAM | driessen@iram.fr |
| Ermakov, Andrey | Chalmers University | andreye@chalmers.se |
| Eskilsson, Robert | Rohde & Schwarz Sverige AB | robert.eskilsson@rohde-schwarz.com |
| Ezawa, Hajime | NAOJ | h.ezawa@nao.ac.jp |
| Falk, Kambiz | Keysight Technologies | kambiz.falk@keysight.com |
| Finkel, Matvey | SRON | m.finkel@sron.nl |
| Fredrixon, Mathias | Chalmers University | mathias.fredrixon@chalmers.se |
| Gan, Yuner | SRON | Y.N.Gan@sron.nl |
| Gao, Jian-rong | SRON-TUDelft | j.r.gao@tudelft.nl |
| Garrett, John | Oxford | john.garrett@astro.ox.ac.uk |
| Goldsmith, Paul | JPL | paul.f.goldsmith@jpl.nasa.gov |
| Goltsman, Gregory | MSPU | goltsman@mspu-phys.ru |
| Gordeeva, Anna | Nizhny Novgorod State Technical University | a.gordeeva@nntu.ru |
| Graf, Urs | University of Cologne | graf@ph1.uni-koeln.de |

| | | |
|---|---|---|
| Groppi, Christopher | Arizona State University | cgroppi@asu.edu |
| Hagelschuer, Till | German Aerospace Center (DLR) | Till.Hagelschuer@dlr.de |
| Hawkins, Coy (Bert) | NRAO | bhawkins@nrao.edu |
| Heinerås, Daniel | Wasa Millimeter Wave AB | daniel@wmmw.se |
| Henke, Doug | NRC Herzberg | doug.henke@nrc-cnrc.gc.ca |
| Hesler, Jeffrey | Virginia Diodes Inc. | Hesler@vadiodes.com |
| Hesper, Ronald | NOVA / Univ. of Groningen | r.hesper@astro.rug.nl |
| Hills, Richard | University of Cambridge | reh1@cam.ac.uk |
| Hoh, Jonathan | ASU | jonathan.hoh@asu.edu |
| Hovey, Gary | OSO/Chalmers | Gary.Hovey@chalmers.se |
| Hübers, Heinz-WIlhelm | DLR | heinz-wilhelm.huebers@dlr.de |
| Irimajiri, Yoshihisa | NICT | irimaji@nict.go.jp |
| Jayasankar, Divya | Chalmers Tekniska Högskola | divyaj@chalmers.se |
| Jung-Kubiak, Cecile | NASA/Jet Propulsion Laboratory | cecile.d.jung@jpl.nasa.gov |
| Justen, Matthias | University of Cologne | justen@ph1.uni-koeln.de |
| Kasalynas, Irmantas | FTMC | irmantak@ktl.mii.lt |
| Kawakami, Akira | NICT | kawakami@nict.go.jp |
| Khudchenko, Andrey | University of Groningen/NOVA | A.Khudchenko@sron.nl |
| Kinev, Nikolay | IREE RAS, Moscow | nickolay@hitech.cplire.ru |
| Klein, Bernd | MPI für Radioastronomie | bklein@mpifr-bonn.mpg.de |
| Knee, Lewis | NRC Canada | lewis.knee@gmail.com |
| Kojima, Takafumi | NAOJ | t.kojima@nao.ac.jp |
| Koshelets, Valery | Koytel'nikov IREE RAS | valery@hitech.cplire.ru |
| Kotiranta, Mikko | Univ. Bern | mikko.kotiranta@iap.unibe.ch |
| Kroug, Matthias | NAOJ | matthias.kroug@nao.ac.jp |
| Lambert, Joseph | NRAO | jlambert@nrao.edu |
| Lapkin, Igor | Chalmers University | igor.lapkin@chalmers.se |
| LI, Jing | PMO | lijing@pmo.ac.cn |
| López, Cristian Daniel | Chalmers University | Cristian.Lopez@chalmers.se |
| López-Fernández, Isaac | Observatorio de Yebes | i.lopez@oan.es |
| Lubenchenko, Aleksandr | MPEI | LubenchenkoAV@mpei.ru |
| Mahdizadeh, Sajjad | University of Cologne | mahdizadeh@ph1.uni-koeln.de |
| Mahieu, Sylvain | IRAM | mahieu@iram.fr |
| Matsuo, Hiroshi | NAOJ | h.matsuo@nao.ac.jp |
| Mehdi, Imran | JPL | imran.mehdi@jpl.nasa.gov |
| Meledin, Denis | Chalmers University | denis.meledin@chalmers.se |
| Mena, Patricio | University of Chile | fpmena@uchile.cl |
| Mertens, Marc | University of Cologne | mertens@ph1.uni-koeln.de |
| Miao, Wei | Purple Mountain Observatory | weimiao@pmo.ac.cn |
| Mirzaei, Behnam | SRON | b.mirzaei@tudelft.nl |
| Monasterio, David | Universidad de Chile | david.monasterio@raig.uchile.cl |
| Montofre, Daniel | Kapteyn Institute | d.a.montofre@rug.nl |
| Moro Melgar, Diego | ACST GmbH | diego.moro-melgar@acst.de |
| Murk, Axel | Uni Bern | murk@iap.unibe.ch |
| Murtagh, Donal | Chalmers | donal.murtagh@chalmers.se |
| Neric, Marko | Arizona State University | mneric@asu.edu |
| Ochiai, Satoshi | NICT | ochiai@nict.go.jp |
| Oshima, Tai | NAOJ | tai.oshima@nao.ac.jp |
| Otto, Ernst | CHALMERS | otto@chalmers.se |

| | | |
|---|---|---|
| Pavolotsky, Alexey | Chalmers University | alexey.pavolotsky@chalmers.se |
| Pettersson, Lars | OSO/Chalmers | Lars.Pettersson@chalmers.se |
| Qian, Yuan | PMO,CAS | yuanqian@pmo.ac.cn |
| Rahiminejad, Sofia | JPL | sofia.rahiminejad@jpl.nasa.gov |
| Reck, Theodore | Virginia Diodes Inc. | theodore.reck@vadiodes.com |
| Ren, Xiaodong | University of Cologne | ren@ph1.uni-koeln.de |
| Ren, Yuan | Purple Mountain Observatory | yren@pmo.ac.cn |
| Richter, Heiko | DLR | Heiko.Richter@dlr.de |
| Risacher, Christophe | IRAM | risacher@iram.fr |
| Rudakov, Kirill | Kapteyn Astronomical Institute/NOVA | rudakov@astro.rug.nl |
| Saenz, Elena | ESA/ESTEC | elena.saenz@esa.int |
| Santamaria Botello, Gabriel | UC3M | gasantam@pa.uc3m.es |
| Schuster, Karl-Friedrich | IRAM | schuster@iram.fr |
| Shan, Wenlei | National Astronomical Observatory of Japan | wenlei.shan@nao.ac.jp |
| SHI, Sheng-Cai | PMO | scshi@pmo.ac.cn |
| Siles, Jose | Jet Propulsion Laboratory | Jose.V.Siles@jpl.nasa.gov |
| Silva, Jose | SRON | j.r.g.d.silva@sron.nl |
| Sjögren, Paulina | Chalmers University | paulina.sjogren@chalmers.se |
| Skyrme, Tess | University of Cambridge | ts676@cam.ac.uk |
| Smirnov, Andrei | ASC LPI | asmirn@asc.rssi.ru |
| Sobis, Peter | Omnisys Instruments AB | ps@omnisys.se |
| Stake, Jan | Chalmers tekniska högskola | jan.stake@chalmers.se |
| Strandberg, Magnus | Chalmers University | magnus.strandberg@chalmers.se |
| Sundin, Erik | Chalmers University | erik.sundin@chalmers.se |
| Takakura, Hayato | Univ. of Tokyo, JAXA/ISAS | takakura@astro.isas.jaxa.jp |
| Tan, Boon Kok | University of Oxford | boonkok.tan@physics.ox.ac.uk |
| Testi, Leonardo | ESO | ltesti@eso.org |
| Tong, Edward | Harvard Smithsonian CfA | etong@cfa.harvard.edu |
| Treuttel, Jeanne | CNRS - OBSERVATOIRE DE PARIS - LERMA | jeanne.treuttel@obspm.fr |
| Trifonov, Andrey | Moscow Pedagogical State University | andrey89trifonov@gmail.com |
| Uno, Shinsuke | The University of Tokyo | uno@ioa.s.u-tokyo.ac.jp |
| Valenzuela, Daniel | Universidad de Chile | daniel.valenzuela@ug.uchile.cl |
| Vdovin, Vyacheslav | IAP RAS | vdovin_iap@mail.ru |
| Widdig, Sina | I. Physics Institute, Cologne | swiddig@ph1.uni-koeln.de |
| Wiedner, Martina | Paris Observatory, CNRS | martina.wiedner@obspm.fr |
| Wienold, Martin | DLR | martin.wienold@dlr.de |
| Wylde, Richard | Thomas Keating Ltd / St Andrews University | r.wylde@terahertz.co.uk |
| Yang, Fei | SEU | yangfei@seu.edu.cn |
| Yoo, Changyun | UCSB | cyoo@ucsb.edu |
| Zhang, Wen | Purple Mountain Observatory | wzhang@pmo.ac.cn |
| Zhu, Haotian | Observatoire de Paris | haotian.zhu@obspm.fr |